Smart Innovation, Systems and Technologies

Volume 336

The Smart Innovation, Systems and Technologies book series encompasses the topics of knowledge, intelligence, innovation and sustainability. The aim of the series is to make available a platform for the publication of books on all aspects of single and multi-disciplinary research on these themes in order to make the latest results available in a readily-accessible form. Volumes on interdisciplinary research combining two or more of these areas is particularly sought.

The series covers systems and paradigms that employ knowledge and intelligence in a broad sense. Its scope is systems having embedded knowledge and intelligence, which may be applied to the solution of world problems in industry, the environment and the community. It also focusses on the knowledge-transfer methodologies and innovation strategies employed to make this happen effectively. The combination of intelligent systems tools and a broad range of applications introduces a need for a synergy of disciplines from science, technology, business and the humanities. The series will include conference proceedings, edited collections, monographs, handbooks, reference books, and other relevant types of book in areas of science and technology where smart systems and technologies can offer innovative solutions.

High quality content is an essential feature for all book proposals accepted for the series. It is expected that editors of all accepted volumes will ensure that contributions are subjected to an appropriate level of reviewing process and adhere to KES quality principles.

Indexed by SCOPUS, EI Compendex, INSPEC, WTI Frankfurt eG, zbMATH, Japanese Science and Technology Agency (JST), SCImago, DBLP.

All books published in the series are submitted for consideration in Web of Science.

John Littlewood · Robert J. Howlett ·
Lakhmi C. Jain
Editors

Sustainability in Energy and Buildings 2022

Springer

Editors
John Littlewood
Cardiff School of Art and Design, The Sustainable and Resilient Built Environment Research Group
Cardiff Metropolitan University
Cardiff, UK

Robert J. Howlett
KES International Research
Shoreham-by-Sea, UK

'Aurel Vlaicu' University of Arad
Arad, Romania

Lakhmi C. Jain
KES International
Selby, UK

Liverpool Hope University
Liverpool, UK

ISSN 2190-3018 ISSN 2190-3026 (electronic)
Smart Innovation, Systems and Technologies
ISBN 978-981-19-8768-7 ISBN 978-981-19-8769-4 (eBook)
https://doi.org/10.1007/978-981-19-8769-4

This Springer imprint is published by the registered company Springer Nature Singapore Pte Ltd.
The registered company address is: 152 Beach Road, #21-01/04 Gateway East, Singapore 189721, Singapore

Organization

International Programme Committee

Kouzou Abdellah	Djelfa University, Algeria
Abdel Ghani Aissaoui	University of Bechar, Algeria
Mahmood Alam	University of Brighton, UK
Ricardo Almeida	Polytechnic Institute of Viseu, Portugal
Hasim Altan	Arkin University of Creative Arts and Design, Cyprus
Martin Anda	Murdoch University, Australia
Maria Beatrice Andreucci	Sapienza University of Rome, Italy
Touraj Ashrafian	Ozyegin University, Turkey
Ahmad Taher Azar	Prince Sultan University, Saudi Arabia
Jiping Bai	University of South Wales, UK
Eva Barreira	University of Porto, Portugal
Umberto Berardi	Ryerson University, Canada
Gabriele Bernardini	Università Politecnica delle Marche, Italy
Francesco Calise	University of Naples FedericoII, Italy
Rosa Caponetto	University of Catania, Italy
Kate Carter	University of Edinburgh, UK
Stefano Cascone	Mediterranea University of Reggio Calabria, Italy
Christopher Chao	The Hong Kong Polytechnic University
Abdellah Chehri	University of Quebec—UQAC, Canada
Giacomo Chiesa	Politecnico di Torino, Italy
Dorota Chwieduk	Warsaw University of Technology, Poland
Paolo Civiero	Catalonia Institute for Energy Research, Spain
Nicola Colaninno	Polytechnic of Milan, Italy
Vincenzo Costanzo	University of Catania, Italy
Alessandro D' Amico	Sapienza Università di Roma, Italy
Elisa Di Giuseppe	Università Politecnica delle Marche, Italy
Miriam Di Matteo	Sapienza University of Rome, Italy

Moustapha Doumiati	ESEO School of Engineering—IREENA UR 4642, France
Mahieddine Emziane	University of Birmingham, UK
Diana Enescu	Valahia University of Targoviste, Romania
Youssef Errami	Chouaib Doukkali University, Morocco
Najib Essounbouli	Université de Reims Champagne-Ardenne, France
Fatima Farinha	Universidade do Algarve, Portugal
Tiago Miguel Ferreira	University of the West of England—UWE Bristol, UK
Antonio Gagliano	University of Catania, Italy
George Georghiou	University of Cyprus
Giada Giuffrida	University of Catania, Italy
Cheng Siew Goh	Heriot-Watt University, UK
Gwanggil Jeon	Incheon National University, Korea
Hong Jin	Harbin Institute of Technology, China
Mohammad Arif Kamal	Aligarh Muslim University, India
Prasad Kaparaju	Griffith University, Australia
George Karani	Cardiff Metropolitan University, UK
Khalil Kassmi	Mohamed Premier University, Morocco
Mohan Lal Kolhe	University of Agder, Norway
Angui Li	Xi'an University of Architecture and Technology, China
John Littlewood	Cardiff Metropolitan University, UK
Alessandro Lo Faro	University of Catania, Italy
Judit Lopez	UPC Barcelona Tech, Spain
Giuseppe Mangano	Mediterranea University of Reggio Calabria, Italy
Simona Mannucci	Sapienza University of Rome, Italy
Gianluca Maracchini	Università Politecnica delle Marche, Italy
Ahmed Mezrhab	Mohamed Premier University, Morocco
Michele Morganti	Sapienza University of Rome, Italy
Benedetto Nastasi	Sapienza University of Rome, Italy
Consuelo Nava	Mediterranea University of Reggio Calabria, Italy
Francesco Nocera	University of Catania, Italy
Masa Noguchi	University of Melbourne, Australia
Sonja Oliveira	University of Strathclyde, UK
Emeka Efe Osaji	Leeds Beckett University, UK
Paul Osmond	University of New South Wales, Australia
Anna Pellegrino	Politecnico di Torino, Italy
Abdelhamid Rabhi	Université of Picardie Jules Verne, Amiens, France
Fernanda Rodrigues	University of Aveiro, Portugal
Federica Rosso	Sapienza University of Rome, Italy
Rachid Saadane	Hassania School of Public Works, Morocco
Atul Sagade	Energy Center, FCFM, University of Chile, Santiago, Chile
Francesca Scalisi	DEMETRA Euro-Mediterranean Documentation and Research Center, Italy
Lloyd Scott	Technological University Dublin, Ireland

Preface

The 14th International Conference on Sustainability and Energy in Buildings 2021 (SEB22) is a significant international conference organised by a partnership made up of KES International and The Sustainable and Resilient Built Environment research group, Cardiff Metropolitan University.

SEB-22 invited contributions on a range of topics related to sustainable and resilient buildings and renewable energy and explored innovative themes regarding building adaptation responding to climate change mitigation and other local, national and global challenges.

The aim of the conference was to bring together University researchers, Government and Scientific experts and Industry professionals to discuss the minimisation of energy use and associated carbon emissions in buildings, neighbourhoods, cities in the urban context but also rurally; from a theoretical, practical, implementation, modelling and simulation perspective. The conference formed an exciting chance to present, interact, and learn about the latest research and practical developments on the subject with real world impact. SEB22 will be held in a hybrid form with physical and virtual attendance, in response to agile work patterns following the global COVID-19 pandemic.

The conference featured two General Tracks chaired by experts in the fields:

- Sustainable and Resilient Buildings
- Sustainable Energy Technologies.

In addition, there were eight Invited Sessions proposed and organised by prominent researchers.

It is important that a conference provides high quality talks from leading-edge presenters. SEB-22 featured the keynote speaker Prof. Pete Walker, Centre for Innovative Construction Materials, Department of Architecture and Civil Engineering at the University of Bath UK.

The conference attracted submissions from around the world. Submissions for the Full-Paper Track were subjected to a two-stage blind peer-review process. With the objective of producing a high-quality conference, only the best of these were selected for presentation at the conference and publication in the Springer as book chapters. Submissions for the Short Paper Track were subjected to a 'lighter-touch' review and published in an online medium, but not in the Springer book.

Thanks are due to the very many people who have given their time and goodwill freely to make SEB-22 a success. We would like to thank the members of the International Programme Committee who were essential in providing their reviews of the conference papers, ensuring appropriate quality. We thank the high-profile keynote speakers for providing interesting talks to inform delegates and provoke discussion. Important contributors to the conference were made by the authors, presenters, and delegates without whom the conference could not have taken place, so we offer them our thanks. Finally, we would like to thank the administrative staff of KES International.

It is hoped that you find the conference an interesting, informative, and useful experience; and remain connected through the KES International Virtual Conference Experience.

Cardiff, Wales, UK
Shoreham-by-Sea, UK
Selby, UK

John Littlewood
Robert J. Howlett
Lakhmi C. Jain
SEB-22 Conference Chairs

Contents

Impact of Climate on Building Energy Performance, Urban Built Form and Urban Geometry 1
Ehsan Ahmadian, Amira Elnolaky, Behzad Sodagar, and Ivan Verhaert

A Decision Support Tool for the Co-design of Energy and Seismic Retrofitting Solutions Within the e-SAFE Project 12
G. Evola, G. Margani, V. Costanzo, A. Artino, D. L. Distefano, G. Semprini, M. Lazzaro, and D. Arnone

Comparison of the Morphological Influence of Canopy Roughness Space in Shenzhen and Harbin Main Urban Areas 22
Ming Lu, Di Song, Jun Xing, Jing Liu, and Lu Wang

The Impact of a Vertical Greening System on the Indoor Thermal Comfort in Lightweight Buildings and on the Outdoor Environment in a Mediterranean Climate Context 37
Grazia Lombardo, Angela Moschella, Francesco Nocera, Angelo Salemi, Gaetano Sciuto, Alessandro Lo Faro, Maurizio Detommaso, and Vincenzo Costanzo

Recycling Volcanic Ash and Glass Powder in the Production of Alkali Activated Materials 47
Loredana Contrafatto, Daniele Calderoni, Salvatore Gazzo, and Enrico Bernardo

Thermal Environment Retrofitting of Outdoor Activity Spaces in Old Settlements in Severe Cold Regions of China 56
Yujing Liu and Jin Hong

A Novel Laboratory Procedure to Determine Thermal Conductivity of Green Roof Substrates 66
Stefano Cascone and Antonio Gagliano

Study on Residents' Perception of Low-Carbon Policy and Its Influence on Low-Carbon Behavior Intention 76
Alin Lin, Jiankun Lou, and Ran Yue

The Techno-Economic Effect of PV-BSS Size Under Various Supporting Schemes: A Replicable Method for Buildings 86
Nikolas G. Chatzigeorgiou and George E. Georghiou

Environmental Assessment, Cost Assessment and User Experience of Electric Excavator Operations on Construction Sites in Norway 97
Marianne Kjendseth Wiik, Kristin Fjellheim, Kamal Azrague, and Jon Are Suul

A Rapid Survey Form for Users' Exposure and Vulnerability Assessment in Risk-Prone Built Environments 109
Enrico Quagliarini, Guido Romano, Gabriele Bernardini, and Marco D'Orazio

Assessing People's Efficiency in Workplaces by Coupling Immersive Environments and Virtual Sounds 120
Arianna Latini, Samantha Di Loreto, Elisa Di Giuseppe, Marco D'Orazio, Costanzo Di Perna, Valter Lori, and Fabio Serpilli

Assessment Analysis of BEV/PHEV Recharge in a Residential Micro-Grid Based on Renewable Generation 130
Dario Pelosi, Linda Barelli, Michela Longo, and Dario Zaninelli

Impact of Using Phase Change Materials with Different Wall Orientations in a Classroom Building Under a Warm Temperate Climate 140
Mohammed Amin Nassim Haddad, Hamza Semmari, Khaled Imessad, Mohammed Cherif Lekhal, Lotfi Derradji, D. Rouag-Saffidine, and Mohamed Amara

Applications of Thermoelectricity in Buildings: From Energy Harvesting to Energy Management 152
Diana Enescu

Natural and Recycled Stabilizers for Rammed Earth Material Optimization 164
Giada Giuffrida, Vincenzo Costanzo, Francesco Nocera, Massimo Cuomo, and Rosa Caponetto

Driving a Photovoltaic Panel in Manhattan with Well-Chosen Projections of the Reflections of the Mirror City 175
Benoit Beckers, Jairo Acuña Paz y Miño, and Inès de Bort

One Stop Shops on Housing Energy Retrofit. European Cases, and Its Recent Implementation in Spain 185
Rolando Biere-Arenas and Carlos Marmolejo-Duarte

Mitigating Multi-risks in the Historical Built Environment: A Multi-strategy Adaptive Approach 197
Federica Rosso, Letizia Bernabei, Gabriele Bernardini, Juan Diego Blanco Cadena, Martina Russo, Alessandro D'Amico, Graziano Salvalai, Edoardo Currà, Enrico Quagliarini, and Giovanni Mochi

Sensorial Design—A Collaborative Approach for Architects and Engineers 208
P. Grant, J. R. Littlewood, R. Pepperell, and F. Sanna

Hemplime Blocks: Innovative Solution for Green Buildings in Italy 218
Chiara Moletti, Patrizia Aversa, Bruno Daniotti, Giovanni Dotelli, Vincenza A. M. Luprano, Anna Marzo, Sergio Sabbadini, and Concetta Tripepi

Have Population Growth and Economic Activity Converged Towards the Same Pattern in the Spanish Metropolises? 228
César Costa Costa, Carlos Marmolejo Duarte, and Rolando Biere Arenas

A Community Housing Association's Strategy for the Benchmarking, Reduction and Sequestration of Carbon Towards a Resilient and Globally Responsible Wales (UK) 240
K. Stevens-Wood, J. R. Littlewood, and F. Sanna

Management of Indoor Thermal Conditions in Heavy and Lightweight Buildings: An Experimental Comparison 249
Nicola Callegaro, Luca Endrizzi, Luca Zaniboni, and Rossano Albatici

Comparison of the Thermal Effect of Two Automatic Controls of Roller Shutters in an Academic Space 261
Marc Roca-Musach, Elena Garcia-Nevado, Carlos Alonso-Montolio, Isabel Crespo Cabillo, and Helena Coch Roura

Energy Poverty and Heatwaves. Experimental Investigation on Low-Income Households' Energy Behavior 271
Gianluca Maracchini, Elisa Di Giuseppe, and Marco D'Orazio

Sensitivity and Uncertainty Analysis on Urban Heat Island Intensity Using the Local Climate Zone (LCZ) Schema: The Case Study of Athens 281
Gianluca Maracchini, Fatemeh Salehipour Bavarsad, Elisa Di Giuseppe, and Marco D'Orazio

Challenges in BiPV/PCM Façade System: Pathways Towards Numerical Modelling and Simulation Approaches 291
Jakub Čurpek, Miroslav Čekon, Ondřej Šikula, and Muhammad Faisal Junaid

Integrating Vegetation and Cities: A Review of the Applicative Solutions from Technical Component to Planning Scale 301
Arianna Peduzzi and Carlo Cecere

RECsim—Virtual Testbed for Control Strategies Implementation in Renewable Energy Communities 313
Antonio Gallo, Marco Savino Piscitelli, Lorenzo Fenili, and Alfonso Capozzoli

Energy Retrofit and Fire Protection in Existing High-Rise Residential Buildings: A Case Study in Modena (Italy) 324
Luca Guardigli and Fausto Barbolini

A Pilot Study to Evaluate Occupant Quality of Life in Optimised Retrofit Dwellings in Wales, UK 337
J. R. Littlewood, X. Zhang, and G. Karani

BIM-Based Workflow for Managing Multi-risk Factors of Open Spaces in Historical Built Environment 347
M. Angelosanti, M. Russo, A. D'Amico, M. Pugnaletto, C. Paolini, E. Quagliarini, and E. Currà

Experimental Assessment of a Preliminary Rule-Based Data-Driven Method for Fault Detection and Diagnosis of Coils, Fans and Sensors in Air-Handling Units 359
Mohammad El Youssef, Francesco Guarino, Sergio Sibilio, and Antonio Rosato

Bridging the Flexibility Concepts in the Buildings and Multi-energy Domains 371
Gianfranco Chicco, Diana Enescu, and Andrea Mazza

Offsite Manufacturing of Timber-Frame Woodfibre Insulated Construction Systems for Nearly-to-Zero Carbon Dwellings in Wales, UK 386
J. R. Littlewood, R. J. M. Hawkins, N. I. Evans, and C. Hale

Energy Communities: The Concept of Waste to Energy-CHP Based District Heating System for an Italian Residential District 397
L. Pompei, F. Nardecchia, V. Lanza, L. M. Pastore, and L. de Santoli

Renewable Energy System Applied to Social Housing Building in Mediterranean Climate 407
Andrea Vallati, Stefano Grignaffini, Costanza Vittoria Fiorini, Simona Mannucci, and Miriam Di Matteo

Cyclic Lateral Load Test of a Wall with Timber Frame Structure and Lightearth Envelope 418
G. Becerra, S. Onnis, G. Meli, M. Wieser, and J. Vargas-Neumann

Takagi-Sugeno Fuzzy Control of an Interleaved DC-DC Boost Converter 431
M. Nachidi, K. Khennoune, I. Ouachani, and A. Rabhi

Author Index 441

About the Editors

John Littlewood is a Professor of Sustainable and Resilient Buildings, and holds a Ph.D. in Building Performance Assessment. He is the head of the Sustainable and Resilient Built Environment research group in Cardiff School of Art and Design at Cardiff Metropolitan University (UK). He has been General Chair for the Sustainability in Energy and Buildings international conference in 2014, 2017 and from 2019 to date. He coordinates three professional doctorates in Art and Design Practice, Engineering, and Sustainable Built Environment. He is a Chartered Building Engineer and his innovation, and research expertise is industry focused, identifying and improving fire and thermal performance in existing and new dwellings to enable occupant quality of life. In addition, to helping organisations use innovative materials in offsite manufacturing and in construction to deliver a sustainable and resilient built environment. He has authored, co-authored and co-edited over 160 peer-reviewed articles and book volumes for Springer.

Robert J. Howlett is the Academic Chair of KES International a non-profit organisation which facilitates knowledge transfer and the dissemination of research results in areas including Intelligent Systems, Sustainability, and Knowledge Transfer. He is Visiting Professor at 'Aurel Vlaicu' University of Arad, Romania, and has also been Visiting Professor at Bournemouth University, UK. His technical expertise is in the use of artificial intelligence and machine learning for the solution of industrial problems. His current interests centre on the application of intelligent systems to sustainability, particularly renewable energy, smart/micro grids and applications in housing and glasshouse horticulture. He previously developed a national profile in knowledge and technology transfer, and the commercialisation of research. He works with a number of universities and international research groups on the supervision teams of Ph.D. students, and the provision of technical support for projects.

Professor Lakhmi C. Jain Ph.D., Dr. H.C., M.E., B.E.(Hons), Fellow (Engineers Australia), is with the Liverpool Hope University and the University of Arad. He was formerly with the University of Technology Sydney, the University of Canberra and Bournemouth University.

Professor Jain serves the KES International for providing a professional community the opportunities for publications, knowledge exchange, cooperation and teaming. Involving around 5000 researchers drawn from universities and companies world-wide, KES facilitates international cooperation and generate synergy in teaching and research. KES regularly provides networking opportunities for professional community through one of the largest conferences of its kind in the area of KES.

Impact of Climate on Building Energy Performance, Urban Built Form and Urban Geometry

Ehsan Ahmadian[1(✉)], Amira Elnolaky[2], Behzad Sodagar[2], and Ivan Verhaert[1]

[1] University of Antwerp, Groenenborgerlaan 171, 2020 Antwerp, Belgium
ehsan.ahmadian@uantwerpen.be

[2] University of Lincoln, Brayford Way, Lincoln LN6 7TS, UK

Abstract. The study investigates the impact of climate on the residential building energy performance and its relationship with urban built form and geometry of the built environment. It aims to identify the most energetically sustainable urban built form and optimal urban geometry in different climates that results in higher energy performance of buildings. Geometrical models of four urban built forms are developed, and a simulation method is used to conduct sensitivity analyses over the four case studies (cities of London, Singapore, Helsinki and Phoenix) that are selected based on specific climatic criteria. The Energy Equity (EE) indicator is used for demonstration of the results, which simultaneously considers the amount of building energy demand as well as energy generation by building-mounted PVs. The results show that increasing the cut-off angle (i.e., reducing buildings distance) reduces building energy demand in cooling-dominated buildings (i.e., in Singapore and Phoenix) between 6% and 56% while increases building energy demand in heating-dominated buildings (i.e., in London and Helsinki) between 2% and 16.5%. Hence, the impact of distance between buildings on building energy demand is more significant in hot climates. In general, building energy demand in London is the lowest among the case studies, while it is the highest in Singapore (up to 219% higher than London). London also shows the highest value of EE (demonstrating the best energy performance) and Helsinki shows the lowest (up to 51% lower than London). It is recommended to use the tunnel-court built form to have a more energy-efficient buildings, specifically in hot climates.

Keywords: Building energy performance · Urban built form · Climate

1 Introduction: The Importance of Design with Climate

It has been well-established by different studies that building energy performance correlates with urban built form and density [1–3], while urban density is directly related to the geometry of the built environment [4]. The correlation itself is influenced by climate and dependent on the geographical location of a city. For instance, both the magnitude and type of building energy demand and the potential of renewable energy generation, specifically solar energy, depend on climate and location. They consequently influence

J. Littlewood and R. J. Howlett (Eds.): SEB 2022, SIST 336, pp. 1–11, 2023.
https://doi.org/10.1007/978-981-19-8769-4_1

the relationship of energy with urban form and geometry [5]. This makes it vital to design buildings according to climatic conditions during the early stages of the design process [6]. Climatic variables must be known to predict the thermal behavior of the building envelope [7].

In contemporary building designs and with the use of mechanical equipment (e.g., air conditioning system) to provide satisfying thermal conditions, less attention has been paid to climatic conditions. Built forms have become very similar in every corner of the world regardless of climate, reflecting the loss of traditional skills with respect to a climate-sensitive design. More recently, with more focus on sustainability, we have begun to consider climate conditions for achieving sustainable building/urban designs. For instance, Dursun and Yavas [8] emphasized that to have a sustainable urban development, a climate-sensitive urban design guideline is urgently needed and the urban built environment should be consistent with climatic conditions. Muhaisen [9] suggested general rules and guidelines for the design of courtyards in four different climatic regions. Kocagil and Oral [10] showed that building form and settlement texture are influential parameters for heating/cooling loads of buildings in a hot-dry climate zone to provide optimum conditions. Khalili and Amindeldar [11] identified that traditional courtyards have emerged in the hot-arid regions of Iran to reduce the detrimental aspects of the climate providing better microclimatic conditions for occupants. Strømann-Andersen and Sattrup [12] argued that in northern European cities with high latitudes and low solar inclinations, urban density is of particular concern since urban geometry affects solar access more than in other urban centers around the world.

Therefore, climate not only influences building energy demand but also determines suitable built forms and density of urban areas. Although previous studies have investigated the impact of climate on building energy demand, few have considered the impact of climate on the energy performance of buildings with different built forms and urban geometric variables. The aim of this study is identification of the most energetically sustainable urban built form and optimal urban geometry in different climates that results in higher energy performance of buildings. Building energy performance includes energy demand along with solar energy generation from roof-mounted PV panels that is necessary to be considered to achieve sustainable cities of future. Four case studies from different climate zones are selected and for each case, the correlation of building energy performance with urban geometric variables and the selected built forms is investigated. Simulation method is adopted for energy simulation of the built form models, and consequently, a comparative analysis suggests the most energy-efficient urban built forms in the different climates.

2 Methodology

The study initially develops geometric models of the four selected built forms using three influential geometric parameters. Secondly, case studies from different climate zones are selected. Finally, simulation trials are performed to obtain building energy demand and solar energy generation from roof-mounted PVs. The results from different case studies are compared to identify the impact of climate on building energy performance, urban built form and urban geometry.

2.1 Developing Geometrical Models of Different Built Forms

Geometrical models of four urban built form, namely, pavilion, terrace, court and tunnel-court form are developed (see Fig. 1) using three geometrical parameters, namely, the cut-off angle (θ), the plan depth (x) and the number of floors (n) [4]. These three variables explain the whole geometry of a built environment and have a significant effect on building energy performance [13]. As shown in Fig. 1 (right), the cut-off angle represents the distance between buildings (L) in the site plan.

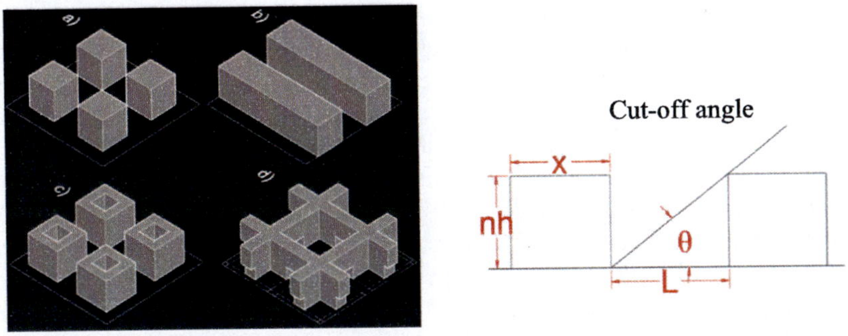

Fig. 1. Generic urban built forms **a** pavilion, **b** terrace, **c** court, and **d** tunnel-court (left), section showing cut-off angle (right).

2.2 Case Study Selection and Energy Simulation

Case studies from different climatic conditions are selected using the Köppen climate classification system (also known as the Köppen–Geiger) [14]. It divides the earth into five main zones, Group A: tropical (mega thermal) climates, Group B: dry (arid and semiarid) climates, Group C: temperate (mesothermal) climates, Group D: continental/cold (microthermal) climates, Group E: polar and alpine (montane) climates. Four large metropolitan cities are selected as the case studies based on their diverse climatic conditions to represent each of the main climate zones. Their great populations show their significant contribution to overall urban energy consumption. Hence, providing guidelines for the optimization of energy with respect to their built form and geometry is beneficial for future developments of these cities that can conserve significant amounts of energy and prevent high levels of carbon emissions.

Group A: Singapore (tropical hot and humid climate): The metropolitan City of Singapore, located at the latitude of 1.3521° N and longitude of 103.8198° W, is an equatorial city with a hot, humid and rainy climate. Energy consumption of buildings contributes about a third of Singapore's total electricity production [15]. Although using passive design strategies are encouraged in 80% of the residential built area, the energy performance of a building is measured according to active mechanical systems [16].

Group B: Phoenix (hot and arid climate): The metropolitan City of Phoenix as the capital of the state of Arizona in the USA, located at the latitude of 33.4484° N and longitude of 112.0740° W. Phoenix has a long, hot summer and short, mild winter. It is one of the sunniest cities in the world (in a desert location) with approximately 300 days of sunshine per year. It makes this city a suitable candidate for this study since the potential of PV energy harvesting is being considered.

Group C: London (temperate climate): The metropolitan City of London as the capital of England, located at the latitude of 51.5074° N and longitude of 0.1279° W. It has a temperate climate with warm summer and without dry season.

Group D: Helsinki (continental cold climate): The metropolitan City of Helsinki as the capital city of Finland, located at the latitude of 60.1699° N and longitude of 24.9384° W. It has a continental cold climate with warm summer and without a dry season. Its intense winters impose a significant heating load on buildings.

This study does not find any necessity to analyse a city from group E because there are no large metropolitan urban areas in these parts of the World; and the outcome would be identical to the continental cold climate with similar (but sharper) trends.

The simulation method is adopted for the energy analysis of the case studies. An urban energy simulation software, CitySim, is used to perform an energy analysis on the geometrical models of the chosen built forms. CitySim considers parameters such as the shadowing effect of adjacent buildings, radiative inter-reflection between external surfaces, and the Urban Heat Island (UHI) effect [17], which are important features for investigating the impact of the geometry of the built environment on building energy performance. UHI effect is considered by calculating the surface temperature of all the surfaces existing in the site plans on an hourly basis. The climate files of each of the case studies are derived from the Meteonorm database, which contains 10 years of average data for each location plus their horizon files [18]. Theoretical site plans of buildings are developed for each built form to be fed into CitySim for energy analysis, which includes heating/cooling, lighting and appliances energy demands. Each site plan is composed of a 5 * 5 grid of similar buildings while only the energy performance of the central block is taken into account to not only limit the edge effect, but also, provide a more realistic microclimatic condition of a built environment composed of a specific form of buildings. Simulation trials are repeated by changing the geometrical variables to identify the impact of each variable on the building energy performance. Subsequently, the whole process is repeated separately for each case study using its relevant climate data. To ensure a like-for-like comparison between built areas with different geometries, the parameters such as building materials, insulation, infiltration rate (0.5 ACH), glazing ratio (40%), occupant density (35 m^2/person) and room setpoint temperature (20 °C for heating and 24 °C for cooling) are kept constant. All buildings are assumed to be highly insulated with wall and roof U-values of 0.18 and 0.13 W/m^2K, respectively. In practice, the physical characteristics of a building envelope might be influenced by climatic conditions. For instance, the value of glazing ratio in Helsinki and Singapore should be different since solar gains have an opposite impact on their building energy demand. However, in this study, to be consistent in all case studies and to focus the study on the impact of climate, these parameters are kept constant for all climatic zones.

In addition to climate and horizon files, the other input data for the simulation that is variable for different case studies is the heating/cooling period considered for energy simulations. This factor is the direct offspring of the climate that is varied for different climates. In Singapore, the average temperature during the day and the night is almost constant throughout the year. Therefore, buildings require only cooling related energy all year round, which is defined as the annual electrical energy consumption of the air-conditioning system [15]. Cooling energy is considered to be supplied by a heat pump in the simulations for this study. Therefore, it increases the total electricity consumption of buildings. Looking at historical climate data for Phoenix [19] and following information provided by authors of previous studies on this city [20, 21], the typical building cooling period is considered to be from April to October and the heating period from November to March. Phoenix and Singapore both have a hot climate, however, they possess considerably different climatic conditions that create different building energy requirements. Singapore requires 12 months of cooling while Phoenix requires seven months of cooling and five months of heating. Due to the desert location of Phoenix, there is normally a substantial change in temperature between daytime and nighttime, therefore, the thermal behavior of the hot-dry climate is very distinctive due to wide daily and seasonal fluctuations [10]. In London, the heating season begins in October and lasts until the end of May according to SAP [22]. Due to the temperate climate of the UK and its mild summers, normally no cooling load is considered for residential buildings [23–25]. Helsinki has a cold climate with a long heating season. To be consistent with London case study, only the heating period is considered for simulation trials of Helsinki, which is similarly the period between October and May. For the purpose of this study, gas is used for preparation of heat for homes.

For each case study, 216 simulation trials are conducted for different building plans. These site plans are obtained by combining the changes in the geometric variables that means altering the number of floors (from 1 to 30), cut-off angle (25°, 45° and 65°), and plan depths (from 6 to 60m with 6m intervals). The selection of 6m interval is based on the passive to non-passive area ratio determined in the LT method [26]. The resulting values of building energy demand are given in kWh/m^2/year for each plan. Meanwhile, it is assumed that 90% of all building roofs are covered by PVs to obtain the solar energy potential of buildings in different climates. A dimensionless energy indicator termed Energy Equity [13] is used, which is defined as the ratio of the yearly energy generation by building-mounted PVs over building energy demand. It is an indication of building energy self-sufficiency. Please note that if seasonal self-sufficiency is accounted (instead of yearly one), the outcomes might be different.

3 Results and Discussion

In this section, initially the impact of cut-off angle on building energy demand is investigated, and subsequently, the comparative analysis of building energy performance in different climates is illustrated.

3.1 Impact of Cut-Off Angle in Different Climates

To investigate the impact of cut-off angle on building energy demand in each case study, building plans composed of similar buildings but different cut-off angles are simulated. The results for different climates are collected, and exemplar cases are shown in Fig. 2, which represents the general trends of all cases.

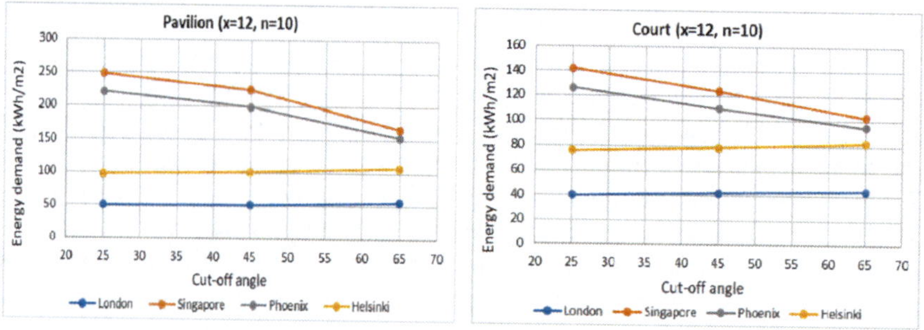

Fig. 2. Impact of cut-off angle on building energy demand in different climates (exemplar cases).

It can be seen from Fig. 2 that, in both London and Helsinki, greater cut-off angle results in higher building energy demand. It means that, for all the built forms, energy demand of buildings is the highest for built environment with a cut-off angle of 65°, while having cut-off angle of 25° leads to the lowest building energy demand. For instance, for pavilion buildings with plan depths of 12 m and 10 number of floors in London, energy demand is equal to 50, 51 and 56 (kWh/m^2) for the $\theta = 25°$, $\theta = 45°$ and $\theta = 65°$ cases, respectively. In Helsinki, varying the cut-off angle from 25° to 65° can increase building energy demand between approximately 2% and 12%, depending on the plan depth. The main reason for this outcome is the shadowing effect of the neighbor buildings. Higher cut-off angles mean building are closer to each other, which blocks a larger portion of sunlight. This not only reduces the solar gain of buildings through glazing, but also decreases the amount of energy stored in building thermal mass. As a result, buildings need more energy to satisfy their heating energy demand [27]. It means that a higher urban density is not advantageous for continental/cold/temperate climates. Considering urban energy planning targets for these cities, this may encourage urban planners to plan new urban built areas to have lower cut-off angles by increasing the distances between buildings.

The results show an opposite trend for the cities of Singapore and Phoenix. In Singapore, changing the cut-off angle from 25° to 65° can diminish building energy demand from approximately 8–56% depending on the plan depth. Therefore, higher density reduces building energy demand in hot climates that buildings are cooling-dominated. The reason is that by increasing the cut-off angle the buildings become closer and therefore the shadowing effect of adjacent buildings protects them from intense solar radiation (which reduces solar heat gain) that consequently decreases the cooling energy requirement of a building [28–30]. However, the trend of this reduction in Phoenix is not as pronounced as for Singapore due to the fact that the buildings in Phoenix demand heating load in wintertime, while cooling is required for buildings in Singapore all year round. This heating load is the element that mitigate the sharpness of this trend. Changing cut-off angle from 25° to 65° in Phoenix can reduce building energy demand between approximately 6% and 47% (depending on the plan depth).

In general, the change in the building energy demand by altering cut-off angle is significantly smaller in heating-dominated buildings than the change it imposes to the cooling-dominated buildings. The analysis of this section emphasizes that the impact of urban density on building energy demand definitely depends on the climate and geographical location.

3.2 Comparison of Different Climates

Here, the results obtained from all case studies are aggregated to make a comparison between the energy performance of the studied built forms in the different climates. The resulting values of building energy demands from the four case studies are compared, and eight exemplar cases are shown in Fig. 3. These cases are selected in the way to represent the whole range of values for the geometric variables including high and low-rise buildings, small and great depth buildings, and high/low cut-off angles. They are the similar cases from the different case studies (shown by different colors), that have been chosen among more than 200 datasets obtained from the simulation trials, where the general trend of all of them are similar. In each case, built form, cut-off angle (θ), plan depth (x) and number of floors (n) are kept constant, which means the density is constant too as a result of the similarity of all parameters considered. Therefore, the only variable in each case is climate.

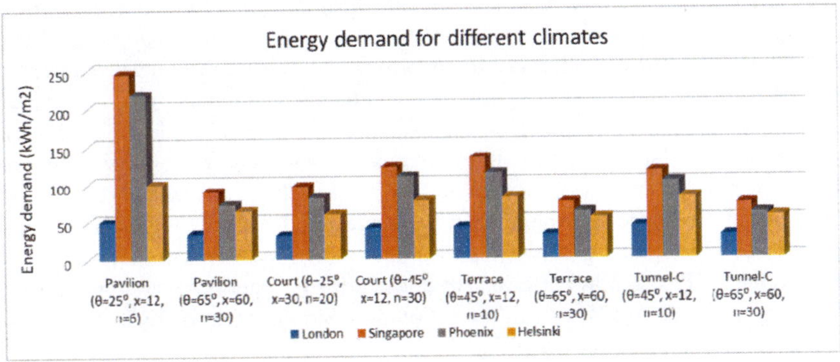

Fig. 3. Comparison of the energy demand of the built forms with similar geometric parameters in different climates (representative selection of 8 cases out of 216 datasets).

It can be observed that the lowest energy demand belongs to London, having a significant difference compared to the others. The next lowest energy demand is associated with Helsinki and is followed by Phoenix. Finally, the highest energy demand belongs to Singapore. The low energy demand of London is due to its temperate climate which necessitates less heating energy to reach the thermal comfort temperature of occupants. Due to the cold climate of Helsinki, the outside temperature has a larger divergence from the inside setpoint room temperature. Phoenix and Singapore mainly require cooling demand that itself requires more energy compared with heating demands. Moreover, according to their climatic conditions, they demand energy 12 months of a year, while it is only eight months for London and Helsinki. Therefore, these two case studies show higher energy demand. Notably, the weather in Phoenix is harsher and hotter in the summer period which requires higher cooling demand to the buildings but requires less total energy than Singapore which requires cooling all year round (Phoenix buildings require heating for five months of a year).

To investigate the scale of these differences, the case of terrace form with $\theta = 45°$, $x = 12$ m and $n = 10$ is analysed here. The resulting energy demand of buildings in London, Helsinki, Phoenix and Singapore are 42, 81, 114 and 134 kWh/m^2/year, respectively. This shows that yearly building energy demand in Helsinki, Phoenix and Singapore are 93%, 171% and 190% higher than in London. This highlights the significant impact of climate on building energy demand.

Among the cases shown in Fig. 3, the first case that is composed of pavilion form with $\theta = 25°$, $x = 12$ m and $n = 6$ shows a relatively abnormal high energy demand for the Phoenix and Singapore case studies. In this specific instance, the energy demands for these two case studies are unexpectedly much higher than in London and Helsinki, and their percentage differences are not following the above-mentioned trend. In fact, the energy demand for Phoenix and Singapore are 338% and 392% higher than London, respectively, while they have only a 12% difference between each other. This substantial difference is due to a combination of three features, (i) it is a pavilion, (ii) it has a small plan depth, and (iii) it has a low cut-off angle. The pavilion built form consists of smaller internal space compared with other built forms [4], therefore, its envelope energy efficiency is more vulnerable to outside weather conditions. In addition, it has a small plan depth that makes it even more sensitive to the changes happening outside the building, and finally (and more importantly), the low cut-off angle increases the cooling load of the building in hot climates (i.e., Phoenix and Singapore). As previously demonstrated in Fig. 2, in hot climates, the increasing cut-off angle would decrease cooling demand of buildings. Therefore, in plans with a low cut-off angle, the difference between energy demand in hot climates and the cities such as London and Helsinki (that require heating load) are very significant. By increasing the cut-off angle, the difference is significantly reduced (e.g., $\theta = 65°$).

A similar analysis is now performed by considering PV energy generation in addition to building energy demand for the different climates. Similar cases to Fig. 3 are compared using their EE values, as shown in Fig. 4.

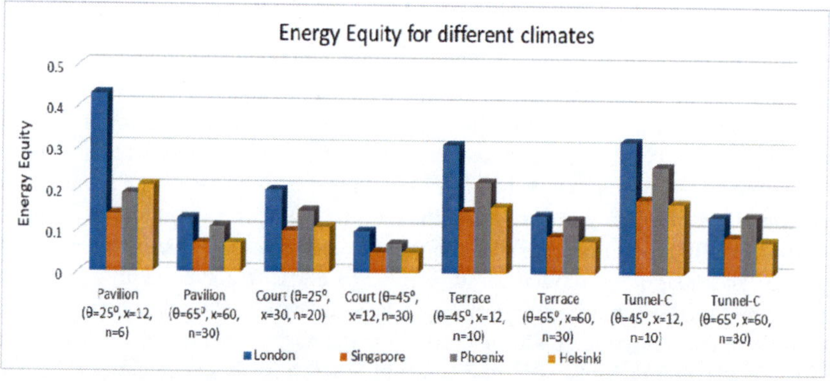

Fig. 4. Comparison of the Energy Equity (EE) of the built forms with similar geometric parameters in different climates.

Having similar geometry, density and built form in each case, Fig. 4 shows only the impact of climate on the EE indicator. It can be seen that the EE of London is higher than the others in all cases except with $\theta = 65°$, where the domination of the London case study, with respect to Phoenix, is not very significant (the reason will be discussed in the last paragraph of this section). Phoenix is ranked second in this comparison, achieving higher values than Singapore and Helsinki except in the first case. As explained when considering the results of Fig. 3, in that exceptional case, the cooling load in Phoenix and Singapore is very high which creates a substantial reduction in their EE. In this case, Helsinki, despite its low solar potential, acquires a higher value of EE than those. By way of a holistic comparison of the lowest-ranked case studies, Helsinki and Singapore, it is seen that Helsinki has greater EE than Singapore in site plans with low cut-off angles, while it is opposite in cases with large cut-off angles. This is connected to their energy demand (the denominator of the EE equation). It is shown in Fig. 2 that increasing the cut-off angle increases the energy demand of Helsinki (and decrease Singapore's) that reduced its EE value (and magnifies Singapore's). Therefore, although the amount of solar radiation in Singapore is substantially greater than in Helsinki, their EE values are relatively similar. According to the results of the simulation trials of PV energy generation, London PV generation is 1% more than Helsinki, Singapore is 54% more than London and Phoenix is 26% more than Singapore. Therefore, although there is a 55% difference between the PV generation potential of Helsinki and Singapore, their EE values remain similar.

As discussed above, for the cases with $\theta = 65°$, the EE of London is very close to that of Phoenix (and in the last case they are almost equal). The reason again is that in plans with high cut-off angle, building energy demand in London is increased while in Phoenix it is decreased (Fig. 2), which causes an opposite impact on the EE. Moreover, the reason that in the last case their EE is equal is that this is a tunnel-court form with $\theta = 65°$. For the tunnel-court form the roof surface area available for PV installation is greater than in other built forms, and in Phoenix, the intensity of solar radiation is greater than the other studied cities, especially London. These two features combined considerably increase the EE of Phoenix which results in equality of its value with London's.

4 Conclusion

In this study, four cities are analysed to investigate the impact of climates on their building energy performance and its relationship with urban geometric variables and built forms. The results show that by increasing the cut-off angle, the energy demand of buildings in London and Helsinki rise while it reduces building energy demand in Singapore and Phoenix. The reason is that energy demand in London and Helsinki is heating-dominated while in Singapore and Phoenix is cooling-dominated. The findings show that closely packed buildings provide shade for their neighbours, resulting in cooler environments that increases the heating load while decreases cooling load. The impact of cut-off angle on the building energy demand of cooling-dominated buildings is significantly higher than on heating-dominated buildings.

The direct comparison of the studied built forms in the chosen case studies shows that yearly building energy demand is a minimum in London while it is maximum in

Singapore. Helsinki and Phoenix are in the middle, though Phoenix shows higher energy demand than Helsinki. Building EE is the highest for London (i.e., buildings in London can achieve energy self-sufficiency easier than in the other case studies) that is followed by Phoenix because of their higher potential for solar energy generation with respect to their building energy demand. The value of this indicator is low for Singapore and Helsinki with approximately similar values. When the cut-off angle of the building plan is low, Helsinki acquires higher EE while Singapore shows higher values in case of having a greater cut-off angle.

In general, pavilion form acquires highest energy demand in all case studies. Tunnel-court form shows the lowest energy demand in Singapore and Phoenix, while the terrace and court forms show the lowest energy demand in Helsinki and London. Meanwhile, the tunnel-court form achieves the highest value of EE in all case studies, where the lowest value belongs to the pavilion. The magnitude of difference between EE of tunnel-court and pavilion forms is significantly higher for cooling-dominated buildings. The tunnel-court form performs between 7% and 32% higher than pavilion form in London and Helsinki, while it performs between 27% and 67% higher than pavilion form in Singapore and Phoenix. It demonstrates the higher importance of choice of built form in hotter climates. Hence, although the tunnel-court form is the best choice in all climates, it is specifically recommended to be used in hot climates. This built form together with a low cut-off angle is the best choice for cold climates while it should be planned with a large cut-off angle in hot climates to achieve the energetically sustainable solutions in the built environments around the world.

It should be noted that energy performance is not the only variable to be considered while designing a building, and there are other priorities such as social and economic aspects that should be considered at the same time. Hence, this study suggests design recommendations to identify the highest energy-performance built forms and urban geometry for different climates, and the main variables and design criteria affecting it.

References

1. Ahmadian, E., Byrd, H., Sodagar, B., Matthewman, S., Kenney, C., Mills, G.: Energy and the form of cities: the counterintuitive impact of disruptive technologies. Archit. Sci. Rev. **62**(2), 145–151 (2019)
2. Ji, Q., Li, C., Makvandi, M., Zhou, X.: Impacts of urban form on integrated energy demands of buildings and transport at the community level: a comparison and analysis from an empirical study. Sustain. Cities Soc. **79**, 103680 (2022)
3. Mangan, S.D., Oral, G.K.: Impacts of future weather data on the energy performance of buildings in the context of urban geometry. Cogent Eng. **7**(1), 1714112 (2020)
4. Ahmadian, E., Sodagar, B., Mills, G., Byrd, H., Bingham, C., Zolotas, A.: Sustainable cities: the relationships between urban built forms and density indicators. Cities 95 (2019)
5. Tsirigoti, D., Tsikaloudaki, K.: The effect of climate conditions on the relation between energy efficiency and urban form. Energies **11**(3), 582 (2018)
6. Heidari, S.: A deep courtyard as the best building form for desert climate an introduction to effects of air movement (Case study: Yazd). Desert. **15**(1), 19–26 (2010)
7. Oral, G.K., Yilmaz, Z.: Building form for cold climatic zones related to building envelope from heating energy conservation point of view. Energy Build. **35**(4), 383–388 (2003)

8. Dursun, D., Yavas, M.: Climate-sensitive urban design in cold climate zone: the City of Erzurum, Turkey. Int. Rev. Spat. Plann. Sustain. Dev. **3**(1), 17–38 (2015)
9. Muhaisen, A.S.: Shading simulation of the courtyard form in different climatic regions. Build. Environ. **41**(12), 1731–1741 (2006)
10. Kocagil, I.E., Oral, G.K.: The effect of building form and settlement texture on energy efficiency for hot dry climate zone in turkey. Energy Proc. **78**, 1835–1840 (2015)
11. Khalili, M., Amindeldar, S.: Traditional solutions in low energy buildings of hot-arid regions of Iran. Sustain. Cities Soc. **13**, 171–181 (2014)
12. Strømann-Andersen, J., Sattrup, P.A.: The urban canyon and building energy use: urban density versus daylight and passive solar gains. Energy Build. **43**(8) (2011)
13. Ahmadian, E., Sodagar, B., Bingham, C., Elnokaly, A., Mills, G.: Effect of urban built form and density on building energy performance in temperate climates. Energy Build. 110762 (2021)
14. Peel, M.C., Finlayson, B.L., McMahon, T.A.: Updated world map of the Köppen-Geiger climate classification (2007)
15. Chua, K., Chou, S.: Energy performance of residential buildings in Singapore. Energy **35**(2), 667–678 (2010)
16. GM, R.B.: Green mark for residential buildings. Technical Guide and Requirements. In: Authority BaC (Ed.). Singapore, pp. 30, 75 (2016)
17. Dorer, V., Allegrini, J., Orehounig, K., Moonen, P., Upadhyay, G., Kämpf, J., et al.: Modelling the urban microclimate and its impact on the energy demand of buildings and building clusters. Proc. BS. **2013**, 3483–3489 (2013)
18. Meteonorm. Meteonorm Software: Meteotest. https://meteonorm.com/en/
19. U.S. Climate Data. Climate Phoenix—Arizona (2021). https://www.usclimatedata.com/climate/phoenix/arizona/united-states/usaz0166
20. Guhathakurta, S., Williams, E.: Impact of urban form on energy use in central city and suburban neighborhoods: lessons from the phoenix metropolitan region. Energy Proc. **75**, 2928–2933 (2015)
21. Sailor, D.J., Elley, T.B., Gibson, M.: Exploring the building energy impacts of green roof design decisions–a modeling study of buildings in four distinct climates. J. Build. Phys. **35**(4), 372–391 (2012)
22. SAP: The government's standard assessment procedure for energy rating of dwellings (2012)
23. Palmer, J., Cooper, I.: United Kingdom housing energy fact file. Department of Energy and Climate Change London (2013)
24. Rode, P., Keim, C., Robazza, G., Viejo, P., Schofield, J.: Cities and energy: urban morphology and residential heat-energy demand. Environ. Plann. B. Plann. Des. **41**(1), 138–162 (2014)
25. Steemers, K.: Energy and the city: density, buildings and transport. Energy Build. **35**(1), 3–14 (2003)
26. Ratti, C., Baker, N., Steemers, K.: Energy consumption and urban texture. Energy Build. **37**(7), 762–776 (2005)
27. Coccolo, S., Monna, S., Kaempf, J.H., Mauree, D., Scartezzini, J.-L.: Energy demand and urban microclimate of old and new residential districts in a hot arid climate. In: 36th International Conference on Passive and Low Energy Architecture, Los Angeles (2016)
28. Chan, A.: Effect of adjacent shading on the thermal performance of residential buildings in a subtropical region. Appl. Energy **92**, 516–522 (2012)
29. Nikoofard, S., Ugursal, V.I., Beausoleil-Morrison, I.: Effect of external shading on household energy requirement for heating and cooling in Canada. Energy Build. **43**(7), 1627–1635 (2011)
30. Numan, M., Almaziad, F., Al-Khaja, W.: Architectural and urban design potentials for residential building energy saving in the Gulf region. Appl. Energy **64**(1–4), 401–410 (1999)

A Decision Support Tool for the Co-design of Energy and Seismic Retrofitting Solutions Within the e-SAFE Project

G. Evola[1(✉)], G. Margani[2], V. Costanzo[2], A. Artino[2], D. L. Distefano[2], G. Semprini[3], M. Lazzaro[4], and D. Arnone[4]

[1] Department of Electric, Electronic and Computer Engineering (DIEEI), University of Catania, Viale A. Doria 6, 95125 Catania, Italy
gevola@unict.it

[2] Department of Civil Engineering and Architecture (DICAR), University of Catania, Via Santa Sofia 64, 95123 Catania, Italy

[3] Department of Industrial Engineering, University of Bologna, Viale Risorgimento 2, 40136 Bologna, Italy

[4] Research and Development Department, Engineering Ingegneria Informatica S.p.A., Viale Della Regione Siciliana N.O. 7275, 90122 Palermo, Italy

Abstract. The innovation project e-SAFE, funded by the EU under the H2020 Programme, is developing a new deep renovation system for non-historical reinforced concrete (RC) framed buildings, which combines energy efficiency and improved seismic resistance. The present paper describes the main functionalities of a Decision Support System (e-DSS) that is being developed by e-SAFE experts, aimed at guiding the technicians and the building owners through a conscious preliminary co-design activity, and leading to the choice of the most suitable renovation solution amongst those envisaged by the e-SAFE portfolio. The e-DSS allows assessing—with a reasonable degree of approximation—the energy performance of the building before and after the proposed renovation action, the environmental benefits in terms of decarbonization (i.e. reduction in CO_2 emission for space heating, space cooling and DHW preparation), the expected costs and time for the building renovation and the expected time of Return of the Investment (ROI), based also on the savings in the annual operating costs. The paper explains the criteria used by the tool to identify those solutions that are not suitable for the selected building, and discusses the degree of approximation behind the calculation of energy, cost and environmental performance.

Keywords: Energy renovation · Seismic renovation · Decision support · Energy saving · Decarbonization

1 Introduction

The topic of combined energy and seismic upgrading of buildings has become increasingly important because of the growing attention to the economic, social and environmental sustainability in the real estate sector. However, frequently retrofit actions are not

J. Littlewood and R. J. Howlett (Eds.): SEB 2022, SIST 336, pp. 12–21, 2023.
https://doi.org/10.1007/978-981-19-8769-4_2

chosen based on a detailed evaluation and comparison of the several possible alternatives, but rather on the designer's experience and widespread best practices. In addition, the building process can be particularly complex from both a technical (e.g. because of the low number of companies specialized in combined seismic and energy retrofitting) and an administrative point of view (e.g. because of bottlenecks in the approval process for renovation actions in apartment buildings).

For these reasons, there has been a growing interest in the development of Decision Support Systems (DSS) to guide the decision process of various retrofit interventions. As an example, some authors [1–3] analyzed and grouped the most common decision-making methods and found that multi-criteria approaches are widely used in DSSs within the construction sector. In addition, some companies and universities have already started developing decision support tools themselves. Amongst them, Kamari et al. [4, 5] applied a hybrid approach based on a genetic algorithm able to define several scenarios and to evaluate their performances in terms of energy consumption, thermal comfort, and investment costs. Another interesting reference can be found in the RENO-EVALUE tool [6], which is meant as a basis for dialogue among building professionals and building users while also supporting the formulation of specific objectives for renovation projects. This system can also be used for comparing alternative project proposals and to follow-up on a project and assess its actual performance. Furthermore, Campos and Neves-Silva developed a DSS tool called EnPROVE ("Energy consumption prediction with building usage measurements for software-based decision support") that supports investors in the selection of the most suitable renovation scenarios by considering budget, technical, and usage constraints [7, 8]. Although being a powerful tool for ranking energy-efficient long-term projects, it needs a technical consultant to define legislation and incentive schemes that can be applied in the specific location where the renovation should take place.

What emerges from the review of existing DSS tools is that several renovation scenarios are first generated through genetic algorithms or user-defined schemes, and then they are assessed and finally ordered according to the stakeholders' priorities.

Differently from such approaches, this paper introduces a new Decision Support System called e-DSS, developed within the H2020 project e-SAFE. Indeed, this tool is specifically designed to support professionals, building managers and residents in choosing amongst the different technologies made available by the e-SAFE project. The main outcome of the tool is the comparison between energy, environmental and economic performance of the building in its current state and after the renovation: these results are helpful to the designer during the preliminary design process, since it allows him/her to show the residents all the potential benefits of the selected solution. Furthermore, the e-DSS guides the designer in the selection of the most appropriate renovation solution amongst those envisaged in e-SAFE, based on a series of checks regarding the shape of the building, the nearby context and the presence of balconies and large glazed surfaces.

The paper describes the main features of the e-DSS in its first release, its current limitations and the criteria behind the selection process for the most suitable renovation solutions. Further developments and functionalities are being implemented and will be available in 2023 in the second release of the tool.

2 The e-SAFE Renovation System and the Role of e-DSS

In e-SAFE, the energy and seismic retrofit of the existing RC-framed buildings will be primarily achieved through two different envelope solutions: (i) timber-based panels including a wood-based insulating material (e-PANEL), and (ii) structural panels made of Cross Laminated Timber (e-CLT) that increase seismic performance through specifically designed friction dampers attached to the existing RC beams. The e-CLT will also include an outer insulation whose thickness is calculated in order to get the same thermal transmittance of the e-PANEL. The two types of panels will externally clad the existing walls seamless and, as a general rule, the e-PANEL will be applied on those walls including openings while the e-CLT to the remaining façade surfaces (see Fig. 1). Both panels are customizable in terms of size, thermal transmittance, and finishing material; they will be prefabricated through BIM-based design procedures and installed through cranes, without the need of scaffoldings.

Finally, the e-SAFE project envisages also the possibility of applying a metallic exoskeleton (e-EXOS) made of bi-dimensional bracings equipped with dampers and connected to the existing RC frame for increasing its seismic resistance.

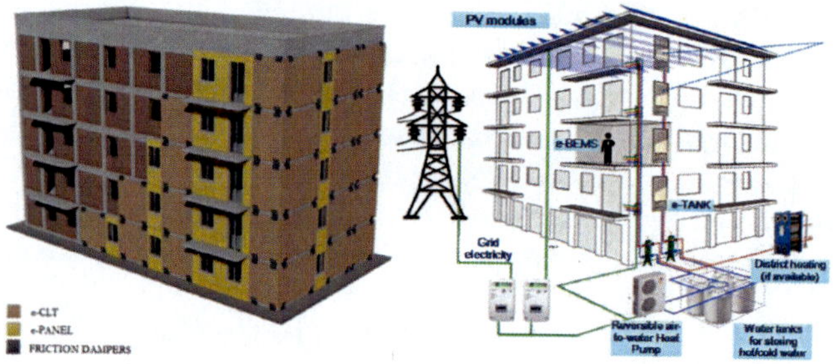

Fig. 1. Left: proposed envelope retrofit solutions. Right: concept for thermal systems (e-THERM)

Apart from acting on the thermal insulation of the building shell, energy savings are also achieved by renovating the technical systems. In e-SAFE, the selected technical solution is named e-THERM: this provides space heating and cooling, as well as Domestic Hot Water (DHW), through highly efficient electricity-driven centralized reversible air-to-water heat pumps (Fig. 1). Fan-coils for space heating and cooling are installed in each dwelling, which will also be equipped with a modular plug-and-play small-size tank to store DHW (called e-TANK) specifically developed within the project. Roof-mounted PV panels are also included for on-site electricity generation. In order to make full profit of PV-based electricity production, the e-THERM concept appoints a central role to heat storage. A first storage level is provided by large centralized water tanks: the system will be equipped with a programmable control system in order to use the water tanks as a buffer and let the heat pump operate in the most convenient conditions (e.g. when PV electricity is available, or the outdoor air temperature exceeds a minimum threshold). A second level of energy storage is provided by the decentralized e-TANKs. The final aim is reducing peak energy demand and increasing the self-consumption rate by the heat pumps.

However, the **e**-DSS is not just a calculation tool for the technician: indeed, it is also conceived as a means of communication between the technician and the building owners/residents during the co-design stage (see Fig. 2). Thanks to the **e-**DSS, the owners are made aware of the potential benefits of the renovation solutions, but also of the costs and the potential disruption. Through a specific functionality implemented in the **e-**DSS, and with the support of the Building Manager, they can interact with the technician and express their point of view, their doubts, and their requirements in terms of cladding, colour, windows and cost of the solutions. Based on this interaction, the technician can refine the preliminary design in order to meet the expectation and the needs of owners and residents.

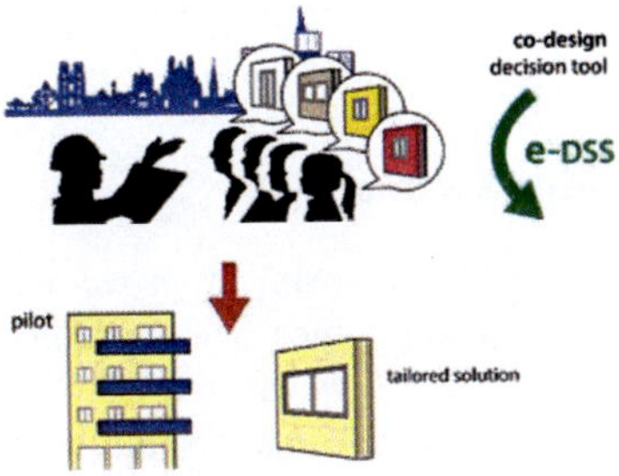

Fig. 2. The e-DSS software as a co-design tool

3 Looking Inside the e-DSS

3.1 Calculating the Energy Needs

The e-DSS estimates the energy needs of both the current and renovated building configurations within the framework of the quasi steady-state energy balance approach described in the European Norm EN ISO 13790:2008 [9]. Specifics concerning the heat transfer through the envelope and the nominal efficiency value of various mechanical and domestic hot water (DHW) production systems are instead gathered from the Italian Norm Series UNI 11300:2014 [10–12]. Other Italian technical norms are recalled also for the calculation of air change rates and endogenous heat contribution (UNI 10339:1995 [13]), as well as for the outdoor weather data (UNI 10349:2016 [14]). This is done because the demo building of the e-SAFE project is located in Italy, and as such it has to comply with the local building codes and prescriptions. Nevertheless, the quasi steady-state calculation approach is commonly adopted by many European Countries for the assessment of the energy performance of buildings, so the e-DSS tool can be easily adapted to different contexts by simply allowing the users to change the values of some parameters and by selecting different weather data based on a GIS tool: these features will be implemented in the second release, available in 2023. Since the tool does not aim to provide an extremely precise estimation of the energy needs, and cannot be used for official energy certification purposes, some simplifications are introduced in order to speed up the data input and make the tool user friendly. For instance, heat transfer through thermal bridges is taken into account by a coefficient that multiplies the heat transfer rate through the opaque envelope components. Coming to the technical systems, the e-DSS is able to determine their global efficiency as the product of the efficiency of several sub-systems: generation, distribution, emission, and control. Each efficiency has default values depending e.g. on the water temperature, the presence of thermal insulation in the distribution system, the type of emission terminal. The COP of air-to-water heat pumps is calculated on a monthly basis as a function of the mean outdoor air temperature, according to UNI 11300:4 [12]. Finally, the e-DSS considers the possible presence of solar thermal systems and PV systems, and subtracts from the energy needs the corresponding thermal/electric energy production, in order to assess the non-renewable primary energy demand by considering appropriate primary energy conversion factors [15].

3.2 Choosing the Renovation Solutions

The e-DSS does not aim at assessing the seismic improvement provided by the proposed renovation solutions, since this result comes from a complex structural analysis that goes beyond the scopes of the tool. However, the e-DSS supports the technician in the decision-making process of the most suitable solution for seismic improvement, in relation to the seismic zone, the height and the shape of the building, and the current state of conservation of the reinforced concrete structures.

The seismic risk of the city where the building is located is defined based on the Peak Ground Acceleration (PGA), automatically retrieved through the EFEHR web service [16]. According to the seismic risk, the **e**-DSS suggests either a simple energy refurbishment or a combined seismic and energy refurbishment. In the latter case, a series of questions are posed by the e-DSS, such as:

- level of degradation of the existing structure (RC frame)
- percentage of perimeter occupied by balconies in a typical floor
- number of facades attached to other buildings (see Fig. 3);
- presence of constraints that prevent altering the appearance of the building (e.g., local regulations, cultural heritage restrictions);
- availability of sufficient space for crane's operations around the building.

Based on the responses, but also on the seismic risk and the number of floors, the e-DSS attributes a score to the building, and determines the most suitable combined solution (e-CLT or e-EXOS). In some cases, the e-DSS might exclude one or even both solutions, which happens for instance if the number of floors is above twelve, or when local regulations forbid altering the façade.

3.3 Calculating Energy and Cost Savings

In the e-DSS, energy savings are first determined by the reduction of heat losses through the envelope because of the application of e-CLT and e-PANEL to the existing façades.

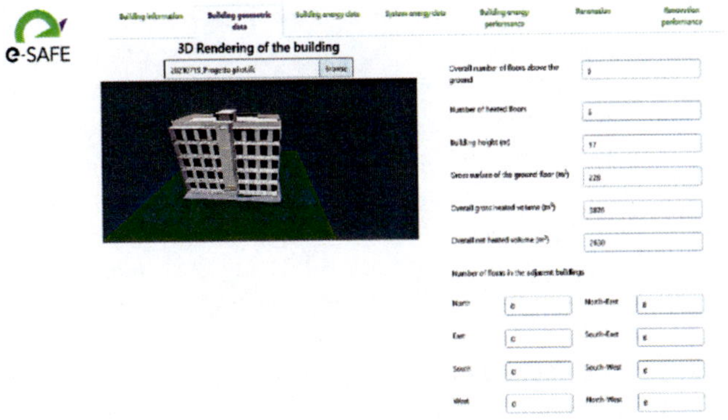

Fig. 3. Screenshot of the e-DSS user interface

The tool calculates the minimum insulation thickness that must be adopted in order to get a certain target U-value, inserted by the user. The target U-value may come from national regulations, or other performance targets identified by the technician himself. In doing this calculation, the tool considers default values for the thermal conductivity of the various layers of materials composing both panels. The user can then select a specific insulation thickness (higher than the minimum value), and the e-DSS defines the new U-value for the renovated buildings, taking also into account the different surfaces covered with e-PANEL or e-CLT, if this is the case. If the user selects e-EXOS in place of e-CLT as the seismic upgrade technology, the new U-value is estimated for the only e-PANEL. In any case, the user is also allowed to renovate the roof: after selecting the desired insulating material and its thickness, the **e-**DSS calculates the new thermal transmittance of the roof. As far as the windows are concerned, there is no calculation performed by the **e-**DSS but rather a unique U-value input by the user. The shutter boxes (if any) are assumed to be thermally insulated by default and the windows to be provided with shadings whose shading factor is set by default as well. On the other hand, energy savings come also from the upgrade of the technical systems. The data regarding the new technical systems (e-THERM) are specified by the user through a series of questions posed by the e-DSS. Many parameters are defined by default according to the specific **e-**THERM concept described in Sect. 2.2, while others (e.g. type of heat pump, COP value in standard conditions, SEER value, number and type of PV modules) can be inserted by the user. The energy savings achievable through the renovation are then estimated by subtracting the non-renewable energy needs of the retrofit scenario from those estimated for the building in its current state, as already described in Sect. 3.1. The e-DSS then computes the renovation costs of both the envelope and technical systems separately. As concerns the envelope costs, they are estimated by multiplying the base unit costs of e-CLT, e-PANEL, windows and roof (€$\cdot$m^{-2}) retrofit by their surface extension (m^2). All unit costs depend on the materials selected during the renovation co-design process, and are reported in the database from which the user selects the materials. The unit costs will be regularly updated in order to follow market prices. Instead, the costs of the heat pump, storage tanks and auxiliary systems are considered proportional to the thermal power of the heat pump, whose size is specified by the technician during the renovation co-design process. Similarly, the size and the cost of the circulation pumps are proportional to the water flow rate, which in turn depends on the thermal power of the heat pump. A similar approach is used for estimating the costs related to the PV system (PV panels and batteries namely), which are considered proportional to the peak power installed. As a further step, the **e-**DSS also assesses the renovation costs associated to each apartment by splitting the overall installation costs proportionally to the net surface of the dwellings. This information is very useful during the co-design stage, since the building manager can show it to the residents and discuss with them the convenience of the proposed solutions.

Finally, the Simplified Payback Period of the investment is computed by dividing the total costs for the renovation (envelope+technical systems) by the annual cost savings. The last ones are estimated as the annual savings on the energy bill due to the energy savings and are calculated considering the constant unit costs for electricity and natural gas embedded in the database (regularly updated for reflecting cost variations through time). The second release of the tool will include the Compound Payback Period and will quantify the potential benefits arising from the improved seismic resistance.

3.4 e-DSS Architecture, Data Model and Protocols

e-DSS is a web application implemented according to the Model-View-Controller (MVC) [17] architectural design pattern separating an application into three main logical components: the Model, the View and the Controller. The Model component corresponds to the e-SAFE data model, which represents the knowledge base of the building renovation process. The View component is what is presented to the end user, and as such, it is the e-DSS Graphical User Interface (GUI) responsible for the visualization of building's relevant information and for guiding the end user into the building renovation co-design process. Finally, the Controller component is the brain of the application since it implements the needed business logic for the calculation of building's energy needs and the algorithms supporting the co-design and renovation process. The e-DSS tool takes advantage of the main benefits associated to the MVC like the efficient code reuse and parallel development of the application, faster development process, easy modification of the entire application and a simplified testing process. The e-DSS architecture is shown in Fig. 4, where the main functionalities and architectural choices are shown. As it is possible to see, the View component consists in the web application front-end addressing the user interaction whereas the Model and the Controller components belong to the application back-end, manly related to the e-DSS data model and the e-DSS business logic. The front-end relies on HTML, CSS, JavaScript and Vue framework [18], while the back-end is implemented in NodeJS [19]. The Renovation Space Representational Model is delivered as a MySQL database [20], while the interaction between the Client and the Server side is addressed through REST services based on Express framework [21] exploiting the HTTP protocol.

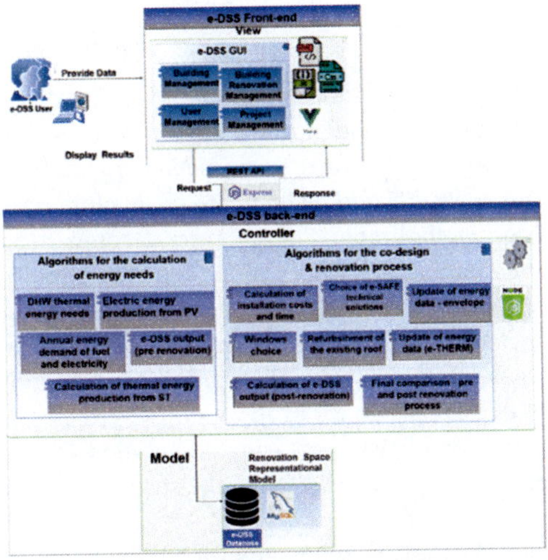

Fig. 4. e-DSS tool architecture.

4 Limitations and Further Developments

The first release of the e-DSS has several limitations, which will be tackled in the second release ready in the summer 2023. One of the main limitations concerns the monthly weather data used for calculating the energy needs: now, they are taken from the Italian Standard UNI 10349:2016, and just refer to Catania (Italy), i.e. the city hosting the pilot building. In the second release, the weather data will be extracted from PVGIS EU web-service, starting from latitude and longitude of the site, and after some simple processing by the same e-DSS. One more limitation affects the data input process for the building geometry: in the first release, the size of the building, including net/gross volume and surfaces, is assigned manually by the user, while in the second release these data will be automatically derived from a BIM-based file. To this aim, the e-SAFE partners are investigating the possible use of IFC and/or gbXML format. Furthermore, the e-DSS considers a standard duration of the heating and cooling seasons in the calculation process, depending on the climatic zone as established by Italian regulations. The second release will adopt a more general approach, where the duration depends on the ratio of the monthly heat gains to the monthly heat losses, but also with the possibility for the user to freely assign these periods.

Finally, the current release of e-DSS calculates the installation costs according to the Italian market as for December 2021. The second release will include updated costs extended to other European countries.

Further information about limitations and future development, together with more details about the e-DSS tool, can be found in the Deliverable D4.2 of the e-SAFE project, available in the ZENODO open access repository [13].

5 Conclusions

A new Decision Support System developed within the H2020 project e-SAFE and called e-DSS is presented with its main features and limitations. The tool is specifically designed to support all the stakeholders involved in combined energy and seismic building renovations in choosing the most appropriate technological solution made available by the e-SAFE project. Indeed, the tool reports as an output a comparison in terms of energy, environmental and economic performances of the building in both its current state and after the renovation, thus allowing for a quick scenario analysis.

The e-DSS does not address the seismic improvement provided by the chosen renovation solutions because this result would require complex and cumbersome structural analysis that are out of the scopes of the tool. Nevertheless, the e-DSS supports the decision-making process of the most suitable solution for seismic upgrading by accounting for the seismic zone, the height and the shape of the building, and the state of conservation of the bearing structure.

In its current release, the tool can only work for building renovations taking place in Catania (Italy), where the pilot building of the e-SAFE project is located and for which weather data and renovation costs are provided in the library of the software.

However, an upcoming release of the tool due on summer 2023 will overcome these limitations and allow the users to analyse the performances of the e-SAFE renovation solutions for buildings located everywhere in the EU by extracting the relevant weather data from PVGIS EU web-service and considering nation-specific renovation costs. Further improvements are also planned concerning the use of the gbXML format for automatically deriving the main geometric features of the building and thus simplifying the users' modelling tasks.

References

1. Nielsen, A.N., Jensen, R.L., Larsen, T.S., Nissen, S.B.: Early stage decision support for sustainable building renovation—a review. Build. Environ. **103**, 165–181 (2016). Elsevier. https://doi.org/10.1016/j.buildenv.2016.04.009
2. Marcher, C., Giusti, A., Matt, D.T.: Decision support in building construction: a systematic review of methods and application areas. Buildings **10**(10) (2020). https://doi.org/10.3390/BUILDINGS10100170
3. Ferreira, J., Duarte Pinheiro, M., de Brito, J.: Refurbishment decision support tools: a review from a Portuguese user's perspective. Constr. Build. Mater. **49**, 425–447 (2013). https://doi.org/10.1016/j.conbuildmat.2013.08.064
4. Kamari, A., Kirkegaard, P.H.: Development of a rating scale to measuring the KPIs in the generation and evaluation of holistic renovation scenarios. IOP Conf. Ser. Earth Environ. Sci. **294**(1) (2019). https://doi.org/10.1088/1755-1315/294/1/012043
5. Kamari, A., Kirkegaard, P.H., Leslie Schultz, C.P.: PARADIS—a process integrating tool for rapid generation and evaluation of holistic renovation scenarios. J. Build. Eng. **34** (2021). https://doi.org/10.1016/j.jobe.2020.101944
6. Jensen, P.A., Maslesa, E.: Value based building renovation—a tool for decision-making and evaluation. Build. Environ. **92**, 1–9 (2015). https://doi.org/10.1016/j.buildenv.2015.04.008

7. Campos, A.R., Marques, M., Neves-Silva, R.: A decision support system for energy-efficiency investments on building renovations. In: 2010 IEEE International Energy Conference and Exhibition, EnergyCon 2010, 2010, pp. 102–107. https://doi.org/10.1109/ENERGYCON.2010.5771656
8. Campos, A.R., Neves-Silva, R.: Decision support system for building renovation. In: Proceedings Campos 2012 DECISIONS
9. ISO 13790:2008 Energy performance of buildings—calculation of energy use for space heating and cooling [Online]. https://www.iso.org/standard/41974.html. Accessed 03 Nov 2021
10. UNI 11300-1: Prestazioni energetiche degli edifici—parte 1: Determinazione del fabbisogno di energia termica dell'edificio per la climatizzazione estiva ed invernale. Ente Italiano di Normazione (in Italian) (2014). Accessed 03 Nov 2021
11. UNI 11300-2: Prestazioni energetiche degli edifici—parte 2: Determinazione del fabbisogno di energia primaria e dei rendimenti per la climatizzazione invernale, per la produzione di acqua calda sanitaria, per la ventilazione e per l'illuminazione in edifici non residenziali. Ente Italiano di Normazione (in Italian) (2019). Accessed 03. Nov 2021
12. UNI 11300-4: Prestazioni energetiche degli edifici - Parte 4: Utilizzo di energie rinnovabili e di altri metodi di generazione per la climatizzazione invernale e per la produzione di acqua calda sanitaria. Ente Italiano di Normazione (in Italian) (2016). Accessed 03 Nov 2021
13. UNI 10339: Impianti aeraulici a fini di benessere. Generalità, classificazione e requisiti. Regole per la richiesta d'offerta, l'offerta, l'ordine e la fornitura. Ente Italiano di Normazione, 2016 (in Italian) (1995). Accessed 03 Nov 2021
14. UNI 10349-1: Riscaldamento e raffrescamento degli edifici—dati climatici—parte 1: Medie mensili per la valutazione della prestazione termo-energetica dell'edificio e metodi per ripartire l'irradianza solare nella frazione diretta e diffusa e per calcolare l'irradianza solare su di una superficie inclinata. Ente Italiano di Normazione (in Italian) (2016). Accessed 03 Nov 2021
15. Deliverable D4.2—Decision Support System (e-DSS)—Renovation space representational model. https://doi.org/10.5281/zenodo.6496938. Accessed 28 Apr 2022
16. http://www.efehr.org/earthquake-hazard/hazard-map/. Accessed 28 Apr 2022
17. Developing GUI Applications: Architectural Patterns Revisited—A survey on MVC, HMCV and PAC Patterns. http://citeseerx.ist.psu.edu/viewdoc/summary?doi=10.1.1.232.3191
18. Vue.js—a progressive framework for building user interfaces. Available from: https://vuejs.org/v2/guide/. Accessed 18 Oct 2021
19. Node.js—JavaScript runtime. https://nodejs.org/en/. Accessed 18 Oct 2021
20. MySQL Database service. https://www.mysql.com/. Accessed 18 Oct 2021
21. Expressjs framework. https://expressjs.com/. Accessed 18 Oct 2021

Comparison of the Morphological Influence of Canopy Roughness Space in Shenzhen and Harbin Main Urban Areas

Ming Lu[1,2], Di Song[1,2](✉), Jun Xing[1,2], Jing Liu[1,2], and Lu Wang[1,2]

[1] School of Architecture, Harbin Institute of Technology, Harbin 150006, China
553586585@qq.com

[2] Key Laboratory of Cold Region Urban and Rural Human Settlement Environment Science and Technology, Ministry of Industry and Information Technology, Harbin 150006, China

Abstract. In this paper, Shenzhen and Harbin, cities with large climate differences, are selected as the research object to explore the relationship between canopy roughness and spatial morphological parameters. Respectively for the spatial configuration parameters of different land use types, build relationships on the analysis of the influence degree, after perspective in order to make clear the climate differences between roughness and spatial form of city space effect difference, facilitate subsequent city get better ventilation effect conforms to the climate characteristics of statistical reference.

Keywords: Spatial morphology influence · Canopy roughness: land type · Comparative analysis · Major urban areas

1 Introduction

At present, there are buildings of different uses and heights in the city, which make the artificial underlying surface have an increasing influence on the vertical ventilation of the canopy [1, 2]. Therefore, it is necessary to analyze the degree of urban roughness and its influence on ventilation, so as to guide how to obtain an appropriate spatial layout to meet the needs of ventilation in urban development [3]. In addition, the climate difference caused by the regional geographical position of the city also makes the difference of the influencing factors, and the comparative study of these influencing factors is also very necessary.

2 Literature Review

At present, the commonly used urban ventilation impacts mainly rely on wind rose maps, numerical simulation, wind tunnel tests, geographic information system (GIS) and remote sensing technology (RS) [4]. Wind rose chart is the most intuitive method

J. Littlewood and R. J. Howlett (Eds.): SEB 2022, SIST 336, pp. 22–36, 2023.
https://doi.org/10.1007/978-981-19-8769-4_3

to describe ventilation in urban meteorology [5, 6]. The frequency distribution of wind speed and direction can be used as the main basis for urban layout. But at the same time, the measurement site is far from the high-density area, and the local universality is insufficient. Numerical simulation and wind tunnel test can express the influence of buildings on air flow [7, 8] and obtain fine wind field. However, due to the limitation of computing equipment and preparation cycle, the cycle is long, the cost is high, and the scope is small. Therefore, it cannot be applied to large-scale overall urban space. GIS and RS technologies make it possible to study the wind environment at the urban scale, and usually combine the meteorological data of the city [9]. The major representative cities are Hong Kong [10, 11], Jinan [12], Guangzhou [13], Wuhan [14] and Beijing [15], which involve urban terrain, water body, greening, building height and land use, etc.

In recent years, the urban canopy space form under the influence of ventilation is valued gradually, on the basis of the above factors, the increase in considering building volume [16],windward acreage [17], streets ratio [18], the sky open degree, aerodynamic roughness parameters [19, 20] to evaluate the roughness effects on ventilation of the city. It can be seen that the indicators of urban ventilation influence are multiple and complex, and the corresponding parameters are often selected according to the evaluation objectives and. At the same time, because different cities are located in different geographical locations, the relationship between spatial roughness and urban layout parameters is different, and the influencing factors and degrees are also different. Therefore, it is necessary to analyze the roughness characteristics of urban canopy in different climates. In addition, there are also limitations of analysis and spatial scale for urban ventilation [21]. For example, the scale of ventilation corridors in Beijing, Wuhan, Fuzhou and Guangzhou varies from 80 to 150 m [22, 23], so it is necessary to conduct a fine analysis on the influence of ventilation.

In this study, GIS and RS technologies were combined to carry out rapid modeling at the urban scale [24, 25], and corresponding roughness parameters and spatial morphological parameters were extracted. The central urban areas of Shenzhen and Harbin, which have large climate differences between the north and the south, were selected as examples to analyze and compare the roughness of urban canopy and its influence on ventilation, so as to provide statistical reference for the subsequent relationship between urban spatial planning and ventilation influence.

3 Methodology

3.1 Data Extraction

The building data of this study came from the supplier of electronic map, and the three-dimensional building form was reviewed and simplified through field investigation [26]. Land use parameters of a city determine the boundary of each land by referring to the overall planning of the city where it is located, and land and above-ground buildings are assigned with reference to the medium nature listed in national standard GB 50352-2019

[27]. The meteorological data of the city came from the National Climatic Data Center (NCDC) of the United States, with a statistical unit of 3 h per time, covering the period 1974–2020 [28]. Finally, ARCGIS is used to extract point, line and plane information to form the required urban spatial database.

3.2 Morphological Parameters

In the vertical direction of the city, the basic non-uniform representative parameter is the building height. In order to express the complex urban morphological characteristics from the perspective of wind environment, the maximum value (H_{max}) and the weighted mean height ($\overline{z_h}$) and standard deviation (σh) of the building plane area are selected as the characteristic parameters of the height. Since the space scale of buildings is different from the land area, the plane area fraction (λ_p) and surface area fraction (λ_b) are selected to represent the influence of space scale on the roughness. As for the blocking effect of each building on incoming flow, the windward area fraction weighted by 16 wind directions ($\overline{\lambda_f}$)) is selected to express it [29].

3.3 Roughness Parameters

Roughness parameters selected in this study are roughness length (Z_O) and displacement height (Z_d), which are used to express the height of wind speed attenuation to 0 affected by roughness elements such as buildings and the initial height of cutting effect on incoming flow [30, 31]. The two-parameter operation formula is as follows:

$$Z_0 = \left[20.21 * \left(\frac{\lambda_p \sigma_h}{\overline{Z_h}}\right) - 0.77 * \left(\frac{\lambda_p \sigma_h}{\overline{Z_h}}\right) + 0.71\right] * \left(1 - \frac{Z_{dm}}{\overline{Z_h}}\right) EXP\left\{-\left[0.5\beta \frac{C_{1b}}{k^2}\left(1 - \frac{Z_{dm}}{\overline{Z_h}}\right)\lambda_f\right]^{-0.5}\right\} \quad (1)$$

$$\text{where: } Z_{d_m} = \left[1 + 4.43^{-\lambda_p}(\lambda_p - 1)\right] * \overline{Z_h} \quad (2)$$

$$Z_d = -0.17 * \left(\frac{\sigma_h + \overline{Z_h}}{H_{max}}\right)^2 + \left(1.29 * \lambda_p^{0.36} + 0.17\right) * \left(\frac{\sigma_h + \overline{Z_h}}{H_{max}}\right) \quad (3)$$

In the formula, Z_{dm} is the calculation method proposed by Macdonald [32, 33], $\beta = 1.0$ is the drag correction parameter [34], $C_{Ib} = 1.2$ is the drag coefficient [35]. Excel function and Grasshopper script were used to calculate the operation of the above parameters, and SPSS was used to conduct regression analysis on the influence relation between roughness and spatial morphology [36].

4 Comparative Analysis of Canopy Roughness and Spatial Morphology

4.1 Overview of the Study Area

The main urban areas of Harbin and Shenzhen are both urban areas with a population of more than 5 million, and they are the main political and economic areas of their cities, with significant differences in geographical location and climate [37]. As shown in Figs. 1, 2 and 3, Harbin is the capital of Heilongjiang Province and a cold megalopolis with the highest latitude in China. Throughout the year, the dominant wind direction is SSW-S, the probability of calm wind is 7.17% and the average wind speed is 3.23 m/s. Shenzhen is a large city located in Guangdong Province, South China. The annual dominant wind direction is NNE-NE, the probability of calm wind is 5.88%, and the average wind speed is 3.13 m/s. The overall wind speed decreased year by year in Harbin and increased year by year in Shenzhen. The number of buildings and land involved in the main urban areas of the two cities is shown in Table 1. In order to ensure the consistency of statistical caliber, the land such as mountains in the main urban area of Shenzhen is removed.

Table 1. Statistical table of construction land in Harbin/Shenzhen central area

Land use nature	Land use code	Shenzhen (SZ)			Harbin (HRB)		
		Land area (km^2)	Building number	Plots number	Land area (km^2)	Building number	Plots number
Administration	A1	2.89	1627	207	2.36	817	275
Culture	A2	1.02	365	82	1.84	222	70
Education and research	A3	10.77	4748	501	30.60	5978	1250
Physical education	A4	1.51	346	62	0.72	139	38
Medical hygiene	A5	1.19	831	103	2.88	1022	228
Social welfare, historic sites and religion	A679	0.13	160	14	0.78	234	40
Business	B1	6.36	4197	463	12.23	3859	796
Business	B2	11.40	5571	698	2.95	1188	346
Entertainment, public outlets and other services	B349	7.01	1917	320	3.07	1042	280

(continued)

Table 1. (*continued*)

Land use nature	Land use code	Shenzhen (SZ)			Harbin (HRB)		
		Land area (km^2)	Building number	Plots number	Land area (km^2)	Building number	Plots number
Green space and square	G123	33.00	2207	3324	36.47	1403	2070
Regional traffic, public facilities, special facilities and border ports	H2349	8.00	1168	161	10.44	996	122
Reserved, to be built	E9	14.77	2619	470	22.09	1763	309
Industry	M	11.52	6016	331	41.33	5515	423
Roads and traffic	S2349	2.00	398	133	2.43	454	290
1st residential building	R1	0.58	709	18	0.00	0	0
2nd residential building	R2	44.08	35269	2173	91.17	32410	5037
Village in the City	R3	7.87	27689	276	15.32	5958	372
Public facilities, logistics and warehousing	UW	6.12	2360	139	14.84	2916	393
Total	–	170.21	98197	9475	291.52	65916	12339

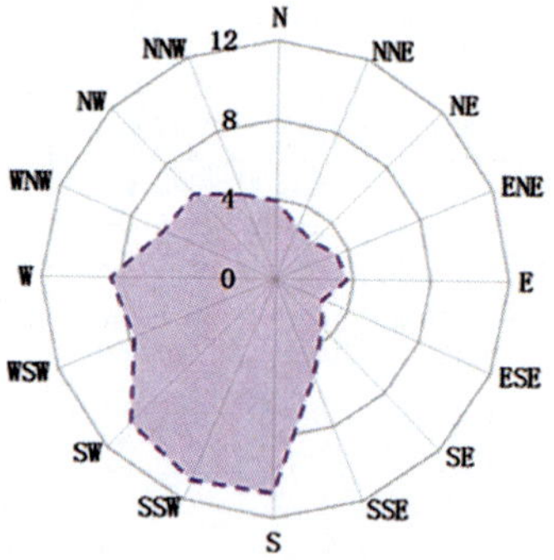

Fig. 1. Harbin wind frequency

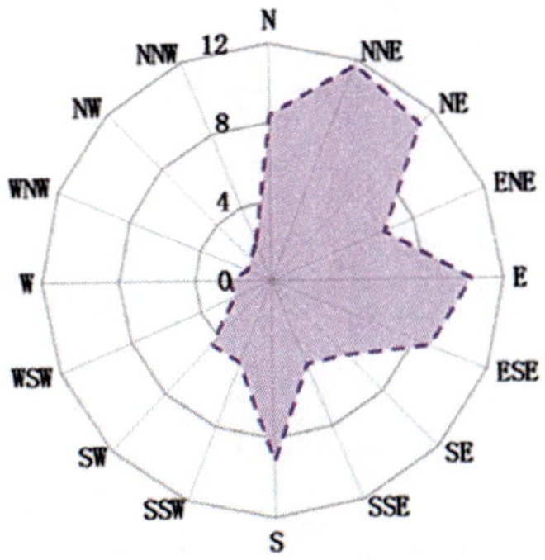

Fig. 2. Shenzhen wind frequency

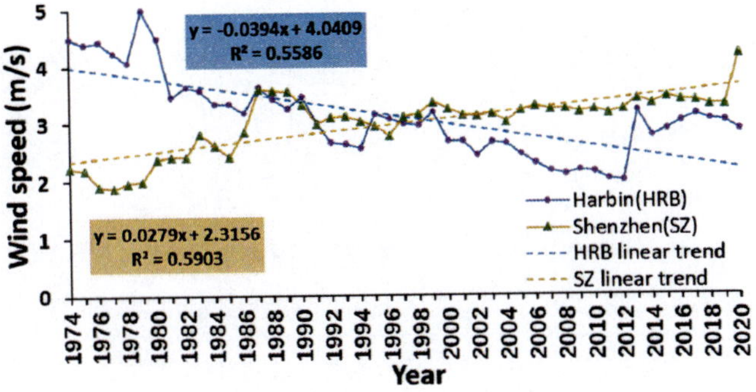

Fig. 3. Building height distribution in central area

4.2 Planar Spatial Analysis

The statistical data of land use in Shenzhen and Harbin are shown in Table 2.

(1) Highly weighted average ($\overline{z_h}$), standard deviation (σh), the maximum (H_{max}): Fig. 4 shows the height of the Shenzhen and Harbin weighted average distribution, Shenzhen's overall average (29.39 m) above the Harbin (19.35 m), from the land distribution, health care, business buildings, 2nd residential building, urban villages far from the average. The further gap is reflected in extreme value and standard deviation (Fig. 5). The huge gap between commercial buildings in the two cities reflects that super high-rise buildings in Shenzhen mainly focus on business functions. The standard deviation of 2nd residential building fluctuates greatly (37.76 in Shenzhen and 26.38 in Harbin), reflecting the wide distribution range of residential building heights in the two cities.

Table 2. Statistical table of vertical morphological parameters of urban space.

Land use code	Shenzhen (SZ)						Harbin (HRB)					
	H_{max} (m)	σh	$\overline{Z_h}$ (m)	λp	λb	$\overline{\lambda_f}$	H_{max} (m)	σh	$\overline{Z_h}$ (m)	λp	λb	$\overline{\lambda_f}$
A1	199.38	21.67	21.91	0.24	1.20	0.29	21.79	24.37	21.79	0.23	1.19	0.27
A2	94.70	21.67	11.64	0.39	1.13	0.22	170.60	22.96	13.98	0.15	0.40	0.07
A3	120.00	19.81	12.40	0.19	0.97	0.22	226.00	17.36	13.35	0.16	0.59	0.12
A4	84.80	18.90	9.21	0.23	0.55	0.10	33.34	17.02	7.45	0.29	0.69	0.12
A5	116.30	27.60	20.08	0.26	1.73	0.43	102.70	14.83	14.83	0.22	0.93	0.20
A679	35.20	10.36	9.10	0.32	1.30	0.29	54.00	8.99	6.28	0.17	0.43	0.07
B1	375.60	23.20	23.42	0.39	1.71	0.39	288.00	18.79	19.95	0.32	1.06	0.21
B2	599.10	55.40	45.91	0.32	2.68	0.71	180.00	27.08	26.34	0.30	1.53	0.35
B349	89.14	11.69	7.29	0.11	0.33	0.07	99.00	10.78	8.13	0.19	0.52	0.10
G123	86.60	8.57	6.41	0.01	0.04	0.01	32.83	6.85	4.06	0.01	0.03	0.01
H2349	104.60	13.34	9.49	0.08	0.25	0.05	105.00	11.00	11.41	0.06	0.16	0.03
E9	120.00	10.18	8.74	0.06	0.18	0.04	86.32	9.02	7.03	0.06	0.14	0.02
M	304.30	17.48	13.07	0.27	0.99	0.22	104.03	9.45	7.19	0.13	0.28	0.05
S2349	54.00	14.76	10.61	0.20	0.47	0.08	56.01	14.00	6.95	0.22	0.46	0.07
R1	34.60	11.81	5.52	0.23	1.00	0.24	0.00	0.00	0.00	0.00	0.00	0.00
R2	388.10	37.76	28.65	0.28	2.69	0.67	150.00	26.83	22.43	0.26	1.49	0.32
R3	63.51	20.59	8.51	0.43	3.61	0.99	144.48	6.30	4.44	0.30	0.58	0.07
UW	54.00	14.21	7.83	0.22	0.60	0.12	67.01	10.61	6.14	0.21	0.46	0.08

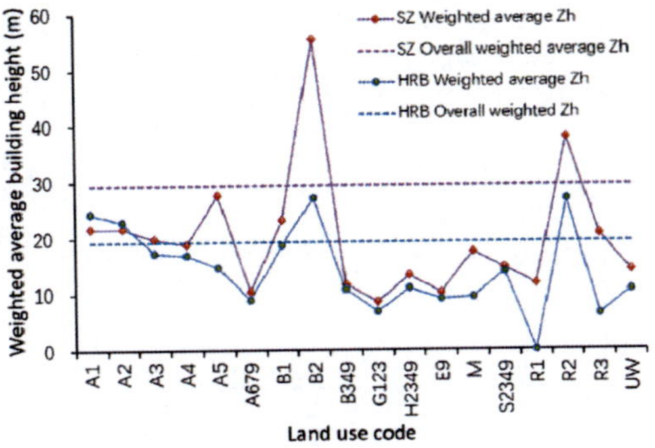

Fig. 4. Building height weighted mean line chart

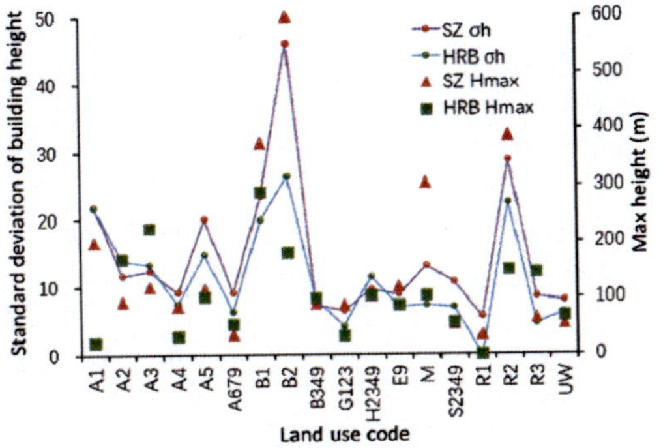

Fig. 5. Standard deviation/maximum building height broken line chart

(2) Plane area fraction (λ_p) and surface area fraction (λ_b): according to the statistical results (Fig. 6), the overall density of Shenzhen is 0.198 and Harbin is 0.176, both of which belong to the region with disturbance influence on downwind direction [38]. Different land use has different degree of disturbance to downwind, and the green space, regional traffic and the area to be built in the two cities have isolated flow, which is conducive to obtaining a good ventilation environment. On the whole, the downwind is sheltered by commercial business and urban village. Shenzhen has a high overall density due to its large cultural buildings.

Figure 7 shows the statistical results of surface area fraction. As there are more high-rise buildings in Shenzhen, the λ_b value is mostly higher than Harbin, and the commercial and second-class residential areas are significantly higher. In addition, in the urban village land, the λ_b value in Shenzhen is significantly higher than that in Harbin due to different building forms and higher density. The λ_b value of green area traffic in the two cities is relatively low.

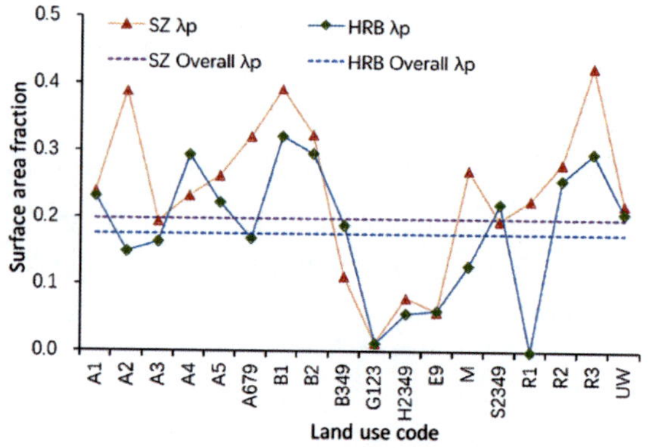

Fig. 6. Plane area fraction broken line chart

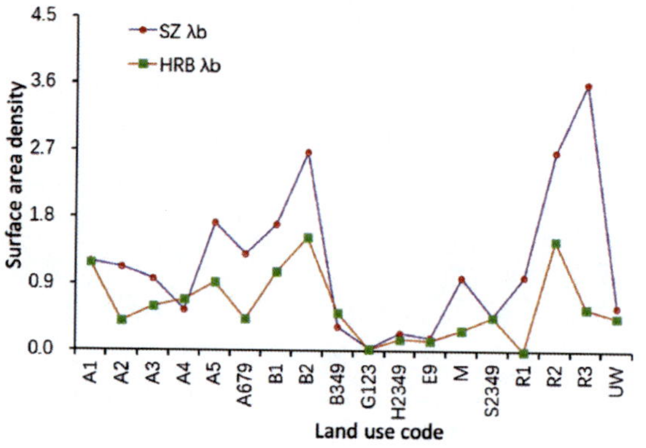

Fig. 7. Surface area fractional broken line chart

(3) Weighted windward area fraction ($\overline{\lambda f}$):Fig. 8 shows the statistical result of 16 $\overline{\lambda f}$ weighted wind direction. The total $\overline{\lambda f}$ of Shenzhen is 0.338 and that of Harbin is 0.166. This also reflects that due to the different distribution of high-rise buildings, Shenzhen is higher than Harbin in terms of blocking the coming current. In the statistical results of different land uses, commercial and commercial buildings, second-class residential buildings and urban village buildings all have high windward side, which is very different between the two cities. This is mainly because high-rise buildings are distributed in commercial and second-class residential land, while multi-storey buildings in urban villages in Shenzhen are more and densely distributed. The low value area mainly includes green space, various traffic stations and sports venues.

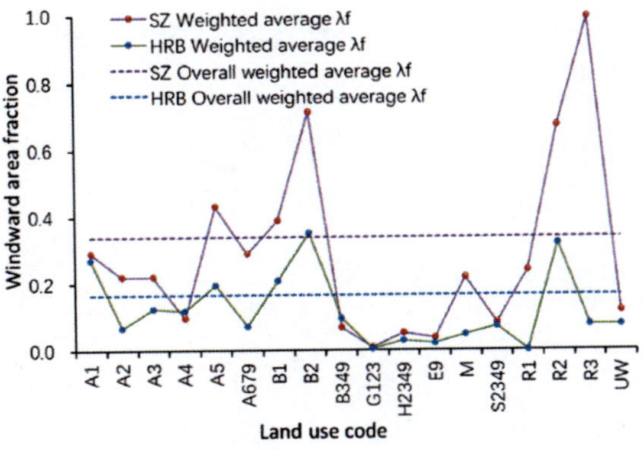

Fig. 8. Weighted windward area fraction broken line chart

4.3 Roughness Parameter Analysis

Figure 9 shows the distribution of roughness length (Z_0) and displacement height (Z_d) of the two cities. The Z_0 and Z_d values of Shenzhen are higher than those of Harbin on the whole, reflecting that the roughness of Shenzhen is higher on the whole. From the perspective of various land uses, the Z_0 values of most of the land uses in the two cities were below 9 m. In Shenzhen, the building heights of commercial buildings and urban villages were 17.56 m and 10.47 m, respectively, and began to influence the wind speed. When the Z_d value is below 60 m, the height (101.42 m and 63.61 m) of wind speed attenuation to 0 of buildings in commercial and urban villages in Shenzhen is higher. The lower roughness of green Spaces, traffic stations and urban roads in the two cities will promote ventilation.

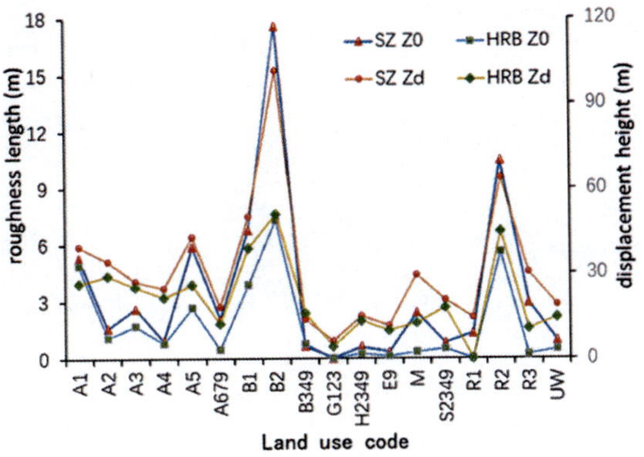

Fig. 9. Roughness length/ Displacement height broken line chart

4.4 Correlation Between Spatial Morphology and Roughness

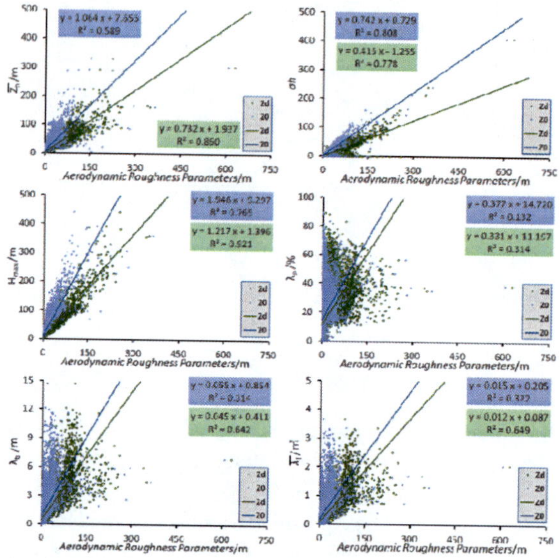

Fig. 10. Correlation between roughness parameters and vertical parameters (Shenzhen)

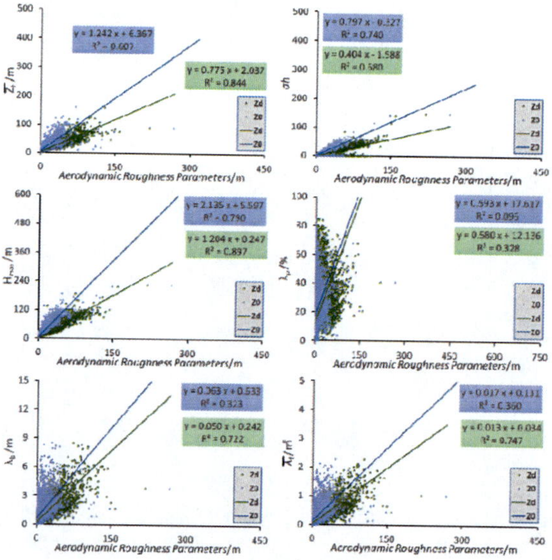

Fig. 11. Correlation between roughness parameters and vertical parameters (Harbin)

The roughness parameters and morphology parameters of Shenzhen and Harbin are analyzed by unary regression analysis (Figs. 10 and 11). The results show that the roughness length Z_0 is significantly correlated with the standard deviation, extreme value and weighted average height of the two cities. The displacement height Z_d is significantly correlated with weighted average height, standard deviation and extreme value, and is well correlated with surface area fraction and weighted windward area fraction.

Multiple regression analysis was conducted on roughness parameters and morphology parameters in Shenzhen and Harbin, and the importance distribution of morphology parameters to roughness parameters was obtained (Formula 4, 5, 6, 7):

$$Z_{dHRB} = -0.175 + 0.137H_{max} + 0.692\sigma_h + 1.894\lambda_p + 3.755\lambda_b + 0.355\overline{Z_h} + 4.989\overline{\lambda_f} \quad (4)$$

$$R^2 = 0.983 \quad \beta_{H_{max}} = 0.176 \quad \beta_{\sigma_h} = 0.338 \quad \beta_{\lambda_p} = 0.020$$
$$\beta_{\lambda_b} = 0.219 \quad \beta_{\overline{Z_h}} = 0.300 \quad \beta_{\overline{\lambda_f}} = 0.074$$

$$Z_{0HRB} = 0.139 + 0.125H_{max} + 0.486\sigma_h - 1.743\lambda_p - 1.376\lambda_b + 0.241\overline{Z_h} - 0.479\overline{\lambda_f} \quad (5)$$

$$R^2 = 0.866 \quad \beta_{H_{max}} = 0.306 \quad \beta_{\sigma_h} = 0.454 \quad \beta_{\lambda_p} = -0.035$$
$$\beta_{\lambda_b} = -0.153 \quad \beta_{\overline{Z_h}} = 0.390 \quad \beta_{\overline{\lambda_f}} = -0.014$$

$$Z_{dSZ} = -1.447 + 0.104H_{max} + 0.827\sigma_h + 8.349\lambda_p + 0.096\lambda_b + 0.537\overline{Z_h} + 7.473\overline{\lambda_f} \quad (6)$$

$$R^2 = 0.984 \quad \beta_{H_{max}} = 0.132 \quad \beta_{\sigma_h} = 0.389 \quad \beta_{\lambda_p} = 0.049$$
$$\beta_{\lambda_b} = 0.005 \quad \beta_{\overline{Z_h}} = 0.424 \quad \beta_{\overline{\lambda_f}} = 0.112$$

$$Z_{0SZ} = -0.292 + 0.079H_{max} + 0.769\sigma_h - 0.621\lambda_p - 0.925\lambda_b + 0.217\overline{Z_h} - 2.644\overline{\lambda_f} \quad (7)$$

$$R^2 = 0.862 \quad \beta_{H_{max}} = 0.176 \quad \beta_{\sigma_h} = 0.635 \quad \beta_{\lambda_p} = -0.006.$$
$$\beta_{\lambda_b} = -0.091 \quad \beta_{\overline{Z_h}} = 0.300 \quad \beta_{\overline{\lambda_f}} = -0.069$$

It can be seen from the above formula that spatial morphological parameters can explain more than 86% of Z_0 and Z_d changes in the main urban areas of the two cities. Among the morphological parameters that have important influences on Z_d values, Shenzhen has a higher degree of influence on weighted height mean, standard deviation and weighted windward area, and Harbin has a higher degree of influence on extreme values. Among the most important morphological parameters in Z_0 value, standard deviation and extreme value are more important in Shenzhen, and weighted height mean is more important in Harbin. The weighted windward area and the area fraction of the two cities are negatively correlated.

5 Discussion

Current city there are many factors influencing the distribution of wind environment, and the effect of height are also gradually improving. Selection of parameters in this study after building area weighted can more accurately explain the effects of roughness, by comparing the various parameters in Harbin and Shenzhen, can clear under climate differences between roughness and shape parameters of the relationship.

The limitation of this stage is that there are altogether seven types of climate zones in China, and only two types of urban samples are selected at present. In future work, more urban samples should be selected for analysis according to the typicality of roughness, so as to further clarify the characteristics of urban roughness in different climates.

6 Conclusions

In this study, the correlation between Shenzhen and Harbin, which are representative cities under different climate conditions, is analyzed by using the vertical form parameters and roughness parameters of urban space.

The results show that there are more high-rise buildings in Shenzhen, which can be reflected from the height extremum and standard deviation. Thus, the surface area fraction of Shenzhen is higher, and the weighted windward area fraction is higher. Furthermore, through roughness analysis, the overall roughness of Shenzhen is higher than Harbin.

The results of regression model show that roughness parameters can explain more than 86% of the influence of morphology parameters on ventilation in the two main urban areas. In terms of importance, the roughness of the two cities is basically the same, Harbin is less affected than Shenzhen. During construction, the change of building volume and height difference should be controlled, and the plane layout of the highest building should be controlled.

Acknowledgments. This study was supported by the National Natural Science Foundation of China (No. 51878208).

References

1. Yuan, C., Ng, E.: Building porosity for better urban ventilation in high-density cities - A computational parametric study. Build. Environ. **50**, 176–189 (2012)
2. Ragheb, A.A., El-Darwish, I.I., Ahmed, S.: Microclimate and human comfort considerations in planning a historic urban quarter. Int. J. Sustain. Built Environ. **5**(1), 156–167 (2016)
3. Santiago, J.L.: Urban obstacles influence on street canyon ventilation: a brief review. Environ. Sci. Proc. **8** (2021)
4. Toparlar, Y., Blocken, B., Maiheu, B. et al.: A review on the CFD analysis of urban microclimate. Renew. Sustain. Energy Rev. **80**, 1613–1640 (2017)
5. Mikhailuta, S.V., Lezhenin, A.A., Pitt, A., et al.: Urban wind fields: phenomena in transformation. Urban Clim. **19**, 122–140 (2016)

6. Xiaoyi, F., et al.: Review and prospect of climate information application in urban planning in my country. Adv. Earth Sci. **30**(04), 445–455 (2015)
7. Kato, S., Hong, H.: Ventilation efficiency of void space surrounded by buildings with wind blowing over built-up urban area. J. Wind Eng. Ind. Aerodyn. **97**(7–8), 358–367 (2009)
8. Li Lei, W., Di, Z.L., Lei, Y.: Research on ventilation assessment for detailed urban block planning based on numerical simulation. J. Environ. Sci. **32**(04), 946–953 (2012)
9. Li, Z., Qingming, Z., Wanlu, O.: A GIS-based study on the summer monsoon environment in Wuhan. Landscape Archit. **03**, 89–97 (2017)
10. Man, S.W., Niched, J.E., To, P.H., et al.: A simple method for designation of urban ventilation corridors and its application to urban heat island analysis. Build. Environ. **45**(8), 1880–1889 (2010)
11. Liang, C., Ng, E.: Quantitative urban climate mapping based on a geographical database: a simulation approach using Hong Kong as a case study. Int. J. Appl. Earth Observ. Geoinf> **13**(4), 586–594 (2011)
12. Yonghong, L., Shuo, Z., Pengfei, C., Peng, C., Lai, W., Xiaoyi, F.: Research and application of thermal environment and wind environment assessment for urban planning: taking Jinan Central City as an example. Ecol. Environ. J. **26**(11), 1892–1903 (2017)
13. Mingrui, L., Qin, Z., Wei, D., Huixia, O.: Research on wind environment assessment and planning control of urban design in key areas: taking guangzhou as an example. J. Urban Plan. **04**, 35–42 (2021)
14. Xie, P., Liu, D., Liu, Y., et al.: A least cumulative ventilation cost method for urban ventilation environment analysis. **2020**, 1–13 (2020)
15. Tushi, Y., Weiwen, W., Ming, C., Xuemei, W.: Numerical simulation and comprehensive identification of potential air ducts in Beijing. J. Geo-Inf. Sci. **22**(10), 1996–2009 (2020)
16. Jie, Y., Qingming, Z.: Research on the excavation of urban ventilation corridor in Wuhan. J. Mod. Urban Res. **10**, 58–63 (2017)
17. Chao, Y., Chao, R., Edward, Ng.: GIS-based surface roughness evaluation in the urban planning system to improve the wind environment—a study in Wuhan, China. Urban Clim. **10**, 585–593 (2014)
18. Ng, E., Chao, Y., Liang, C., et al.: Improving the wind environment in high-density cities by understanding urban morphology and surface roughness: a study in Hong Kong. Landsc. Urban Plan. **101**(1), 59–74 (2011)
19. Hsieh, C.M., Huang, H.C.: Mitigating urban heat islands: a method to identify potential wind corridor for cooling and ventilation. Comput. Environ. Urban Syst. **57**, 130–143 (2016)
20. Liu, Y., Fang, X., Cheng, C., et al.: Research and application of city ventilation assessments based on satellite data and GIS technology: a case study of the Yanqi Lake Eco-city in Huairou District Beijing. Meteorol. Appl. **23**(2), 320–327 (2016)
21. Huanchun, H., Yuan, M., Hailin, Y., Xin, D., Xinhui, Z.: Exploration of geographical design methods for ventilation corridors in megacities. Planners **37**(13), 66–71 (2021)
22. Xiang, C., Sun, W., Zitong, S., Linlin, Z., Zhang Jiabin, X., Wei.: Characteristics of air ducts and evaluation of ventilation efficiency in the main urban area of Guangzhou. Acta Geogr. Sin. **76**(03), 694–712 (2021)
23. Qingming, Z., Wanlu, O., Zhicheng, J., Li, Z.: Research and planning guidelines for urban ventilation potential based on RS and GIS. Planners **31**(11), 95–99 (2015)
24. Xiao, X., Zhu, Q., Du, Z., et al.: A semantics-constrained profiling approach to complex 3D city models. Comput. Environ. Urban Syst. **41**(09), 309–317 (2013)
25. Luo, Y., Jiang, H., He, Y.: A rule-based city modeling method for supporting district protective planning. Sustain. Cities Soc. **28**, 277–286 (2017)
26. Baidu Map-Baidu online panoramic map. http://quanjingbaidu.com/#/
27. Uniform standard for design of civil buildings (GB 50352-2019). National Standards of People's Republic of China (2019)

28. The National Climatic Data Center (NCDC) public FTP server. ftp://ftp.ncdc.noaa.gov/pub/data/noaa/isd-lite/
29. Song, D., Ming, L., Xing, J., Jing Liu, L., Wang.: Analysis of the influence of aerodynamic roughness on urban vertical space form: an example of Shenzhen central area. Sustain. Energy Build. **2021**(263), 29–43 (2021)
30. Yingqin, J.: Research on kinetic roughness and thermodynamic roughness of different underlying surface types master thesis, Nanjing University of Information Science & Technology (2012)
31. Yuqi, T.A.N.G., Bingxin, M.A., Xiaofeng, W.A.N.G., et al.: Analysis on applicability of morphology methods in the estimation of aerodynamic parameter of urban areas in Shanghai. Torrent. Rain Disast. **39**(4), 382–391 (2020)
32. Hawkes, D.J., Macdonald, R., Walkter, S., et al.: Measurements of dispersion within simulated urban arrays—a small scale wind tunnel study. J. Appl. Meteorol. **80**, 178–196 (1996)
33. Grimmond, C.S.B., King, T.S., Roth, M., Oke, T.R.: Aerodynamic roughness of urban areas derived from wind observations. Bound.-Layer Meteorol. **89**(1), 1–24 (1998). https://doi.org/10.1023/A:1001525622213
34. Kent, C.W., Grimmond, S., Barlow, J., et al.: Evaluation of urban local-scale aerodynamic parameters: implications for the vertical profile of wind speed and for source areas. Bound.-Layer Meteorol. **164**(2), 183–213 (2017)
35. Kanda, M., Inagaki, A., Miyamoto, T., et al.: A new aerodynamic parametrization for real urban surfaces. Bound.-Layer Meteorol. **148**, 357–377 (2013)
36. Song, D., Ming, L., Xing, J.: Analysis of the spatial morphology facing wind environment in Harbin central area. Sustain. Energy Build. **2020**(203), 1–15 (2021)
37. Circular of the State Council on Adjusting the Criteria for Urban Scale Division, No. 51. http://www.gov.cn/zhengce/content/2014-11/20/content_9225.htm.2014.10.29
38. Oke, T.R.: Street design and urban canopy layer climate. Energy Build. **11**(1–3), 103–113 (1988)

The Impact of a Vertical Greening System on the Indoor Thermal Comfort in Lightweight Buildings and on the Outdoor Environment in a Mediterranean Climate Context

Grazia Lombardo, Angela Moschella, Francesco Nocera, Angelo Salemi, Gaetano Sciuto, Alessandro Lo Faro(✉), Maurizio Detommaso, and Vincenzo Costanzo

University of Catania, Via S. Sofia 64, 95125 Catania, Italy
{grazia.lombardo,angela.moschella,francesco.nocera, angelo.salemi,gaetano.sciuto,alessandro.lofaro, vincenzo.costanzo}@unict.it, maurizio.detommaso@phd.unict.it

Abstract. Vertical Greening Systems (VGSs) represent an effective solution for new building design and for existing building retrofitting in order to improve the indoor thermal comfort, decrease the building cooling energy needs and mitigate the Urban Heat Island (UHI) phenomenon. VGSs can reduce the outer surface temperature of walls, and internal temperatures in buildings through shading and evapotranspiration effects due to the foliage. In this study, the authors evaluate the effectiveness of a green façade on the indoor thermal comfort in a lightweight building prototype, as well as its effects on the outdoor surrounding microclimate, through Computational Fluid Dynamic simulations performed in ENVI-met. With this aim, two prefabricated modules, one with a vertical vegetation layer and one not, located at the University Campus of Catania (Italy) are investigated. The CFD analysis is carried out for a preliminary assessment of the effects of a vertical greening layer with Trachelospermum jasminoides, which will be installed on the west-oriented wall of one of the two prefabricated modules. The simulation results have revealed a relevant decrease in the peak of outer surface temperature, and a reduction in the internal air temperature in the hottest hours of a typical sunny day. These outcomes will be further studied by means of measurements campaigns on the VGS during next summer period.

Keywords: Vertical greening system · Microscale analysis · CFD simulations · Thermal comfort in buildings · Outer surface temperature

1 Introduction

The continuing urbanization process is subtracting big quantities of natural vegetation and replacing it with impervious low-albedo surfaces [1, 2]. This process has reduced the evapotranspiration rate in urban areas, leading to a phenomenon known as the urban heat island (UHI) [1, 2].

J. Littlewood and R. J. Howlett (Eds.): SEB 2022, SIST 336, pp. 37–46, 2023.
https://doi.org/10.1007/978-981-19-8769-4_4

In recent decades, there has been an increasing pressure towards the application of mitigation and adaption strategies to climate change [3]: sustainable building designs have been proposed as an approach that can contribute to the reduction of cooling energy demand while also mitigating the UHI effects [1].

Since vegetation plays an important role in the reduction of UHI, some of the best performing and interesting solutions for this purpose are greenery systems for buildings [4]. There are basically two ways to integrate vegetation into a building: green roofs and green walls [5, 6]. Nowadays, green wall technologies can be divided in Green façades and Living walls. Green façades can be further divided into two different systems: traditional green façades and double-skin green façades [7]. Factors such as the cover percentage, density and width of plant foliage have an important bearing because their extent defines the effectiveness of plant cover [5]. In this regard, the optimal efficiency of a vertical vegetation system can only be achieved through high leaf area index (LAI) per façade unit [5].

Nevertheless, the variable growth along the time and space of the vegetation and its variation in shape, weight and foliage density involves higher maintenance costs. Consequently, these drawbacks may compromise the implementation of green façades in buildings and the quantification of the thermal behavior of green façades [5]. Thus, the choice of the suitable type of plants to use in green façades and its maintenance is a crucial factor. This affects the typology of supporting structure of the creeping vegetation that has a great importance in the façades' thermal behavior. Several computer simulation studies have investigated the temperature reductions on the building envelope and within buildings when green façades are applied. More recently, some studies have investigated the cooling performance achievable with green façades by means of CFD simulations carried out by means of ENVI-met tool [8]. Most studies have investigated the performances of green facades predominantly using specific perennial and evergreen species such as ivy (Hedera helix and Parthenocissus tricuspidata). One only study has investigated the trend of outer surface temperature of a bare wall covered with Rhyncospermum jasminoides [9]. Therefore, a lack of application of different species of plants suitable green wall in hot summer climate conditions was found [10]. The present paper aims at advancing this knowledge by evaluating the effectiveness of a green façade on the indoor thermal conditions in well insulated and lightweight building structure and on its surrounding outdoor microclimate by means of CFD simulations. A VGS realized using the Trachelospermum jasminoides plant species, which will be installed in the next months on the west-oriented wall of a prefabricated module as part of the "BETA" Intradepartmental Project at the University of Catania (Italy), is preliminary analyzed and discussed.

2 Methodology

The effectiveness of a green façade in terms of indoor thermal comfort in buildings and outdoor surface temperature is investigated through Computational Fluid-Dynamics (CFD) simulations. With this aim, a prefabricated lightweight module—already installed at the University Campus of Catania (Italy)—is simulated in ENVI-met with the addition of a Vertical Greening System made up of a climbing evergreen plant supported by lightweight structure anchored to the west oriented wall of the prefabricated module. In addition, an identical prefabricated module is placed in close proximity to the one simulated: this will allow future experimental comparisons between the module equipped with the VGS and the one that is not.

The ENVI-met model is calibrated with meteorological data recorded at a weather station placed nearby the prefabricated modules. Finally, the thermal behavior of the prefabricated module with and without the VGS is appraised by comparing the inner and outer surface temperatures of the west oriented wall, as well as the indoor air temperatures achieved.

2.1 Experimental Measurements

The weather station "LSI Lastem" is used to measure dry bulb air temperature, relative humidity, direct and diffuse solar radiation, and wind speed. The following sensors' features are declared by the manufacturer: (1) wind speed: measurement range = 0–50 m s^{-1}; threshold = 0.36 m s^{-1}; uncertainty = 1% below 3 m s^{-1} and 1.5% above 3 m s^{-1}; resolution = 0.06 m s^{-1}; (2) radiometer for solar irradiance: spectral response = 300–3000 nm; operative temperature = −40 °C /+80 °C; uncertainty = ±4 W m^{-2} (according to ISO 9060). The sampling time of one minute is chosen in order to keep a high temporal granularity for the experimental measurements.

2.2 ENVI-Met Calculations: Heat Transfer Through Multi-layer Walls

ENVI-met adopts a multiple transient state model to calculate the surface temperatures of walls and roofs [11, 12]. The building indoor temperature is handled as a prognostic variable progressing with respect to the calculated energy fluxes at the building envelope [13]. A wall (or roof) is treated as a uniform structure built out of one homogenous material that featured physical nodes, one on the inner surface, one on the outer surface and one in between. In its current state, the wall can be set of up to three different layers which can vary in width and materials used. Thus, a wall with three materials consists of seven nodes, and every node has its own physical properties (absorption, transmission, reflection, emissivity, thermal conductivity, specific heat capacity, and density) [11, 12].

3 Case Study

The case study is represented by two prefabricated modules located in Catania, a metropolitan city along the Mediterranean coast in Southern Italy (latitude 37°30' North and longitude 15°04' East). According to the international Köppen-Geiger climate classification, Catania is characterized by a warm and temperate climate (Csa), with warm and humid summers and moderately cool, and wet winters.

Figure 1 depicts the 2D horizontal view of the investigated area, the two analyzed prefabricated modules and the placement of the LSI—Lastem weather station.

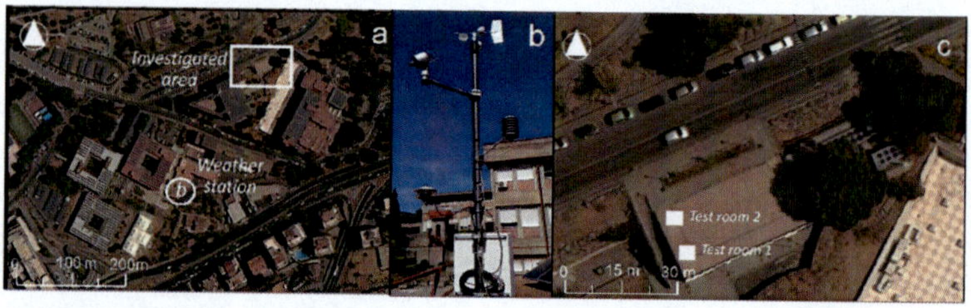

Fig. 1. **a** 2D View of urban neighborhood; **b** weather station; **c** view of the investigated area.

3.1 Description of the Prefabricated Module

The two prefabricated modules are placed on an open-space platform sited in an open area free from obstacles in the nearby surroundings so that the walls and the roof are not shaded during the year. Each of the prefabricated modules has a square shape of 2.50 m × 2.50 m plan dimensions and a height of 3.0 m, and is oriented in a way that every wall faces the four main compass directions. The envelope of such modules is made with sandwich panels comprising an insulation layer of polystyrene 6.5 cm thick that is a common thickness of insulation layer used in the new buildings of the climate zone where the investigated area is suited. A finishing layer—both internally and externally—is made of an osb panel with a thickness of 0.6 cm and 0.9 cm, respectively. Consequently, the total thickness of the sandwich panel is constant and equal to 8 cm.

Figure 2 shows the 2D horizontal section of the prefabricated module (a) and a vertical cross section of the wall (b).

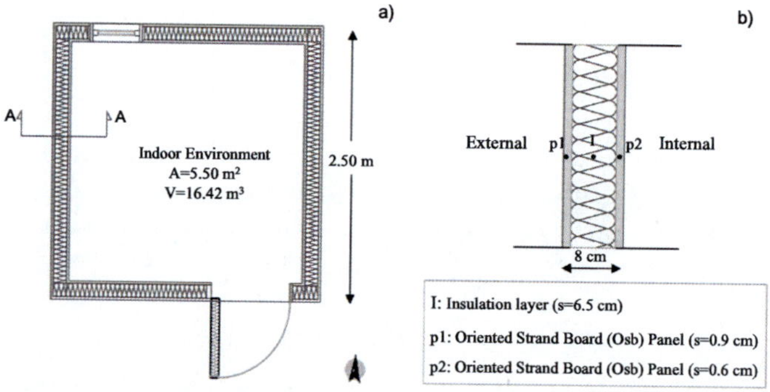

Fig. 2. **a** 2D horizontal section of the prefabricated module; **b** cross-section (A-A) of the wall.

The floor and roof have the same stratigraphy of the wall. In addition, the roof has a metallic corrugated sheet in the outermost face; the floor raised off the ground and is finished in laminate with parquet effects. The external door is made with white honeycomb steel, while the south-facing window is double-glazed with an air gap and PVC frame. The external surface of the walls is finished by means of paint with quartz powder and has an absorption coefficient of 0.30.

The geometrical and thermo-physical properties of each layer of wall, roof and window structures are reported in Table 1.

Table 1. Thermo-physical properties of the walls, roofs and windows [14].

Component	Layer	Thickness s (m)	Thermal conductivity λ (W/mK)	Specific heat Cp (J/kgK)	Density ρ (kg/m^3)
Wall/roof	Osb (p)	0.009	0.12	650	600
	Polystirene (I)	0.065	0.040	1450	15
	Osb (p)	0.006	0.12	650	600
Window	Glass	0.004	1.00	750	2500
	Air Gap	0.015	0.025	1006	1.20
	Glass	0.004	1.00	750	2500

3.2 Features of the Proposed Green Façade

The proposed green façade plans a climbing plant on **supporting structure arranged in trellis**. A single type of evergreen climbing plant, namely the trachelospermum jasminoides, is selected because it is an endemic species and a fast-growing plant that can climb up a trellis or a wall with ease until to 6.0 m, forming an eye-catching 'living screen' effect [15]. The jasminoides species is characterized by lai values varying in the range of 2–4 m^2/m^2 [15]. At the time of maximum growth of the plant, the plant is able to realize an almost total shading of a building façade. The west-facing wall of dimensions of 2.50 m × 3.00 m will house 6 plants, planted at the ground level in pots. Figure 3 shows the place of the pots in the 2D horizontal section and west oriented wall of the prefabricated module (test room 1) where it is planned the installation of the VGS.

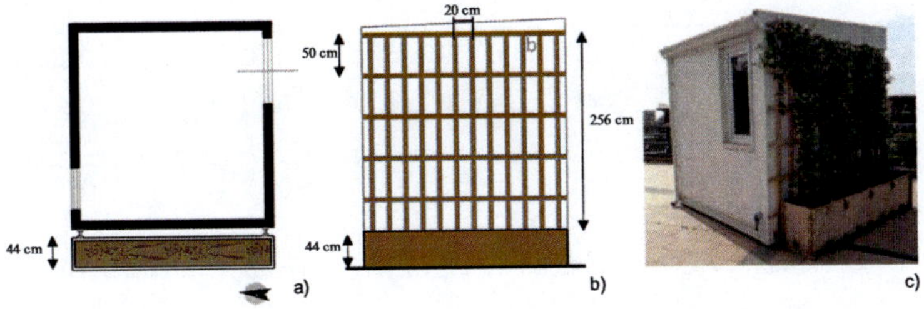

Fig. 3. 2D horizontal section at 1.50 m a.g.l. and west oriented wall of the prefabricated module; **b** West facing wall: frame of the support structure for the VGS; **c** current state

The framework is anchored in the lower section of the pot and to the top of the wall of the prefabricated module. The plants are arranged in order to achieve a dense distribution of foliage in the moment of maximum development of the plants.

3.3 Simulation Model

The investigated area is implemented in ENVI-met using a grid of size 120 × 120 × 17 featured by a mesh of 1.0 m × 1.0 m × 1.0 m and resulting in a total area of 120 × 120 m^2 in the horizontal dimension. Along the vertical z-axis, a telescoping grid with a factor set to 3% is adopted. Ten grid cells for every dimension in space are set as nesting grids along the lateral domain borders. Figure 4 shows the 3D view of the ENVI-met model of investigated area.

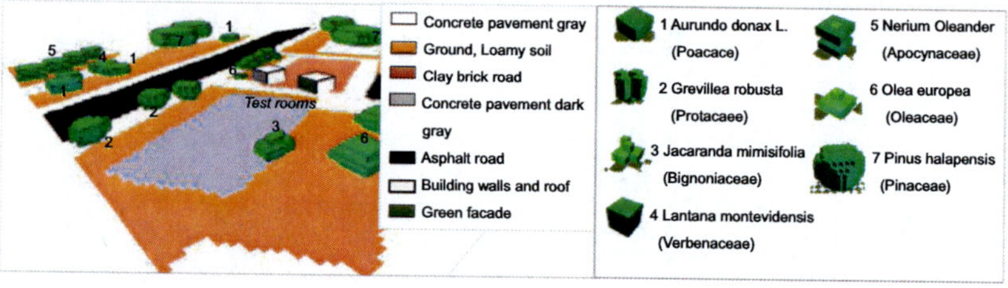

Fig. 4. 3D view of the ENVI-met model for the investigated area.

The materials making up the surrounding context, such as soil, paved areas and buildings, are implemented in the model using the tool called "Database Manager". The input data for each material are thickness, absorption coefficient, reflection (albedo), emissivity, heat capacity, thermal conductivity, and density. International literature sources were adopted for thermal and optical properties of such materials [12].

Table 2 reports the values of albedo and emissivity of the materials of urban surfaces and prefabricated modules adopted.

Table 2. Optical properties of soil and module structures.

	Material	Surface	Colour	Albedo r (−)	Emissivity ε (−)
Prefabricated modules	Osb panel	Wall	White	0.70	0.90
	Coating Aluminum	Roof	White	0.70	0.90
	Concrete pavement	Roads & pavements	Gray	0.50	0.90
	Loamy soil	Natural surface	Brown	0.20	0.98
Soil	Brick road	Decorative	Red	0.30	0.90
	Concrete Pavement	Roads & pavements	Dark gray	0.30	0.90
	Asphalt road	Roads & pavements	Black	0.20	0.90

The modeling of the green façade in ENVI-met is carried out adding a greening layer on the outer surface of the existing wall of prefabricated module. The "Trachelospermum Jasminoides" species described above is represented through a greening layer with an albedo value of 0.2 [15]. The greening layer was implemented under the hypothesis of maximum growth development of plants, with a dense distribution of the foliage and a LAI value of 4.0 m^2/m^2 [15].

3.4 Investigated Wall Scenarios

Figure 5 shows the investigated wall configurations: the bare wall and the one with the green facade implemented on the west oriented wall of test room 1.

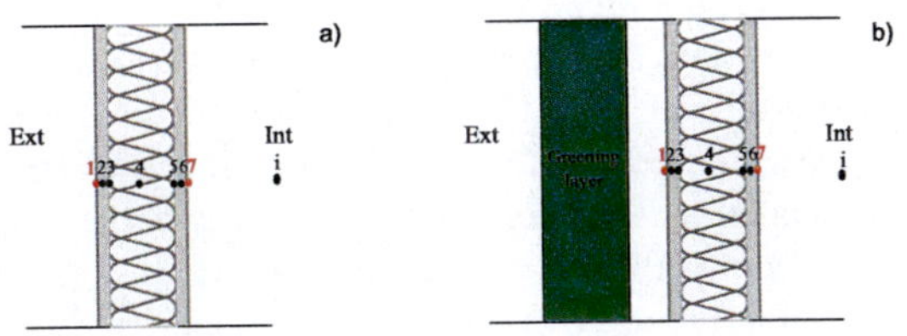

Fig. 5. Investigated wall scenarios: **a** bare wall; **b** green wall.

CFD simulations are carried out during two summer days (from 6 a.m. on 7th August to 6 a.m. on 9th August); these days are selected because they are among warmest days of the summer period recorded in 2021. Figure 6 shows the hourly profiles of global solar irradiance on the horizontal surface ($G_{h,I}$) and outdoor air temperature (T_o) from 6 a.m. on 7th August to 6 a.m. on 9th August.

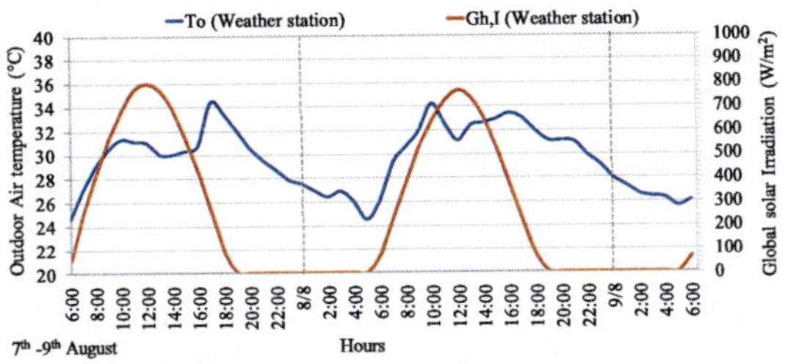

Fig. 6. Profiles of global horizontal irradiation ($G_{h,I}$) and outdoor air temperature (T_o) from 6 a.m. on the 7th August to 6 a.m. on the 9th August.

4 Results and Discussions

4.1 Bare Wall Versus Green Wall: Outer and Inner Surface Temperature

Figure 7 depicts a comparison in terms of outer surface temperature between bare wall and green façade.

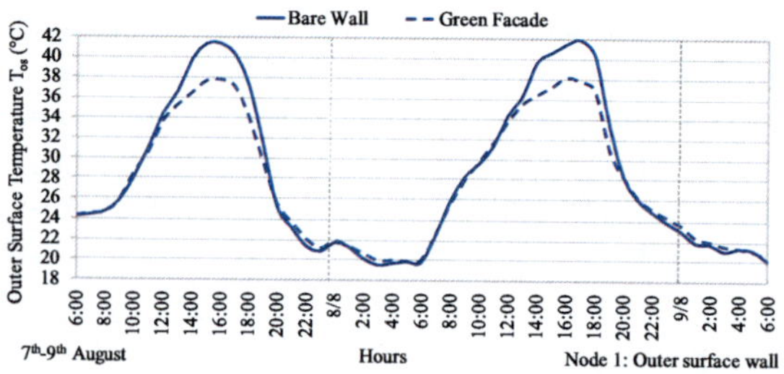

Fig. 7. Outer surface temperature (T_{OS}) on the west oriented wall: comparison between bare wall and green facade.

As it is easy to observe, the green façade is able to reduce the T_{os} value (node 1) almost throughout a sunny day, while during nighttime both façade solutions have the same hourly trend. The profile of T_{os} is always below 38 °C and a decrease of 4.10 °C in the peak values of T_{os} is achieved at around 16:00 in presence of the green façade. The extent of temperature reduction is mainly due to the density of the foliage cover and the consequent shading effect of the leaves on the bare wall.

Figure 8 depicts the hourly profile of inner surface temperature (T_{is}) of the bare wall and green façade respectively.

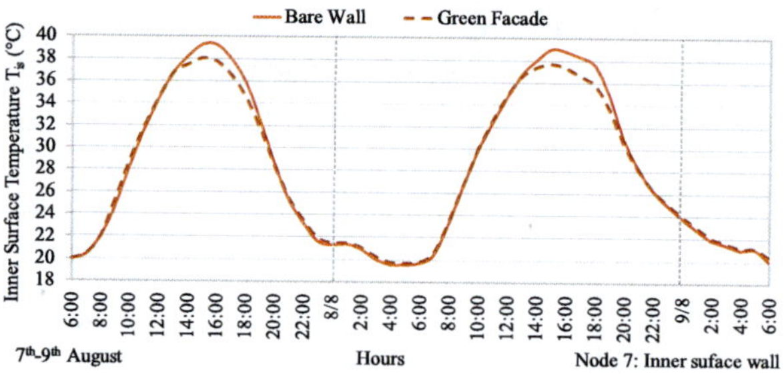

Fig. 8. Inner surface temperature (T_{IS}) on the west oriented wall: comparison between bare wall and green facade.

Comparing the hourly path line of T_{is} of the bare wall and green façade respectively, it emerges that the addition of the vegetation layer allows reducing the T_{is} value only during afternoon hours. The hourly trend of T_{is} of the bare wall has a peak value of 40.0 °C around 16:00, whereas the wall covered by vegetation layer reaches a peak value of 38.0 °C at the same time. Thus, a reduction about 2.0 °C in peak value of T_{is} is achieved thanks to the addition of vegetation layer.

4.2 Bare Wall Versus Green Wall: Indoor Air Temperature

Figure 9 shows the hourly profile of indoor air temperature calculated in a reference prefabricated module (test room 2) made by bare walls and the prefabricated module (test room 1) equipped with a green façade on the west oriented wall.

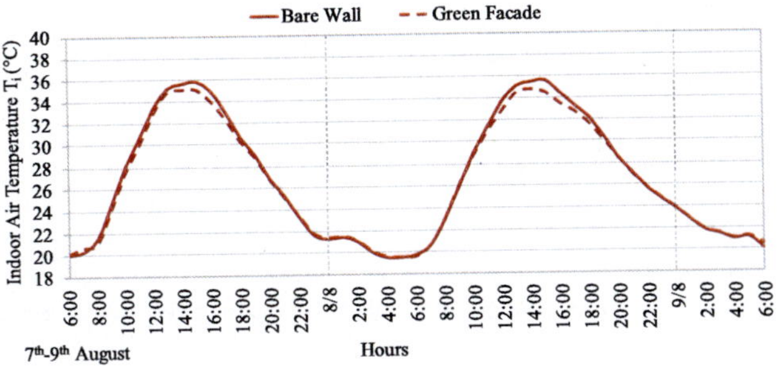

Fig. 9. Indoor air temperature (T_i) on the west oriented wall: comparison between bare wall and green facade.

The reference prefabricated module shows that the indoor air temperature (T_i) ranges from a minimum value of 20.0 °C to a maximum value of 36.1 °C. The peak value of T_i occurs at around 16.00, the same time when the peak of surface temperature is achieved. This is because the lightweight structure of the modules does not allow to shift the heat flux incoming during the hottest hours of the day.

5 Conclusions

The present study dealt with the preliminary assessment of the thermal behavior of a Vertical Greening System (VGS) in improving indoor thermal comfort conditions while also positively influencing the immediate outdoor surroundings.

To this purpose, a CFD micro-scale model of two lightweight prefabricated modules installed at the University Campus of Catania (Italy) is analyzed by means of Computational Fluid Dynamic simulations in ENVI-met.

Results showed that a decrease in the peak of outer surface temperature on the wall with the vegetation layer installed can be as high as 4.10 °C, thus proving the ability of the VGS in improving microclimate conditions in the close surroundings. On the other hand, a decrease in the peak of inner surface temperature of 2.0 °C has been achieved at the same time.

Green façade proved also to improve the indoor thermal conditions thanks to a reduction of the indoor air temperature of about 1.0 °C in the hottest hours of a sunny day.

Future investigations are planned to monitor the effectiveness of the green façade by an experimental measurement campaign lasting one year. The experimental measurements will thus help refining the CFD model presented in this research and allow running a parametric analysis to optimize the features of the VGS.

Acknowledgements. The present study has been conducted within the frame of the "BETA Intradepartmental Project: Thermo-hygrometric well-being in internal and external environment and energy saving through vertical greenery systems (VGS)". We would like to thank the National Association of Building Constructors "ANCE" of Catania for its contribution to the purchase of the prefabricated modules; the company "DOMUS—Prefabbricati Mobili s.r.l." for providing us with the prefabricated modules; the company "Piante Faro" for its support in the selection of suitable plant species to install for the VGS.

References

1. Mohajerani, A., Bakaric, J., Jeffrey-Bailey, T.: The urban heat island effect, its causes, and mitigation, with reference to the thermal properties of asphalt concrete. J. Environ. Manage. **197**, 522–538 (2017)
2. Wong, N.H., et al.: Thermal evaluation of vertical greenery systems for building walls. Build. Environ. **45**, 663–672 (2010)
3. Santamouris, M., Synnefa, A., Karlessi, T.: Using advanced cool materials in the urban built environment to mitigate heat islands and improve thermal comfort conditions. Sol. Energy **85**, 3085–3102 (2011)
4. Kontoleon, K.J., Eumorfopoulou, E.A.: The effect of the orientation and proportion of a plant-covered wall layer on the thermal performance of a building zone. Build. Environ. **45**, 1287–1303 (2010)
5. Pérez, G., Coma, J., Martorell, I., Cabeza, L.F.: Vertical greenery systems (VGS) for energy saving in buildings: a review. Renew. Sustain. Energy Rev. **39**, 139–165 (2014)
6. Seyam, S.: The impact of greenery systems on building energy: systematic review. J. Build. Eng. **26**, 100887 (2019). https://doi.org/10.1016/j.jobe.2019.100887
7. Pérez, G.: Green vertical systems for buildings as passive systems for energy savings. Appl. Energy **88**, 4854–4859 (2011)
8. Morakinyo, T.E., Laib, A., Ka-Lun Laua, K., Ng, E.: Thermal benefits of vertical greening in a high-density city: case study of Hong Kong. Urban Forest. Urban Green. (2017)
9. Blanco, F.I., Scarascia, G., Vox, G.: Modeling of the thermal effect of green façades on building surface temperature in Mediterranean climate. Innov. Biosyst. Eng. Sustain. Agric. Forest. Food Prod. **67**, 179–188 (2019)
10. Blanco, I., Schettini, E., Scarascia, G., Vox, G.: Thermal behaviour of green façades in summer. J. Agric. Eng. **49**(3), 183–190 (2018)
11. ENVI-met V4.4, 2020. Urban Environment Through Holistic Microclimate Modeling. https://www.envi-met.com/. Accessed 13 June 2020
12. Huttner, S.: Further Development and Application of the 3D Microclimate Simulation ENVI-met, Dissertation. Johannes Gutenberg-Universitat, Mainz, Germany (2012)
13. Simon, H.: Modelling urban microclimate Development, implementation and evaluation of new and improved calculation methods for the urban microclimate model ENVI-met, Dissertation. Johannes Gutenberg-Universitat, Mainz, Germany (2016)
14. Evola, G., Costanzo, V., Magrì, C., Margani, G., Marletta, L., Naboni, E.: A novel comprehensive workflow for modeling outdoor thermal comfort and energy demand in urban canyons: results and critical issues. Energy Build. **216**, 109946 (2020)
15. Blanco, I., Convertino, F., Schettini, E., Vox, G.: Wintertime thermal performance of green façades in a Mediterranean climate, WIT transactions on ecology and the environment. Urban Agric. City Sustain. **II**(243), 47–56 (2020)

Recycling Volcanic Ash and Glass Powder in the Production of Alkali Activated Materials

Loredana Contrafatto[1](✉), Daniele Calderoni[1], Salvatore Gazzo[1], and Enrico Bernardo[2]

[1] University of Catania, Department of Civil Engineering and Architecture, Catania, Italy
loredana.contrafatto@unict.it, daniele.calderoni@phd.unict.it, salvatore.gazzo@dica.unict.it

[2] University of Padova, Department of Industrial Engineering, Padova, Italy
enrico.bernardo@unipd.it

Abstract. The environmental sustainability of new materials and production processes is of crucial importance nowadays. The mechanical properties of a new binder obtained by light activation of waste pozzolanic materials are investigated. The compound consists of volcanic ash from street sweeping after pyroclastic eruptions of Mt. Etna and waste glass fine dust from the recycling process of glass containers. A single component NaOH solution at low molarity (3M) is used. The mechanical strength is determined through of three-point bending tests and compression tests. The results of the performed basic physical and mechanical tests show that the material is promising to be used in the construction field, with compressive strength between 15 and 20 MPa.

Keywords: Alkali activated material · Glass powder · Volcanic ash · Recycled waste

1 Introduction

The reuse of waste materials, both of natural origin and deriving from industrial production processes nowadays plays a key role in international environmental protection policies. On the one hand, the use of waste materials reduces the exploitation of depleting natural resources, on the other hand it aims at creating production processes that require less energy, thus obtaining a reduction in greenhouse gas emissions. Therefore, the concept of sustainability relies on both re-usability and simple processing.

The use of volcanic ash as replacement of Portland cement in mortars and concretes is widely documented in the literature [1], both in mortars and concrete. The reuse of glass powders has also been extensively studied. Glass powder is used both as Supplementary Cementitious Material (SCM), replacing Portland

J. Littlewood and R. J. Howlett (Eds.): SEB 2022, SIST 336, pp. 47–55, 2023.
https://doi.org/10.1007/978-981-19-8769-4_5

cement in mortar mixes [2], and as alternative fine aggregate in replacement up to 100% of natural sand in geopolymer materials [3]. In [4] it is shown that glass powder also introduces a beneficial effect on durability of glass powder mortars, especially on resistance to acid and sulfate attacks as well as to the alkali-silica reaction.

More recently, new potentials of these materials as binders not containing Portland cement have been investigated [5]. Different polymerization and activation processes in an alkaline environment have been studied [6,7]. For instance, Djobo in [8] studied the mechanical properties and durability of geopolymer mortars based only on volcanic ash. However, in almost all cases high molarity alkaline solutions are used, based on not eco-friendly alkali.

The paper deals in particular with the reuse of the specific type of volcanic ash erupted by Mt. Etna, that has less pronounced pozzolanic properties than the volcanic ash of other volcanoes on the planet. The reuse of Etna volcanic ash in traditional construction materials, such as cement, concrete, insulating mortar, has been largely investigated in [9–11]. The application proposed in the present paper concerns a new sustainable binder, which falls in the context of Alkali Activated Materials. Different mixtures, containing Etna volcanic ash and waste glass powder in the same proportion, are described. The main mechanical properties of the designed mixtures will be discussed with reference to the results of the experimental testing, carried out varying the alkaline solution-to-solid ratio.

2 Waste Materials and Sustainability Issues

In this study the attention was focused on two specific waste materials.

1. Volcanic Ash (VA) from street cleaning. A very significant natural production characterizes the world regions with active volcanoes. However, a limited exploitation of the resource is currently documented.
2. Soda-Lime Glass (SLG) fine powders from the crushing of glass containers. A significant fraction remains un-recycled, according to enhanced contamination from other materials. The fraction is discarded to reduce the risk of degradation of the production quality.

Specifically, ashes erupted by Mt. Etna are considered. The eruptive activities of the Etna volcano, located in the eastern part of the island of Sicily in the Mediterranean Sea, are of two different types. The effusive activity is characterized by the continuous emission of lava flows, in the absence of explosive phenomena. In the paroxysmal activity explosive manifestations generated by the violent and sudden expansion of volcanic gases contained in the magma are added to the emission of typically very viscous lava, causing it to fragment. During the fragmentation process, the expansion of gases produces vesicles inside the pyroclastic products. Due to these cavities the volcanic material present high porosity.

The product of the explosion are classified according to their size in: bombs (size greater than 64 mm); lapilli (size between 64 and 2 mm); ashes (dimension less than 2 mm). The pyroclasts, transported by air currents, fall and settle on urban soil, even tens of kilometers away from the emission. Based on the extent and nature of the phenomenon, even very large grain sizes are able to deposit at a great distance from the active volcanic vent. As the distance travelled by the pyroclasts increases, the size usually decreases, reaching even very fine dimensions. Figure 1 shows, for example, the products fallen on 21 September 2021 at about 25 km from the South East Crater.

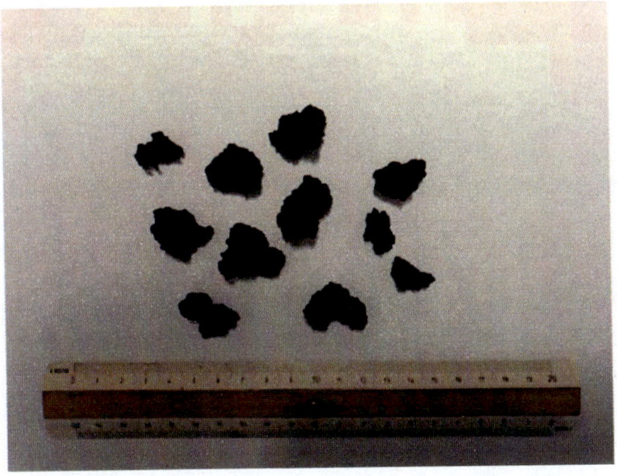

Fig. 1. Pyroclastic products from the event on 21 September 2021, Milo, Catania

The glass powder comes from a treatment plant of waste glass located in the northern Italy. It is the residue from the crushing process which cannot be further reused for the production of recycled glass.

The ongoing sustainability challenges are thus the reuse of both these materials with limited energy/material inputs and the design of marketable products (up cycling). Moreover, a further end-of-life option can be addressed: the waste-derived new products should possibly be reused as feed-stock for a second generation of products (up cycling the up cycled).

3 Alkali Activated Composite with Volcanic Ash and Recycled Glass Powder

The 1st generation of products which is proposed consists of cement-free binder materials, obtained by light activation of VA-SLG mixture. Usually, in the synthesis of volcanic ash based geopolymer pastes strong activation are performed, with high molarity NaOH solution and the introduction of synthetic additives such as sodium silicate or alumina-oxide (besides NaOH) and metakaolin [12,13]. Alternatively, phosphoric acid activation of volcanic ash is proposed in [6]. A review of new eco-friendly alternative activators in alkali activated materials has been recently reported in [14]. Strong activation of VA based geopolymer pastes containing only sodium hydroxide solution is proposed in [15].

In the present study for the activation process a light alkaline sodium hydroxide solution with concentration of 3 mol/L was prepared. A 99 % by mass of purity flaked sodium hydroxide (Marten s.r.l., Italy) was mixed with distilled wate r to obtain a single-source alkaline activator. The solution was cooled to room temperature before use.

Three alkali activated pastes with a constant VA/SLG ratio of 1:1 were prepared by varying the water-to-solid ratio w/s (0.35, 0.39, 0.42), with fixed concentrations of the NaOH solution (3M). Table 1 provides the characteristics of the designed mixtures.

Table 1. Designed mixtures, characterized by different water-to-solid ratio w/s.

Series	VA [%]	SLG [%]	w/s
A-B, H-G-L, M-P-Q	50	50	0.42
E-F, R	50	50	0.39
C-D	50	50	0.35

3.1 Paste Preparation

First, Etna volcanic ash was washed with distilled water in the volume ratio ash-to-water equal to 1:2, to remove impurity. Then, VA was then dried in oven at 100 °C for 24 h, finely milled and sieved at < 75 µm. SLG powder (Savel CS60 type) was sieved at < 75 µm too.

Mechanical stirring of each mix was carried on at 400 rpm for 30 min.

For each type of mixture not less than 6 prismatic samples 40 × 40 × 160 mm^3 were cast in three-gang moulds following standard procedures (UNI EN 196-1:2016), vibrated using a shaking table for 3 min to remove air bubbles, and cured at 75 °C for 72 h. Before being placed in the oven, the moulds were covered with a thin film of polyethylene to avoid sudden water evaporation. The samples were demolded and tested after 24 h they had been cured.

The nth specimen of the X series is labeled as Xn.

In the case of the water-to-solid ratios 0.39 and 0.42 the mixtures were prepared twice and thrice, respectively, due to uncertainties in the casting phase influencing the repetitiveness of the testing.

3.2 Test Method

The mechanical properties of hardened pastes were evaluated according to UNI EN 196-1:2016. Three-point bending tests on the 40 × 40 × 160 mm prisms and compression tests on the resulting half prisms after the breakage were carried on. The loading rate was controlled at 50 N/s ± 10 N/s and 2400 N N/s ± 200 N/s in flexural and compressive strength test, respectively. Linear shrinkage upon drying was evaluated according to EN 12617-4.

Figure 2 shows the setting for three point bending test on prismatic sample, (sample H2, $w/s = 0.42$), for the compression test on half-prism sample (sample R1, $w/s = 0.39$), and the shrinkage of prismatic samples (series A and B, $w/s = 0.35$).

(a) (b) (c)

Fig. 2. **a** Three point bending test on prismatic sample, series H; **b** compression test on half-prism sample, series R and **c** shrinkage of prismatic samples, series A-B

4 Results and Discussion

4.1 Mass Density

Figure 3 reports the mass density of all the tested specimens. From the pictures it is evident that the variation of the water-to-solid ratio does not significantly affect the mass density. The values vary from a minimum value of 1549.32 kg/m^3 to a maximum value of 1607.18 kg/m^3 in the case of mixtures with $w/s = 0.35$; from a minimum value of 1494.95 kg/m^3 to a maximum value of 1686.38 kg/m^3 in the case of mixtures with $w/s = 0.39$; from a minimum value of 1502.64 kg/m^3 to a maximum value of 1630 kg/m^3 in the case of mixtures with $w/s = 0.42$.

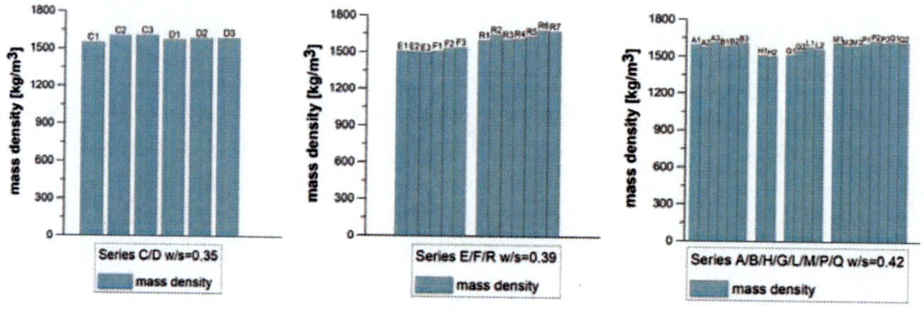

(a) (b) (c)

Fig. 3. Mass density. **a** $w/s = 0.35$; **b** $w/s = 0.39$ and **c** $w/s = 0.42$

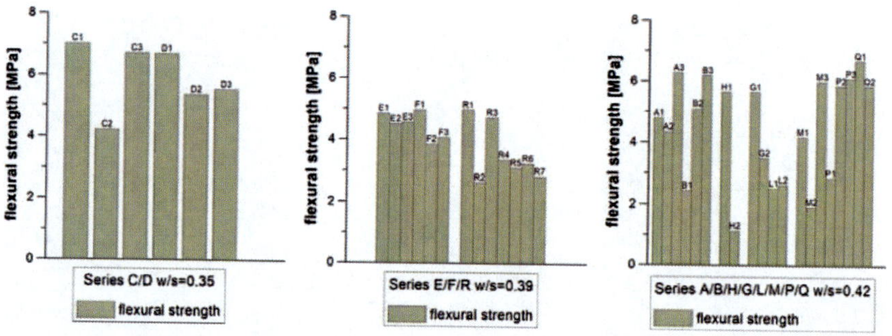

(a) (b) (c)

Fig. 4. Flexural strength. **a** $w/s = 0.35$; **b** $w/s = 0.39$ and **c** $w/s = 0.42$

4.2 Flexural and Compressive Strength

Figures 4, 5 show the results of the bending tests and of the compressive tests, respectively.

In the case $w/s = 0.35$ the flexural strength vary in the range $[4.22, 7.02] \pm 1.08$ MPa while the compressive strength is in the range $[14.25, 16.51-7.02] \pm 1.53$ MPa. In the case $w/s = 0.39$ the flexural strength vary in the range $[2.64, 4.97] \pm 0.68$ MPa while the compressive strength is in the range $[11.20, 17.77] \pm 1.89$ MPa. In the case $w/s = 0.42$ the flexural strength vary in the range $[1.14, 6.71] \pm 1.95$ MPa while the compressive strength is in the range $[12.14, 20.87] \pm 2.12$ MPa.

The higher values of the standard deviation in the case of the mixtures of the batch $w/s = 0.42$ are to be attributed to slight variations in the procedures during the casting phase which led, in some specimens, to the formation of micro-cracks on the surface of the specimens. These cracks are responsible for the greater instability of the bending test results, as they drive the initiation of the specimen fracture during the test, while they does not affect the compression test.

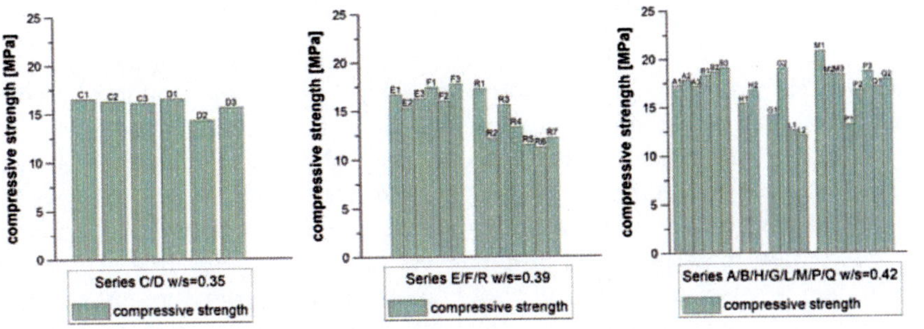

(a) (b) (c)

Fig. 5. Compressive strength. **a** $w/s = 0.35$; **b** $w/s = 0.39$ and **c** $w/s = 0.42$

4.3 Free Shrinkage

The minimum shrinkage was exhibited by the mixtures with the lowest solution content ($w/s = 0.35$), whereas the maximum value was observed in series E and F ($w/s = 0.39$) (Fig. 6).

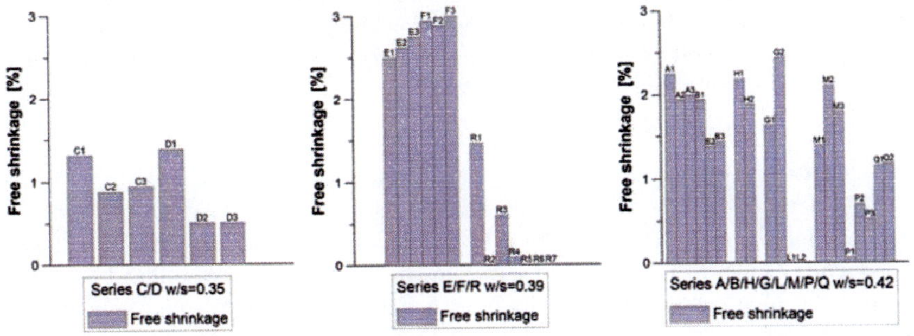

(a) (b) (c)

Fig. 6. Free shrinkage. **a** $w/s = 0.35$; **b** $w/s = 0.39$ and **c** $w/s = 0.42$

The variability of the shrinkage values can be attributed to the sequence followed in the casting phases of the different batches. Furthermore, the workability of the mixtures, which is extremely low in the case $w/s = 0.35$, certainly played a fundamental role. The paste in this case is very dry. For $w/s = 0.42$ a significant bleeding phenomenon was observed. Bleeding was not present for w/s=0.35 and not very pronounced in the case w/s = 0.39.

5 Concluding Remarks

Volcanic Ash has a great potential for the obtainment of new products based on (light) alkali activation in combination with Soda-Lime Glass.

Preliminary results obtained through basic physical and mechanical tests on the proposed new binder, which is part of the Alkali Activated Materials (AAM) framework, show that the compound appears to be promising for use in construction materials. The proposed binder is completely free of Portland cement and consists exclusively of waste materials. The production process does not require additional resources, neither natural nor artificial, except for the use of caustic soda in low concentration. Caustic soda is a low cost material, easily usable at low molarity. The new composite therefore respects a sustainability principle based on the principles of re-usability and simple processing.

Further developments are directed towards the study of the mineralogical and chemical characteristics of the new compounds and towards the design of new mixtures capable of developing better mechanical performance, even with maturation at room temperature. Finally, a further sustainability feature of the proposed products is their re-usability as glass-ceramic foams, by thermal treatment (thermal foaming) [16,17].

Acknowledgements. The experimental study was conducted with the financial support of University of Catania, Italy, Program *Piano di incentivi per la ricerca di Ateneo 2020-2022 (Pia.ce.ri.)*, Research Line 2, Rewards D and E to the Department of Civil Engineering and Architecture. The authors are grateful to Eng. E. Mangano for his contribution in the material testing activity.

References

1. Contrafatto, L.: Volcanic Ash. In: Siddique, R., Belarbi, R. (Eds.) Sustainable Concrete Made with Ashes and Dust from Different Sources, Woodhead Publishing Series in Civil and Structural Engineering, pp. 331–418. Woodhead Publishing (2022). https://doi.org/10.1016/B978-0-12-824050-2.00011-5
2. Bostanci, L.: Effect of waste glass powder addition on properties of alkali-activated silica fume mortars. J. Build. Eng. **29**, 101154 (2020)
3. Mejía de Gutiérrez, R., Villaquirán-Caicedo, M.A., Guzmán-Aponte, L.A.: Alkali-activated metakaolin mortars using glass waste as fine aggregate: mechanical and photocatalytic properties. Constr. Build. Mater. **235**, 117510 (2020)
4. Idir, R., Cyr, M., Pavoine, A.: Investigations on the durability of alkali-activated recycled glass. Constr. Build. Mater. **236**, 117477 (2020)
5. Samarakoon, M., Ranjith, P., De Silva, V.: Effect of soda-lime glass powder on alkali-activated binders: rheology, strength and microstructure characterization. Constr. Build. Mater. **241**, 118013 (2020)
6. Djon Li Ndjock, B., Robayo-Salazar, R., Mejía de Gutiérrez, R., Baenla, J., Mbey, J., Cyr, M., Elimbi, A.: Phosphoric acid activation of volcanic ashes: influence of the molar ratio R=(MgO + CaO) / P2O5 on reactivity of volcanic ash and strength of obtained cementitious material. J. Build. Eng. **33**, 101879 (2021)

7. Lemougna, P.N., MacKenzie, K., Chinje Melo, U.: Synthesis and thermal properties of inorganic polymers (geopolymers) for structural and refractory applications from volcanic ash. Ceram. Int. **37**, 3011–3018 (2011)
8. Djobo, J., Elimbi, A., Tchakoute, H.K., Kumar, S.: Mechanical properties and durability of volcanic ash based geopolymer mortars. Constr. Build. Mater. **124**, 606–614 (2016)
9. Contrafatto, L.: Recycled Etna volcanic ash for cement, mortar and concrete manufacturing. Constr. Build. Mater. **151**, 704–713 (2017)
10. Contrafatto, L., Lazzaro Danzuso, C., Gazzo, S., Greco, L.: Physical, mechanical and thermal properties of lightweight insulating mortar with recycled Etna volcanic aggregates. Constr. Build. Mater. **240**, 117917 (2020). https://doi.org/10.1016/j.conbuildmat.2019.117917
11. Contrafatto, L., Gazzo, S., Purrazzo, A., Gagliano, A.: Thermo-mechanical characterization of insulating bio-plasters containing recycled volcanic pyroclasts. Open Civil Eng. J. **14**(1), 66–77 (2020)
12. Tchakoute, H., Elimbi, A., Mbey, J., Ngally Sabouang, C., Njopwouo, D.: The effect of adding alumina-oxide to metakaolin and volcanic ash on geopolymer products: a comparative study. Constr. Build. Mater. **35**, 960–969 (2012)
13. Tchakoute, H., Elimbi, A., Yanne, E., Djangang, C.: Utilization of volcanic ashes for the production of geopolymers cured at ambient temperature. Cement Concr. Compos. **38**, 75–81 (2013)
14. Mendes, B.C., Pedroti, L.G., Vieira, C.M.F., Marvila, M., Azevedo, A.R., Franco de Carvalho, J.M., Ribeiro, J.C.L.: Application of eco-friendly alternative activators in alkali-activated materials: a review. J. Build. Eng. **35**, 102010 (2021)
15. Zhou, S., Lu, C., Zhu, X., Li, F.: Upcycling of natural volcanic resources for geopolymer: comparative study on synthesis, reaction mechanism and rheological behavior. Constr. Build. Mater. **268**, 121184 (2021)
16. Marangoni, M., Secco, M., Parisatto, M., Artioli, G., Bernardo, E., Colombo, P., Altlasi, H., Binmajed, M., Binhussain, M.: Cellular glass ceramics from a self foaming mixture of glass and basalt scoria. J. Non-Cryst. Solids **403**, 38–46 (2014)
17. Bernardo, E., Contrafatto, L.: Double-life construction materials from discarded glass and volcanic ash. In: 46th International Conference and Expo on Advanced Ceramics and Composites (ICACC2022) Virtual Conference. 16th International Symposium on Advanced Processing and Manufacturing Technologies for Structural and Multifunctional Materials and Systems (2022)

Thermal Environment Retrofitting of Outdoor Activity Spaces in Old Settlements in Severe Cold Regions of China

Yujing Liu and Jin Hong(✉)

School of Architecture, Key Laboratory of Cold Region Urban and Rural Human Settlement Environment Science and Technology, Ministry of Industry and Information Technology, Harbin Institute of Technology, Harbin 150001, China

442620874@qq.com

Abstract. Outdoor environmental problems have long been common in old settlements in severe cold regions. In particular, the narrow outdoor activity space and the poor thermal environment seriously impair the community environment and residents' outdoor comfort. Therefore, it is necessary to transform the outdoor space environment of the old settlements in severe cold regions. This paper selects the old residential area of Wenlin as the research object, which is located in Harbin, a typical severe cold city. Based on the current situation of the community, we comprehensively analyze the spatial thermal comfort level of outdoor activities through investigation, actual measurement, and numerical simulation. Thus, the renovation scheme is put forward. It is found that the most effective measures to improve the thermal environment of outdoor activity space are some technical measures such as demolishing illegal private buildings, changing ground material, optimizing residential greening, and increasing wind walls. This study will provide a reference for the renovation of old urban settlements in severe cold regions and the improvement of human living environment performance.

Keywords: Severe cold regions · Renovation of old settlements · Outdoor spaces · Thermal environment

1 Introduction

Currently, China is in a general environment of promoting the renovation of old urban settlements in a comprehensive manner. On the one hand, the era of big demolition and big construction of renewal has basically passed. On the other hand, the high property prices in big cities make many people unable to improve their living conditions by choosing their homes. With the improvement of economic level and the arrival of new crown epidemic emergencies, people's demand for living environment is getting higher and higher, which constitutes a source of motivation for the renovation of the stock of residences. Therefore, the renovation of old settlements is inevitable [1]. The development of urban settlements has prompted the need for higher quality outdoor spatial environments in settlements [2]. Outdoor space in older settlements is an important spatial vehicle to

J. Littlewood and R. J. Howlett (Eds.): SEB 2022, SIST 336, pp. 56–65, 2023.
https://doi.org/10.1007/978-981-19-8769-4_6

address the growing demand for a better life, and is an important component to improve the quality of urban space. Therefore, such renovation should take into account both the residential comfort of the residents and the benefit of their outdoor activities. Due to the frequent influence of cold Siberian air, winters in cold cities are cold and long with strong winds [3]. There is also a very hot period in the summer.Such special climatic conditions can directly affect the comfort of people during outdoor activities and the use of space. As the outdoor activity space in old settlements is an important activity place for residents, designers should use microclimate design means to improve the thermal environment in order to stimulate the vitality of the space and improve the space utilization [4].

With the in-depth research on the microclimate of urban outdoor public spaces, researchers found that the thermal environment has a great influence on the comfort of outdoor crowd activities [5–7]. Crowd preferences in summer and winter are different, the parameters affecting crowd activity are mainly air temperature, solar radiation, wind speed, while the influence of relative humidity is weaker [8, 9]. In addition, a large number of researchers have studied the factors affecting the thermal environment of public spaces, showing that the layout of buildings, the greening of spaces, the choice of water bodies and substrates, as well as the choice of materials and colors of leisure facilities, can directly affect the outdoor thermal environment [10–12]. For severe cold regions, complex greening structure is beneficial to regulate the microclimate of outdoor space [13]. The floor material will also directly affect the thermal environment of the space, high heat absorption and low reflectivity flooring material is ideal for underlayment in cold regions [14].

Although the research on outdoor thermal environment has been more comprehensive, the research on outdoor thermal environment in old urban settlements in cold regions is still less. Studies on old settlements have also mostly been conducted from the planning and design levels such as land use and traffic layout. Therefore, the renovation design strategy proposed in this paper from the perspective of thermal environment is innovative and can make up for the lack of research in this field. In order to improve the vitality and efficiency of outdoor spaces in old urban settlements in cold regions, this paper focuses on the Wenlin old settlement in Harbin, a typical city in severe cold regions. The thermal comfort level of different outdoor activity spaces is studied and evaluated by using the methods of investigation, actual measurement, and numerical simulation. Exploring design strategies to enhance the exterior activity space environment of old settlements in cold cities. In recent years, China has attached great importance to the renovation of old urban communities. It is hoped that this study has some theoretical and practical significance for the implementation of outdoor environment improvement projects in old settlements in severe cold regions.

2 Field Measurements and Simulation Software Validation

2.1 Study Site and Methods

Wenlin Settlement is a very representative enclosed old settlements in severe cold regions, built in 1996, with 11 residential buildings, and most of the community residences are arranged in a north-south direction. The settlement was included as one of the key renovation projects for old settlements in Harbin City. The current situation of the settlement is shown in Fig. 1. The roads are made of concrete and the activity sites are made of red bricks. There is a serious lack of outdoor public activity space in the settlement, and a large number of unauthorized buildings make the outdoor public space layout crowded. The amount of greenery in the settlement is very small, and the public greenery is mainly trees, shrubs and grasses, mainly distributed around the activity site.

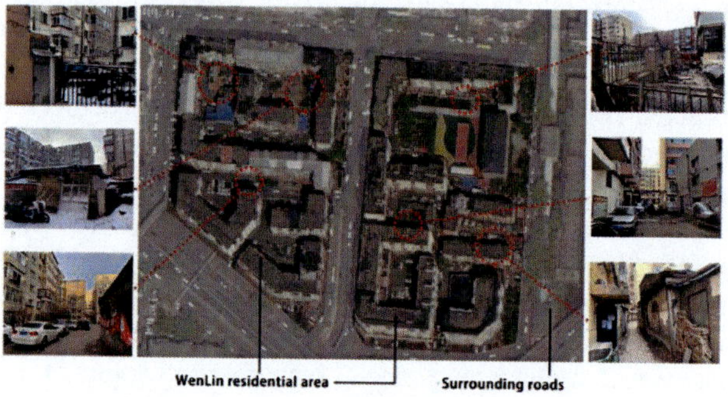

Fig. 1. Study area plan and node status

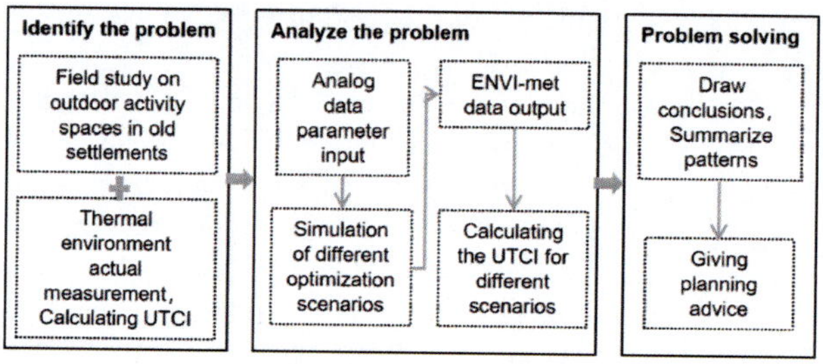

Fig. 2. Research framework

Figure 2 shows the methodology of this study. Field measurements of the thermal environment at the study site were carried out using technical equipment to measure factors including temperature, humidity, wind speed and solar radiation. The Universal Thermal Climate Index (UTCI) was integrated and calculated to explore the current state of thermal comfort for residents' outdoor activities. Combined with previous studies, different optimization schemes were designed and analyzed using numerical simulations to make an objective analysis of the UTCI regulation effects produced by different optimization schemes, leading to research conclusions.

2.2 Evaluation of the Thermal Environment of the Outdoor Activity Space

Harbin is one of the representative cities of typical cold regions in China. According to the statistics from the website of National Meteorological Information Center. The ground climate background information of the last ten years is selected to analyze the climate characteristics of the severe cold region in Northeast China. The average temperature of the coldest month (January) in the coldest region of Northeast China is about −30 to −10 °C, and the average relative humidity is between 60 and 80%; the average temperature of the hottest month (July) is about 15–26°C, and the average relative humidity is mainly between 70 and 90%. Comprehensive meteorological data from weather stations were used to select good weather with few clouds and no rain or snow for the actual test. Test days were chosen: summer July 22, 2021 with an average temperature of 27.0 °C and an average relative humidity of 81%. In winter, the average temperature on December 25, 2021 was −23.7 °C and the average relative humidity was 70.5%. The test day had the typical climate characteristics of winter and summer seasons, and the test results were highly representative. This study focused on the time period when people are more likely to participate in outdoor activities. Therefore, the thermal environment was measured from 8:00 to 18:00 in the center of a small public activity square in the residential area. The air temperature and relative humidity were recorded with a BES-02 temperature and humidity collector, and the temperature of the fog black sphere with a diameter of 0. 08 m was recorded with a BES-01 temperature collector. Wind speed was recorded using a Kestrel 5500 mini weather station with the temperature and humidity collector placed inside a foil radiation shield to protect it from direct sunlight and to maintain good ventilation (Fig. 2), and test data were recorded at 1 min intervals.

The Universal Thermal Climate Index (UTCI) was used for the thermal environment evaluation index. There are differences in the acceptance and assessment of thermal environment in different regions of the population, so it is necessary to correct the range of UTCI corresponding to the level of thermal stress in different regions. The corrected UTCI ranges for the study site Harbin are shown in Table 1 [15].

Table 1. UTCI ranges for different levels of thermal stress

Thermal stress levels	Standard UTCI/°C	Calibration UTCI/°C
Extreme thermal stress	>46	>49. 4
Very strong thermal stress	38 to 46	40. 9 to 49. 4
Strong thermal stress	32 to 38	29. 1 to 40. 9
Mild thermal stress	26 to 32	23. 0 to 29. 1
No thermal stress	9 to 26	−3. 8 to 23. 0
Slight cold stress	0 to 9	−7.2 to −3.8
Gentle cold stress	−13 to 0	−18.3 to −7.2
Strong cold stress	−27 to −13	−25.6 to −18.3
Very strong cold stress	−40 to −27	−30.2 to −25.6
Extreme cold stress	<−40	<−30.2

As shown in Fig. 3, which shows the UTCI thermal comfort index for the active site in the Wenlin Old Settlement during the hours of 08:00 to 18:00 in summer and winter

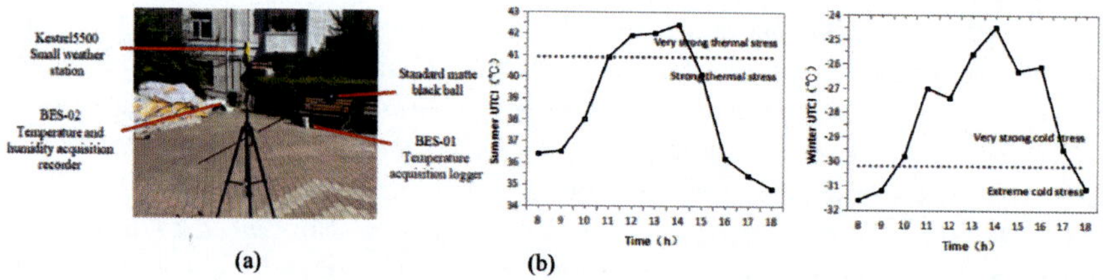

Fig. 3. **a** Experimental apparatus set-up; **b** settlement activity site UTCI (°C) in summer and winter

respectively. It shows that the duration of the very strong heat stress level reaches about 3.5 h in summer and the other time periods are at strong thermal stress levels. There is a duration of extreme cold stress levels of about 2 h in winter, and very strong cold stress levels for the other time periods. Therefore, it is necessary to improve the comfort level of residents' activities in different seasons.

2.3 Evaluation of the Thermal Environment of the Outdoor Activity Space

The ENVI-met V4 model was used for all simulations in this study. The corresponding model is established in the software based on the actual measured settlement planning layout of the field research. The main model domain area representing the current situation of the study area was established in a 150 × 120 × 25 grid, where the horizontal and vertical spatial resolutions were dx = 2 m, dy = 2 m and dz = 2 m, respectively, and the height of the study model was ≥ 2 times the maximum building height. In order to reduce the influence of the model boundary and improve the stability and accuracy of the simulation, 10 nested grids were added at the periphery of the main model area. The simulation location is Harbin, Heilongjiang Province, with coordinates 126.57, 45.93. The simulation time is set from 0:00 to 20:00 on July 22, 2021 and December 25, 2021, with a total simulation time of 20 h. The hourly air temperature, relative humidity, wind direction and wind speed are obtained from the meteorological station in Harbin City as the meteorological boundary conditions, according to Some of the model parameters were set as shown in Table 2.

2.4 Simulation Software Validation

The variation of ambient air temperature is controlled by different factors, such as the thermal characteristics of surrounding natural and artificial surfaces, ventilation conditions and ambient shading rate. It is reasonable to use the air temperature as a composite index to represent the thermal state of the test site. Therefore, the measured and simulated obtained air temperatures are compared. Figure 4 shows that the root mean square error (RMSE) is 1.05 °C and 1.52 °C , the mean error (MAE) is 0.99 °C and 1.38 °C for summer and winter, respectively, both close to 1. The index of agreement (d) is 0.84 and 0.82, respectively, both close to 0. Figure 5 shows that the results of the variables are linearly correlated. The R-squared values close to 1 indicate a strong correlation between simulated and measured air temperatures. The validation shows that the simulated air temperature at 1.5 m above the ground level is basically consistent with the field

Table 2. Model parameter settings

Simulation date	2021.7.22	2021.12.25
Maximum air temperature	32.8 °C	−19.6 °C
Minimum air temperature	21.2 °C	−27.8 °C
Maximum relative humidity	98%	86%
Minimum relative humidity	64%	55%
Wind speed at 10 m height	1.44 m/s	3.30 m/s
Wind direction	202.5	225
Solar radiation conditioning factor	0.71	0.96
Specific humidity of air at 2500 m height	1.35 g/kg	10.17 g/kg
Temperature/humidity of the upper layer of soil (0–20 cm)	26.1 °C/30.8%	−18.5 °C/33.5%
Temperature/humidity in the upper layer of the soil (20–50 cm)	22.2 °C/31.8%	−10.0 °C/40.5%
Temperature/humidity in deeper layers of soil (>50 cm)	21.4 °C/33.2%	−6.0 °C/37.5%

measurements, and the model can simulate the thermal environment of the settlement in summer and winter more accurately.

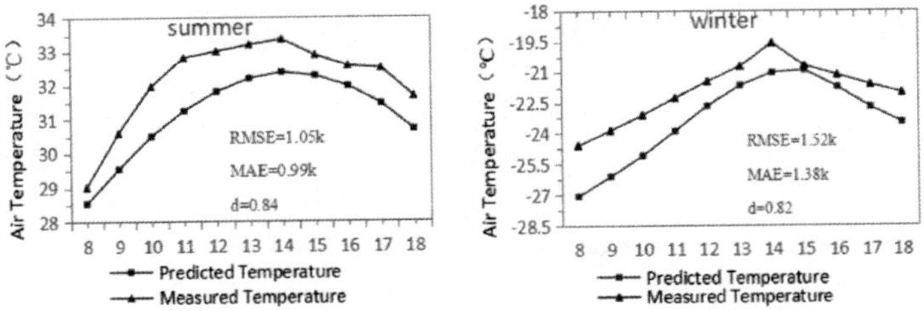

Fig. 4. Comparison of measured and simulated air temperature values in summer and winter

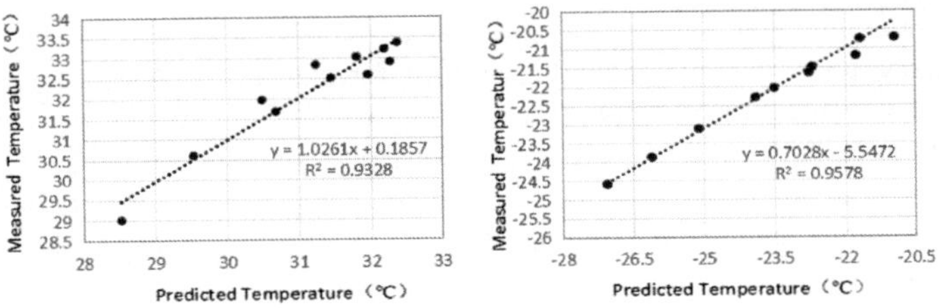

Fig. 5. Correlation analysis between measured and simulated values in summer and winter

3 Transformation Strategy

3.1 Transformation Programme

First, the original scenario was established by removing the private construction violations in the settlement and expanding the public activity space (Fig. 6c, d). After performing the removal of private building violations to expand the activity area, three improvement scenarios were simulated for the Wenlin settlement in order to study the contribution of greenery and surface materials to the outdoor thermal comfort. The three improvement scenarios are as follows: 1. Use ground materials with high heat absorption and low reflectivity (Fig. 6e); 2. Increase the greening area of road space and public activity space and add gardens. Increase the complexity of greening structure, such as shrubs, trees and evergreen plants, and add wind-blocking walls in winter (Fig. 6f); 3. Combination of scheme 1 and scheme 2 (Fig. 6g). The measurement points were chosen in the road, small square and garden regions, respectively.

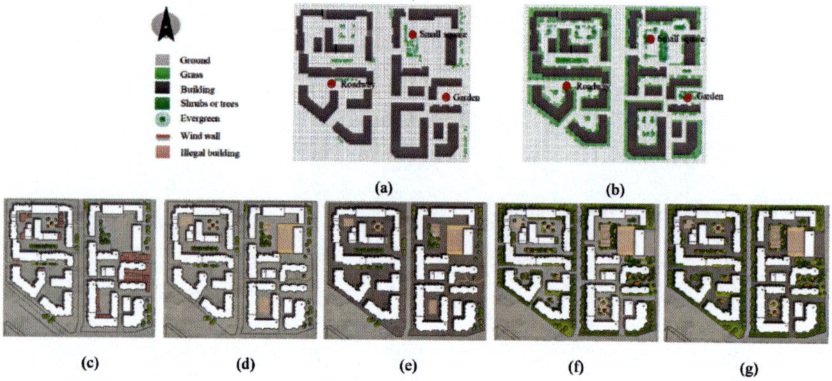

Fig. 6. **a** Model construction I; **b** model construction II; **c** field scene; **d** original scene; **e** ground material scheme scene; **f** greening scheme scene; **g** combined scheme scene

3.2 Summer UTCI Analysis

Figure 7 shows the change of outdoor thermal comfort state for four scenarios in summer. It was observed that after the modification of the low-reflectivity ground material, the UTCI in the road, active small square and garden study areas were all reduced in thermal stress levels compared to the original scenario. However, the decrease in UTCI was very small, at most reducing the peak UTCI by 0.82 °C. The duration of the very strong thermal stress level in both the small square and garden areas was reduced by 2 h. The addition of the greening scenario significantly reduces the peak UTCI, changes the heat stress level (see Table 1) and improves outdoor thermal comfort in summer. For example, compared to the original scenario, the duration of the very strong heat stress level in the small square after improving the greening of the settlement was reduced by 5.5 h, significantly reducing the peak UTCI by up to 4.5 °C. The garden and road scenes, on the other hand, reduced the duration of very strong thermal stress levels by 5 and 4 h, respectively, with maximum differences in UTCI changes of 4.1 °C and 4.4 °C , respectively. Although the impact of changing the ground material on improving outdoor thermal comfort is currently found to be very small, this combined scenario was still found to minimize the peak UTCI (up to 4.8 °C) and the duration of very strong thermal stress. This implies that greenery is a major factor in improving UTCI and should be given sufficient attention when improving the summer thermal environment or planning the settlement environment.

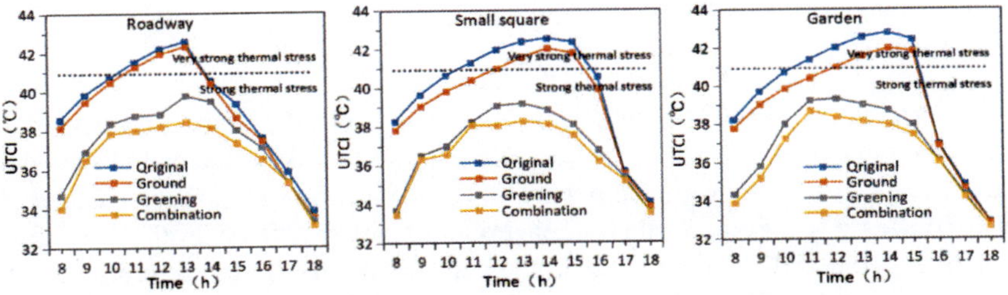

Fig. 7. UTCI for four scenes in winter in the road, small square and garden area

3.3 Winter UTCI Analysis

Figure 8 shows the UTCI for the four winter scenarios. it can be observed that the heat stress level increases for all three retrofit scenarios compared to the initial scenario, which indicates that all three scenarios have a positive impact on the UTCI. But in the overall view, the UTCI of the low-reflectivity ground material scenario and the greenery and wind walls scenario do not change much compared to the original scenario. The ground material scenario increases the peak UTCI by 0.6 °C at most, and has almost no effect on the duration of the thermal stress level. In contrast, the greenery and wind walls scenario raised the peak UTCI by at most 1.9 °C. The duration of extreme cold stress levels in both the garden and road areas was reduced by 0.5 h. The duration of the very strong cold stress level in the garden and small square areas was also reduced by 0.5 h. This means that greenery has a significant positive effect on improving UTCI. Although most greenery has lost its leaves in winter, dense shrub branches or tree trunks can block cold winter winds. The combination of greenery and the use of wind walls can effectively improve outdoor thermal comfort in winter. In addition, although the improvement in cold stress levels was not significant for the ground material scenario, the combined solution still maximizes the UTCI peak (up to 2.4 °C) and reduces extreme or very strong cold stress levels for the longest duration as well.

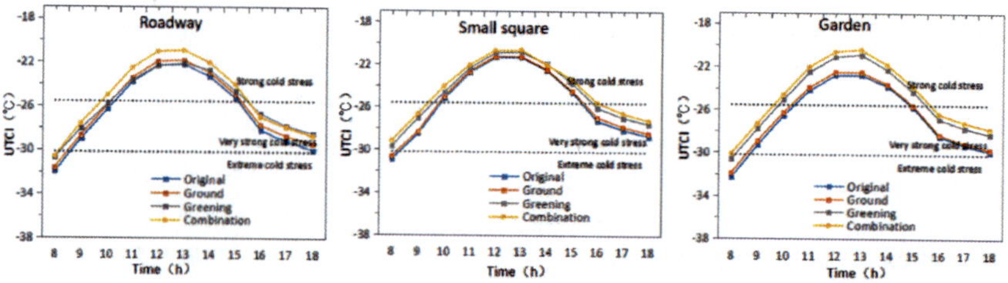

Fig. 8. UTCI for four scenes in winter in the road, small square and garden area

4 Conclusion

In the renovation design of old settlements, it is recommended to demolish all the illegal private buildings that affect the residents' life, and to transform the demolished open space into an activity site with appropriate greening planting and facility vignettes. For severe cold regions, the outdoor thermal environment is very different from other regions. Winters are cold and more prolonged, while summers are short. Therefore, targeted solutions should be proposed, with winter as the main consideration. The simulation shows that different ground materials will directly affect the air temperature and average radiation temperature of the outdoor space. In both summer and winter, low-reflectivity ground material can improve the outdoor thermal environment in severe cold urban settlements, but the improvement is not significant compared to the greening scheme. It is worth noting that low-reflectivity flooring materials can increase the ambient air temperature after solar exposure. Therefore, shading treatment should be given to the lower mat surface to prevent the lower mat surface from absorbing large amounts of heat causing urban heat island problems. In addition, greening can significantly improve the environmental comfort of outdoor space microclimate in old settlements. Therefore, the renovation should enrich the existing greenery in the settlement, make the green structure more complex, increase the green area in the vacant public space and add small gardens. Since most of the greenery in the severe cold regions is deciduous in winter, it can be used in combination with wind walls to improve the effect.

In addition to ground materials, greenery and wind walls, the choice of materials and colors for facility vignettes and building facades should also take into account the climatic conditions of the cold city and meet the comfort of use. In short, the outdoor activity space of old settlements should fully consider the outdoor thermal environment. Traditional settlement renovation usually focuses only on the replacement of old facilities. However, in today's social development, attention should also be paid to the spiritual aspect and health needs of residents. Old settlements renovation should create a healthy, harmonious and beautiful living environment while promoting sustainable urban development.

References

1. Wei, Z., Chen, Y.: Demand, implementation dilemma and planning countermeasures for ageing-friendly retrofitting in old urban settlements: a synergistic governance discussion between endogenous dynamics and external interventions. Senior Housing **40**(12), 27–34 (2020)

2. RuiXuan, L., XiangHui, L.: Spatial environment of urban settlements under the influence of modern planning theory. In: 2008 China Urban Planning Conference. Dalian, Liaoning, China (2008)
3. Hong, L., Qing, Y.: Scientific concept of building a livable urban environment in the context of severe cold climate. Urban Plan. **10**, 26–31 (2008)
4. Leng, H., Liang, S., Yuan, Q.: Outdoor thermal comfort and adaptive behaviors in the residential public open spaces of winter cities during the marginal season. Int. J. Biometeorol. **64**(2), 217–229 (2019). https://doi.org/10.1007/s00484-019-01709-x
5. Nikolopoulou, M., Lykoudis, S.: Use of outdoor spaces and microclimate in a Mediterranean urban area. Build. Environ. **42**(10), 3691–3707 (2007)
6. Chen, H., Ooka, R., Harayama, K., et al.: Study on outdoor thermal environment of apartment block in Shenzhen, China with coupled simulation of convection, radiation and conduction. Energy Build. **36**, 1247–1258 (2004)
7. Yumeng, J., Hong, J., Jian, K., et al.: Study on thermal environment and thermal comfort of public space in urban settlements in severe cold regions. Build. Sci. **37**(12), 29–37 (2021)
8. Hong, J., Peng, C., Liang, Q.: Study on the correlation between microclimate comfort and number of participants in a square in a severe cold region. Build. Sci. **33**(10), 1–7 (2017)
9. Lin, T., Tsai, K., Liao, C., et al.: Effects of thermal comfort and adaptation on park attendance regarding different shading levels and activity types. Build. Environ. **59**, 599–611 (2013)
10. Huang, Q., Meng, X., Yang, X., et al.: The ecological city: considering outdoor thermal environment. Energy Proc. **104**, 177–182 (2016)
11. Huang, J., Zhou, C., Zhuo, Y., et al.: Outdoor thermal environments and activities in open space: an experiment study in humid subtropical climates. Build. Environ. **103**, 238–249 (2016)
12. Lai, D., Liu, W., Gan, T., et al.: A review of mitigating strategies to improve the thermal environment and thermal comfort in urban outdoor spaces. Sc. Total Environ. **661**, 337–353 (2019)
13. Hong, J., Huangyu, L., Yujie, L.: Study on the influence of vegetation structure on winter and summer microclimate in urban settlements in severe cold regions. Scenic Garden **25**(10), 12–15 (2018)
14. Xiaowei, J.: Impact of urban subsurface on outdoor radiation field and thermal environment. Xi'an University of Architecture and Technology (2019)
15. Jin, H., Siqi, L., Jian, K.: Thermal comfort range and influence factor of urban pedestrian streets in severe cold regions. Energy Build. **198**, 197–206 (2019)

A Novel Laboratory Procedure to Determine Thermal Conductivity of Green Roof Substrates

Stefano Cascone[1(✉)] and Antonio Gagliano[2]

[1] Mediterranea University of Reggio Calabria, 89124 Reggio Calabria, RC, Italy
stefano.cascone@unirc.it
[2] University of Catania, 95124 Catania, CT, Italy

Abstract. Green roofs are recognized as a solution able to reduce building energy consumption and to increase envelope thermal performance. Nowadays, new materials are continuously proposed and investigated for green roof layers to optimize environmental and economic sustainability. However, in absence of specific national and international standards for laboratory calculations of the thermal performance of substrate for green roofs, thermophysical characterization of substrates generally excluded that thermal conductivity should be modified as water content changes from dry to saturated. In this paper, a novel laboratory testing method was proposed to assess thermal conductivity of substrates for green roofs. The tests were performed by varying the water content into substrates and saturated, dry, and moist conditions were analyzed. Three commercial substrates different in composition were investigated. Thermal conductivity measurements were conducted using the HFM 436/3/1 Lambda and the TLS-100 devices to cross-referencing the measurements obtained. The results for all the substrates demonstrated that thermal conductivity fluctuates by varying water content. At moist conditions, the thermal conductivity value is an average between the one at saturated and dry conditions. The lower thermal conductivity is obtained in dry condition while the higher one is obtained in saturated condition due to water influence. Since the thermal conductivity was consistent with those provided by the company manufacturing commercial substrates, the proposed testing method could be considered reliable and it could be applied to different substrates, especially concerning the HFM Lambda device. This laboratory testing can potentially provide a useful guideline for assessing thermal conductivity of green roof substrates.

Keywords: Growing media · Testing method · Water content

1 Introduction

Nowadays, global warming and rising energy prices impacts daily lives of citizens from environmental and economic point of view [1]. Because the built environment accounts for two-thirds of all greenhouse gas emissions, requiring a huge quantity of resources, it is necessary finding sustainable alternatives to the traditional construction techniques [2–4].

J. Littlewood and R. J. Howlett (Eds.): SEB 2022, SIST 336, pp. 66–75, 2023.
https://doi.org/10.1007/978-981-19-8769-4_7

Among such solutions, green roofs represent one of the most widely investigated and acknowledged sustainable approaches able to reduce building energy consumption and CO_2 emissions and to provide many ecosystem services, such as reduction in urban heat island effect, decrease in storm water runoff, optimization of biodiversity, cleaning of runoff water and surrounding air and increase in building energy-savings [5].

Green roof is a roof that is covered with substrate layer planted with vegetation, as well as other functional layers (e.g., drainage and filter layer) [6]. Among all these layers, the substrate and drainage layer are the most important ones in water runoff reduction, peak flow reduction, water quality improvement, and thermal benefits [7]; thus, they need to be strategically selected to maximize the benefits of green roofs [8].

In recent years, several energy balance models were developed and refined to estimate heat fluxes through green roofs and building energy savings from green roofs. These savings can be predicted from models if the specific properties of the main layers which form a green roof are known. Therefore, thermal conductivity data was investigated for several green roof substrates. Sailor and Hagos [9] conducted a laboratory study to measure thermal conductivity, volumetric heat capacity, and thermal diffusivity of 12 green roof soil samples of varying composition. Ouldboukhitine et al. [10] investigated the thermo-physical properties of green roofs by correlating the thermal conductivity of the substrate with the water content for different substrates and maximum water capacities. Next, the authors characterized the moisture storage using the dynamic vapor sorption technique to determine both the sorption and desorption isotherms for three different temperatures as well as the moisture buffer capacity. Finally, they characterized the microstructural properties of green roof substrate using mercury intrusion porosimetry to measure the porosity range of the substrate and to compare this porosity range with different concrete porosities. Pianella et al. [11] measured thermal conductivity values of three substrates with steady-state and transient techniques under three moisture conditions. In addition, the authors measured specific heat capacities of the green roof substrates with a transient technique. Steady-state measurements were performed with a "k-Matic" apparatus while transient measurements with KD2 Pro needles. Barozzi et al. [12] carried out a multi-year laboratory testing campaign on 8 samples which aimed to define the thermal resistance reference values of growing media as a function of density and water content. Fabiani et al. [13] investigated the impact of water content on the thermal performance of different disaggregated materials from green roofs substrates commonly used in Mediterranean climates. The authors used the TPS method to assess the effect of humidification and raining processes on the final performance of the considered samples. Finally, Shao et al. [14] investigated the influence of moisture content on substrate thermal conductivity at different temperatures for four different commercially available substrates for green roofs.

These studies underlined the importance of studying thermal performance but in absence of specific national and international standards for laboratory calculations of the thermal performance of substrate for green roofs, previous research mainly focused on thermal conductivity results to characterize substrate materials, create heat transfer models, and simulate green roof performance without establishing common and replicable methods to easily measure substrate thermal conductivity. Therefore, there is the need to establish common procedures that companies and researchers can easily follow to determine thermal conductivity of granular materials, such as green roof substrates by varying water content, also validating them by comparing results from different laboratory procedures.

The research presented in this paper aims to establish a laboratory testing method that can be followed to evaluate thermal properties of green roofs substrates, such as thermal conductivity by varying water content. To this end, the novel method was applied to two devices measuring thermal conductivity and the results were compared. In addition, thermal conductivity was compared to technical data provided by companies for commercial substrates. In the following sections, the proposed method is described and the results of its application to three commercial substrates are presented.

2 Materials and Methods

The method proposed for testing substrate thermal conductivity was set-up at the Building Physic Laboratory of the University of Catania, Italy, by using two different devices, the HFM 436/3/1 Lambda (Heat Flow Meter) and the TLS-100 (Transient Line Source), to cross-referencing the measurements obtained (Fig. 1).

Because substrate is a granular material, the samples were enclosed into containers, based on the requirements of the devices. It was not mandatory using a specific material for the containers, but they have a regular shape to easily determine their volume.

Fig. 1. Measurements with TLS-100 (on the left) and HFM Lambda (on the right)

Before each step, the sample was weighed through a precision scale to determine material physical properties such as mass, volume, and density of the material.

First, the measurements using the TLS-100 device with a 100 mm needle probe (suitable for granular materials) were carried out. The substrate was placed in a rectangular plastic container perforated at the base to allow water drainage when saturated conditions were reached and a geotextile sheet as filter layer was placed on the bottom surface of the container, avoiding the fine particles of the substrate to flow out through container holes. The measurements were repeated in six different points close to the centre of the sample to validate data.

Before starting, the needle probe was placed into the sample for about 20 min until the thermal equilibrium between the probe and material was reached. After that, making sure a 3 cm layer of material was left between the probe and the container as suggested by the device technical guide to avoid external environmental condition effects on results, the measurements started. For each of the six points, three measurements were taken with a time-out of 10 min between a test and the next one. This procedure allowed the cooling of the device leaving the probe in the same position to re-establish thermal equilibrium conditions. The average of these three measurements was assumed as the value of thermal conductivity for each of the six points. The average of these six points was assumed as the thermal conductivity of the substrate. The medium square deviation (or standard deviation) was assessed to estimate the accuracy of the measurements.

The measurements with the HFM 436/3/1 Lambda equipment required the preparation of the sample due to its granular structure as the sample have to be placed in direct contact with the two heated plates of the HFM. The sample was placed into a high-conductivity aluminium container to create a continuous surface. The contribution of this aluminium container to the measured thermal conductivity can be neglected since its thermal conductivity is much higher than the tested substrates. The heat flux measured by the HFM, according to the Fourier postulate, depends on the thermal conductivity, the thickness, and the surface area of the sample, as well as the temperature difference between the two plates.

The tests with HFM 436/3/1 Lambda device were carried out by setting as average temperatures 20 °C and 30 °C by means of the Q-Lab software. Thus, a temperature of 30 °C and 40 °C for the upper plate and 10 °C and 20 °C for the lower plate were set to maintain a difference of temperature of 20 °C between the plate at 20 °C and 30 °C average temperature, respectively. These temperature differences between the plates generated the heat flow from the hotter to the colder plate and its value depended on the thermal conductivity of the substrate. When the steady-state condition was achieved, the thermal conductivity of the sample was determined. The higher the recorded heat flux, the higher will be the thermal conductivity of the substrate. In addition to thermal conductivity, HFM 436/3/1 Lambda provided a series of information such as the duration of the measurements, sample thickness, plate temperatures, temperature average and difference between the two plates and heat flow.

Globally, three sets of measurements were carried out for each reference temperature (i.e. 20 °C and 30 °C) with the HFM 436/3/1 Lambda without moving the material once it was placed into the device. A period of about two hours from the end of the test and the stat of the new one was appointed to allow for cooling the device. Six thermal conductivity values were obtained, three values for both 20 °C and 30 °C average temperatures. Finally, the average of the three measurements for each temperature set-up was assumed as the thermal conductivity of the sample.

The above-mentioned procedure using HFM 436/3/1 Lambda and TLS-100 was carried out on substrates with different water contents:

1. Moist: voids inside the material are filled with water and air in different percentage. This condition occurs after rain or irrigation events. It is intermediate between saturated and dry conditions.
2. Saturated: voids inside the material are filled with water. This condition occurs during rain or irrigation events until the water starts evaporating. When the water starts evaporating, substrate becomes moist.

3. Dry: voids inside the material are filled air. This condition occurs during drought periods when rain or irrigation events does not occur for a certain period allowing all the water to evaporate.

The tests were conducted on three commercial substrates. Substrate S1 consisted of lapilli, pumice, zeolites, peat, and slow-release fertilizers; Substrate S2 was a mixture of mineral volcanic materials mixed with organic substances, Substrate S3 was made with higher percentage of organic matter compared to the other substrates to increase water retention and it is made with local-available materials.

Only for the Substrate S1, commercial specifications include the thermal conductivity in dry, moist, and saturated conditions (Table 1). This allowed comparing these values with the experimental values obtained through the novel laboratory testing procedure described above to validate it.

Table 1. Physical properties of the substrates.

Conditions	m (g)	V (m^3)	ρ (kg/m^3)	ω (%)
Moist	2540.00	2.56×10^{-3}	991.37	31.57
Saturated	2860.20	2.39×10^{-3}	1355.55	–
Dry	1930.50	1.93×10^{-3}	1000.21	–

The first set of measurements was executed using the substrates in the unaltered state as they were in moist conditions. The next set of measurements was carried out on substrates in saturated conditions. The samples were placed in a perforated container, and they were saturated by pouring tap water. Water was inlet into the sample through subsequent steps until it cannot be further stored in the sample.

When the saturating process was completed, the excess water flows out through the container. Afterwards about 15 min, the measurements procedure was started with the TLS 100 mm and HFM Lambda.

Once the measurements in saturated conditions were completed, the material was oven-dried to assess the thermal properties in dry condition. The perforated plastic container with inside the saturated substrate was placed into a laboratory oven at a temperature of 80 °C for several days to allow the water inside the substrate evaporating. Intermediate weighing was carried out during the drying process and the final weight was obtained when the same readings was measured in two subsequent weighing. In this case, the voids are filled only with air as the water evaporated during the drying process. Finally, thermal conductivity measurements were performed in the laboratory, using the testing procedure previously described with the HFM and the TLS equipment.

3 Results and Discussion

In this section, the substrate thermal conductivity in moist (Fig. 2), saturated (Fig. 3) and dry conditions (Fig. 4), with HFM Lambda at 20 °C and 30 °C average temperatures are presented. Tables 2, 3 and 4 show the thermal conductivity and the standard deviation for the three substrates by using TLS-100 and HFM (both at 20 °C and 30 °C) in moist, saturated, and dry conditions, respectively.

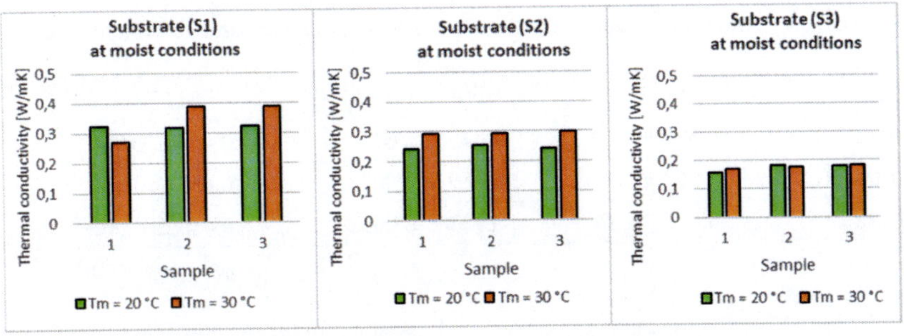

Fig. 2. Substrates measurements in moist conditions by HFM Lambda for T = 20 and 30 °C

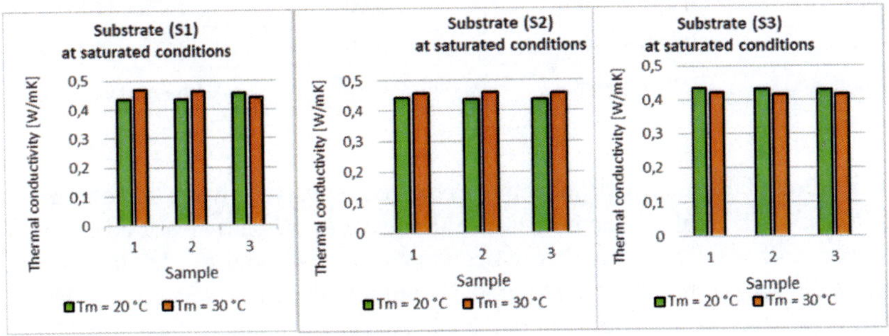

Fig. 3. Substrates measurements in saturated conditions by HFM Lambda for T = 20 and 30 °C

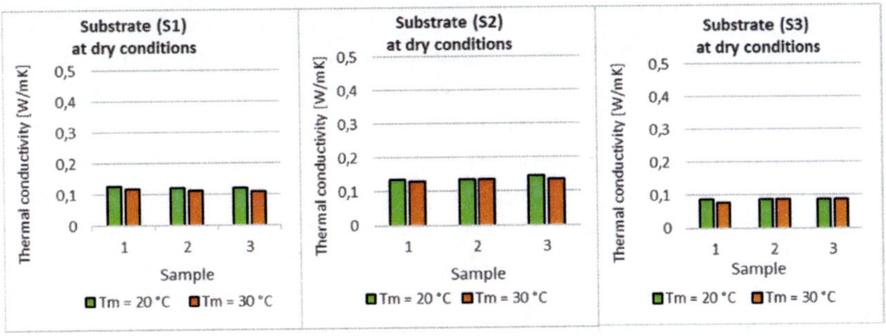

Fig. 4. Substrates measurements in dry conditions by HFM Lambda for T = 20 and 30 °C

Table 2. Thermal conductivity and standard deviation in moist conditions.

Sample	TLS—100	HFM (T = 20 °C)	HFM (T = 30 °C)
Substrate (S1)	0.231 ± 0.049	0.321 ± 0.003	0.347 ± 0.068
Substrate (S2)	0.210 ± 0.045	0.246 ± 0.006	0.295 ± 0.004
Substrate (S3)	0.204 ± 0.024	0.173 ± 0.013	0.176 ± 0.006

Table 3. Thermal conductivity and standard deviation for in saturated conditions.

Sample	TLS—100	HFM (Tm = 20 °C)	HFM (Tm = 30 °C)
Substrate (S1)	0.572 ± 0.111	0.443 ± 0.011	0.463 ± 0.006
Substrate (S2)	0.568 ± 0.065	0.439 ± 0.003	0.458 ± 0.003
Substrate (S3)	0.659 ± 0.060	0.433 ± 0.003	0.418 ± 0.003

Table 4. Thermal conductivity and standard deviation for the three in dry conditions.

Sample	TLS—100	HFM (T = 20 °C)	HFM (T = 30 °C)
Substrate (S1)	0.091 ± 0.002	0.124 ± 0.004	0.113 ± 0.005
Substrate (S2)	0.080 ± 0.010	0.139 ± 0.006	0.134 ± 0.003
Substrate (S3)	0.063 ± 0.004	0.088 ± 0	0.084 ± 0.006

As expected, the thermal conductivity changed by varying water content. In moist conditions, the thermal conductivity had values between the ones measured in saturated and dry conditions. The lower thermal conductivity was found for the dry conditions while the higher was found for the saturated conditions due to the water inside the voids, having higher thermal conductivity compared to the air. In addition, the thermal conductivity also varied with the temperature, as observed looking at the HFM measurements for the 20 °C and 30 °C temperature.

Comparing the thermal conductivity of the three substrates, it is possible to underline that Substrate S3 has thermal conductivity lower than both Substrates S1 and Substrate S2. Therefore, Substrate S3 allows to achieve greater roof thermal insulation compared to the other two substrates. This better performance is confirmed for the different water contents of the substrates, especially in moist conditions.

This result is significant because Substrate S3 was developed by using local-available materials, therefore, when formulating substrates, there is the need to consider the specific application, the climatic and territorial context, and the expected level of maintenance. Furthermore, building substrates using local materials allow also the reduction of costs.

By comparing the thermal conductivity determined with the HFM Lambda to the thermal conductivity provided by the manufacturer for Substrate S1, it can be observed that the results are similar (Table 1). Therefore, the proposed testing method may be considered reliable, and it could be used for the evaluation of the thermal conductivity of granular materials. On the other hand, the measurements obtained by the TLS-100 showed less reliability. It is possible to argue some possible causes of this scarce trustworthiness of the measurements carried out with the TLS 100. For saturated conditions, the TLS measurements indicated thermal conductivity of about 0.568 W/m^{2}K, while for dry conditions thermal conductivity of about 0.08 W/m^{2}K was obtained. These thermal conductivities are quite similar to the thermal conductivity of water and air. Therefore, the content of air or water in the voids influenced the results carried with TLS for granular materials.

Finally, Table 5 reports the physical parameters determined for the three substrates. Results show that these parameters increase going from moist to saturated conditions and decrease at dry conditions.

Table 5. Physical parameters of all tested materials at different conditions.

Moist conditions			
m (g)	V (m^3)	ρ (g/m^3)	ω (%)
2540.0	2.56 × 10^{-3}	991.37	31.5
2900.0	2.56 × 10^{-3}	1131.8	31.05
1684.8	2.56 × 10^{-3}	657.5	59.7
Saturated conditions			
m (g)	V (m^3)	ρ (kg/m^3)	m (g)
2860.2	2.11 × 10^{-3}	1355.5	2860.2
3246.8	2.39 × 10^{-3}	1358.4	3246.8
2273.1	1.92 × 10^{-3}	1183.9	2273.1
Dry conditions			
1930.5	1.93 × 10^{-3}	1000.2	1930.5
2212.8	2.41 × 10^{-3}	919.4	2212.8
1054.6	1.74 × 10^{-3}	605.4	1054.6

4 Conclusions

This study focused on proposing a novel methodology to evaluate thermal conductivity of green roof substrates by varying water content. This methodology allows easy comparison of the results to establish the substrate offering the best thermal performance. In addition, the physical properties could help finding substrates that comply with the weight limit value of the building structure in case of renovation. The substrates were tested at three different water content conditions: moist, saturated, and dry.

By comparing the thermal conductivity determined with the HFM Lambda to the thermal conductivity provided by the manufacturer for Substrate S1, it can be observed that the results are in good agreement. Therefore, the proposed testing method can be considered reliable, and it could be used for the evaluation of the thermal conductivity of granular materials. On the contrary, the measurements delivered by the TLS-100 showed less reliability.

This novel methodology could help companies and researchers for the thermal characterization of green roof substrates by varying water content since they do not provide this information because the manufacturers usually are more focused on the aesthetic outcome rather than the thermal insulation and energy performance. Therefore, this laboratory testing procedure can potentially provide a useful guideline for assessing thermal conductivity of green roof substrates.

The significance of the results has prompted researchers and industrial partners to engage in further investment and ongoing tests in this area, targeting the definition of a standard laboratory method to be presented worldwide.

References

1. Hanemann, M., Labandeira, X., Loureiro, M.L.: Climate change, energy and social preferences on policies: Exploratory evidence for Spain. Clim. Res. **48**, 343–348 (2011)
2. Lotteau, M., Loubet, P., Pousse, M., Dufrasnes, E., Sonnemann, G.: Critical review of life cycle assessment (LCA) for the built environment at the neighborhood scale. Build. Environ. **93**, 165–178 (2015)
3. Cascone, S.M., Cascone, S., Vitale, M.: Building insulating materials from agricultural by-products: a review. In: Littlewood, J., Howlett, R., Capozzoli, A., Jain, L. (eds.) Sustainability in Energy and Buildings. Smart Innovation, Systems and Technologies, vol. 163, pp. 309–318 (2020)
4. Contrafatto, L., Gazzo, S., Purrazzo, A., Gagliano, A.: Thermo-mechanical characterization of insulating bio-plasters containing recycled volcanic pyroclasts. Open Civil Eng. J. **14**, 66–77 (2020)
5. Chen, C.F.: Performance evaluation and development strategies for green roofs in Taiwan: a review. Ecol. Eng. **52**, 51–58 (2013)
6. Vijayaraghavan, K.: Green roofs: a critical review on the role of components, benefits, limitations and trends. Renew. Sustain. Energy Rev. **57**, 740–752 (2016)
7. Vestrella, A., Biel, C., Savè, R., Bartoli, F.: Mediterranean green roof simulation in Caldes de Montbui (Barcelona): thermal and hydrological performance test of Frankenia laevis L., Dymondia margaretae Compton and Iris lutescens Lam. Applied Sciences (Switzerland) **8** (2018)
8. Cascone, S., Gagliano, A., Poli, T., Sciuto, G.: Thermal performance assessment of extensive green roofs investigating realistic vegetation-substrate configurations. Build. Simul. **12**(3), 379–393 (2018)
9. Sailor, D.J., Hagos, M.: An updated and expanded set of thermal property data for green roof growing media. Energy Build. **43**, 2298–2303 (2011)
10. Ouldboukhitine, S.E., Belarbi, R., Djedjig, R.: Characterization of green roof components: measurements of thermal and hydrological properties. Build. Environ. **56**, 78–85 (2012)
11. Pianella, A., Clarke, R.E., Williams, N.S.G., Chen, Z., Aye, L.: Steady-state and transient thermal measurements of green roof substrates. Energy Build. **131**, 123–131 (2016)

12. Barozzi, B., Bellazzi, A., Maffè, C., Pollastro, M.: Measurement of thermal properties of growing media for green roofs: assessment of a laboratory procedure and experimental results. Buildings **7**, 99 (2017)
13. Fabiani, C., Coma, J., Pisello, A.L., Perez, G., Cotana, F., Cabeza, L.F.: Thermo-acoustic performance of green roof substrates in dynamic hygrothermal conditions. Energy Build. **178**, 140–153 (2018)
14. Shao, B., Valeo, C., Mukhopadhyaya, P., He, J.: Influence of temperature and moisture content on thermal performance of green roof media. Energies **14**, 1–21 (2021)

Study on Residents' Perception of Low-Carbon Policy and Its Influence on Low-Carbon Behavior Intention

Alin Lin[1](✉), Jiankun Lou[2], and Ran Yue[3]

[1] School of Civil Engineering and Architecture, Zhejiang Sci-Tech University, Hangzhou 310018, People's Republic of China
linalin@zstu.edu.cn

[2] College of Architecture and Urban Planning, Tongji University, Shanghai 200092, People's Republic of China

[3] School of Architecture, Harbin Institute of Technology, Harbin 150006, People's Republic of China

Abstract. Energy consumption and carbon emissions by residents in China are increasing in recent years. Promoting residents to participate in low-carbon behavior practice is an effective way to reduce energy consumption and achieve the "double carbon goal". This study summarizes the policy development process of residents' energy-saving awareness and low-carbon behavior in China from four aspects: information policy, economic policy, technical policy and administrative policy. By using statistical methods such as Structural Equation Model, taking Hangzhou as an example, this paper analyzes residents' low-carbon behavior intention and low-carbon policy perception level, and explores the impact of low-carbon policy perception on low-carbon behavior intention. The results show that residents' participation and satisfaction of low-carbon policy have a positive and significant impact on low-carbon behavior intention, but the impact of understanding of low-carbon policy on low-carbon behavior intention is insignificant. Moreover, this paper also proves that gender and education level have a positive relationship with low-carbon behavior intention. In the future, government should pay close attention to residents' participation of low-carbon policy. Meanwhile, the improvement of residents' satisfaction with low-carbon policy can also promote the formation of residents' low-carbon behavior awareness.

Keywords: Low-carbon behavior intention · Low-carbon policy · Residents' perception · Structural equation modeling · Hangzhou

1 Introduction

Climate change has brought serious impact on human life [1]. According to the data from International Energy Agency, China's total carbon dioxide emission ranks first in the world, while the per capita carbon dioxide emission of United States is the highest (14.4 t CO_2/person). The energy consumption of China in the field of residents' lives

J. Littlewood and R. J. Howlett (Eds.): SEB 2022, SIST 336, pp. 76–85, 2023.
https://doi.org/10.1007/978-981-19-8769-4_8

is increasing year by year (see Fig. 1) [2]. By 2019, the per capita energy consumption has reached 438 kg of standard coal [3]. China promises that carbon emissions peak by 2030 and carbon neutralization by 2060, which is called "double carbon goal" in China. Therefore, it is of great significance to take the residents oriented, make them participate in the practice of low-carbon behavior, to realize the "double carbon goal".

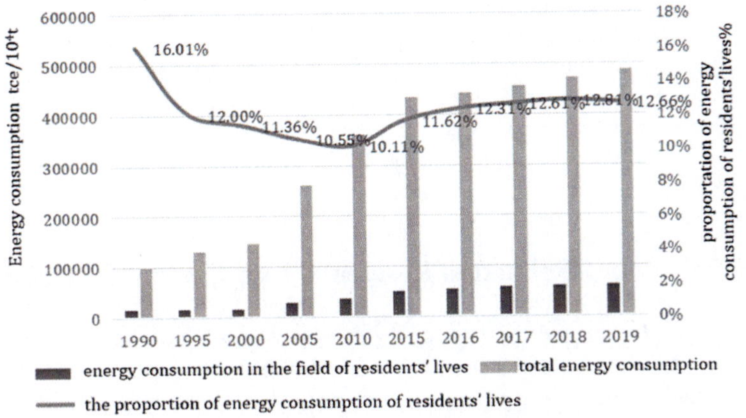

Fig. 1. Energy consumption of China in the field of residents' lives and its proportion

Many studies have been carried out on the determinants of low-carbon behavior. Psychological factors such as values, perception of environmental problems, environmental knowledge affect low-carbon behaviour [4]. Situational factors such as policy, social norm and demographic factors also have an impact on low-carbon behaviour [5, 6]. The theory of planned behavior (TPB) indicates that subjective norm, attitudes and perceived behavioral control affect the low-carbon behavior intention [7]. Attitude-Behavior-external Conditions Model (ABC) believes that the occurrence of behavior is the influence of psychological factors and external factors [8]. As an external factor affecting low-carbon behavior, low-carbon policy can effectively promote residents to adopt low-carbon behavior.

Based on previous studies [4, 9], policy tools related to residents' low-carbon behavior can be divided into four aspects: information policy, economic policy, technical policy and administrative policy. Information policy refers to information feedback and information cue, which can significantly change individuals' intention to save energy and adopt low-carbon behaviour [10, 11]. However, the long-term use of information policy will make people feel indifferent and ignore the importance of choice. Economic policy includes positive (e.g. financial subsidies, tax relief) and negative incentive policy (e.g. raising prices, increasing taxes). Generally, economic policy forms external incentives for low-carbon consumption behaviour [12, 13], and has little effect on daily habitual low-carbon behavior. Technical policy focuses on the research and development of energy-saving products and technology, and plays an indirect role in promoting residents' low-carbon behaviour [14]. Administrative policy refers to mandatory regulations that have a direct impact on low-carbon behavior, but the disadvantage of administrative regulations is that low-carbon behavior may be implemented out of fear of authority and punishment rather than active motivation, so it's difficult to maintain for a long time [15].

The implementation effect and resident's perception of low-carbon policy are different in different regions [16]. Meanwhile, the intervention and guidance measures of residents' low-carbon behavior between China and other countries have a little difference. Therefore, this study analyzes the development of China's low-carbon policy related to residents' low-carbon behavior, and summarizes Chinese low-carbon policy from the above four aspects. Taking Hangzhou city as an example, this paper evaluates the residents' perception level of low-carbon policy, and explores the impact of residents' perception on low-carbon behavior intention. The results can provide reference for low-carbon policy making, especially for the promotion of residents' low-carbon behavior intention.

2 CHina's Policy Related to Residents' Low Carbon Behavior

In 1998, China formulated the Law of the People's Republic of China on Conserving Energy, which established that energy conservation is China's basic national policy. Since then, relevant policies to promote energy conservation and emission reduction have been gradually formulated from the aspects of saving resources, protecting the environment, reducing energy consumption and improving energy efficiency. Table 1 shows the representative policies related to residents' energy conservation from the aspects of information policy, economic policy, technical policy and administrative policy [17, 18].

The emphasis on residents' awareness of energy conservation and low-carbon behavior from Chinese government began in 1991. Since then, information policy are gradually formulated and continuously improved, such as energy efficiency label, intelligent electric meter, tiered pricing for electricity etc. In 2007–2012, the government gave financial subsidies to energy saving products. In terms of economic and information policy, governments and communities mainly focus on publicity, guidance and education, residents have a weak sense of participation. There are few technical policy and administrative policy directly related to residents' energy saving behavior. Most of technical policy are aimed at the development and progress of key energy-saving technology and energy-efficient products. And most of the relevant regulations involved in administrative policy are encouragement or restraint, there are few legally mandatory ones. Meanwhile, the implementation effect of Chinese low-carbon policy needs to be evaluated.

3 Methodology

Hangzhou is a new first-tier city in China, its economic level is relatively high and the policy implementation is better. This study selects Gongshu District, Xihu District, Binjiang District and Shangcheng District within the main city of Hangzhou, Zhejiang Province as the study area (see Fig. 2). According to the amount of residential areas in each district and their built ages (four ages: before the year of 1989, 1990–1999, 2000–2009, after 2010, respectively), based on stratified sampling method, a total of

Table 1. Policies related to residents' low-carbon behavior in China [17, 18].

Year	Name
Administrative policy	
1998	Law of the People's Republic of China on conserving energy
2000	Administrative measures for electricity saving
2004	Outline of energy medium-and long-term development plan
2008	Regulations on energy saving in public institutions
2015	The overall plan for the reform of the ecological civilization system
2021	Outline of the 14th five-year Plan and long-term goals for 2035
Technical policy	
2006	Technical catalogue of application of energy-saving and land-saving buildings
2008	Regulations on energy efficiency of civil buildings
2008	Notice on further strengthening the work of fuel saving and electricity saving
2009	Demonstration and implementation of renewable energy building application city
Economic policy	
2007	Financial subsidy funds for the promotion of high efficiency lighting products
2009	Energy-saving products benefit the people project
2009	Demonstration and popularization of energy saving and new energy vehicles
2009	Financial subsidy funds for the application of solar photovoltaic buildings
2011	Tiered pricing for electricity
2014	Tiered pricing for water
Information policy	
1991	National energy conservation week
2004	Energy efficiency labeling system
2007	National action on energy conservation and emission reduction
2007	Air-conditioning temperature control standards in public buildings
2014	Criteria of conduct for citizens of *Breathe together and Strive together*
2017	Waste segregation system
2018	Criteria of conduct for citizens' ecological environmental behavior
2021	Criteria of conduct for water saving

17 residential areas were selected. Multi-storey and high-rise residential buildings were included.

A questionnaire survey was conducted from September 2021 to January 2022. The questionnaire was constructed using an online-survey-system of Wenjuanxing and face to face interviews in residential areas. The questionnaire was randomly answered by 500 respondents, then we collected 500 questionnaires, of which 475 were valid and could be used for further statistical analysis.

Three parts were included in the questionnaire. (1) Respondents' demographic characteristics: age, gender, monthly income and their education level (see Table 2). (2) Low carbon behavior intention, which is measured by behavioral attitudes (BA), subjective norm (SN) and perceived behavioral control (PBC). (3) Perception level of low-carbon policy, which is evaluated evaluated by three variables, the understanding of low-carbon policy, the participation of low-carbon policy and the satisfaction of low-carbon policy. Low carbon behavior intention and perception level of low-carbon policy were evaluated by 5 likert scale.

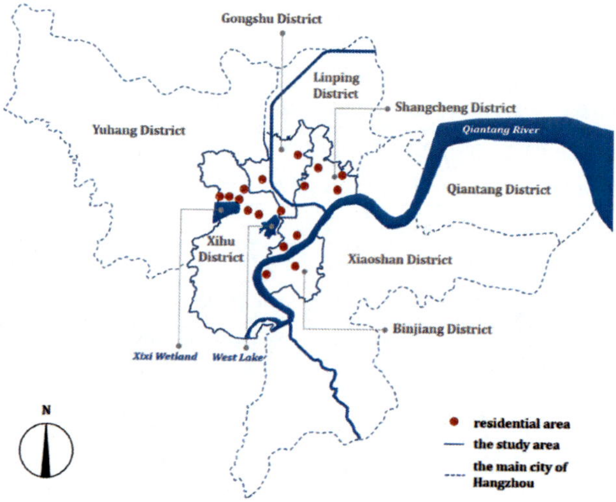

Fig. 2. The study area.

Table 2. Demographic characteristics of the participants.

Characteristic	Category	n(%)	Characteristic	Category	n(%)
Gender	Male	39	Monthly income (Chinese yuan)	≤4000	18
	Female	61		4000–8000	33
Age (years)	≤29	18		8000–12000	21
	30–39	43		≥12000	28
	40–59	29	Education	High school or below	24
	≥60	10		Bachelor	60
				Master	16

Structural Equation Modeling (SEM) [19] method was used in this study. A conception model was built [20], of which the understanding, participation and satisfaction of low-carbon policy were independent variables and low-carbon behavior intention was dependent variable. Figure 3 shows the structural equation model. The calculation was completed by SPSS 22.0 and AMOS 23. The results indicated high reliability and validity. The structural equation model provided a satisfactory goodness of fit to the sample data (χ^2/df = 4.793, RMSEA = 0.089, GFI = 0.712, CFI = 0.807, TLI = 0.794). Table 3 shows the hypothesis test results.

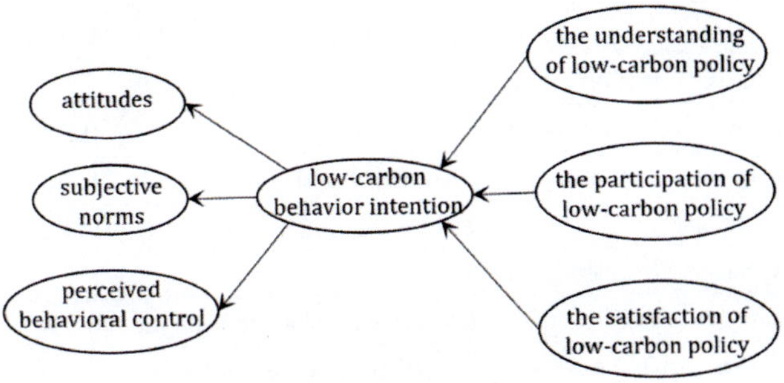

Fig. 3. The conception model.

4 Results and Discussion

4.1 Low-Carbon Behavior Intention of Residents

The mean score of low-carbon behavior intention of respondents is 4.14. The mean score of behavioral attitude is highest (M = 4.40), followed by the perceived behavioral control (M = 4.01) and the subjective norm (M = 3.97). It can be found that respondents have a strong psychological tendency to adopt low-carbon behavior. However, the belief of overcoming the difficulties encountered in the implementation of low-carbon behavior still need to be further improved.

As shown in Table 3, behavioral attitude, perceived behavioral control and subjective norm hold direct positive relationship with low-carbon behavior intention. Subjective norm has the largest factor load ($\beta = 0.893$, $p = 0.000$), which is consistent with the existing research [21]. It shows that the current low-carbon behavior intention of respondents mainly comes from compliance with social norm. This may be because Chinese people have habits of collecting and are easily influenced by around people's beliefs or motivations about a certain behavior. Generally, environmental policy that emphasize ecological value and improve personal environmental awareness and responsibility can strengthen subjective norm and enhance people's low-carbon intention. The coefficients of behavioral attitude and perceived behavioral control were 0.776 ($p = 0.000$) and 0.763 ($p = 0.000$), respectively. Behavioral attitude is an endogenous factor that promotes low-carbon behavior intention, while perceived behavioral control is the exogenous factors along with subjective norm.

Additionally, the correlation result of demographic characteristics with low-carbon behavior intention shows that gender ($\beta = 0.121$, $p < 0.01$) and education level (β

= 0.120, $p < 0.05$) of respondents have a positive influence on low-carbon intention. This finding is consistent with previous conclusion. Women's energy-saving behavior intention are significantly higher than those of males, and the degree of education plays a positive role in adjusting residents' saving energy level [4].

Table 3. Hypothesis test results.

Path	Coefficient	*P* value
Low-carbon behavior intention		
Low-carbon behavior intention → behavioral attitude	0.776	0.000
Low-carbon behavior intention → perceived behavioral control	0.763	0.000
Low-carbon behavior intention → subjective norm	0.893	0.000
Policy → Low-carbon behavior intention		
Understanding of low-carbon policy → low-carbon behavior intention	0.036	0.318
Participation in low-carbon policy → low-carbon behavior intention	0.857	0.000
Satisfaction to low-carbon policy → low-carbon behavior intention	0.238	0.000

4.2 The Residents' Perception of Low-Carbon Policy

The perception level of low-carbon policy is determined by the understanding, participation and satisfaction of low-carbon policy. The mean score of low-carbon policy perception level is 3.65. The higher perception level low-carbon policy, the more understanding of the detail and implementation purpose of low-carbon policy. Thus residents tend to have a strong willingness to conduct low-carbon behavior. Respondents reported higher score on the participation of low-carbon policy (M = 3.71) than satisfaction (M = 3.60) and understanding (M = 3.58) of low-carbon policy.

Table 4 reports the descriptive statistic result of low-carbon policy. The mean scores of understanding of low-carbon policy show that Hangzhou has a better effect on the information measures than other aspects, including normative cue, publicity and education advertising on low-carbon knowledge. But the awareness of respondents related to low-carbon regulations needs to be further improved. Also, we find that the subsidy policy made by the government on energy-saving household appliances is not better perceived by residents. At present, the importance of energy-saving and environmental protection technologies isn't attached by residents, the energy-saving technological innovation of alternative fuel vehicles and solar water heaters isn't perceived either. Safe energy-saving technology is a key factor affecting the willingness of low-carbon consumption.

The mean scores of participation of low-carbon policy show that respondents are actively participating in publicity and education activities and knowledge learning activities related to low-carbon behavior and have reached a high willingness to participate in low carbon activities. Residents consider that the existing technology is not mature, the government should enhance the credibility of technology. Financial subsidy will encourage residents to conduct low-carbon behavior. And low-carbon behavior of respondents may rely too much on administrative regulations. The mean scores of satisfaction of low-carbon policy indicate that the satisfactory to low-carbon policy still needs to be improved. The difference of satisfaction values between different policy tools is not obvious.

Table 4. The mean scores of perception of low-carbon policy.

	Information policy	Technology policy	Economic policy	Administrative policy
Understanding of low-carbon policy	3.73	3.37	3.47	3.61
Participation of low-carbon policy	3.25	4.17	3.97	3.84
Satisfaction of low-carbon policy	3.60	3.62	3.53	3.64

4.3 The Influence of Low-Carbon Policy Perception on Behavior Intention

Positive associations were found among most of the variables in this model. The effect of the participation of low-carbon policy on low-carbon behavior intention is significant ($\beta = 0.857, p = 0.000$), indicating that the higher degree of participation of residents in low-carbon policy, the stronger desire to engage in low-carbon behavior. The satisfaction of low-carbon policy is positively associated with low-carbon behavior intention ($\beta = 0.238$, $p = 0.000$), suggesting that respondents who have higher satisfaction on low-carbon policy may express greater low-carbon behavior intention.

The understanding of low-carbon policy has no significant impact on low-carbon behavior intention ($p = 0.318$). It shows that recognition of policy cannot affect residents' low-carbon behavior intention. Therefore, it is more necessary to further strengthen residents' participation in low-carbon policy, and try to improve the satisfaction on low-carbon policy.

Based on previous studies, we find that residents' low-carbon behavior intention is not only depends on values, beliefs, environmental knowledge and environmental awareness, but also depends on the benign guidance of the government and society. Therefore, low-carbon policy system making should be further improved by putting forward information tools, economical measures, technical measures and administrative regulations.

5 Conclusions

Low-carbon behavior intention is affected by self-factors (subjective norm, behavioral attitude, perceived behavioral control, gender and education level) and external factors (policy perception). The participation and satisfaction of low-carbon policy have a positive significant impact on low-carbon behavior intention, but the understanding of low-carbon policy on low-carbon behavior intention is not significant. Therefore, we can promote the residents' low-carbon behavior intention by improving the residents' satisfaction with the low-carbon policy, and the government must clarify the importance of public participation (participate in making policy and policy that can be involved in low-carbon practice).

In terms of low-carbon policy: (1) Strengthen the implementation of information policy. The individual low-carbon awareness can be raised by information cue, through which the low-carbon behavior of personal life with ecological value and consumer prices can be closely linked. In addition, real-time electricity price display of household electricity meters should be used for urban household. (2) Enhance the sense of security of low-carbon technology in the minds of Chinese residents. An in-depth understanding of low-carbon technical policy can be got through information tools (e.g. information cue and activity to popularize low-carbon technology). (3) Improve the subsidy mechanism and give a clear explanation in the sales process of product to encourage low-carbon consumption behavior. Our results provide a reference for the improvement of low-carbon behavior intention and its transformation to low-carbon behavior. Also, we suggest local government attaches the importance of public participation in making low-carbon policy.

Acknowledgments. This work is the research findings of Zhejiang Federation of Humanities and Social Sciences (2022N29, 浙江省社会科学界联合会研究课题成果 in Chinese).

References

1. Lin, A.L., Lu, M., Lou, J.K.: The assessment of urban solar energy's available space based on energy landscape. Planners **1**, 77–83 (2021)
2. International Energy Agency. Data and statistics. https://www.iea.org/data-and-statistics/data-products/?filter=emissions. Last accessed 21 Apr 2022
3. National Bureau of Statistics: China Statistical Yearbook 2021. China Statistics Press, Beijing (2021)
4. Ding, Z., Jiang, X., Liu, Z., et al.: Factors affecting low-carbon consumption behavior of urban residents: a comprehensive review. Resour. Conserv. Recycl. **132**, 3–15 (2018)
5. Wang, T., Shen, B., Springer, C.H., et al.: What prevents us from taking low-carbon actions? A comprehensive review of influencing factors affecting low-carbon behaviors. Energy Res. Soc. Sci. **71**(2), 101844 (2021)
6. Liu, D., Du, H., Southworth, F., et al.: The influence of social-psychological factors on the intention to choose low-carbon travel modes in Tianjin, China. Transp. Res. Part A Policy Pract. **105**, 42–53 (2017)
7. Hu, X., Wu, N., Chen, N.: Young people's behavioral intentions towards low-carbon travel: extending the theory of planned behavior. Int. J. Environ. Res. Public Health **18**(5), 2327 (2021)
8. Guagnano, G.A., Stern, P.C., Dietz, T.: Influences on attitude-behavior relationships: a natural experiment with curbside recycling. Environ. Behav. **27**(5), 699–718 (1995)
9. Lindén, A., Carlsson-Kanyama, A., Eriksson, B.: Efficient and inefficient aspects of residential energy behaviour: what are the policy instruments for change? Energy Policy **34**(14), 1918–1927 (2006)

10. Du, L., Guo, J., Wei, C.: Impact of information feedback on residential electricity demand in China. Resour. Conserv. Recycl. **125**, 324–334 (2017)
11. Nilsson, A., Stoll, P., Brandt, N.: Assessing the impact of real-time price visualization on residential electricity consumption, costs, and carbon emissions. Resour. Conserv. Recycl. **124**, 152–161 (2017)
12. Li, G., Liu, W., Wang, Z., Liu, M.: An empirical examination of energy consumption, behavioral intention, and situational factors: evidence from Beijing. Ann. Oper. Res. **255**(1–2), 507–524 (2016). https://doi.org/10.1007/s10479-016-2202-8
13. Maki, A., Burns, R.J., Ha, L., et al.: Paying people to protect the environment: a meta-analysis of financial incentive interventions to promote pro-environmental behaviors. J. Environ. Psychol. **47**, 242–255 (2016)
14. Zhao, D., Mccoy, A.P., Du, J., et al.: Interaction effects of building technology and resident behavior on energy consumption in residential buildings. Energy Build. **134**, 223–233 (2017)
15. Zhang, B., Chen, H., Du, Z., et al.: Does license plate rule induce low-carbon choices in residents' daily travels: motivation and impacts. Renew. Sustain. Energy Rev. **124**, 109780 (2020)
16. Bergquist, M., Nilsson, A.: I saw the sign: promoting energy conservation via normative prompts. J. Environ. Psychol. **46**, 23–31 (2016)
17. National Energy Administration. Compilation of energy laws, regulations and policies. www.nea.gov.cn/n_home/nyflfghb/index.htm. Last Accessed 05 March 2022
18. Long, R.Y., Yue, T.: Influencing Factors and Guiding Policy of Energy-Conservation Behavior of Urban Residents. Science Press, Beijing (2017)
19. Dong, W., Lou, J.K., Dong, Y.: Impact of community walkability on the elderly's place attachment and willingness to aging in place: an empirical study of Xiangfang old industrial zone in Harbin. Shanghai Urban Plann. Rev. **155**(6), 30–35 (2020)
20. Li, W., Zhao, S., Ma, J., et al.: Investigating regional and generational heterogeneity in low-carbon travel behavior intention based on a PLS-SEM approach. Sustainability **13**, 3492 (2021)
21. Mi, L.Y., Gu, M., Yang, J., et al.: Empirical research on the psychological motivation factors of urban residents' low carbon oriented energy consumption behavior in Xuzhou city. Resour. Sci. **38**(4), 609–621 (2016)
22. Lin, B., Wang, X.: Does low-carbon travel intention really lead to actual low-carbon travel? Evidence from urban residents in China. Econ. Anal. Policy **72**, 743–756 (2021)

The Techno-Economic Effect of PV-BSS Size Under Various Supporting Schemes: A Replicable Method for Buildings

Nikolas G. Chatzigeorgiou(✉) and George E. Georghiou

PV Technology Laboratory, Department of Electrical and Computer Engineering, University of Cyprus, Nicosia, Cyprus
nchatz05@ucy.ac.cy

Abstract. As the competitiveness of Photovoltaic (PV) systems increased in recent years, mainly due to their vast cost reduction, PVs penetrated power systems and already reached Grid Parity in numerous countries. Nowadays, policymakers consider Battery Storage Systems (BSS) as an integral part of distributed systems in buildings, in order to limit technical counter effects. As a result, a shift to more self-consumption promoting supporting schemes is observed in recent years. Given the recent introduction of domestic BSS in Cyprus, this study aims to provide guidelines for PV-BSS sizing and to evaluate their economic viability under various supporting schemes. The study uses a holistic methodology and an energy-flow simulation for domestic buildings, which can be replicated across different countries and building types. The results showed that for the case of Cyprus, such systems are already a viable investment, even under the most pessimistic scenario and verified that PV-BSS size has a significant effect on the addressed techno-economic indexes.

Keywords: Photovoltaics · Battery storage systems · Techno-economic analysis · Self-consumption

1 Introduction

The vast reduction in Photovoltaics (PVs) cost, driven mainly by their continuous technological advancement and maturity, resulted in a notable global growth in grid-connected PV systems in recent years [1]. PVs have already reached Grid Parity in many countries [2, 3] and hence, policymakers are in the process of modifying current supporting schemes to meet the needs of modern power systems more precisely. Specifically, similarly to previous experience when (the more necessary at the time) Net-Metering replaced Feed-in Tariffs (FiTs) in early 2010s (as the competitiveness of PVs made their support through FiTs no longer necessary), the dawn of 2020s sees a shift to supporting schemes promoting self-consumption, e.g. Net-Billing [4]. In any case, this process is still directly focused on buildings and mostly, domestic premises.

Concerning Cyprus, the high solar resource [5] results in high PV generation, which in combination with the presence of favourable supporting schemes (Net-Metering),

J. Littlewood and R. J. Howlett (Eds.): SEB 2022, SIST 336, pp. 86–96, 2023.
https://doi.org/10.1007/978-981-19-8769-4_9

resulted in ≈63 MWp of PVs in buildings by 2021, having a significant share (≈25%) of the island's total installed PV capacity [4].

Due to their positive effect on modern power systems [6–8], Battery Storage Systems (BSS) are nowadays considered a possible part of PVs distributed in end-user sites, specifically for domestic and other buildings, in an effort to solve the arising security and stability issues, which stem from the mismatch of PV generation and building consumption. Coupling PVs with BSS decreases both energy import/export by storing and then using excess PV generation. Thus, building self-consumption is enhanced, leading to more self-sufficient and sustainable buildings [6, 9].

Cyprus enabled the integration of BSS with domestic PVs in 2022. To date, there is a limited number of PV-BSS, which were implemented as pilots in domestic buildings in recent years. Their operation has been studied regarding their sizing for self-consumption/-sufficiency maximisation and for grid export control to avoid PV curtailment or affecting building consumption [4, 6, 7]. Various studies explored the operation of domestic PV-BSS for sizing focusing on their payback [10, 11] or analysed their competitiveness under a specific operation [3]. Yet, none of these addressed the techno-economic effect of PV-BSS size in buildings considering supporting schemes.

Given the recent enablement of BSS in Cyprus, the present study was conducted aiming to serve as (a) a guideline for PV-BSS size and (b) an evaluation of the economic viability of such systems under various supporting schemes. The study performed an energy-flow simulation for (but not limited to) domestic buildings, replicable for different countries and building types, realising the need for effective supporting schemes to further increase PVs in buildings, while enhance their sustainability.

2 Methodology

A Techno-economic Analysis was performed with dedicated simulation to evaluate the viability of domestic PV-BSS under various supporting schemes, considering Cyprus as a case study. The evaluation used the following techno-economic indexes: Net Present Value (NPV), Internal Rate of Return (IRR), Discounted Payback Period (DPP) and Levelised Cost of Electricity (LCOE). Consistent technical information (e.g. system characteristics, operational data, etc.) and updated economic information (e.g. system costs, electricity prices, etc.) under a holistic methodology are required for a detailed Techno-economic Analysis, consisting of the following four steps:

Project Characterisation: Implementation of energy-flow simulation considering power profiles and PV-BSS operation.

Cost Estimation: Quantification of total economic costs during lifetime.

Benefit Estimation: Assessment of total economic benefits during lifetime.

Cost & Benefit Comparison: Contrast of economic costs and benefits.

2.1 Techno-Economic Indicators

The four techno-economic indicators used for the analysis are:

NPV (€): Difference of present value of cash inflows and outflows over lifetime (1).

IRR (%): Discount rate to break-even (NPV = 0) the investment (2).

DPP (years): Time to break-even the investment (3).

LCOE (€/kWh): Minimum sale price to break-even the investment (4).

$$NPV = I_0 + \sum_{n=1}^{20} \frac{I_n}{(1+d)^n} \tag{1}$$

$$NPV = I_0 + \sum_{n=1}^{20} \frac{I_n}{(1+IRR)^n} = 0 \tag{2}$$

$$NPV = I_0 + \sum_{n=1}^{DPP} \frac{I_n}{(1+d)^n} = 0 \tag{3}$$

$$LCOE = \frac{\sum_{n=1}^{20} \frac{I_n}{(1+d)^n}}{\sum_{n=1}^{20} \frac{P_n}{(1+d)^n}} \tag{4}$$

For (1)–(4), I_0 = Initial Investment Cost, I_n = Annual Profit or Loss, n = Year, d = Discount Rate and P = PV Production.

Generally, a positive NPV shows a profitable investment. Also, the higher the IRR, the higher the net cash flows and the shorter the DPP, the more attractive the investment to the investor, while Grid Parity occurs when the LCOE is less than the retail electricity price [3].

Moreover, the monthly Self-Consumption (SCR) and Self-Sufficiency (SSR) Rates were calculated. The former regards the percentage of PV generation consumed by the building, while the latter concerns the percentage of building consumption that is covered by the PV-BSS. Clearly both indexes are higher compared to a no-BSS case (i.e. PV system only), as a higher percentage of PV generation is used by the building.

2.2 Simulation Specifications

By considering the domestic power consumption profile for a typical working and non-working day for each month of the year and the corresponding average daily PV generation profile, an energy-flow simulation was performed assuming a 20-year system lifetime. Specifically, dedicated profiles[1] (acquired in a 15-min timestamp from 300

[1] Due to the absence of official domestic consumption and PV generation profiles, data already in the disposal of the PV Technology Laboratory from previous research work were used.

domestic buildings in Cyprus) were used as inputs to the simulation. It must be noted that both generation and consumption data were comparable to ones found in the literature [12, 13], confirming the validity of the approach.

A BSS control aiming to maximise the building's self-consumption was considered (i.e. charge BSS with surplus PV generation when PV generation > domestic consumption, else discharge BSS to cover the deficit and meet building demand, thus avoiding energy imports), as energy arbitrage and the provision of ancillary services to the power system are not allowed in Cyprus, currently. No BSS component (e.g. battery unit, converter etc.) replacement was examined due to relevant BSS service life [3]. Finally, constant domestic consumption level and retail electricity price (i.e. no increase/decrease during the analysis period) were assumed. Any possible increase in the building consumption level or retail electricity price is expected to enhance the viability of the PV-BSS.

2.3 Analysis Types and Parameters Used

Two types of analysis were performed, i.e. Single Analysis and Parametric Analysis. Specifically, for Single Analysis, results were extracted for a specific PV-BSS size. For Parametric Analysis, two further cases were applied, particularly a calculation for PV size range with constant BSS size and a calculation for BSS size range with constant PV size. Tables 1 and 2 list the technical and economic parameters considered for the analysis, respectively.

Table 1. Technical parameters considered for the analysis.

Description	Value	Comment
PV degradation	0.5%	Yearly decrease of nominal PV power
PV power	5 kWp	Nominal PV system power for single analysis
PV power min	3 kWp	Min nominal PV system power for parametric analysis
PV power max	10 kWp	Max nominal PV system power for parametric analysis
BSS energy	10 kWh	Nominal BSS capacity for single analysis
BSS energy min	5 kWh	Min nominal BSS capacity for parametric analysis
BSS energy max	15 kWh	Max nominal BSS capacity for parametric analysis
BSS level min	5%	Min allowable BSS charging level as % of nominal capacity

Table 2. Economic parameters considered for the analysis.

Description	Value	Comment
Tariff structure	Flat tariff	Regular tariff structure in Cyprus
Electricity price	0.2304 €/kWh	Incl. 19% Value Added Tax (VAT)[a,b]

(continued)

Table 2. (*continued*)

Description	Value	Comment
PV cost	1,100 €/kWp	Excl. VAT[c]
BSS cost	500 €/kWh	Excl. VAT[d]
O&M cost	2%	Operation and maintenance expense rate
Insurance cost	0.5% per year	Insurance expense rate
Discount rate	4%	Discount rate value
Inflation	1%	Inflation rate value
Tariff inflation	2%	Inflation rate value
Energy sales tariff	0.08 €/kWh	Price for sale of exported PV generation

[a]EUROSTAT, Electricity Price Statistics Explained. *Link.* [1/5/2022]
[b]SCOPULUS, European Vat Rates. *Link.* [1/5/2022]
[c]JRC (2019). PV Status Report 2019. Publications Office of the EU, Luxembourg, 2019
[d]IRENA (2019). Innovation landscape brief: Behind-the-meter batteries, IRENA, Abu Dhabi

2.4 Supporting Schemes

Finally, the following supporting schemes (SS) were examined:

- SS1: Net-Metering (1-year netting period)
- SS2: Net-Billing (self-consumption with energy sales)
- SS3: Full Self-Consumption (self-consumption without energy sales)
- SS4: Impact of future reduced BSS cost (200 €/kWh[5]) on SS3.

3 Results

3.1 Single Analysis

This simulation considered a 5 kWp/10 kWh PV-BSS size. Figure 1 illustrates the building's energy analysis during the PV-BSS lifetime, Fig. 2 demonstrates the cash flow under SS1 (similar pattern for SS2-SS4), while Table 3 summarises the estimated economic indexes for each SS.

For each SS, a positive NPV is derived for this PV-BSS size. Moreover, SS1 is the most suitable compensation mechanism for such systems, as grid import/export are valued the same. Yet, even SS3 (the most pessimistic case) derives a profitable investment, with slightly increased DPP compared to the others. The decrease in BSS cost has a vast effect on all indexes, resulting also in the lowest DPP. Finally, Grid Parity is reached under each SS, as the estimated LCOE is less that the electricity price, evincing the system's economic viability. Overall, this specific PV-BSS size is a profitable investment for domestic buildings under all cases.

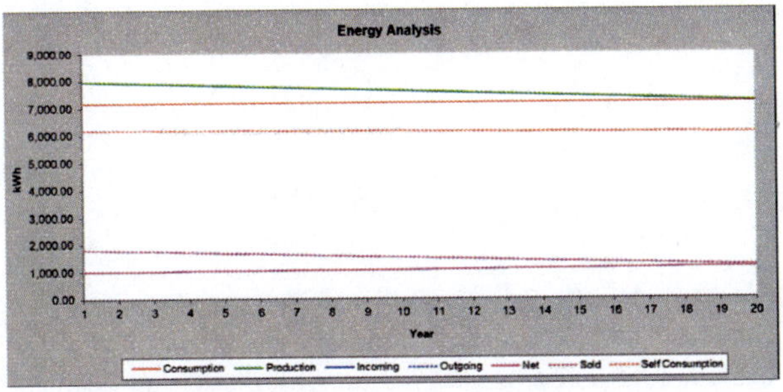

Fig. 1. Energy analysis during the PV-BSS lifetime.

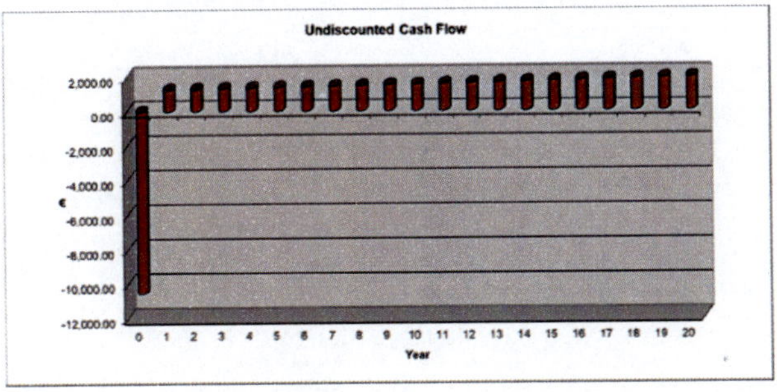

Fig. 2. Cash flow during the PV-BSS lifetime for SS1 (similar pattern for SS2-SS4).

3.2 Parametric Analysis: PV Size

This simulation considered a 10 kWh BSS with a 3–10 kWp PV. Figure 3 illustrates the energy analysis for the first year of operation and the impact of PV size on SCR and SSR. The higher the PV size, the higher the PV generation, resulting in higher gird export. This in turn results in decreased SCR and diminishing SSR.

Figures 4, 5 and 6 demonstrate the impact of PV size on NPV, IRR and DPP, respectively. NPV is positive under each SS. Yet, there is a certain PV size for each BSS size that maximises it for every SS. Similar observation for IRR. The PV size which results to the highest NPV and IRR, gives the lowest DPP, calculated at ≈10 years for SS1-SS3 and even less for SS4, as BSS cost decrease has a vast effect on all economic indexes.

Table 3. Estimated economic indexes of the investment for each SS.

Index	SS1	SS2	SS3	SS4
NPV (€)	10,430.68	8,772.93	6,802.00	10,920.88
IRR (%)	12.61	11.50	9.93	16.07
DPP (years)	9.17	9.87	11.25	7.26
LCOE (€/kWh)	0.1393	0.1393	0.1393	0.0996

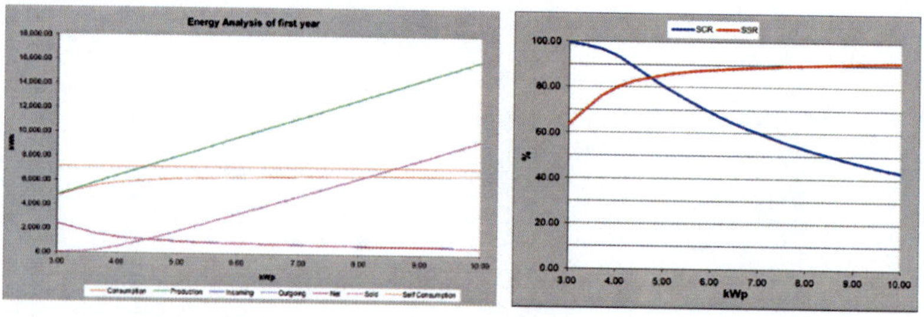

Fig. 3. Energy analysis for the first year of operation (L); PV size impact on SCR & SSR (R).

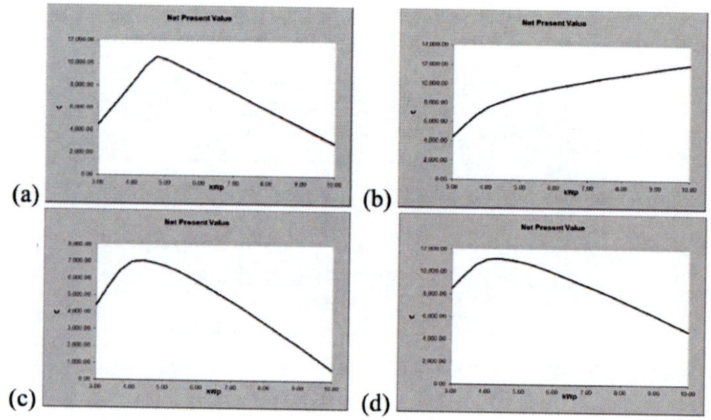

Fig. 4. NPV of the PV-BSS for **a** SS1, **b** SS2, **c** SS3, **d** SS4.

3.3 Parametric Analysis: BSS Size

This simulation considered a 5 kWp PV with a 5–15 kWh BSS. Figure 7 illustrates the energy analysis for the first year of operation and the impact of BSS size on SCR and SSR. The higher the BSS size, the higher the PV self-consumption, which results in reduced grid import/export and thus, increased SCR and SSR.

Figures 8, 9 and 10 demonstrate the impact of BSS size on NPV, IRR and DPP, respectively. NPV is positive under each SS. Yet, there is a certain BSS size for each PV size that maximises it for every SS. Similar observation for IRR. Increasing the BSS size slightly decreases the DPP due to unnecessary BSS capacity. Finally, the BSS cost decrease has a vast effect on all economic indexes.

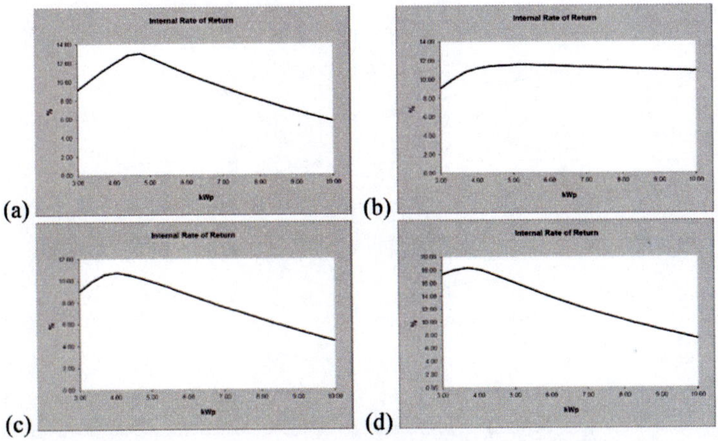

Fig. 5. IRR of the PV-BSS for **a** SS1, **b** SS2, **c** SS3, **d** SS4.

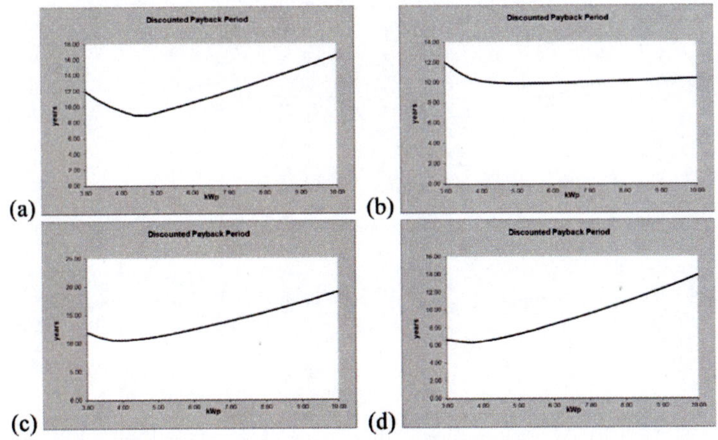

Fig. 6. DPP of the PV-BSS for **a** SS1, **b** SS2, **c** SS3, **d** SS4.

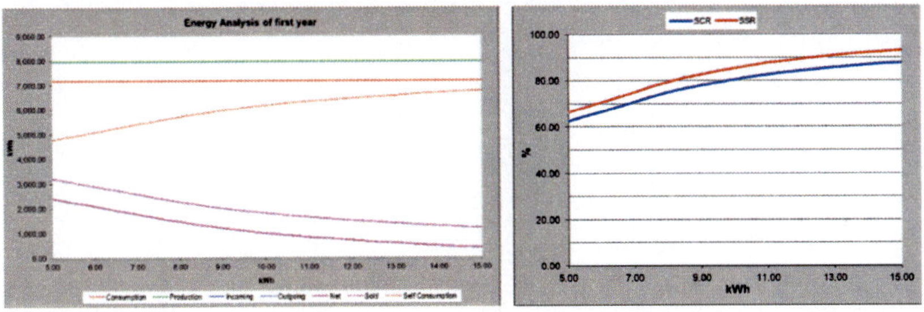

Fig. 7. Energy analysis for the first year of operation (L); BSS size impact on SCR & SSR (R).

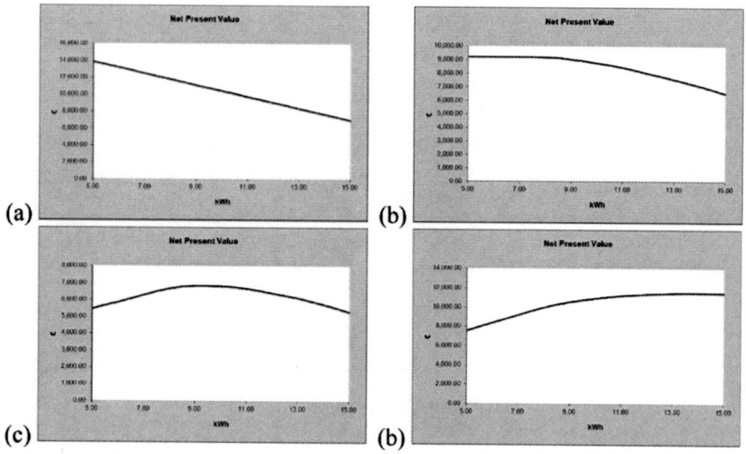

Fig. 8. NPV of the PV-BSS for **a** SS1, **b** SS2, **c** SS3, **d** SS4.

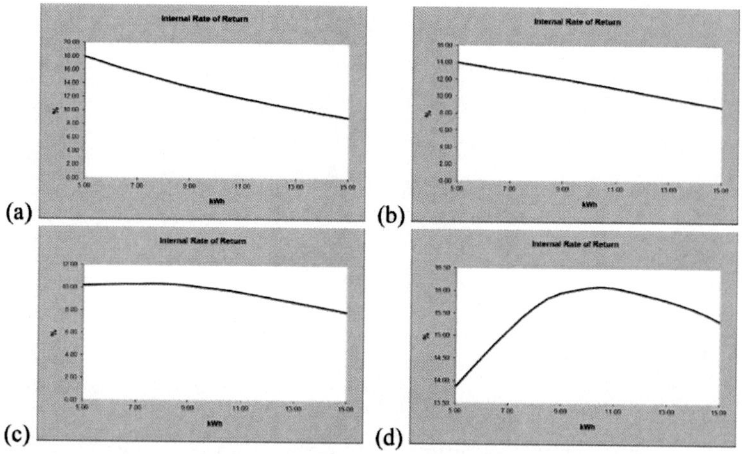

Fig. 9. IRR of the PV-BSS for **a** SS1, **b** SS2, **c** SS3, **d** SS4.

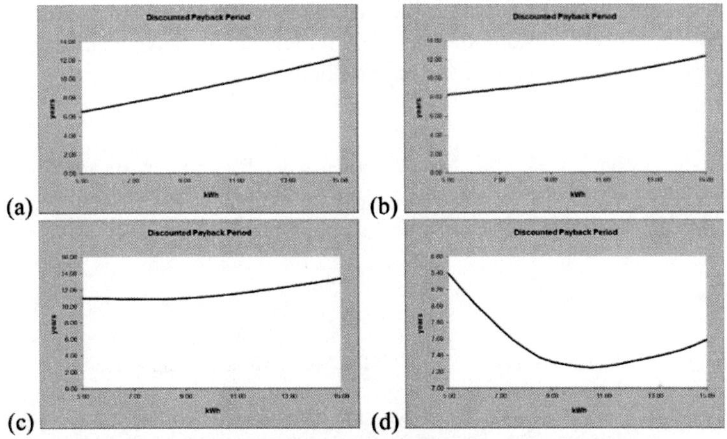

Fig. 10. DPP of the PV-BSS for **a** SS1, **b** SS2, **c** SS3, **d** SS4.

4 Conclusion

This study described an energy-flow simulation for the techno-economic evaluation of PV-BSS in domestic buildings in Cyprus under different SSs. Following a holistic methodology, which includes 4 different steps (Project Characterisation, Cost Estimation, Benefit Estimation and Cost & Benefit Comparison), the simulation can be replicated for any country and building type.

The Techno-economic Analysis demonstrated the viability of PV-BSS in domestic buildings under different SSs. Specifically, even with the current economic conditions and imposed barriers (such as restriction to energy arbitrage, etc.), domestic PV-BSS investments are profitable in Cyprus, while future BSS cost reductions will impact positively all addressed economic indexes. Furthermore, the PV-BSS size must be consciously considered during investment decisioning, as the results clearly showed its significant influence on investment viability. The main outcomes derived by the Techno-economic Analysis are:

- Further PV utilisation and more sustainable buildings due to SCR/SSR increase achieved by PV-BSS, considering also electricity as the main energy carrier in buildings in Cyprus due to cooling needs (mainly use of air-conditioning units).
- Installing a BSS under a Net-Metering scheme is still more attractive than the other addressed policies, due to the equal valuation of grid import/export.
- Self-consumption with sales (Net-Billing) promotes more self-consumption and self-sufficiency maximisation when compared to Net-Metering.
- Self-consumption without sales (most pessimistic case) can still be an attractive SS, especially in cases of high PV generation and electricity prices (like Cyprus).
- PV-BSS size is of great importance, as the purchase of unnecessary PV or BSS capacity results in decreased profitability.

- Net-Metering is the most attractive SS under flat-tariffs, which block further benefits from increasing self-consumption and self-sufficiency. A move from flat to non-flat pricing is required, as it can further exploit the BSS features.

Funding Information. This work has been funded by the Interreg Balkan-Mediterranean Programme, under the project "Enhancing storage integration in buildings with Photovoltaics (PV-ESTIA)" and national funds of Cyprus.

References

1. Briano, J.I., Pardo, I.P., Denoyes, C.R.: PV Grid Parity Monitor Commercial Sector. 5th issue. Creara Energy Experts (2018)
2. Chatzigeorgiou, N.G., Afxentis, S., Panagiotou, K., Georghiou, G.E.: Analysis of 'Increase self-consumption' battery energy storage system use—A residential case study in cyprus [Paper presentation]. In: The 1st International Conference on Energy Transition in the Mediterranean Area 2019 (SyNERGYMED2019), 28–30 May, Cagliari, Italy (2019)
3. Barzegkar-Ntovom, G.A., et al.: Assessing the viability of battery energy storage systems coupled with photovoltaics under a pure self-consumption scheme. Renew. Energy **152**, 1302–1309 (2020)
4. Chatzigeorgiou, N.G., Florides, M.A., Georghiou, G.E.: The financial impact of policy schemes on PV+Battery systems in residential buildings: a case study in Cyprus [Paper presentation]. In: The 13th International Conference on Sustainability in Energy and Buildings 2021 (SEB-21), 15–17 September, KES Virtual Conference Centre (2021)
5. Jäger-Waldau, A.: PV Status Report 2019—EU Science Hub—European Commission. EU Science Hub—European Commission (2019)
6. Chatzigeorgiou, N.G., Yerasimou, Y.P., Florides, M.A., Georghiou, G.E.: Grid export reduction based on time-scheduled charging of residential battery energy storage systems—A case study in Cyprus. In: Littlewood, J., Howlett, R.J., Jain, L.C. (eds.) Sustainability in Energy and Buildings 2020. Smart Innovation, Systems and Technologies, vol. 203. Springer, Singapore (2021)
7. Chatzigeorgiou, N.G., Poize, N., Florides, M.A., Georghiou, G.E.: Analysing the operation of residential photovoltaic—Battery storage systems in cyprus [Paper presentation]. In: The 12th Mediterranean Conference on Power Generation, Transmission, Distribution and Energy Conversion 2020 (MEDPOWER2020), 9–12 November, Online Conference (2020)
8. Santos, J.M., Moura, P.S., Almeida, A.T.D.: Technical and economic impact of residential electricity storage at local and grid level for Portugal. Appl. Energy **128**, 254–264 (2014)
9. Wang, Y., Lin, X., Pedram, M.: Adaptive control for energy storage systems in households with photovoltaic modules. IEEE Trans. Smart Grid **5**(2), 992–1001 (2014)
10. Khalilpour, K.R., Vassallo, A.: Technoeconomic parametric analysis of PV-battery systems. Renew. Energy **97**, 757–768 (2016)
11. Marchi, B., Pasetti, M., Zanoni, S.: Life cycle cost analysis for BESS optimal sizing. Energy Procedia 127 (2017)
12. Venizelou, V., Makrides, G., Efthymiou, V., Georghiou, G.E.: Methodology for deploying cost-optimum price-based DSM for residential prosumers. Renew. Energy **153**, 228–240 (2020)
13. EU Science Hub - European Commission. (2020). Photovoltaic geographical information system (PVGIS)—EU Science Hub. https://ec.europa.eu/jrc/en/pvgis. Accessed 1 May 2022

Environmental Assessment, Cost Assessment and User Experience of Electric Excavator Operations on Construction Sites in Norway

Marianne Kjendseth Wiik[1](✉), Kristin Fjellheim[1], Kamal Azrague[1], and Jon Are Suul[2]

[1] SINTEF Community, Børrestuveien 3, 0373 Oslo, Norway
marianne.wiik@sintef.no
[2] SINTEF Energy, Sem Sælands Vei 11, 7034 Trondheim, Norway

Abstract. This paper measures the environmental impacts and life cycle costs associated with electrifying diesel excavators (8.5, 17.5 and 38t) and summarizes experiences from the pilot testing of these electric excavators in Norway. The demonstration and pilot testing of the studied excavators is due to requirements set by Oslo municipality for all public construction sites to be emission-free by 2025. The results show that electrified operation leads to much lower total GHG emissions than diesel excavators, and that electrified excavators can have lower costs over their life span than their diesel-powered equivalents, even if they have a higher investment cost. Experiences from pilot testing show that electric excavators can provide the required performance and that they can be successfully used on construction sites with only minor adjustments to operational procedures. Although the studied cases and pilot tests are motivated by local Norwegian requirements, they imply a large potential for wider application of the technology for fully electrified construction machinery.

Keywords: Environmental assessment · Life cycle assessment (LCA) · Life cycle costing (LCC) · Pilot testing · Electric excavator

1 Introduction

According to the International Energy Agency (IEA), the building and construction sector is responsible for ca. 40% of energy and process-related global greenhouse gas (GHG) emissions [1]. While a large share of these emissions is related to construction materials, maintenance, and heating, the fossil fueled machineries used during the construction process also contribute to emissions. With growing global population and increasing need for buildings and infrastructure combined with the urgent need for reducing GHG emissions, the potential for reducing emissions generated during all stages of construction should be utilized. For instance, in Oslo, Norway, it is estimated that construction machinery is responsible for around 30% of transport sector emissions [2]. In response to the Paris agreement, Oslo municipality has set climate mitigation goals to reduce GHG emissions by 95% by 2030 compared to levels in 1990 [2–5]. In addition,

J. Littlewood and R. J. Howlett (Eds.): SEB 2022, SIST 336, pp. 97–108, 2023.
https://doi.org/10.1007/978-981-19-8769-4_10

the municipality has agreed that all public construction sites shall be emission free by 2025 and all private construction sites shall be emission free by 2030. All public procurement will require fossil free construction operations and award contracts to suppliers with emission free machineries.

In Norway, a fossil free construction site is defined as a construction site that does not use any fossil fuels in any of its onsite construction activities [6]. An emission free construction site does not have any direct or indirect GHG emissions from its construction activities [10]. Construction activities include transport of materials, transport and operation of construction machinery, transport of construction workers, energy use, internal transport, storage, temporary works, additional materials for installation of building materials and components, transport of waste, waste treatment and disposal [7–10]. Fossil fuels (i.e., diesel or propane) are often replaced with bioenergy and fuels (i.e., HVO or wood pellets) or alternative renewable energy resources such as electricity or hydrogen [11]. To achieve this goal, construction machineries are being electrified and tested out on construction sites around the Oslo region. This paper measures the environmental impacts and life cycle costs (LCC) associated with electrifying diesel excavators (8.5, 17.5 and 38t) and summarizes experiences from the pilot testing of these electric excavators on three construction projects in Norway.

2 Method

2.1 Life Cycle Assessment

The goal of the life cycle assessment (LCA) is to ascertain the environmental impacts from an 8.5t, 17.5t, and 38t electric excavator compared to diesel excavators of equivalent size and evaluate potential environmental impacts after electrification. These machines are prototypes converting standard diesel machines to be electric. The comparative LCAs are carried out according to ISO 14044: 2006 and EN 15804: 2012 [7, 12] and include raw material extraction and processing, transport to the manufacturer, manufacturing, transport to the building site, installation on site, maintenance, repair and replacement, operation on site, transport to end of life, waste processing and disposal.

Life cycle inventory data has been collected from the manufacturers [13] and Ecoinvent v3.1 database [14]. The life cycle inventory models are developed in SimaPro Analyst v9.0.0.48 [15]. For impact assessment, the ReCiPe Midpoint Hierarchist v1.13 method is used, whereby all eighteen mid-point indicators have been assessed [16]. The functional unit is one-hour operation given 1800 h of operation per year and a reference study period and lifetime of 10 years. As a starting point, the same material inventory has been used for a diesel excavator. The diesel power train components have then been replaced with the electrical components necessary for an electric excavator.

It is assumed that all the excavators are produced in Japan and transported to Larvik, Norway using containerships. The 15t "hydraulic digger" process in Ecoinvent is used and adjusted to 8.5t, 17.5t and 38t diesel and electric excavators with background information from the manufacturer. This means that the original "hydraulic digger" process is adjusted from a 15t diesel excavator produced in Europe to 8.5t, 17.5t and 38t diesel excavator produced in Japan. The processes are switched from European processes (RER) that use European electricity mixes, to Japanese processes (JP) that use the Japanese electricity mix. When Japanese processes were not available, processes representative for the rest of the world (ROW) are used. The electrical excavators were then rebuilt for electrical operation at a special production facility in Larvik and use processes with the Norwegian electricity mix (NO). During the rebuilding process the electrical excavators all needed extra components such as batteries (8.5t and 17.5t), electrical engines (all excavators), powertrains (all excavators), inverters (all excavators), 230m cable (17.5t and 38t excavator), galvanic isolation (17.5t and 38t excavator), and storage container (17.5t and 38t excavator). These elements are included in calculations, see Fig. 1.

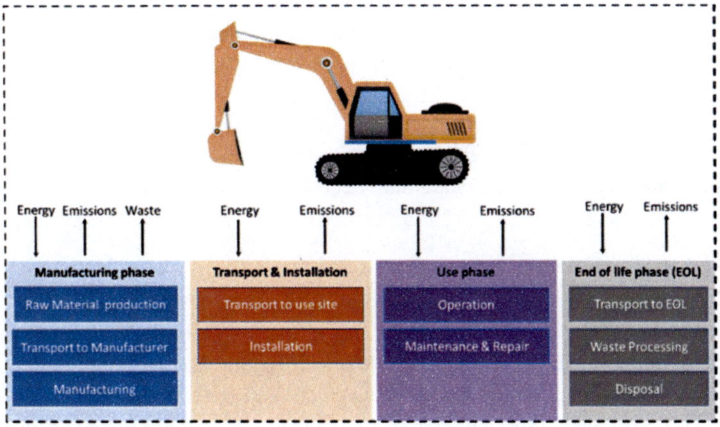

Fig. 1. System boundary for the life cycle assessment of diesel and electric excavators, (authors own).

The construction phase includes transport from Larvik to Oslo (128 km) and installation of the excavators on the construction site. The installation on the construction site includes the charging infrastructure of the charging cable and storage container for the cable roll. It is assumed that the excavators are at the same construction site for approximately one year and then moved to a different site in the Oslo area, the distance from one construction site to the next is negligible.

The operational energy use for each excavator is listed in Table 1 and each excavator has 1800 operating hours per year. For diesel, the emission factor is 3.32 $kgCO_2e$/liter, and for electricity the Norwegian emission factor (0.011 $kgCO_2e$/kWh) is used [14].

All excavators will go through waste processing and disposal stages. The diesel excavators that are produced in Japan consist mainly of steel and will be sent to recycling. For the electrical excavators, the batteries and cables are sent to disposal, the containers are sold at the second-hand market, the electrical engine, powertrain, and aluminum inverters are dismantled and recycled.

Table 1. Energy use per excavator.

Technology	Unit	8.5t	17.5t	38t
Diesel	l/h	5.5	10	30
Electric	kWh/h	13	28	100

2.2 Life Cycle Costing

LCC is a comprehensive decision-making tool that can be used for evaluating total costs generated over the entire lifetime of physical assets. The LCC of any excavator can be computed by calculating the total cost of ownership (TCO) with respect to initial purchase price, depreciation costs of the vehicle over its economic life, operation costs, maintenance costs, financing costs including interest, costs associated with taxes, and insurance costs. All costs are reduced to their present value (PV) which considers the time value of money. PV can be calculated according to the following formula:

$$PV = C/(1+i)^n \tag{1}$$

where C is any cost incurred in year n and i is the discount rate. The present value of every cost can be added to get the full cost of an alternative. The TCO of each alternative is calculated according to Eq. 2, the purchase prices and the electric supply infrastructure are considered together with the yearly costs calculated based on the PV.

$$\begin{aligned} TCO = & \text{Vehicle infrastructure acquisition price} \\ & + \sum_{n=1}^{N} \frac{RC}{(1+i)^n} - \text{residual value}_n \end{aligned} \tag{2}$$

where RC expresses the recurring costs. The planning horizon used in LCC calculations is 6 years. This is shorter than the lifetime used in LCA since it is based on the typical planning horizon used by the industry of the excavator and not the lifetime of the product. A sensitivity analysis has also been performed for changes in energy prices, purchase cost discount rate, and residual value.

2.3 Pilot Testing

All three excavator prototypes are designed to have operating performances equivalent to their diesel equivalents and have been tested under real conditions at Olav Versus Street, Oslo accident and emergency (A&E) and Biri care home. The 8.5t and 17.5t excavators have been tested on Olav Versus Street and Oslo A&E construction sites, and per April 2020 have accrued 1505 and 1071 operational hours, respectively. The 38t excavator has been tested at Biri care home from May 2019 to January 2020 and has accrued 392 operational hours.

Olav Versus Street lies between Oslo's national theatre and city hall and has been refurbished as a pedestrianized street with wider pavements, newly planted trees and charging stations for electric taxis. The construction site is Norway's first emission free site, meaning that all machinery and equipment within the construction site is emission free. The construction phase began in September 2019 and was finished at the end of 2020. In contrast, Oslo A&E consists of 26,000 m^2 heated floor area and 4,400 m^2 basement carparking. Oslo municipality have a clear environmental goal that the building shall have low operational energy use, fulfil passive house standards, achieve BREEAM Excellent and be a fossil free construction site. The construction began in August 2019 and will be completed in 2023. Finally, Biri care home consists of 16 nursing home places and 16 sheltered housing units. The development has environmental goals for fulfilling passive house requirements, energy class A, reducing material GHG emissions by 40%, implementing renewable energy as well as using fossil and emission free construction machinery. The performance assessment of the electric excavators has been carried out through dialogue with important stakeholders, (e.g., contractors, site managers, construction machinery operators and machinery manufacturers), and collecting firsthand experiences from them and comparing these experiences to those of conventional diesel excavators.

3 Results and Discussion

3.1 Life Cycle Assessment

Table 2 displays the impact category results for the different excavator alternatives. The results for global warming potential show a 95% reduction in total GHG emissions for the conversion from diesel to electric excavators. The 38t electric excavator (powered by cable) has lower impacts across all impact categories than its diesel counterpart, except for in the human toxicity category where this is 22% higher. For the 8.5t and 17.5t battery electric excavators, higher impacts are experienced in the freshwater eutrophication (60% and 84% respectively), human toxicity (117% and 169%), freshwater ecotoxicity (50% and 84%), marine ecotoxicity (48% and 82%) and metal depletion (41% and 38%) impact categories. These results are mainly due to the batteries installed in the 8.5t and 17.5t excavators and the dynamic cable installed in the 17.5t and 38t excavators. Environmental impacts associated with batteries are high as reported by several other studies also dealing with electric vehicles (17–19). The 8.5t, 17,5t and 38t electric excavator's contribution to the remaining impact categories are reduced compared to their diesel excavator equivalents.

The same lifetime of 10 years is used for both diesel and electric excavators, however it can be argued that an electric excavator may have a longer lifetime than its diesel equivalent since there are fewer mechanical, movable parts. The battery performance may also affect service life. Service lifetimes of batteries and electric excavators can be better established through further pilot testing and warrants scope for future work. Extending the service life of the electrified excavators will have a positive effect on emissions since emissions from the production processes and relating to batteries will be shared across the longer service life of the excavator. Extending the service life of the electric excavators and associated components (e.g., batteries) will also have a positive impact on the LCC of electric excavators and reduce the pay-back time.

Table 2. Mid-point results for 8.5t, 17.5t and 38t excavators per functional unit.

Impact category	Unit	8.5t		17.5t		38t	
		Diesel	Electric	Diesel	Electric	Diesel	Electric
Global warming potential	kg CO_2e	43.92	2.37	85.94	4.42	214.74	9.35
Ozone depletion	kg CFC-11 e	3.61E−06	2.21E−07	6.62E−06	4.55E−07	1.94E−05	7E−07
Terrestrial acidification	kg SO_2 e	0.35	0.0157	0.68	0.0312	1.70	0.0479
Freshwater acidification	kg P e	1.53E−03	0.00245	2.97E−03	5.46E−03	7.53E−03	0,00692
Marine Eutrophication	kg N e	2.34E−02	0.00151	4.56E−02	3.06E−03	1.15E−01	0,00331
Human toxicity	kg 1,4-DB e	2.0	4.34	3.9	10.5	9.9	12,1
Photochemical oxidant formation	kg NMVOC	0.63	0.0096	1.24	0.0183	3.07	0.0333
Particulate matter formation	kg PM10 e	0.19	0.00771	0.36	0.0154	0.90	0.0271
Terrestrial ecotoxicity	kg 1,4-DB e	1.26E−03	3.49E−04	2.43E−03	7.88E−04	6.26E−03	0.00109
Freshwater ecotoxicity	kg 1,4-DB e	0.07	0.105	0.14	0.257	0.37	0.327
Marine ecotoxicity	kg 1,4-DB e	0.07	0.104	0.14	0.254	0.37	0.321
Ionising radiation	kBq U235 e	1.46	0.148	2.69	0.246	7.86	0,377
Agricultural land occupation	m^2a	0.13	0.0951	0.25	0.185	0.62	0.329
Urban land occupation	m^2a	0.06	0.0312	0.11	0.0573	0.31	0.0971
Natural land transformation	m^2	7.30E−04	0,00038	1.47E−03	7,38E−04	3.40E−03	0.0015

(continued)

Table 2. *(continued)*

Impact category	Unit	8.5t		17.5t		38t	
		Diesel	Electric	Diesel	Electric	Diesel	Electric
Water depletion	m^3	0.05	0.0264	0.10	0.0383	0.28	0.0609
Metal depletion	kg Fe e	1.7	2.4	3.3	4.56	8.2	6.5
Fossil depletion	kg oil e	7.10	0.58	13.06	1.07	38.04	2.05

3.2 Life Cycle Costing

Figure 2 shows the sum of costs in Euros per life cycle category after the end of the 6-year planning horizon. This shows the main cost element for the electrical excavators is acquisition costs, while the operation cost has a higher relative impact on the total LCC for diesel excavators.

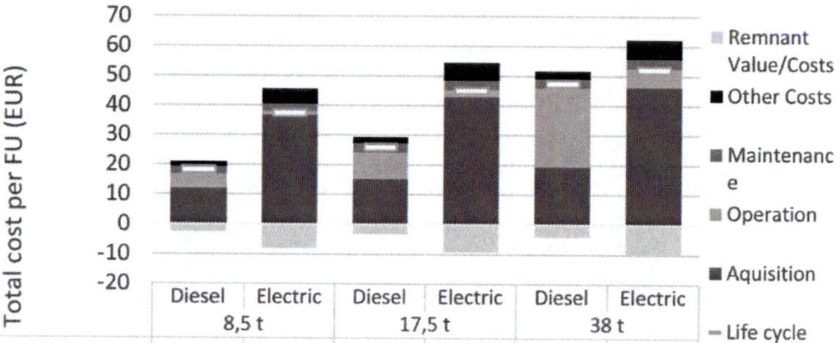

	Diesel 8,5 t	Electric 8,5 t	Diesel 17,5 t	Electric 17,5 t	Diesel 38 t	Electric 38 t
Remnant Value/Costs	- 3	- 8	- 3	- 9	- 4	- 10
Other Costs	2	5	2	6	3	6
Maintenance	2	2	3	3	3	3
Operation	5	1	9	3	27	6
Aquisition	12	37	15	43	19	46
Life cycle cost	18	38	26	45	47	52

Fig. 2. Diagram showing the TCO of the various diesel and electric excavator alternatives.

Several sensitivity analyses have been conducted to see the effects of energy prices, purchase cost, discount rate and residual value. A possible development in the energy market can lead to higher prices for fossil resources and lower prices for renewables.

A sensitivity analysis on increased prices for diesel and decreased prices for electricity by 10% has been performed. The results show that the energy prices will affect the competitiveness of the electrical excavators slightly, but it still does not make any of the electrical excavators cheaper than diesel excavators over a 6-year planning period. It is assumed that the cost of manufacturing the excavators will reduce over time especially considering that the price of batteries ($/kWh) has reduced by 87% since 2010 (20). The concept of economies of scale will also be driving the total cost of electrical excavators down as the first prototypes will always be more expensive due to long delivery times for components and that the order batches are small compared to large scale production. A sensitivity analysis is performed where the cost of electrical excavators is reduced by 37.5% for 8.5t, 26% for 17.5t and 22% for 38t, as it is assumed that the cost of batteries and excavators will become cheaper over time due to the economics of scale. The larger reduction for the 8.5t and 17.5t excavator is because it is assumed that the batteries will have a higher price drop. The results show that the 8.5t and 17.5t electric excavators are still more expensive over the 6-years planning period, but the 38t electric excavator is cheaper. As the discount rate can have a large impact on estimated future expenses a sensitivity analysis has been performed for a lower discount rate of 2.5%. The lower discount rate has little effect on the life cycle cost and only closes the cost gap between the electric and diesel excavators, for all excavator sizes the electric version is still more expensive over the 6-year planning period. However, since most building and construction sites in Norway will be emission free by 2030 it is likely that the rest value of diesel excavators will decrease over time and that there will be a higher demand for electric excavators. There are high uncertainties related to residual value, so this is evaluated at 0% and 10% of the purchase cost instead of the standard 25%.

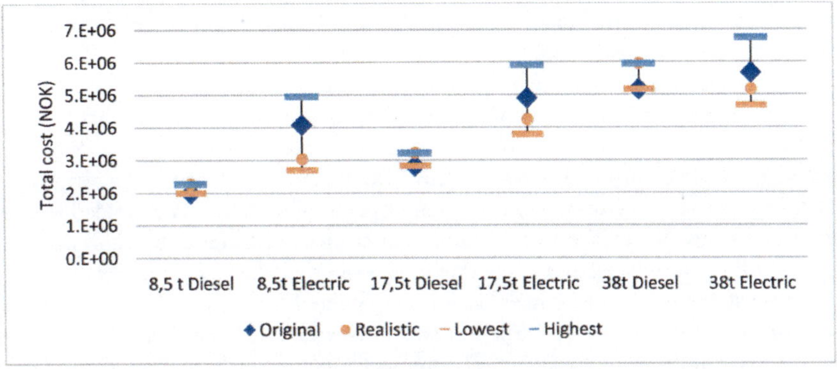

Fig. 3. Total cost after 6 years showing the range of variation from lowest to highest sensitivity scenario.

The results from the sensitivity analysis are shown in Fig. 3. The original scenario shows the results without sensitivity analysis. The lowest scenario shows the original cost for diesel excavators and the reduced cost of electric excavators. The highest scenario shows the optimistic scenario for diesel excavators and the residual value equal to zero for the electric excavators. The realistic scenario shows the cost with reduced discount rate, reduced cost of electric excavators, lower electricity prices and higher diesel prices. The sensitivity analyses show that minor changes are required in the production process, energy market, or economic market to increase the competitiveness of electrical excavators compared to diesel excavators in a life cycle perspective.

3.3 Pilot Testing

The electrical excavators have been tested under real conditions on Olav Versus Street, Oslo A&E and Biri care home. The electrical excavators have operated a total of 6,817 h per April 2020. Altogether the electric excavators have saved over 30,748 L of diesel, 372,000 Norwegian kroner (NOK) in energy costs, and over 91 tCO_2e in GHG emissions. Experiences obtained from prototype testing show that a construction site using new electrical prototype excavators requires more preparation in advance of the excavator arriving on site than when using traditional diesel excavators. In addition to standard construction site preparations, the electricity provider should be included early in the planning and preparation phases. As the excavators are electric there should always be competent personnel on site during the startup phase and under operation. Errors occurring on the construction site can lead to delays and increased cost as the availability of new excavators are small and replacement parts are harder to come by. The 8.5t and 17.5t excavators have built in batteries which makes it easier to move around the construction sites without the issue of always being connected to a cable.

Biri care home was the first construction site to test the 38t electric excavator. The excavator operator felt that the excavator was a bit too large for the work it was contracted for, however they did not experience any delays in operation or damage to the cable. They did experience some electrical interference and flashing from the tube light. The manufacturer has evaluated this feedback and will relook at the way the machine is put together to avoid any further interference. It is believed that having a separate transformer will resolve most of these issues.

At Oslo A&E the 8.5t electric excavator has been used during demolition. The 17.5t electric excavator is planned to be used later during the construction process, however implementation of the 17.5t electric excavator was delayed due to one of the battery sets being damaged during prototype testing. The battery set at the time of writing was repaired by the manufacturer, and the battery operating system was upgraded to avoid any further issues. To avoid any cable damage or electrical interference a cable container solution will be introduced and a separate transformer for the electric excavator.

The most extensive experiences have been gathered from Olav Versus Street since this is the first full-electric construction site in Norway. At Olav Versus Street there was some downtime during start-up. This project was unique since multiple electrified construction machineries were running simultaneously from multiple suppliers. When the electric excavators were first delivered onsite, two of the chargers were burnt out, firstly because of overloading the electric grid by trying to charge too many construction machineries at the same time and secondly because one of the machines (not one of the excavators assessed in this article) did not have electromagnetic noise insulation installed in the transformer. There have been some issues with the destruction of cables, however this has not led to any downtime. These learning experiences have highlighted the need for adjustments in operating routines on construction sites when using electric excavators. This was resolved by charging construction machinery during lunch hours to have full operation during the whole day. There were also some issues with cold temperatures during winter as the battery performance was reduced and a heater was installed to keep adequate temperatures during operations, breaks and at night. Lower outdoor temperatures also led to lower operating hours as some of the battery capacity needed to be used for heating and lighting to maintain good working conditions for the excavator operator. General comments from the excavator operators include less noise and pollution onsite leading to a better working environment. There was also less hazardous waste spill (i.e., diesel) from the operation of electric excavators. The collective experiences from these projects have been invaluable for the further development of electric excavators and transition towards emission free construction sites in Norway.

4 Conclusion

This article presents an assessment of environmental impacts and life cycle costs associated with electrifying diesel excavators and summarizes experiences from the pilot testing of these in Norway. Three different cases of retrofitted excavators with an electric drivetrain are compared to original diesel units. This includes a 38t excavator electrified by cable, a 17.5t excavator with cable and a small onboard battery, and an 8.5t battery-electric excavator. The results show that the electrified excavators have a 95% reduction in total GHG emissions. Electrification also helps reduce other environmental impacts. However, the use of batteries and cables result in higher impacts for human toxicity, freshwater and marine ecotoxicity, freshwater eutrophication and metal depletion. While the TCO for the electrified excavators is higher than in diesel excavators, the operation costs are significantly lower. A sensitivity analysis of the costs demonstrates that limited changes in cost factors can result in conditions where the electrified excavators will have lower lifetime costs. This is most likely to happen for the largest units where the operation costs have the highest impact on the TCO. It is expected that the use of electric excavators will increase in Norway under public requirements for zero emission construction sites, and that this will lead to experiences applicable in a wide range of construction projects both locally and internationally.

Acknowledgements. The authors would like to acknowledge project partners Nasta, Skanska, Omsorgsbygg, Bellona and Difi for their collaboration in the Zero Emission Digger project. Funding: This work was supported by the Norwegian Research Council, Innovation Norway and Enova through the PILOT-E programme 2018–2020 [grant number 281804].

References

1. IEA.: UN Environment programme. 2019 Global status report for buildings and construction. Towards a zero emissions, efficient and resilient buildings and construction sector (2019)
2. Oslo kommune.: Klima- og energistrategi for Oslo. Behandlet av Oslo bystyre 22.06.2016 (sak 195/16). Klimaetaten, Oslo (2016)
3. United Nations General Assembly. United Nations Climate Change Conference. Paris (2015)
4. Oslo kommune.: Det grønne skiftet. Klima- og energistrategi for Oslo. Oslo kommune. Klima- og energiprogrammet, Oslo (2015)
5. Multiconsult.: Review of implementation of fossil free building sites. Document no. 10206471-TVF-RAP-001. 2018 (2018)
6. Fufa, S.M., Wiik, M.R.K., Mellegård, S.E., Andresen, I.: Lessons learnt from the design and construction strategies of two Norwegian low emission construction sites. IOP Conf Ser Earth Environ Sci EES. **352** (2019)
7. EN 15804: 2012 + A1: 2013. Sustainability of Construction Works—Environmental Product Declarations—Core Rules for the Product Category of Construction Products. Vol. CEN/TC 350. Standard Norway, European Standard, Oslo, Norway (2013)
8. EN 15978.: Sustainability of construction works—assessment of environmental performance of buildings—calculation method (2011)
9. NS 3720.: Metode for klimagassberegninger for bygninger/Method for greenhouse gas calculations for buildings (2018)

10. Fufa, S.M., Mellegård, S., Wiik, M.K., Flyen, C., Hasle, G., Bach, L., et al.: Utslippsfrie byggeplasser - State of the art. Veileder for innovative anskaffelsesprosesser. SINTEF Fag rapport nr. 49 (2018). ISBN:978-82-536-1589–9
11. Fasting, G., Lie, A.Ø., Dugstad, E.: Fossil- og utslippsfrie byggeplasser rapport. Energi Norge, Norsk Fjernvarme i samarbeid med Bellona, og Enova SF. DNV GL AS Energy, Oslo. Report No.: 2017–0637 (2017)
12. ISO 14044. Environmental Management—Life Cycle Assessment—Requirements and Guidelines. Geneva, Switzerland: International Organization for Standardization (2006)
13. NASTA. ZAXIS-6 Series Zaxis300 hydraulisk gravemaskin. Hitachi NASTA (2019)
14. Ecoinvent. Ecoinvent database v3.1. Swiss Centre for Life Cycle Inventories, Dübendorf, Switzerland (2014)
15. PRE Consultants.: Simapro. LCA software for fact-based sustainability. Simapro (2019). https://simapro.com/. Accessed 1 Jan 2019
16. Goedkoop, M., Heijungs, R., Huijbregts, M., Schryver, A.D., Struijs, J., Rv. Z.: ReCiPe 2008: a life cycle impact assessment method which comprises harmonised category indicators at the midpoint and the endpoint level. Report I: Characterisation (2013)
17. Booto, G.K., Aamodt Espegren, K., Hancke, R.: Comparative life cycle assessment of heavy-duty drivetrains: a Norwegian study case. Transp. Res. Part Transp. Environ. **95** (2021)
18. Lie, K.W., Synnevåg, T.A., Lamb, J.J., Lien, K.M.: The Carbon footprint of electrified city buses: a case study in Trondheim, Norway. Energies 14 (2021)
19. Puig-Samper Naranjo, G., Bolonio, D., Ortega, M.F., García-Martínez, M.J.: Comparative life cycle assessment of conventional, electric and hybrid passenger vehicles in Spain. J. Clean Prod. **291** (2021)
20. Bloomberg, N.E.F.: Battery pack prices fall as market ramps up with market average at $156/kWh in 2019 (2019). https://about.bnef.com/blog/battery-pack-prices-fall-as-market-ramps-up-with-market-average-at-156-kwh-in-2019/. Accessed 29 Jul 2020

A Rapid Survey Form for Users' Exposure and Vulnerability Assessment in Risk-Prone Built Environments

Enrico Quagliarini(✉), Guido Romano, Gabriele Bernardini, and Marco D'Orazio

DICEA Department, Università Politecnica Delle Marche, Via Brecce Bianche, Ancona, Italy
e.quagliarini@staff.univpm.it

Abstract. The sustainable transition to resilient cities is linked to the evaluation of their citizens' habits. Understanding the Built Environment (BE) use is fundamental to plan effective risk-mitigation strategies, and users' features and behaviors deeply affect the way BEs are used. Recent studies are moving toward the definition of typological (idealized) scenarios—namely Built Environment Typologies (BETs)—for simulation-based analyses aimed at the assessment of real-life BEs safety and resilience. Rapid surveys are available to collect data on typological features and hazards/physical vulnerability factors, but not to adequately assess BEs users' vulnerability and exposure to single/multi natural and human-related risks, and their spatiotemporal variability. Within this framework, this work aims at providing an expeditious survey form to quantify, collect and represent such data. The form is based on remote analyses for rapid evaluations of critical hourly/daily users-related conditions. Among BEs/BETs, the attention is here focused on squares, which represent meeting spaces par excellence and host main functions for communities. A real-world square (Piazza Duomo in Reggio Calabria, Italy) is selected for the form application because of its geomorphological and riskiness characterization in correlation with its previously defined BET type. Results are assessed through Key Performance Indicators (KPIs) resuming daily trends according to the users' age, position and familiarity with outdoor and indoor areas. Promoted in the BE S^2ECURe Italian Research Project, the form can also support safety planners and local administrations in simulation-based assessment and risk-mitigation strategies development.

Keywords: Survey form · Users · Squares · Users' exposure · Users' vulnerability · Built environment · Multi-hazards

1 Introduction

In order to cope with future challenges of resilience and sustainability development, the Built Environment (BE) in which we leave must become increasingly smart, secure, and inclusive to its users [1, 2]. Worldwide, about 70% of cities are already facing the effects of climate changes [2] that every day affect the BE with slow-onset disasters, like

J. Littlewood and R. J. Howlett (Eds.): SEB 2022, SIST 336, pp. 109–119, 2023.
https://doi.org/10.1007/978-981-19-8769-4_11

pollution and heat-waves, to which must be added risks due to sudden-onset disasters, like earthquakes and terrorist attacks [1].

Many previous literature works offer definitions of the concepts of "sustainability", which aims at increasing users' quality of life with respect to economic, environmental, and social systems, and "resilience", which focuses on the response of those systems to disasters [3–5]. In the context of this work, both sustainability and resilience are declined with respect to the BE system intended as "a network of buildings, infrastructures, and open spaces" and its users, whose vulnerability and exposure to risk drastically change depending on the BE features [1]. For instance, users can behave differently in relation to factors such as [6–9]: their position in indoor/outdoor areas depending on the type of hazard, their familiarity with the BE and emergency procedures that can vary depending on the type of area (e.g., residential/commercial/touristic), the demographic characterization in relation to categories at risk (such as children, elderlies, and disabled). Therefore, the design of a resilient, safe, and sustainable BE cannot disregard considerations on how, when, where, and which users occupy the different areas of the BE itself.

Risk assessment methodologies are increasingly emphasizing the importance of a holistic BE characterization by jointly considering its features and those of the hosted users [10]. Idealized scenarios in terms of typological categories of BEs, namely Built Environment Typologies (BETs), can represent relevant conditions to assess communities' risks and test sustainable and resilient solutions against disasters [11]. Furthermore, they can take advantage of rapid survey forms to collect and quantify data on the BE geometry, the exposure to a given type of risk, or the population features [1, 8]. As shown by the results of the Italian PRIN research (Projects of Relevant National Interest) BE S^2ECURe, of which this work is part, such scenarios can take advantage of cluster analyses to characterize and resume morphological, functional, constructive, and physical features of real-world case studies, offering the bases for simulation analyses for the assessment of the users' safety under common and typical conditions [1, 8, 11]. Thus, a correct collection of data plays a critical role in defining effective BETs, representative of actual conditions [11]. Key Performance Indicators (KPIs) can be used to this end, being able to derive typological recurrent conditions by considering a sample of case studies of the same relevant context (e.g., squares, streets, districts), as well as depict local conditions by investigating single cases.

Within this framework, this work aims at providing a rapid survey form for the collection and organization of KPIs on users' exposure and vulnerability (and their spatiotemporal variability) in risk-prone BEs, depending on BEs features. According to the main BE S^2ECURe application context, that is Italian historic cities and, i.e., squares as relevant BEs [11], Piazza Duomo in Reggio Calabria is selected as a study case, also in view of its significant geomorphological and multi-hazards characterization.

2 Methodology

This work is organized in two parts, which respectively concern input data collection for the form and their organization into KPIs. The form is similarly organized in two panels, one for each work part. Thus, the first part/form panel introduces the criteria to quantify the number of users through the breakdown analysis of the BE in outdoor and indoor areas, and the organization of demographic data (Sect. 2.1). As a consequence, users' vulnerabilities are expressed in relation to the *type of area* occupied and users' *age range*, while exposure through the maximum number of users that potentially occupy a certain area during a certain timetable [7, 12]. Afterward, such data are converted into KPIs to jointly consider exposure and vulnerability issues for the application to an idealized scenario for simulation-based analysis (Sect. 2.2). Those data are resumed in the second panel of the form.

The form is applied to Piazza Duomo in Reggio Calabria, Italy. The square has a total gross area of about 5000 m^2 (sides of 64 × 77 m), hosts the main city church and other public buildings (including commercial activities), its sides are crossed by two main urban streets open to vehicular traffic, and includes some green areas (i.e. trees implying shaded areas). The square can be represented as a BET 3-type square [11], that has a high level of compactness and regularity on flat ground. It is potentially affected by multi-hazards conditions, i.e.: heatwaves (geographical localization); pollution (vehicular traffic + urban context); earthquakes (seismic zonation); terrorist acts (presence of the main church and other buildings open to the public).

2.1 Built Environment Breakdown Analysis for the Users' Quantification

The method relies on remote-based quick analysis for quantifying the maximum Number of Users NU [pp] depending on the BE features [1, 12]. Different *types of areas* are distinguished to consider vulnerabilities due to the users' outdoor/indoor position and familiarity/unfamiliarity with the BE (Table 1) [7, 13]. Demographic data are organized into *age ranges* to consider common conditions in motion (Table 2) [6].

Table 1. Summary of the users' vulnerabilities, description, and timetable according to the *type of area.*

Type of users (position, familiarity)	Description (Type of area ID [pp/m^2])	Occupant load [13] [pp/m^2] (timetable)
Only outdoor users OO (outdoor, unfamiliar)	Users occupying public walkable areas like sidewalks, accessible non-fenced green areas, parks (OA_1)	W and H: 0.1 pp/m^2 (7–24) 0.0 pp/m^2 (1–6)
Prevalent outdoor users PO (outdoor, unfamiliar)	Users occupying open-air terraces of restaurants, open markets and other outdoor areas related to specific buildings or intended uses (OA_2)	W and H: 0.4 pp/m^2 (i.u. opening time) 0.0 pp/m^2 (i.u. closing time)

(*continued*)

Table 1. (*continued*)

Type of users (position, familiarity)	Description (Type of area ID [pp/m^2])	Occupant load [13] [pp/m^2] (timetable)
Non-resident users NR (indoor, unfamiliar)	Users occupying non-residential areas such as commercial activities, churches, government building… (IO_1)	W and H: Depending on the intended use* (i.u. opening time)**
Residents users R (indoor, familiar)	Users occupying residential areas (IO_2)	W and H: 0.05 pp/m^2 (0–24)

W is for Working days, H is for Holidays. *0.4 pp/m^2 for intended uses open to public (including restaurants, bars, shops, public offices), 0.7 pp/m^2 for churches, 0.1 pp/m^2 for intended uses close to public. **Churches' opening times refer to Sunday services timetables.

Table 2. Summary of the users' vulnerabilities, timetable, and presence coefficient according to their *age range*.

Type of users [6, 12] (age range, motion conditions)	Familiar users (Timetable // Presence coeff.)	Unfamiliar Users (Timetable // Presence coeff.)
Toddlers T (0–4, assisted); Elderlies E (70 +, assisted)	W and H: (1–24 // 1)—at home	W and H: (i.u. opening time // 1); (i.u. closing time and offices // 0)
Parents-assisted Children PC (5–14, assisted); Young Autonomous YA (15–19, autonomous)	W: (8–13 // 0)—at school (1–7 and 14–24 // 1)—at home H: (1–24 // 1)—at home	W and H: (i.u. opening time // 1); (i.u. closing time and offices // 0)
Adults A (20–69, autonomous)	W: (8–18 // 0.09)—at work/univ (1–7 and 14–24 // 1)—at home H: (1–24 // 1)—at home	W and H: (i.u. opening time // 1); (i.u. closing time // 0)

Intended uses close to the public are considered occupied only by Adults. W is for Working days, H is for Holidays. Familiar users are R, Unfamiliar users are OO, PO, NR.

In view of the users' spatiotemporal distribution variations [7], NU [pp] is evaluated on hourly sampling according to Eq. 1:

$$NU = \sum\nolimits_{i,a} SU_i \cdot OL_i \cdot AP_a \tag{1}$$

where:

- SU_i [m^2] is the available surface of the i-th type of area between those in the second column of Table 1, estimated through freeware online tools like Calcmaps (www.

calcmaps.com/it/map-area—to perform measurements on aerial views) and Google Street Maps (www.google.it/maps/?hl=it—to check intended uses, number of floors, the opening times, and presence of doors, passages, or gates connecting indoor and outdoor areas last access 20/04/2022);

- OL_i [pp/m^2] is the Occupant Load of the i-th type of area (3rd column of Table 1);
- AP_a [%] is the users' age percentage distribution of the a-th range (1st column of Table 2), evaluated through freeware online census databases(www.tuttitalia.it) and multiplied by a "presence coefficient" equal to 1 if users are present, to 0 if users are absent, and to 0.09 to consider unemployed users spending their time at home (www.istat.it/it/archivio/occupati+e+disoccupati—2nd and 3rd columns of Table 2).

According to a conservative approach, the following assumptions were made in this phase [7, 14]: (1) in order to consider the contribution of users who, depending on the hazard, may decide to enter/exit the assessed BE, OA_1 and OA_2 include the square and half the streets linked to it (except those with a slope greater than 8% towards the square, and having access by archways, porticoes, and stairways); (2) IA_1 and IA_2 include those directly connected with OA_1 and/or OA_2 (assuming the possible indoor/outdoor displacement in relation to the type of danger); (3) carriageway, parking lots, monuments, fountains, stairs, and fenced areas were excluded from the presence of users (0.00 pp/m^2); (4) only Resident users are considered familiar with the environment and eventual evacuation plans; (5) working days and holidays were distinguished to respectively compare recurring conditions during the year with Sundays and other national festivities.

2.2 KPIs for the Users' Exposure and Vulnerability Characterization

In view of the general application to a BET-oriented approach, the data from Sect. 2.1 are refined and organized into the form to provide scenario representations for further analyses (including simulation-based). Thus, NU is first organized according to the following specific users' vulnerabilities:

- the type of area occupied: NU_t with t = [OO, PO, NR, R] (Table 1);
- the position: NU_p with p = [indoor, outdoor] (Table 1);
- the familiarity with the BE: NU_f with f = [familiar, unfamiliar] (Table 1);
- the age range, tracing NU_a with a = [T, PC, YA, A, E] (Table 2);
- the motion conditions $\sum NU_m$ with m = [autonomous, assisted] (Table 2).

According to a consolidated literature approach [7, 8, 12], Key Performance Indicators (KPIs) like the users' density per square meters, the normalized number and the percentage of hosted users, and the ratio between specific types of users have been introduced. Such data allow depicting specific square conditions, and the following application to idealized contexts by simply scaling the number of users into the BET depending on its dimensions. Table 3 summarizes the adopted KPIs, evaluated on hourly sampling to trace daily trends (distinguished in working days and holidays) and find out peak conditions of use.

Table 3. KPIs for the users' exposure and vulnerability characterization. The users' outdoor density is assessed including carriageways as available for gathering in case of emergency.

KPIs	Meaning and calculation method
UOd, UId [pp/m^2]	Users' outdoor and indoor density NU/$\sum$OA and NU/$\sum$IA
NUn [-]	Users' normalized number evaluated as the ratio between NU and its maximum daily value NU_{max}
OOp, POp, Rp, NRp [%]	Percentage of users according to the type of area occupied, evaluated as the ratios between NU_t and NU
OIUr, FUUr, AAUr [-]	Ratios between specific type of users according to their vulnerabilities, namely: $\sum NU_{outdoor}/\sum NU_{indoor}$; $\sum NU_{familiar}/\sum NU_{unfamiliar}$; $\sum NU_{autonomous}/\sum NU_{assisted}$

3 Results

Results are herein provided in two visual panels of the form. Figure 1 resumes the breakdown analysis of the analyzed BE in terms of: (1) *type of areas* (walkable areas OA_1 are in orange, dehors areas OA_2 in striped orange, non-residential areas IA_1 in yellow, and residential areas IA_2 in striped yellow); (2) users' *age range* percentage distribution AP_a, and their motion conditions (autonomous users in shades of green, assisted users in shades of brown); (3) the type of BET associated for the simulation purposes [11]. Results evidence that the ratio between outdoor and indoor areas is roughly 1:2, while residential and non-residential areas are almost equal. Dehors represent a marginal area. Data on the users' age range are in line with Italian national statistics, as autonomous users represent 70% of the population, while more vulnerable users (who can ask for aid during emergencies) are the remaining 30%. Although a BET is just an idealized scenario, the comparison between the BE aerial view and the BET plan shows also significant analogies given by the square shape and the position of the linked streets, thus remarking the significance of the case study in respect to typological conditions. The church's presence increases the context risk, especially in case of terrorist acts (i.e. for the symbolic value of the buildings), remarking the difference with BET 3 general conditions, that do not include data on sensible functions hosted by the buildings.

Figure 2 resumes the KPIs daily trends evaluated for working days (on the left) and holidays (on the right), and are organized as follows: the first row is for users' densities and normalized number (UOd, UId, and NUn), the second row is for the percentage of users according to the type of area occupied (OOp, POp, Rp, and NRp), and the third row is for the ratios between specific categories of users according to their specific vulnerability (OIUr, FUUr, and AAUr).

Considering peak conditions of BE use, NUn values suggest that the critical hours range between 9 and 13 in the morning both for working days (when the majority of the activities are open) and holidays (when Sunday services are scheduled, as the church is the biggest indoor area with the higher OL_i). Similarly, in the afternoon the daily peaks are observed between 19 and 20 on working days, and 17–18 on holidays. However, even considering peak conditions of use, UOd is always lesser than the critical value

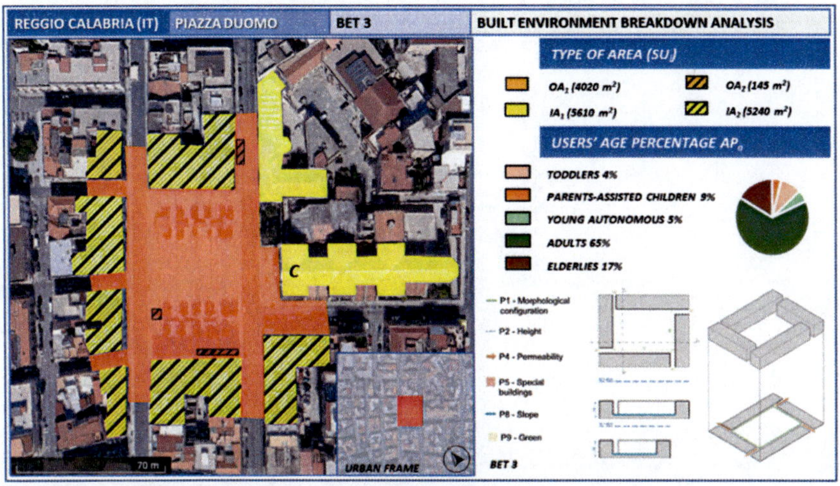

Fig. 1. BE breakdown analysis form panel: left-the case study aerial view (the ground floor of the areas in striped yellow host non-residential areas IA_2); top right-main data on types of areas and users; bottom right-BET 3-type views. *Source* www.bes2ecure.net/). The church is marked with letter C in the aerial view of the case study.

of 3.00 pp/m^2 [15] both on working days and holidays. On the other hand, during the daily peaks UId exceeds the critical value of 0.25 pp/m^2 for indoor areas considering the most severe WHO COVID-19 recommendations (i.e. distancing of about 2 m), reaching 0.31 pp/m^2 in working days and 0.46 pp/m^2 in holidays. However, these values can be limited by adjusting the OL_i adopted for IA_2 areas to the current thresholds (in this work, 0.70 pp/m^2 for churches and 0.40 pp/m^2 for intended uses open to the public).

Indeed, as shown by the percentage of users according to the type of area occupied, during the daily peaks (i.e., the working hours, between 8 and 18), most of the users are NR, ranging between 70–75% both in working days and holidays. On the other hand, in the nighttime R reach over 90% in conjunction with the closure of almost all the activities (the open ones collect the remaining 10%, combing NR and PO), while they represent a small part of the population during the working hours, and reach up to about 30% in the evening (from 21 to 24). Besides, in the hours after meals (7–8, 14–15, and 21–24, where many activities are closed) OO represent the majority of users ranging between 40 and 60%, while PO are always under 5%.

Ratios between specific categories of users show the following outcomes during working days: $OIUr > 1$ only between 7–8 and 14–15, which however correspond to the hours of minimum usage of the OS (NUn between 0.33 and 0.42, UOd about 0.20 pp/m^2, and UId about 0.15 pp/m^2); AAUr is always about 0.5; $FUUr > 1$ whenever but the nighttime, when almost all users are residents (maximum value of about 24 between 11 and 13). On the other hand, during holidays: OIUr is always lesser than 1, with peak values of about 0.8–0.9 in correspondence of the OOp maximum values (after the meals, at 7–8, 14–15, and 21–24); AAUr is always about 0.5; $FUUr > 1$ whenever

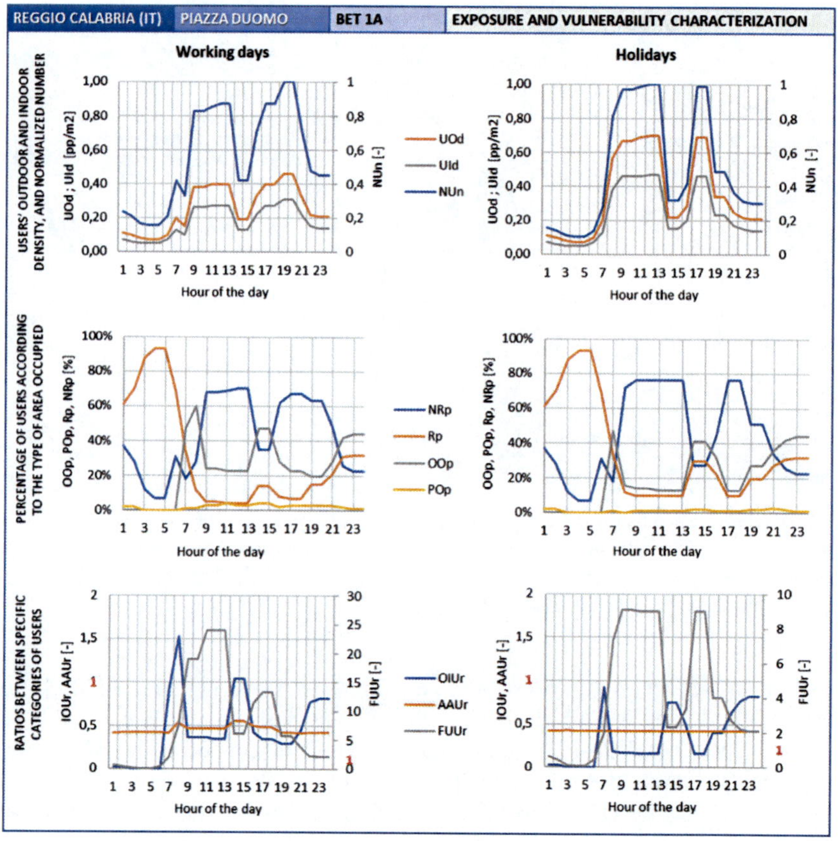

Fig. 2. KPIs form panel: users' exposure and vulnerability characterization.

but the nighttime, when almost all users are residents (maximum value of about 9.0 at 9).

Results seem to suggest that outdoor users play a marginal role in the evaluation of critical conditions, as they represent a large part of the population only in low crowding circumstances (at 7–8 and 14–15). Considering indoor users, although residential and non-residential areas are almost equal (see Fig. 1), both the working day and the holiday conditions point out how the case study is mainly characterized as a commercial area principally occupied by non-resident users, who are potentially unfamiliar with the BE and its eventual evacuation plan. This is true except during the night hours (1–6), when residents represent about 90% of the population.

4 Discussion

This work demonstrates the capabilities of a rapid survey form to collect and manage data on users' exposure and vulnerability in real-world multi risks-prone BEs, which can depict the scenario for further single/multi-risk analysis. In this sense, key findings confirm that the evaluation of the users' spatiotemporal distribution within the analyzed scenario is crucial for such a purpose. In the considered case study, the peak conditions of BE use are gained in the morning between 9–12, and in the afternoon between 17–19, both on working days (that is in line with the opening times of commercial activities) and holidays (due to the presence of a special use like a church hosting a large number of users). Considering specific vulnerabilities, the largest part of the population is composed of users occupying indoor areas, unfamiliar with BE, and their age distribution is in line with the Italian national distributions (due to the absence of high-variation functions, such as schools, nursing houses).

The form can be used to investigate single scenarios (as in this work), and to trace typological conditions between scenarios sharing similar morphological, constructive, and functional features, according to the BET-oriented approach pursued by the BE S^2ECURe project of which this work is part [11]. KPIs are introduced to manage the data and create statistics. Thus, collecting KPIs from more case studies can support the definition of recurring, that is typological, conditions on users' exposure and vulnerability, also depending on the BET to which each assessed square is associated.

Considering applications to both a single case study and a sample of scenarios for BET-oriented assessment, the form allows identifying priority scenarios for users' risks depending on their exposure and vulnerability. These data can be then combined with those from other rapid survey forms on typological hazards, physical vulnerability, and disaster-affected conditions, depending on specific risks-prone scenarios. They offer the bases for: (1) a quick multi-criteria assessment, e.g. through expert judgment or supervised methods (e.g. analytical hierarchy process), by balancing each KPI according to its weighted impact on the whole BE or users' risk; (2) simulation-based methods, which can jointly represent the types of users in the physical scenario and the effects of the emergency in it (or even the combination of more than one disasters, e.g. sudden-onset disasters raising in slow-onset disasters-affected conditions). According to point (2), priority simulations should be correlated to the peaks in users' exposure and vulnerability since these can amplify critical interactions (e.g. in motion). In this sense, the simulation application to BETs-related scenarios can boost the detection of idealized conditions which can represent quick basic standard references for further analysis relating to the single case study specificities.

5 Conclusions

Sustainable and resilient solutions for risk-prone Built Environments (BEs) are a key challenge to guarantee users a safe and full experience of the open spaces and buildings of which they are composed. Anyway, effective solutions should be designed depending on the BEs actual conditions of hazards, physical vulnerability, and users' exposure and vulnerability. Rapid surveys can speed up the application process, reduce economic efforts, and ensure powerful analysis also for non-expert technicians of local authorities in urban areas. In this sense, recent works provided survey forms to collect and organize hazard and physical vulnerability factors, by also moving towards the definition of idealized scenarios which can quickly depict the main BEs recurring features, namely Built Environment Typologies (BETs).

The present work adopts this quick assessment approach, which relies on a remote-based methodology to evaluate and collect users' exposure and vulnerability issues depending on the BE features in which they move. Collected data are organized into a survey form composed of two panels describing (1) the BE composition in terms of the type of areas and demographic data, and (2) Key Performance Indicators (KPIs) useful for the spatiotemporal quantification of the users' exposure and vulnerability. Future researches on risk assessment, including those based on simulations, should be aimed at broadening the current analyses to a greater number of squares prone to multi-risks, to improve the statistical significance of typological scenarios definition. Moreover, this work results encourage future efforts to test similar approaches declined to different types of BEs, such as streets, infrastructural areas and hubs (e.g., airports, stadiums, railway stations), neighborhoods and city districts prone to crowding risks, to investigate peak conditions of use. In this sense, analyses could be also correlated to new pandemic-related requirements and regulations to support local administrations to face new risk challenges for the BE users.

Acknowledgments. This work was supported by the MIUR (the Italian Ministry of Education, University, and Research) Project BE S^2ECURe—(make) Built Environment Safer in Slow and Emergency Conditions through behavioural assessed/designed Resilient solutions (Grant number: 2017LR75XK).

References

1. Angelosanti, M., et al.: Towards a multi-risk assessment of open spaces and its users: a rapid survey form to collect and manage risk factors. In: Littlewood, J.R., Howlett, R.J., Jain, L.C. (eds.) Sustainability in Energy and Buildings 2021. SIST, vol. 263, pp. 209–218. Springer, Singapore (2022). https://doi.org/10.1007/978-981-16-6269-0_18
2. Pirlone, F., Spadaro, I., Candia, S.: More Resilient Cities to Face Higher Risks. The Case of Genoa (2020)
3. Elmqvist, T., et al.: Sustainability and resilience for transformation in the urban century. Nat. Sustain. **2**, 267–273 (2019)
4. Marchese, D., et al.: Resilience and sustainability: Similarities and differences in environmental management applications. Sci. Total Environ. **613–614**, 1275–1283 (2018)
5. Collier, M.J., et al.: Transitioning to resilience and sustainability in urban communities. Cities **32**, S21–S28 (2013)
6. Bosina, E., Weidmann, U.: Estimating pedestrian speed using aggregated literature data. Physica A **468**, 1–29 (2017)
7. Li, J., et al.: Spatiotemporal distribution characteristics and mechanism analysis of urban population density: a case of Xi'an, Shaanxi China. Cities **86**, 62–70 (2019)

8. Cadena, J.D.B., Salvalai, G., Bernardini, G., Quagliarini, E.: Merging heat stress hazard and crowding features to frame risk scenarios within the urban built environment. In: Littlewood, J.R., Howlett, R.J., Jain, L.C. (eds.) Sustainability in Energy and Buildings 2021. SIST, vol. 263, pp. 293–303. Springer, Singapore (2022). https://doi.org/10.1007/978-981-16-6269-0_25
9. Lin, J., et al.: How occupants respond to building emergencies: a systematic review of behavioral characteristics and behavioral theories. Saf. Sci. **122**, 104540 (2020)
10. Arosio, M., Martina, M.L.V., Figueiredo, R.: The whole is greater than the sum of its parts: a holistic graph-based assessment approach for natural hazard risk of complex systems. Nat. Hazard. **20**, 521–547 (2020)
11. D'Amico, A., et al.: Built environment typologies prone to risk: a cluster analysis of open spaces in Italian cities. Sustainability **13**, 9457 (2021)
12. De Lotto, R., Pietra, C., Venco, E.M.: Risk analysis: a focus on urban exposure estimation. In: Computational Science and Its Applications—ICCSA 2019. pp. 407–423. Springer, Cham (2019)
13. Ministry of Interior (Italy): DM 03/08/2015: Fire safety criteria (2015)
14. Hahm, Y., Yoon, H., Choi, Y.: The effect of built environments on the walking and shopping behaviors of pedestrians; A study with GPS experiment in Sinchon retail district in Seoul, South Korea. Cities **89** (2019)
15. Bloomberg, M., Burden, A.: New York City Pedestrian Level of Service Study-Phase 1. New York, NY, USA (2006)

Assessing People's Efficiency in Workplaces by Coupling Immersive Environments and Virtual Sounds

Arianna Latini[1], Samantha Di Loreto[2], Elisa Di Giuseppe[1](✉), Marco D'Orazio[1], Costanzo Di Perna[2], Valter Lori[2], and Fabio Serpilli[2]

[1] Department of Construction, Civil Engineering and Architecture (DICEA), Università Politecnica Delle Marche, Ancona, Italy
e.digiuseppe@staff.univpm.it

[2] Department of Industrial Engineering and Mathematica Sciences (DIISM), Università Politecnica Delle Marche, Ancona, Italy

Abstract. The use of virtual reality (VR) to study the effect of the acoustic environment on performance is still in its infancy, despite its many potentialities due to audio-visual improvement. In this study, a binaural soundtrack was generated and integrated within an immersive virtual environment of an office room, to evaluate the effects of an acoustic ambient on users' cognitive performance and subjective evaluation. To generate the soundtrack, five disrupting sound sources (phone rings, machine noise, mechanical systems, human-based sounds and acoustical effects) were selected. 104 participants performed three productivity tests (working memory, inhibition, task switching) and answered questionnaires under a constant indoor air temperature (24 °C). In particular, an independent measure experimental design was conducted: each group (52 subjects) randomly performed one test session: a no-ambient-noise condition or «quiet environment» was compared to a treated-with-noise condition, or «noisy environment» virtual session. The authors focused on two goals: verifying the external-ecological validity of the virtual model created and evaluating the effect of the sound stimuli on productivity. Findings revealed that the virtual office created an excellent level of presence and immersivity and confirmed, as expected, that work efficiency was negatively influenced by the ambient noise. A decrease in performance was detected in each cognitive test, as the subject evaluated the sound environment to be uncomfortable, chaotic and boring. Hence, the results supported the potentialities of the proposed acoustic virtual reality to study productivity in combination with different stimuli and layouts.

Keywords: Immersive virtual environment · Binaural soundtrack · Work efficiency

J. Littlewood and R. J. Howlett (Eds.): SEB 2022, SIST 336, pp. 120–129, 2023.
https://doi.org/10.1007/978-981-19-8769-4_12

1 Introduction

It is well-established that people's well-being, health, comfort, and productivity are strongly interrelated with the design and characteristics of the built environment in terms of indoor environmental quality and ventilation, thermal environment, natural and artificial lighting, noise and acoustics, biophilia, layout and aesthetics [1]. In particular, «productivity» is the amount of work produced per unit of time [2]. It refers to business-oriented outputs and it is especially relevant in personnel costs, considering that workers' salaries and benefits typically account for 90% of a company's costs [3]. As a consequence, all the aspects influencing staff's productivity should be a major concern for each organisation to ensure the most comfortable, pleasant and efficient workplaces in the long term. Workplaces are becoming smarter in terms of monitoring and controlling light, and temperature, for instance. However, one aspect is very relevant: noise pollution damage users' concentration. Being productive is difficult when noise provides unwanted distractions, leading to stress and dissatisfaction within the office environment [1, 3, 4]. Thus, there is a growing concern to study how the office acoustic environment influences work performance, mainly focusing on sound source type, sound pressure level, and reverberation time [5]. However, occupants are exposed to a plurality of stimuli acting simultaneously that influence comfort, and productivity, not only a single variable at a time. Hence, properly designed indoor workplaces are a crucial component for living and optimising work efficiency. This field of study has been widely addressed in both real-life and laboratory-based studies for years. However, recently, the use of Virtual Reality (VR) for this topic has emerged, thanks to its numerous advantages as low-cost technology, and in terms of speed of execution and repeatability of tests. The aim of this research domain is the application of virtual simulations to study how changes and characteristics of the indoor environment, in terms of layout and stimuli, affect users' well-being. However, to reach high levels of realism and immersivity in virtual environments, it is necessary to improve the sense of hearing with proper acoustics. Several sets of indoor environments can be created and experienced via audio and video using the head-mounted display [6]. The widespread use of VR for the built environment has recently enhanced the interest in generating immersive sound simulation to study the effect of building acoustics on cognitive performance. However, the literature review revealed that a few limited studies investigated the effects of noise, by considering only a single cognitive function at a time [6–8]. Hence, research focusing on the assessment of multiple visual cognitive functions is lacking.

In this study, a multitask cognition experiment was carried out in a virtual office environment to measure the work performance of 104 participants under different acoustic conditions.

2 Material and Methods

The experiment was carried out in an office room located inside the Department of Construction, Civil Engineering and Architecture (DICEA) at Università Politecnica delle Marche (Ancona, Italy). A total of 104 participants were recruited to perform three productivity tests and to answer questionnaires. An independent measure experimental design (52 subjects per group) was carried out in an immersive virtual office environment: each participant was randomly assigned to a no-ambient-noise condition, or «quiet environment» (group 1) or a treated-with-noise condition, or «noisy environment» (group 2) session. All participants were immersed in the same virtual environment with a constant indoor air temperature (mean = 24.45 °C, SD = 0.48) detected with a time-step of 1 s (temperature range: 5°–60° and accuracy ± 0.3 °C).

2.1 IVE Scenario

The generation of the immersive virtual environment involved the creation of an extremely detailed model of a double-occupancy virtual office room (20 m^2). At first, the 3D model was developed using CAD software, and then materials, lights and cameras were applied using the Unity software [9]. To ensure the model an effective realism, the authors wanted to correctly represent the surfaces' colour and materials. The luminance parameter (L*) and chromatic components (a*, b*) of the CIELab model were detected through a spectrophotometer. For each surface (walls, desk, chair, floor) of the real office room, 5 measurements were carried out. Then, the resulting parameters were converted into the RGB coordinates for the Unity model. Two virtual scenarios were created: the first one, corresponding to the adaptation phase, allowing a complete view of the virtual office room, while the second one, necessary for the test operative phase, with the subjects seated at their virtual desks to solve the productivity tests and the surveys (Fig. 1). The performance tasks and questions of the questionnaires were displayed as timed videos and sequences of images through the virtual computer monitor, to avoid a so-called «break-in-presence». The authors developed specific scripts to proceed sequentially to visualise the scenes and then collect the participants' responses by voice. The IVE model was visualized through the HTC Corporation VIVE PRO Eye head-mounted display (1440 × 1600 resolution image per eye) using the SteamVR plugin [10].

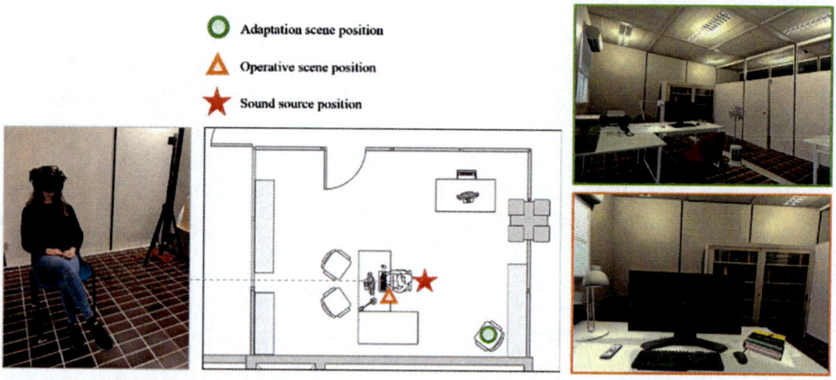

Fig. 1. Virtual environment schematic views and a frame of a participant during the test

2.2 Audio Stimuli Generation

The overall sound stimuli were generated based on computer simulations and reproduced with the headphones of the head-mounted display [11, 12]. In particular, five representative sound sources (phone rings, machine noise, mechanical systems, human-based sounds and acoustical effects) were selected as sound samples due to the disrupting effect on employees' cognitive performance [1, 4–6].

The room is located in a suburban area of the city, away from traffic lines and at the back of the building with an external ambient noise level between 40 and 50 dB(A) during the daytime period. The room has a volume of 78 m^3, an average height of 3 m and a base area of 26 m^2. The room does not have acoustic treatment: the walls are plastered, the floor is tiled, the ceiling is flat, and there is wooden and plastic furniture inside. The calculation of the reverberation time was made according to UNI EN 12354-6:2006 [13], starting from the acoustic absorption of the room (Fig. 2a).

The five sources were mixed into a single stereo soundtrack with a sample rate of 44100 at 16 bit/s, an average LAeq equal to 56.4 dB and loudness of 70.5–50.57 sone for right and left ears, respectively. Then, the physical acoustic characteristics of the sound source were analysed in terms of sound fluctuation strength (0.58–0.48 vacil, for right and left ear respectively) and temporal variation, according to ISO 532-1:2017 [14]. Figure 2b shows the results of the fluctuation responses of the created binaural soundtrack for each channel and the overall frequency of each second.

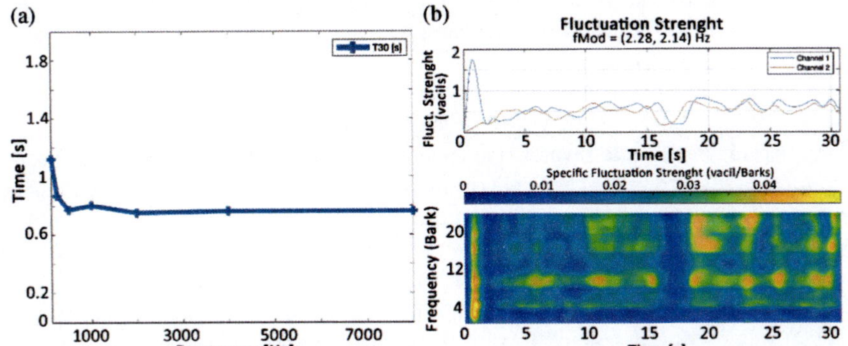

Fig. 2. **a** Calculated reverberation time value in the octave bands 125–4000 Hz; **b** Fluctuation strength of the binaural soundtrack

2.3 Performance Test

The authors assessed objectively the work efficiency through three cognitive tests: the Magnitude-parity test [15] for task switching, the Stroop test [16] for inhibition and the OSPAN test [17] for the evaluation of the working memory.

The Magnitude-Parity test was performed through a timed video (200 ms for each slide). The numbers from "1" to "9" (except "5") were used as stimuli preceded by red and blue circles. Participants were asked to tell whether the displayed number was odd or even after the red circle was shown (parity stimulus), and then if the displayed number was smaller or larger than 5 after the blue circle was displayed (magnitude stimulus). This coloured dot-number combination was presented eight times for each stimulus, for a total of 16 numbers. The Stroop test was presented as a picture with a total of 32 coloured words in red, green, blue, pink and orange on a black background. They were asked to name the colours of the words as fast as they could (i.e. if the word "green" is printed in red ink, the correct answer is "red"), while the researcher detected the execution time. Finally, the OSPAN test consisted of a timed video: in the first slide (3s) a simple math operation was displayed; the second one (3s) showed a possible solution to the previous equation. The participants solved the equation and were asked to tell if the solution was true or false. In the last slide (800 ms) a letter to be memorised was displayed. The combination of the equation—true/false—letter was presented in a set of five items. At the end of each set, the participant was instructed to recall all the letters in the order presented in the video.

2.4 Survey

Before the operative phase, participants completed the pre-experiment survey to retrieve information about demographics (gender, age, height, education level, eyesight problems). Once the participants completed the three cognitive tests, they were provided with the post-experiment questionnaires which included three sections (sound evaluation, cybersickness, sense of presence and immersivity). The noise-related questions were developed according to the standards ISO 12913-2: 2018 [18] and ITU-R BS 1116-3: 2015 [19]. Before performing the three productivity tasks, subjects were asked to rate how much they can distinguish the five sound sources ranging from «not at all» to «a lot». In addition, the virtual office room sound evaluation was carried out by asking participants to rate several characteristics (pleasant, relaxing, peaceful, boring, chaotic) on a seven-point Likert scale (from «totally disagree» to «totally agree»). After the performance tests, the Acoustic Comfort Vote (from «comfortable» to «extremely comfortable») was also investigated through the question "How do you judge this environment on an acoustical personal level?". Two sections were also included to test the ecological validity of the developed model. The Virtual Reality Sickness Questionnaire (VRSQ) was used to assess six motion sickness symptoms [16] on a five-point scale (from «not at all» to «a lot»): general discomfort, fatigue, eye strain, difficulty in focusing, headache, and vertigo. Moreover, the sense of presence and immersivity in IVEs was assessed according to four indicators: Graphical Satisfaction (GS), Spatial Presence (SP), Involvement (INV), and Experienced Realism (REAL), on a seven-point Likert scale.

2.5 Experimental Protocol

At the beginning of each session, all participants signed a consent form and received information about the test procedure. A pre-experimental phase (15 min) was carried out, to allow them to get used to the environmental conditions while responding to a series of demographic questions through an online platform. Then, subjects were instructed to wear and adjust the head-mounted display (and the headphones in the treated-with-noise condition), rest with their eyes closed (30 s) and adapt to the virtual scene for 3 min. These periods allowed, respectively, to reduce any psychological fluctuations related to the virtual environment exposure, and to facilitate immersion (visual and acoustical) [20]. Then, participants were asked to perform the three cognitive tests (3 min) and to complete a survey (2 min). The tests and questions were displayed on the virtual computer screen, simulating a traditional working scenario in an office. Responses to all the three cognitive tests and surveys were given by voice and recorded by the researchers. To reduce fatigue and exposure to the virtual environment, each test session lasted about 25 min (Fig. 3).

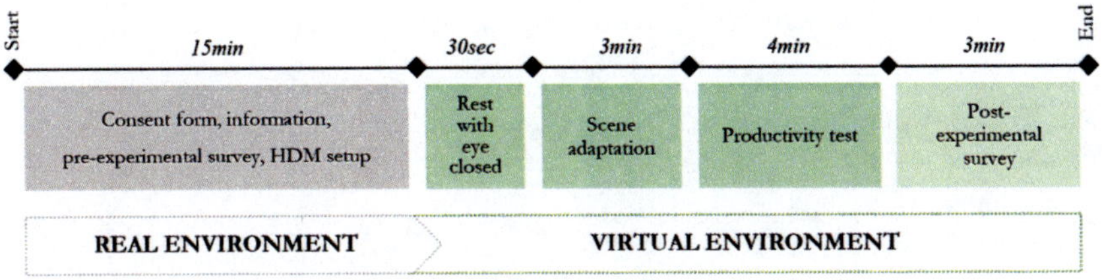

Fig. 3. Experimental schedule

3 Results and Discussion

The following paragraphs present the sample's data analyses and the results according to two objectives: the ecological validity of the modelled IVE, and the evaluation of the effect of the sound stimulus on work efficiency. All datasets were analysed with statistical tests ($\alpha = 0.05$) through RStudio software [21].

3.1 Participants

The sample consisted of 104 participants (52 male, and 52 females) and it was mostly composed of young people as follows: 48% between 20 and 25 years, 35% between 26 and 30, 21% between 31 and 39 and only the 6% over 50 years old. Most of the subjects were already graduated from university (45%), 40% were selected among university students and 14% had a higher educational level (PhD, graduate school). 42% of the participants had eyesight problems, such as myopia and astigmatism, but all of them wore corrective lenses during the tests, not to influence the model visualisation and then the test performance. Moreover, 58% experienced VR tools at least once. A power analysis through the G*Power software [22] confirmed that the sample size (n = 104) was adequate to detect significant effects of statistical power of 0.81 (effect size 0.50).

3.2 Ecological Validity

According to the first aim, the authors analysed data about the sense of presence and immersivity and the cybersickness ratings.

Existing literature [23–26] compared the four indicators scores with the ones from previous studies using the VR tool in the same research domain. The average scores of the present study were at first rescaled from a seven to a five-point scale and then compared with past research. The mean scores (Table 1) are generally higher than a moderate level (i.e. 4). In particular, a very good experienced realism (REAL) was obtained, the participants appreciate the graphics (GS) and felt involved within the IVE (INV). In addition, a very good spatial presence was obtained as the mean value for SP (4.18) is higher than the references [23–25] and almost similar to [26] (4.24). As the difference is negligible (0.06), the virtual environment offered an effective sense of presence and immersivity for the research objectives.

Table 1. Comparison of scores on a five-point scale of the four indicators

Classification		Year	GS	REAL	INV	SP
This study		2022	4.65	4.51	4.29	4.18
Previous studies	[23]	2019	3.65	2.73	3.23	3.39
	[24]	2019	–	3.21	–	3.74
	[25]	2019	–	3.75	–	3.68
	[26]	2020	–	3.54	4.11	4.24

According to the results of the VRSQ, no subject suffered from vertigo since the test was carried out in static conditions. General discomfort, fatigue and headaches symptoms were negligible since between 94% and 100% of the subjects assigned a score of «not at all» and «slightly». Moreover, 12% of them reported «moderate» eye fatigue due to a «difficulty in focusing» (27%), caused by the slightly blurred images presented by the head-mounted display.

3.3 Sound Effects on Work Efficiency

At first, the productivity data from the three cognitive tests were analysed qualitatively (Fig. 4). Then, the related assumptions were tested with parametric statistical analysis. The Magnitude-Parity test analysis involved computing the number of errors expressed as the number of times the subject wrongly classified the digits even/odd and greater/lower than "5". The analysis of the mean and standard deviations (Fig. 4a) revealed approximately double the errors in the noisy condition ($mean_N = 0.98$, $SD_N = 1.16$), compared to the quiet one ($mean_Q = 0.40$, $SD_Q = 0.60$). The t-test was conducted to test the previous assumption. The null hypothesis states that there is no difference between them. The outcome (p-value = 0.002) allowed to reject this hypothesis: the t-value obtained (−3.17) falls outside the critical region (±1.983 for df = 103, $\alpha = 0.05$), therefore the task-switching of the subjects was influenced by the sound environment and, in particular, 59% decrease in performance was registered. For the Stroop test analysis, the authors calculated the times a colour was not correctly recalled by the subjects (number of errors, Fig. 4b), and the time in seconds participants took to pronounce all 32 colours (speed of execution). The qualitative analysis did not reveal difference in the errors between the noisy condition ($mean_N = 0.31$, $SD_N = 0.78$) and the quiet one ($mean_Q = 0.31$, $SD_Q = 0.94$), as confirmed by the t-test results (t-value = 1.00, p-value = 0.00). However, a slight difference was discovered across the speed of execution ($mean_N = 31.97s$, $SD_N = 5.24s$; $mean_Q = 28.85s$, $SD_Q = 5.49s$), with a 10% decrease in the velocity from the quiet to the noisy condition, as also statistically confirmed (t-value = −2.96, p-value = 0.004). In accordance with the automated OSPAN test development [17], the authors computed the number of errors in the true/false strings ($mean_N = 0.33$, $SD_N = 0.62$; $mean_Q = 0.15$, $SD_Q = 0.36$), the correct order of the letters recalled ($mean_N = 2.71$, $SD_N = 1.32$; $mean_Q = 4.48$, $SD_Q = 0.98$), and the OSPAN score, which is the sum of the right true/false and letter correctly reported ($mean_N = 7.38$, $SD_N = 1.52$; $mean_Q = 9.33$, $SD_Q = 1.08$) (Fig. 4c). The absence of significant differences in the mean and standard deviations was detected only across the number of errors in the true/false tasks (t-value = −1.74, p-value = 0.08), while was discovered in the letter recall (t-value = 7.76, p-value = 9.785e−12), and in the overall OSPAN score (t-value = 7.50, p-value = 3.867e−11). A reduction in performance equal to 53% and 65% were detected, respectively.

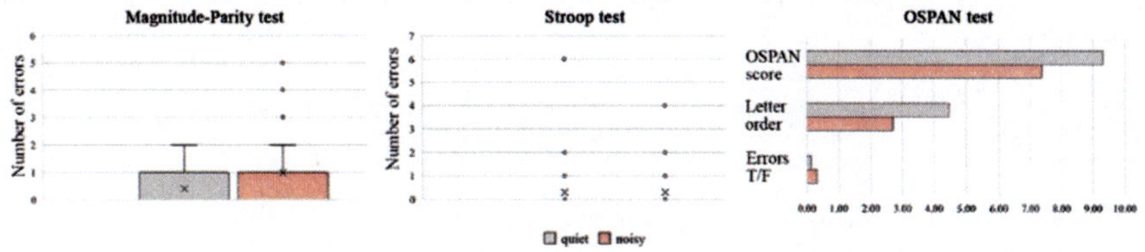

Fig. 4. Results of the three cognitive tests

Finally, the post-experimental survey was analysed concerning the sound characteristics. The majority of the participants (98%) reported the sound environment to be at least «uncomfortable». Indeed, it was assessed as being mostly chaotic (96%) and boring (27%). As expected, none of them assessed the environment to be «pleasant», «relaxing» or «calm» (score from «totally disagree» to «slightly disagree»). All the sample interviewed expressed that the noisy conditions can have an influence (score «a lot» and «extremely») on personal productivity. Thus, a correspondence between the results of the task, the sound environment assessment and the subjects' awareness of its effect on work efficiency was highlighted. Moreover, according to the scores, the prominent sources (score «a lot» and «extremely») were, in relevant order: phone rings (96%), machine noise (58%), mechanical systems noise (17%), human-based sound sources (13%), and acoustical effects (4%).

4 Conclusions

According to the results, the virtual environment created an excellent level of presence and immersivity. The analysis and the comparison of the four indicators (GS, REAL, INV, SP) with similar past studies, allowed the authors to support the ecological validity.

Moreover, the analysis of the cybersickness revealed that the majority of subjects did not report high disorder levels.

As hypothesised, results from the qualitative and statical analyses, showed that cognitive performance was negatively affected by ambient noise. Indeed, noise disturbed most of the task-switching, with a 59% decrease in performance, working memory (53% reduction as in [1, 3–8, 27, 28]) and inhibition (10%). The worst performance observed in the treated-with-noise condition was confirmed by the analysis of the self-reports. Participants evaluated the sound environment to be uncomfortable, chaotic and boring and they were also aware that a noisy environment can negatively affect work efficiency, thus supporting the effectiveness of the proposed acoustic virtual reality.

Future work is needed to perform an acoustical characterization of the room to enhance the productivity evaluation depending on the noise level. The assessment of the indoor soundscape also via self-reports will allow the authors to verify the reliability of the outcomes and then create a robust model to be implemented in other settings, such as other work environments (e.g. open-plan office) with specific layouts (e.g. windows dimensions, walls and light colours) and stimuli (e.g. thermal and visual combined with the acoustical one). Future perspective should also include testing several characteristics of the sources thus analysing the effect on the dependent variables (comfort, work efficiency).

References

1. Al Horr, Y., Arif, M., Kaushik, A., Mazroei, A., Katafygiotou, M., Elsarrag, E.: Occupant productivity and office indoor environment quality: a review of the literature, Build. Environ. **105**, 369–389 (2016)
2. ASHRAE Standard, Journal, June (2019)
3. World Green building Council: Health, wellbeing & productivity in offices. World Green Build. Counc. **1**, 46 (2014)
4. Banbury, S.P., Berry, D.C.: Office noise and employee concentration: identifying causes of disruption and potential improvements. Ergonomics **48**, 25–37 (2005)
5. Meng, Q., An, Y., Yang, D.: Effects of acoustic environment on design work performance based on multitask visual cognitive performance in office space. Build. Environ. **205**, 108296 (2021)
6. Muhammad, I., Vorländer, M., Schlittmeier, S.J.: Audio-video virtual reality environments in building acoustics: an exemplary study reproducing performance results and subjective ratings of a laboratory listening experiment. J. Acoust. Soc. Am. **146**, EL310–EL316 (2019)
7. Doggett, R., Sander, E.J., Birt, J., Ottley, M., Baumann, O.: Using virtual reality to evaluate the impact of room acoustics on cognitive performance and well-being. Front. Virtual Real. **2**, 1–9 (2021)
8. Jeon, J.Y., Jo, H.I., Santika, B.B., Lee, H.: Crossed effects of audio-visual environment on indoor soundscape perception for pleasant open-plan office environments. Build. Environ. **207**, 108512 (2022)
9. Unity. https://unity.com. Last Accessed May 2021
10. SteamVR Plugin. https://bit.ly/38305Ae. Last Accessed May 2021
11. Adobe, Adobe Audition. https://adobe.ly/3FgLwoN. Last Accessed March 2022
12. MathWorks, MATLAB. https://bit.ly/3FgM2mJ. Last Accessed March 2022
13. UNI EN 12354-6:2006 Building Acoustics—Estimation of acoustic performance of buildings from the performance of elements—Part 6: Sound absorption in enclosed spaces (2006)

14. ISO 532-1:2017 Acoustics—Methods for calculating loudness—Part 1: Zwicker method (2017)
15. Wendt, M., Klein, S., Strobach, T.: More than attentional tuning—investigating the mechanisms underlying practice gains and preparation in task switching. Front. Psychol. **8**, 1–9 (2017)
16. Stroop, J.R.: Studies of interference in serial verbal reactions. J. Exp. Psychol. **18**, 643–662 (1935)
17. Unsworth, N., Heitz, R.P., Schrock, J.C., Engle, R.W.: An automated version of the operation span task. Behav. Res. Methods **37**, 498–505 (2005)
18. ISO/TS 12913-2:2018 Acoustics—Soundscape—Part 2: Data collection and reporting requirements (2014)
19. ITU-R BS.1116-3, Methods for the subjective assessment of small impairments in audio systems BS Series Broadcasting service, Int. Telecommun. Union **3** (2015)
20. Li, J., Jin, Y., Lu, S., Wu, W., Wang, P.: Building environment information and human perceptual feedback collected through a combined virtual reality (VR) and electroencephalogram (EEG) method. Energy Build. **224**, 110259 (2020)
21. R Studio. https://www.rstudio.com. Last Accessed May 2021
22. Faul, A., Erdfelder, F.E., Lang, A.-G., Buchner, G.: *Power 3: a flexible statistical power analysis program for the social, behavioral, and biomedical sciences. Behav. Res. Methods **35**, 175–191 (2007)
23. Hong, T., Lee, M., Yeom, S., Jeong, K.: Occupant responses on satisfaction with window size in physical and virtual built environments. Build. Environ. **166** (2019)
24. Abd-Alhamid, F., Kent, M., Bennett, C., Calautit, J., Wu, Y.: Developing an innovative method for visual perception evaluation in a physical-based virtual environment. Build. Environ. **162**, 106278 (2019)
25. Chamilothori, K., Wienold, J., Andersen, M.: Adequacy of immersive virtual reality for the perception of daylit spaces: comparison of real and virtual environments, LEUKOS. J. Illum. Eng. Soc. North Am. **15**, 203–226 (2019)
26. Yeom, S., Kim, H., Hong, T., Lee, M.: Determining the optimal window size of office buildings considering the workers' task performance and the building's energy consumption. Build. Environ. **177**, 106872 (2020)
27. Kaarlela-Tuomaala, A., Helenius, R., Keskinen, E., Hongisto, V.: Effects of acoustic environment on work in private office rooms and open-plan offices—Longitudinal study during relocation. Ergonomics **52**, 1423–1444 (2009)
28. Jahncke, H., Hongisto, V., Virjonen, P.: Cognitive performance during irrelevant speech: effects of speech intelligibility and office-task characteristics. Appl. Acoust. **74**, 307–316 (2013)

Assessment Analysis of BEV/PHEV Recharge in a Residential Micro-Grid Based on Renewable Generation

Dario Pelosi[1](✉), Linda Barelli[1], Michela Longo[2], and Dario Zaninelli[2]

[1] University of Perugia, Via G. Duranti 93, 06125 Perugia, Italy
{dario.pelosi,linda.barelli}@unipg.it
[2] Politecnico di Milano, Via la Masa, 34, 20156 Milano, Italy
{michela.longo,dario.zaninelli}@polimi.it

Abstract. To limit the climate change, a strong evolution on stationary renewable power production and transport sector is needed. More than 34 million of Battery Electric Vehicles (BEVs) and 13 million of Plug-in Electric Vehicles (PHEV) are expected to circulate in Europe by 2030. This negatively affects distribution lines during BEV/PHEV charge, especially in densely populated areas. A possible solution to enhance BEV/PHEV penetration without impact on grid stability is represented by Micro-Grids (MGs), including renewable production, local loads, and energy storage. In this work, a residential MG composed by a Photovoltaic (PV) power generation system, a local load including BEV/PHEV charge, and a Li-ion battery energy storage system is implemented to assess and compare the impact on grid energy independence introducing BEV and PHEV charging load. Four different scenarios are simulated, considering either BEV or PHEV charge, varying the installed PV power (i.e., 3 kWp and 6 kWp) and Li-ion battery nominal capacity (10–20 kWh). The results demonstrate that, in a MG integrating 6 kWp PV and 20 kWh battery, BEV and PHEV can be daily charged by PV energy for 23% and 68% respectively. Hence, enhancing PV installed power and energy storage capacity in the future residential MGs will be needed to avoid stability issues on distribution feeders, in the view of massive BEV penetration for pursuing the limit to global warming.

Keywords: Battery · BEV · Dynamic modeling · Energy storage · Micro-Grid · PHEV · Renewable energy

1 Introduction

Climate change currently represents one of the greatest issues all over the world [1, 2]. To limit the global average temperature below 2 °C at the end of the century, a strong evolution concerning the power generation from decentralized Renewable Energy Sources (RES) is needed. At the same time, the road transport sector constitutes the highest contribution to overall emissions in Europe (in 2019 it emitted 72% of all domestic and international transport greenhouse gas emissions) [3]. Thus, it should make use of

J. Littlewood and R. J. Howlett (Eds.): SEB 2022, SIST 336, pp. 130–139, 2023.
https://doi.org/10.1007/978-981-19-8769-4_13

integrated approaches as enhancing vehicle efficiency, employing alternative fuels as well as different grades of electrification, that can strongly reduce greenhouse effect and pollution, whilst increasing human and environmental health. In such a framework, Battery Electric Vehicles (BEVs) and Plug-in Electric Vehicles (PHEVs) could significantly reduce air pollution caused by transport sector, especially if supplied by renewable electricity [4, 5]. According to [6], more than 34 million of BEVs and 13 million of PHEV are expected to circulate in Europe by 2030. This can negatively affect distribution and transmission lines during BEV/PHEV charge, especially in densely populated areas, as illustrated in [7–9].

Hence, according to what described above, the concept of micro-grid (MG) including renewable production, local loads, and energy storage, will play a key role in the next future. In fact, MG allows to extend energy independence from the main grid, maximize the self-consumption of renewable produced energy and increase BEV/PHEV penetration through home charging, with a low impact on the distribution grid stability [10–12].

In this research work, a residential MG composed by a Photovoltaic (PV) power generation system, a local load including BEV/PHEV charge, and a Li-ion battery energy storage system, is investigated. The main contribution of this work is to assess how BEV and PHEV charge impacts on a typical residential MG. Specifically, four different scenarios are simulated through the implemented dynamic model, considering either BEV or PHEV charge at varying the installed PV power (i.e., 3 kW and 6 kW) and Li-ion battery nominal capacity (10–20 kWh).

The paper is organized as follows: Sect. 2 describes the implemented methodology, including profiles generation, statistical analysis, and dynamic modeling. In Sect. 3 the main obtained results are shown and discussed. Subsequently, Sect. 4 illustrates the conclusions of the work.

2 Methodology

2.1 Input Data and Statistical Analysis

The procedure implemented for the power profile generation is described in the following.

First, the user annual load demand, based on experimental dataset of power consumption relative to one house [13] and gathered with 8 s time step, was expanded in MATLAB® to 1 s time step by means of linear interpolation. Concerning the PV generation, real yearly data (with 1 min time step) of a PV plant sited in the center of Italy were properly scaled to obtain 3 kWp and 6 kWp as maximum instantaneous power and subsequently expanded, according to the same procedure detailed above.

Second, the charging power profiles for BEV and PHEV were generated assuming 3.7 kW as maximum recharging power for a single-phase domestic wall box [14]. Subsequently, such profiles were added to the user load demand, assuming BEV and PHEV charging start at 9 p.m. each day.

Third, aiming at exploiting the available PV power surplus and properly sizing the MG energy storage system, a statistical data elaboration was performed. Thus, the annual load demand and PV generation profiles were divided day by day and subsequently, weekly grouped. Then, considering the instantaneous power difference (*diff*) between PV generation and load demand, the number of daily average hours during which there is a PV power surplus was assessed for each week of the annual profile.

Subsequently, classifying the yearly weeks according to 25 hourly classes (from 0, corresponding to the weeks with null PV surplus, to 24, assumed as the theoretical case of weeks with daily surplus never null), the frequency of the weekly occurrences in terms of daily average hours of PV surplus during the year was determined, as shown in Fig. 1.

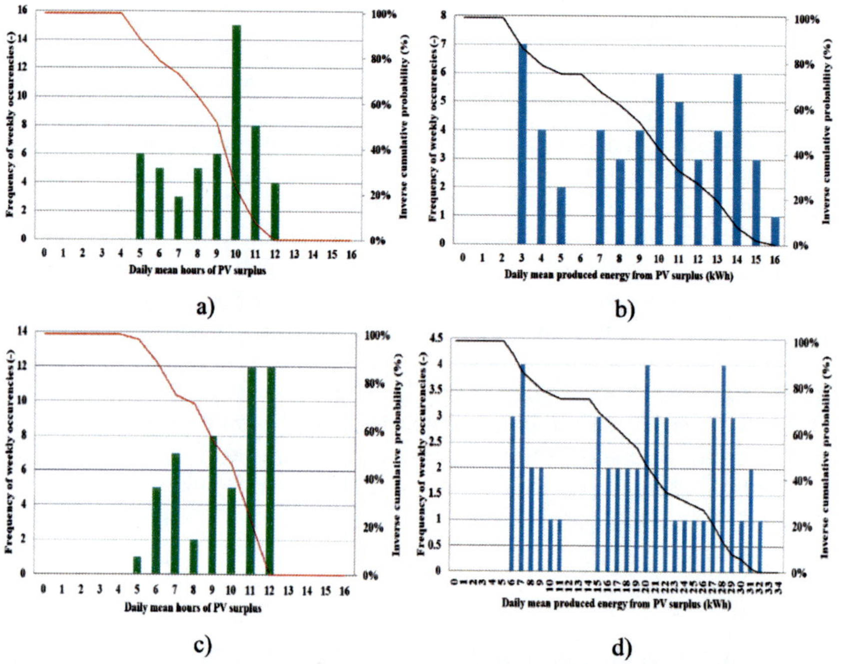

Fig. 1. Weekly inverse cumulative distribution function and occurrences for: **a** daily mean hours of surplus (3 kWp PV), **b** daily produced average energy (3 kWp PV), **c** daily mean hours of surplus (6 kWp PV) and **d** daily produced average energy (6 kWp PV).

Finally, the inverse cumulative probability was computed to define, taking into account the weekly occurrences coverage, the following parameters: the daily average hours of PV surplus (Fig. 1a, c) and the daily average produced energy (Fig. 1b, d).

Specifically, to define a good trade-off between the capacity and the percentage of covered occurrences, the design assumption to cover about 50% of the yearly occurrences, based on the trends depicted in Fig. 1, was chosen. Consequently, the following weeks were selected:

(i) week #40 of the annual diff profile, characterized by daily mean values of 9 h/9 kWh of surplus produced by the 3 kWp PV plant. With reference to Fig. 1a, b, it

means that for more than 54% (52%) of the time over the year, PV surplus energy (hours) is greater than 9 kWh (9 h).
(ii) week #19 of the annual diff profile with a PV surplus, in the case of 6 kWp installation, of 10 h/19 kWh assessed as the daily average values. These data correspond to 46% and 54% of the inverse cumulative probability functions of Fig. 1c, d.

The determined data of PV surplus daily duration and production are then implemented for the energy storage sizing procedure.

2.2 Definition of the Simulated Scenarios

As regards the simulation scenarios, two different installed PV peak powers (i.e., 3 and 6 kWp, as indicated in [15, 16]) were considered. The Li-ion battery was sized to have 10 kWh and 20 kWh as nominal capacity respectively for 3 kWp and 6 kWp PV arrays, according to the daily average energy surplus available in weeks #40 and #19 respectively. This criterion produces a good trade-off to enhance house energy independence from the grid while maintaining adequate storage installation costs. Furthermore, two different vehicles were investigated:

(i) a BEV with a 42-kWh Li-ion battery capacity. It is considered that this vehicle has an average daily route of 100 km. It corresponds to a 50% state of charge (SoC) at the end of the day, i.e., 21 kWh as charging demand.
(ii) a PHEV equipped with a 10-kWh Li-ion battery. For this vehicle a daily operation corresponding to a full-electric average daily distance of about 40 km is supposed. It results in a 20% SoC at the end of the day and about 8 kWh charging demand respectively.

Therefore, four different cases were defined at varying both the PV installed power and BEV/PHEV charging profiles, as follows:

1. 3 kWp PV generation considering a BEV charge each day starting from 9 p.m.
2. 3 kWp PV generation considering a PHEV charge each day starting from 9 p.m.
3. 6 kWp PV generation considering a BEV charge each day starting from 9 p.m.
4. 6 kWp PV generation considering a PHEV charge each day starting from 9 p.m.

As detailed in Sect. 2.1, the maximum power of the wall-box for BEV/PHEV charging is equal to 3.7 kW. Consequently, the charging profiles are determined for the two vehicles and added to the house electric load. The resulting instantaneous power difference (i.e., *diff*) between the PV generation and load demand is illustrated in Fig. 2 relative to the selected weeks, for both BEV (Fig. 2a, c) and PHEV (Fig. 2b, d).

The four scenarios are simulated in MATLAB/Simulink environment through a specific dynamic model described in Sect. 2.3.

2.3 Micro-Grid Dynamic Modeling

According to the statistical analysis described in Sect. 2.1, the simulation time is of one week with 1 s time-step. Figure 3 shows the residential MG layout and the implemented

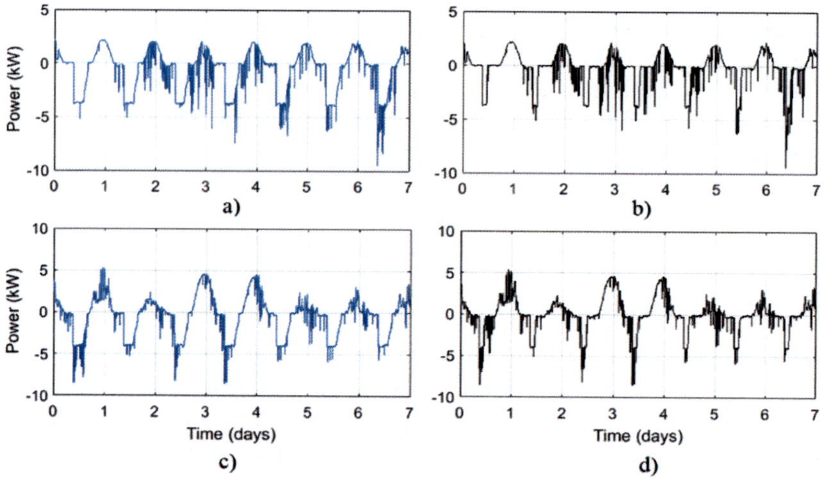

Fig. 2. Instantaneous power profile *diff* for: **a** week #40—BEV charge; **b** week #40—PHEV charge; **c** week #19—BEV charge; **d** week #19 - PHEV charge.

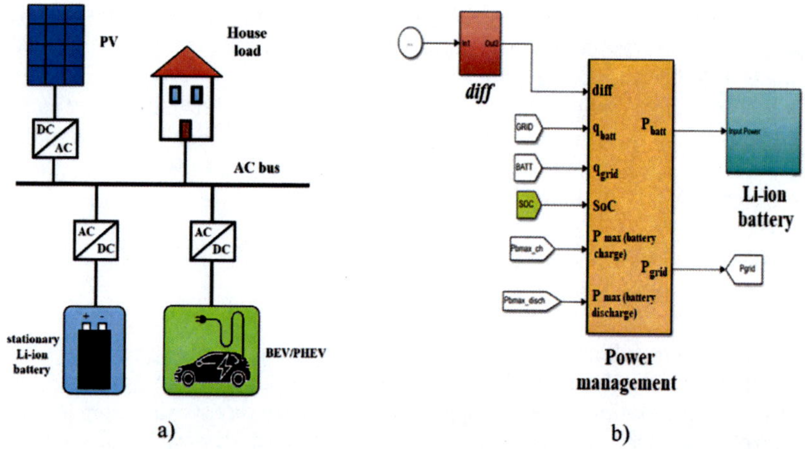

Fig. 3. **a** Layout of the residential micro-grid and **b** the implemented Simulink model.

Simulink model. It is assumed that simulations start at 12 a.m. of the first day and vehicles are charged each day during the night from 9 p.m. To evaluate the impact of a different starting time, simulations were performed also anticipating the BEV/PHEV charging at 7 p.m. No differences were registered; so, the results presented are only referred to charge starting time at 9 p.m. With regards to the stationary Li-ion battery (see Fig. 3a), the initial SoC is set at 50%; moreover, a round trip efficiency of 0.95 and a 90% depth of discharge are implemented. The maximum C-rate for discharge and charge is fixed at C/2. The yellow section in Fig. 3b implements the instantaneous power management strategy that receives as inputs the current *diff* profile (red section). When *diff* is not null, the surplus generated by PV arrays supplies both load and battery (light blue subsystem) charging. Otherwise, load is supplied firstly by the battery discharging and secondly by the grid.

Li-ion battery performance and operating conditions are updated in real-time by the battery subsystem, illustrated in Fig. 4. In detail, it implements, by means of look-up tables, the open circuit voltage and internal resistance measurements reported in Fig. 5. For all details regarding the mathematical models, the Authors refer to their previous works [10–17]. The model instantaneously updates the battery state of charge according to current operating conditions and the battery measured technical features.

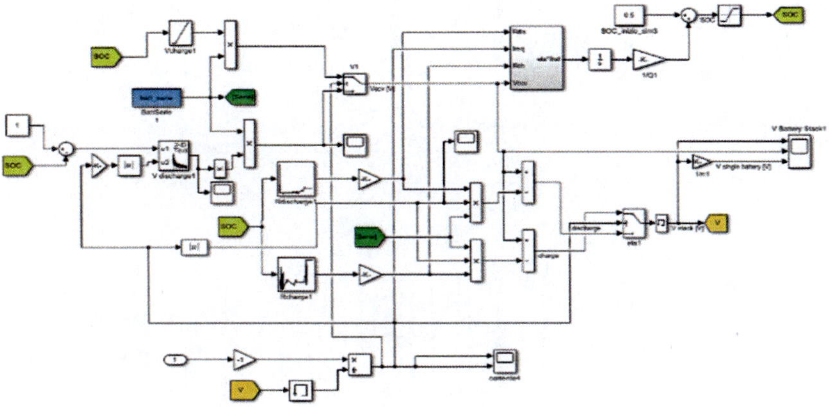

Fig. 4. Li-ion battery subsystem implementation.

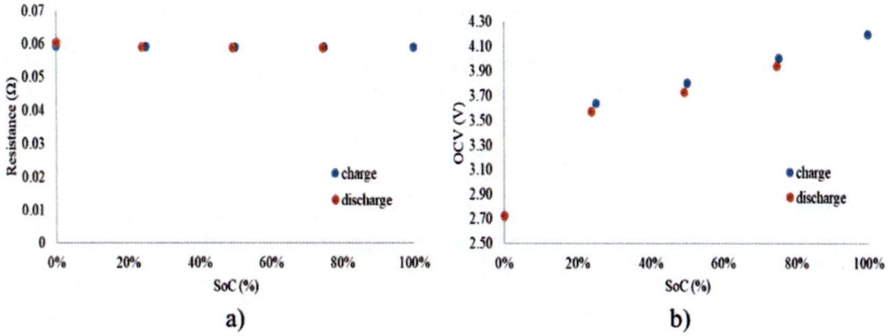

Fig. 5. Open circuit voltages (**a**) and internal resistance values (**b**) at varying the SoC for the implemented Li-ion battery, determined by an experimental campaign.

3 Simulation Results

In this section, the main outcomes of the performed simulations are shown. Specifically, Fig. 6 illustrates the instantaneous evolution of battery SoC and power, the power exchanges with the main grid, PV generation and *diff* profile, over a week with a time step of 1 s.

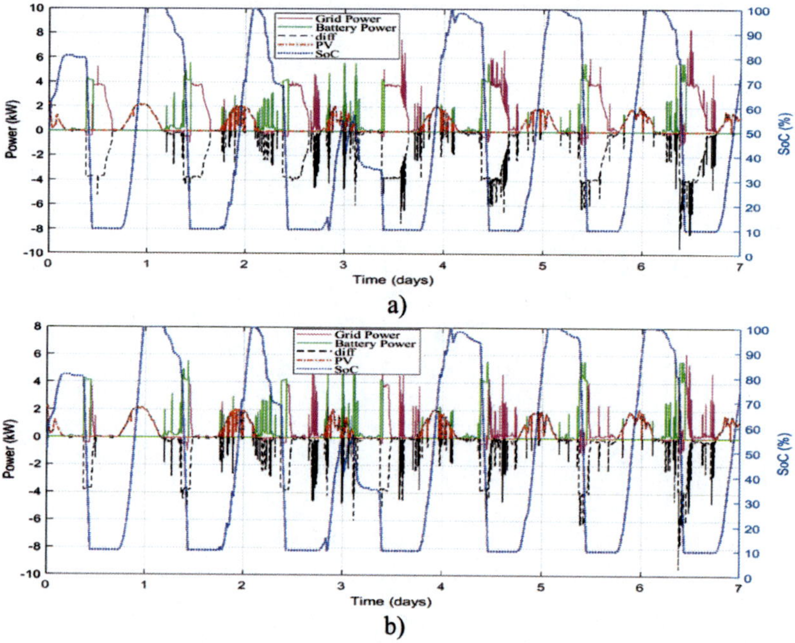

Fig. 6. Trend results of the main considered parameters in the case of PV installed power of 3 kWp and a battery of 10 kWh (week #40): **a** scenario with BEV recharge (case 1) and **b** scenario with PHEV recharge (case 2).

As visible in Fig. 6 for 3 kWp PV power and 10 kWh battery (week #40), BEV charge strongly reduces energy independence from the grid since the amount of required energy for the BEV is more than twice with respect to PHEV (21 kWh vs 8 kWh). SoC evolution of the stationary battery appears to be the same both for BEV and PHEV because the vehicle charging is carried out during the night.

Different parameters evolution can be distinguished from Fig. 7 in the case of 6 kWp of PV installed power and 20 kWh as battery capacity (week #19). Specifically, Fig. 7a (case 3) and Fig. 7b (case 4) refer to BEV and PHEV respectively.

Since the PV generated energy is higher with respect to Fig. 6 (19 kWh vs 9 kWh as daily average surplus energy), the reduction of energy withdrawal from the grid for vehicles charging is significant, as listed in Table 1.

Consequently, the coverage of vehicle charging through PV energy is significant only when installed PV power is increased to 6 kWp and coupled to a 20 kWh battery,

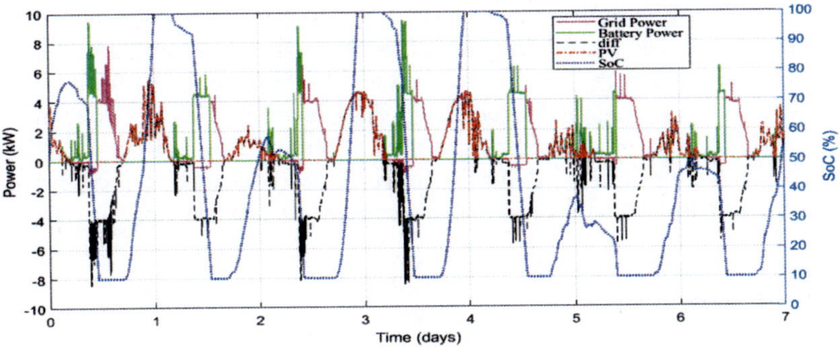

a)

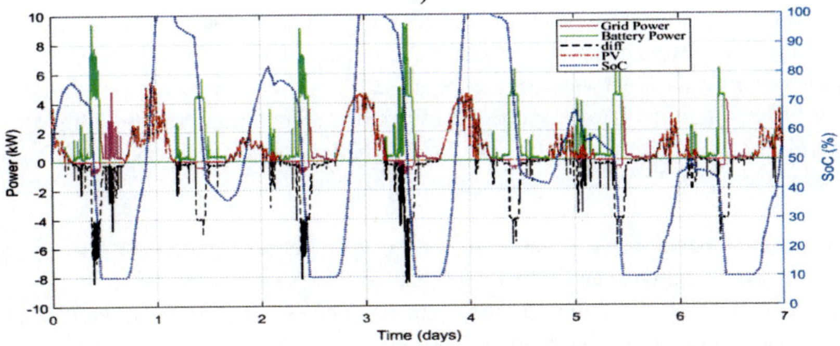

b)

Fig. 7. Trend results of the main considered parameters in the case of PV installed power of 6 kWp and a battery of 20 kWh (week #19): **a** scenario with BEV recharge (case 3) and **b** scenario with PHEV recharge (case 4).

achieving about 23% for BEVs (case 3) and up to about 68% for PHEVs (case 4). Specifically for PHEVs (case 4), PV energy can satisfy the house load and vehicle charge, with 2.54 kWh of daily grid withdrawal.

These data can be referred to about 50% of the weekly occurrences over the year, according to the assumptions made in Sect. 2.1.

Table 1. Weekly storage effects on micro-grid energy independence from the main grid (G_{wd} is the weekly grid withdrawal, $G_{wd,avg}$ is the daily mean grid withdrawal).

Case	G_{wd} (weekly) (kWh)	$G_{wd,avg}$ (daily) (kWh)	BEV/PHEV Charge capacity (kWh)	$G_{wd,avg}$ percentage for BEV/PHEV charge (%)
1	113	19.94	21	95
2	17.79	6.33	8	79.2
3	113	16.14	21	76.9
4	17.79	2.54	8	31.8

Therefore, the MG sized for case 4 can satisfy almost the total energy amount required from the house load and PHEV charge, reaching an energy independence of more than 80% with respect to the grid. This is reduced to 40% for case 3, i.e., for the BEV at parity of PV/battery configuration. Performing the assessment of energy flows of case 4 with respect to the case of storage absence, it results in an increase of self-consumption of about 68%. On the other hand, the PV/battery MG configurations of cases 1–3 are not sufficient to sustain the load and charge requirements, specifically for BEVs.

4 Conclusions

To strongly reduce the emissions, the massive penetration of MGs with renewable generation and electric vehicles could have a great impact in the next years.

This research work aims at evaluating the impact of BEV and PHEV charge on a typical residential MG, at varying the installed PV power (i.e., 3 kW and 6 kW) and Li-ion battery nominal capacity (10–20 kWh). The simulations results highlight that only a MG consisting of 6 kWp PV and 20 kWh battery with PHEV can satisfy both the house load and PHEV charge. Specifically, in this case:

- an energy independence greater than 80% with respect to the grid is achieved considering both house and PHEV charge load.
- PHEV is mainly charged (contribution rate of about 68%) by means of the PV surplus energy stored in the stationary battery.
- if the user owns a BEV, this MG configuration allows to reach less than 40% of energy independence from the grid, taking into account all the loads.

Thus, it is emphasized that enhancing PV installed power and energy storage capacity in the future residential micro-grids should be necessary to avoid stability issues on low voltage feeders, mitigating unbalances, undervoltage and power factor issues, in the view of massive BEV penetration for pursuing the limit to global warming.

References

1. Kumar, V., Ranjan, D., Verma, K.: Global climate change: the loop between cause and impact. Glob. Clim. Chang. 187–211 (2021)
2. European Commission 2050 long-term strategy. https://ec.europa.eu/clima/policies/strategies/2050_en. Last accessed 17 May 2021
3. Greenhouse gas emissions from transport in Europe. https://www.eea.europa.eu/ims/greenhouse-gas-emissions-from-transport. Last accessed 04 May 2022
4. Björnsson, L.H., Karlsson, S.: Electrification of the two-car household: PHEV or BEV? Transp. Res. Part C Emerg. Technol. **85**, 363–376 (2017)
5. Wolfram, P., Lutsey, N.: Electric Vehicles: Literature Review of Technology Costs and Carbon Emissions. International Council on Clean Transportation, pp. 1–23 (2016)
6. Entro il 2030 serviranno 7 milioni di punti ricarica in Europa. https://www.qualenergia.it/articoli/acea-entro-2030-necessari-7-milioni-punti-ricarica-europa/. Last accessed 04 April 2022

7. Gilleran, M., Bonnema, E., Woods, J., Mishra, P., Doebber, I., Hunter, C., Mitchell, M., Mann, M.: Impact of electric vehicle charging on the power demand of retail buildings, vol. 4 (2021)
8. Zhou, C., Wang, H., Zhou, W., Qian, K., Meng, S.: Determination of maximum level of EV penetration with consideration of EV charging load and harmonic currents. IOP Conf. Ser. Earth Environ. Sci. **342** (2019)
9. Ul-Haq, A., Cecati, C., Strunz, K., Abbasi, E.: Impact of electric vehicle charging on voltage unbalance in an urban distribution network. Intell. Ind. Syst. **1**(1), 51–60 (2015). https://doi.org/10.1007/s40903-015-0005-x
10. Barelli, L., Bidini, G., Bonucci, F., Castellini, L., Castellini, S., Ottaviano, A., Pelosi, D., Zuccari, A.: Dynamic analysis of a hybrid energy storage system (H-ESS) coupled to a photovoltaic (PV) plant. Energies **11** (2018)
11. Bayrak, G., Cebeci, M.: Grid connected fuel cell and PV hybrid power generating system design with Matlab Simulink. Int. J. Hydrog. Energy **39**, 8803–8812 (2014)
12. Chen, C., Duan, S.: Optimal integration of plug-in hybrid electric vehicles in microgrids. IEEE Trans. Ind. Inform. **10**, 1917–1926 (2014)
13. Murray, D., Stankovic, L., Stankovic, V.: An electrical load measurements dataset of United Kingdom households from a two-year longitudinal study. Sci. Data **4**, 160122 (2017)
14. Wall-Box 3,7 kW—e-Station Store. https://www.e-station.store/prodotto/wallbox-modo-3/e-line/wall-box-37-kw/. Last accessed 29 April 2022
15. What you need to know about installing solar panels on your home in Italy. https://www.thelocal.it/20210929/what-you-need-to-know-about-installing-solar-panels-in-your-home-in-italy/. Last accessed 04 May 2022
16. How much solar power and solar panels do you need? https://www.choice.com.au/home-improvement/energy-saving/solar/articles/how-much-solar-do-i-need. Last accessed 04 May 2022
17. Barelli, L., Bidini, G., Ottaviano, P.A., Gallorini, F., Pelosi, D.: Coupling hybrid energy storage system to regenerative actuators in a more electric aircraft: dynamic performance analysis and CO2 emissions assessment concerning the Italian regional aviation scenario. J. Energy Stor. **45**, 103776 (2022)

Impact of Using Phase Change Materials with Different Wall Orientations in a Classroom Building Under a Warm Temperate Climate

Mohammed Amin Nassim Haddad[1](✉), Hamza Semmari[1], Khaled Imessad[2], Mohammed Cherif Lekhal[3], Lotfi Derradji[4], D. Rouag-Saffidine[5], and Mohamed Amara[4]

[1] LMSEA Laboratoire de Mécanique et Systèmes Energétiques Avancés, Ecole Nationale Polytechnique de Constantine, BP75, A, Nouvelle Ville Ali Mendjli, 25000 Constantine, Algérie
itshaddad1@gmail.com

[2] Centre de Développement des Energies Renouvelables, Route de l'observatoire, BP 62, BouzareahAlger, Algérie

[3] MSME Laboratoire, UMR-8208 CNRS, Université Gustave Eiffel, 77420 Marne-la-Vallée, France

[4] Centre National d'Etudes et de Recherches Intégrées du Bâtiment, CNERIB, Alger, Algérie

[5] Energy and Environment Laboratory, Faculty of Architecture and Urban Planning, University of Constantine 3, Ali Mendjeli, 25016 Constantine, Algeria

Abstract. In this paper, the effect of the integration of Phase Change Material panels for summer thermal comfort and cooling demand of a classroom under warm temperate climate was numerically assessed. The PCM panels are fixed on the internal faces of the walls (east, west and south) and roof. They are modeled using TRNSYS software, specifically Type 339. A comparison before and after using PCM on the roof and the east, west and south walls as well as on the combination wall-roof was undertaken. The results showed that the cooling demand was reduced from 6.8 kWh/m^2/year to 5.5 kWh/m^2/year with the combined roof-wall configuration of PCM compared to the initial case. Indoor temperature was reduced by 1.13 °C for the roof-wall configuration. Moreover, with regard to the thermal comfort standard that requires an indoor temperature between 21 °C and 27 °C, this combination reduced the discomfort hours by 25% comparing to the base case.

Keywords: PCM panels · Thermal comfort · Classroom building

1 Introduction

The building sector is responsible for nearly one-third of final energy consumption and 40% of CO_2 emissions. Heating and cooling are responsible for 60% of these emissions [1]. This appeals for a back-up of energy efficiency strategies in order to reduce the carbon footprint of building sector. For this purpose, passive energy efficiency strategies have to be adopted as prior upstream approach during the building's design and planning

J. Littlewood and R. J. Howlett (Eds.): SEB 2022, SIST 336, pp. 140–151, 2023.
https://doi.org/10.1007/978-981-19-8769-4_14

steps. Phase change materials (PCM) which form an interesting solution to reduce both indoor air temperature in summer and peak energy demand is among these strategies.

Unlike the conventional materials, PCMs can store energy in two ways, namely sensible and latent. During their phase change, which mostly occurs at a constant temperature, PCMs provide a large energy storage capacity. This advantage can be used to store an important amount of thermal energy across a limited range of temperature. This can be achieved through the design of latent heat energy storage systems (LHESS) [2, 3].

Indeed, during the warm season, the incorporation of PCM improve the absorption of excessive heat gains inside the building during occupancy hours. Theses heat gains can be related to many parameters such as solar radiation through glazed and opaque surfaces, occupancy density and frequency, physical activity of the occupants as well as the use of electrical auxiliaries [4].

2 Literature Review

According to the literature, whether the PCMs are micro or macro encapsulated, different parameters influencing their behavior have been studied. These parameters include, but are not limited to, phase-change temperatures range, location within the wall, heat storage capacity and thickness [5–7]. Usually, the thicknesses of PCM layers incorporated in building walls range from 0.3 cm to 6 cm [8, 9] while the latent heat varies between 70 kJ/kg and 281 kJ/kg [10, 11].

In a study conducted by Salihi et al. [5], the impact of the PCM melting temperature and its thickness for lightweight building walls under semi-arid climate was investigated. The results have shown an energy saving varying between 7.30% to 15.21% in cooling demand and a thickness of 1.5 cm appeared to be the most appropriate. They also reported that the best configuration was a wall with a triple layers of PCM.

For the hot climate regions, Sovetova et al. [12] studied the thermal performance of PCMs integrated in buildings for different cities around the world. The results showed a reduction in energy consumption between 17.97% and 34.36%. In the same work, the application of PCM in a hot climate of Biskra located in Algeria revealed an energy saving of 17.97%.

Moreover, the use of PCM as a passive retrofitting technique to reduce cooling demand in an educational building in a Mediterranean climate in Italy was evaluated by Ascione et al. [13]. A reduction of 11.7% in cooling demand and an increase in thermal comfort by 215 h were found.

Similarly, for a Mediterranean climate in Cyprus Panayiotou et al. [14] evaluated the application of macroincapsulated PCM in a typical dwelling. The authors found that a reduction of 37.4% can be achieved in cooling demand and the mean air temperature can be 3–5 °C lower than the base case.

In another work, Alam et al. [15] inspected different orientations of the PCM layer. The combination between walls + roof was identified as the best configuration reducing the temperature by nearly 3 °C and saving 29% the annual energy was achieved with regard to the climate of Sydney. They reported that although the PCM behavior is climate dependent, the melting point outside the thermal comfort range does not provide a reduction in energy demand. Based on the fact that the PCM melting range is varying throughout the year, the authors suggested the consideration of the lowest annual energy demand as an additional criterion for the PCM selection.

Besides the importance of all the aforementioned parameters that can influence the PCM behavior, the orientation can also play an important role on the thermal performance of the buildings. Therefore, this study aims to investigate the orientation effect of using macro encapsulated PCM panels on thermal comfort and cooling needs of a detached typical classroom building under the warm temperate climate of Constantine (Algeria).

3 Methodology

The present work consists in investigating the orientation of PCM panels. A comparison between the application of the panels on the roof and the east, west and south walls as well as on the combination wall-roof was made. The cooling energy demand of these different orientations is assessed through the simulations.

3.1 Simulation Approach

The study was performed using the dynamic thermal simulation software TRNSYS 17 under hourly time steps [16]. TRNSYS is one of the most suitable tools for the transient thermal systems as is the case for building. In addition to its user-friendly graphic interface, the main advantage of TRNSYS simulation tool lies in its simplicity to link between the outputs and the inputs of the modules (components) required to model the specified project. Consequently, TRNSYS was preferred to perform this study.

The yearly typical weather data file was generated by Meteonorm software and processed in TRNSYS using type 109. The modeling of the PCM panels embedded in the wall was made by type 399 which uses a one-dimensional finite difference method to simulate the heat transfer conduction through the wall [17].

In this study, the methodology used to evaluate the effect of PCMs is similar to that described by Allerhand [18] with some refinement. Allerhand work was focused in the application of PCM panels into the ceiling and a comparison between an all-air system and a thermally active building system (TABS) was made. However, in this work only all-air system was considered while the focus was on the variation of the orientation of PCM panels between walls and the roof.

The simulations have been undertaken from May to October with the exclusion of July and August as the university is in vacation in these two months.

3.2 Climate Description

The geographical location of the expected building is characterized by warm temperate climate. This location is classified as Csa by Koppen-Geiger. Figure 1 shows the monthly outdoor temperatures which are maximum in July and August with an average of 26.3 °C, while the maximum radiation occurs in July with 300 W/m^2 on horizontal surface.

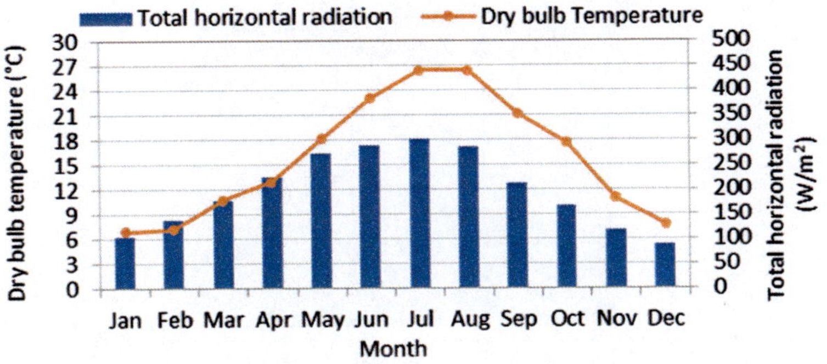

Fig. 1. Monthly average dry bulb temperature and horizontal radiation.

3.3 Building Description

The building has a total area of 106.7 m^2 and is composed of two thermal zones. The conditioned zone represents the classroom with 71.1 m^2 while the non-conditioned part represents the access area and the sanitary with 35.6 m^2. The occupancy capacity of the classroom is 31 (seated) and the attendance schedule is from 9 a.m. to 17 p.m.

The floor plan of the classroom and a 3D model of the building are illustrated in Figs. 2 and 3, respectively. The thermal properties of the building materials are summarized in Table 1 while the configuration of the construction layers is shown in Fig. 4.

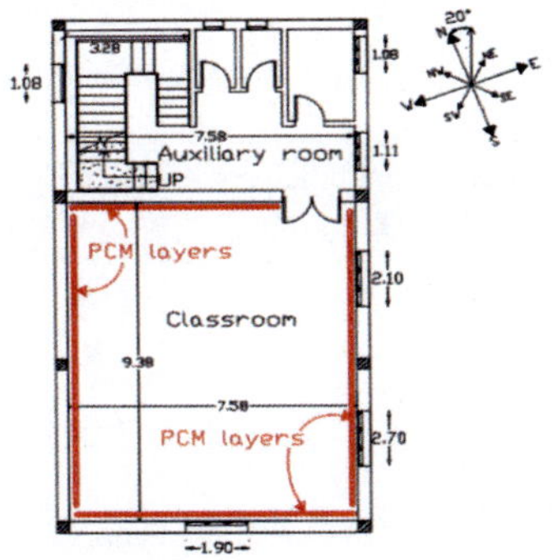

Fig. 2. Floor plan of the classroom.

3.4 PCM Description

Different parameters can influence the thermal behavior of PCM namely: the phase change temperature, the latent heat, the thermal conductivity and PCM thickness, too [7]. In the context of thermal comfort, the PCM selected for this study is Rubitherm RT26 [19]. It is a paraffin-based material with a phase change temperature in the same range as the indoor thermal comfort temperature, as recommended by Alam et al. [15].

Fig. 3. A 3D model of the classroom.

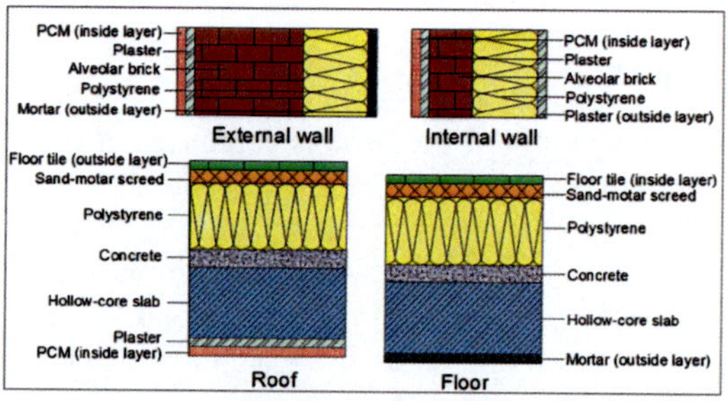

Fig. 4. Configuration of construction layers.

Table 1. Thermo-physical properties of building materials.

Building material	Conductivity [W/m.K]	Specific heat [kJ/kg.K]	Density [kg/m^3]	Thickness [cm]
Plaster	0.35	0.8	800	2
Alveolar brick	0.31	0.79	720	25
Simple brick	0.31	0.79	720	10
Polystyrene	0.04	1.38	25	15
Mortar	1.15	0.8	2000	2
Floor tile	1.71	0.7	2300	2
Sand-mortar	0.93	0.8	1882	3
Concrete	1.76	0.92	2300	4
Hollow-core slab	1.23	0.65	1300	16

The thermal comfort temperature is between 21 °C and 27 °C according to Algerian standards [20]. Table 2 summarizes the properties of the PCM.

The PCM consists in panels of 2 cm thickness, 60 cm length and 60 cm width. These panels are installed side by side and applied on the internal surface of the walls to make

Table 2. Thermal properties of RT26 PCM.

Melting range [°C]	26
Crystalizing range [°C]	25
Thermal conductivity [W/m.K]	1.2
Specific heat [kJ/kg.K]	2
Latent heat [kJ/kg]	180
Maximum operating temperature [°C]	60
Average density [kg/m^3]	880

a total surface of 19 m^2 for every PCM orientation. With the exception for the roof + walls configuration, where 19 m^2 was applied on the walls combined with 19 m^2 applied on the roof, for a total surface of 38 m^2.

It is important to mention that to exploit the full properties of PCM, thermal conductivity should not be low as this will increase the time of the melting and solidification. Otherwise, this will not allow the PCM to melt completely and part of it remains in the solid phase, especially when using a relatively thick layer [15]. Therefore, the thermal conductivity of the PCM used in this study was assumed to have a value of 1.2 W/m.K due to the presence of Aluminum in the PCM panel [18].

4 Results

4.1 Energy Balance

Figure 5 shows the cooling energy balance results for the different PCM panel orientations, including a building without PCM. The building without PCM point out the highest cooling demand. It is noted that June is the most sensitive month for cooling demand with 3.5 kWh/m^2 for a building without PCM compared to 3.1, 3.3, 3.3, 3.2 and 3 kWh/m^2 for the roof, south, east, west and roof + walls, respectively. On the other hand, the lowest demands are associated with the roof + walls configuration. For this latter configuration, the reduction rate varies from 0.17 kWh/m^2 in October to 0.51 kWh/m^2 in June compared to the building without PCM. As well, June is the month with the largest cooling reduction among all the configurations.

Ultimately, the total cooling demand for a building without PCM was the highest with 6.83 kWh/m^2. The reductions achieved after applying PCM are 10%, 11%, 12%, 16% and 19% for the east, south, west, roof and roof + walls respectively.

4.2 Operative Temperature

The operative temperature is defined as the weighted temperature of both the air and the internal surface temperatures of the zone [16]. It can be expressed as follows:

$$T_{op} = A_{op} * T_{air} + \left(1 - A_{op}\right) * T_{surf} \quad (1)$$

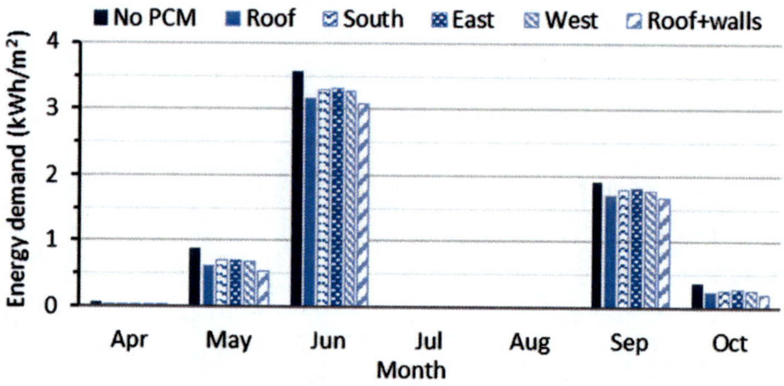

Fig. 5. Monthly cooling demand for several PCM orientations.

where A_{op} represents the weighting factor of the operative temperature, T_{air} and T_{surf} represent the air and internal surface temperatures, respectively.

Figure 6 illustrates the variation of the operative temperature during 3 days (from October 7 to October 9) for the different PCM orientations. As can be seen, the use of PCM panels has reduced the temperatures only in small amounts. For instance, the day of October 7 represents the most interesting reduction of peak indoor operative temperature. It decreases from 26.2 °C for a building with no PCM to 25.1 °C for the roof + walls PCM configuration. This reduction of 1.13 °C was the largest comparing the other configurations.

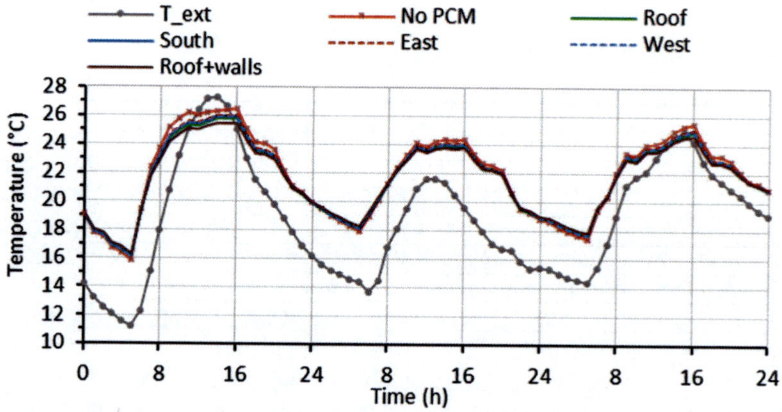

Fig. 6. Operative temperature for a classroom with different PCM orientations.

Additionally, from another perspective, it is known that the incorporation of PCMs in small amounts will have the same effect of a building with high thermal mass and considerably thick opaque construction layers [21]. Thus, the impact of increasing this parameter on the indoor operative temperature can be noticed in the same figure. For the day of October 7, the comparison between a building with no PCM and the other configurations reveals that the reduction of peak temperature starts from 0.59 °C for the east PCM configuration and can reach 1.13 °C for the roof + walls PCM configuration representing thereby the best case.

4.3 Cooling Power Distribution

Figure 7 represents the utilization rate (in number of hours) of cooling power divided into intervals from 0 to 5 W/m^2, from 5 to 20 W/m^2, from 20 to 40 W/m^2 and from 40 to 60 W/m^2, respectively. As can be observed, the case without PCM requires the most important utilization rate independently of the cooling power intensity. It is also noticed that the maximum utilization rate with 179 cooling hours occurs for the cooling intensity of 20 W/m^2.

Additionally, the roof + walls represents the best configuration in terms of reducing the cooling hours. Moreover, regardless of the cooling power intensity, the roof + walls configuration presented the lowest number of total cooling hours with 326 h.

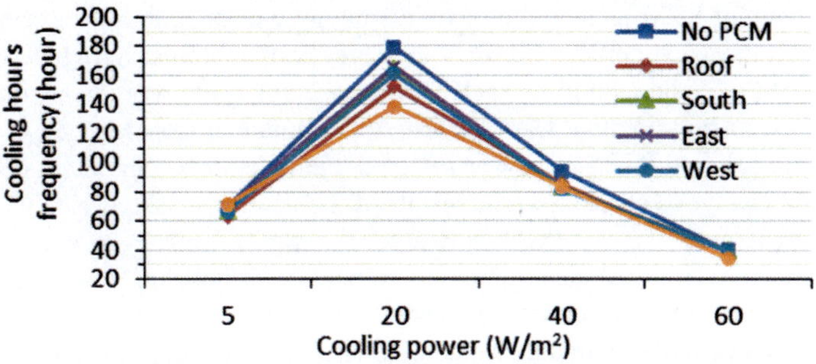

Fig. 7. Cooling power frequency distribution for the PCM orientations.

4.4 Cooling Discomfort Hours

Figure 8 shows the discomfort hours frequency during the cooling season. The discomfort hours are identified when the operative temperature of the classroom exceeds the upper thermal comfort limit of 27 °C. The percentage of reduction of discomfort hours for each PCM configuration compared to a building without PCM is also presented in the same figure.

The results show that the most discomfort hours are associated with a building without PCM with 150 h. A significant reduction was achieved after applying PCM panels. The lowest reduction was observed for the south and east PCM configurations with 16% (24 h) while the west configuration was slightly better with 17% (26 h).

Meanwhile, applying PCM panels on the internal surface of the classroom roof increased this reduction to 22%, corresponding to 33 h. Lastly, the highest amount of reduction was obtained for the roof + walls configuration with 25% reducing, then, discomfort to 113 h against 150 h for a building without PCM. This makes it the optimum PCM configuration.

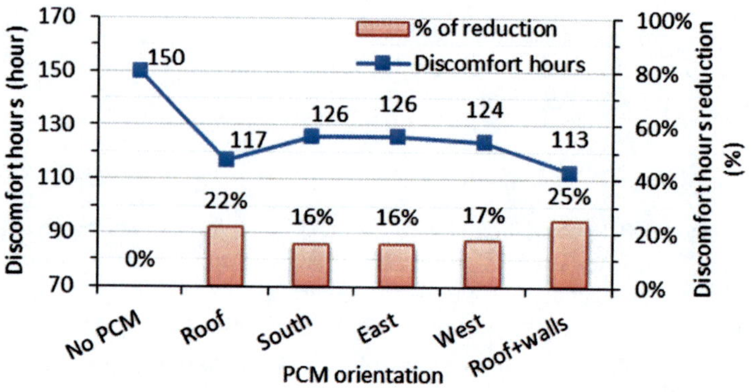

Fig. 8. Discomfort hours frequency and percentage of reduction.

5 Discussion

The integration of PCM panels as a passive strategy with the aim of reducing cooling demand and improving thermal comfort was achieved. Few studies have investigated the orientation of PCMs on a building scale. Therefore, this study was intended to explore this concern for a building in a warm temperature climate of Constantine (Algeria).

The paraffin based RT26 PCM was selected due to its important latent heat storage and its phase change temperature. The results of the cooling needs revealed a reduction ranging from 10% for the east PCM to 19% for the roof + walls PCM combination.

Additionally, an improvement in indoor temperature is noticed. However, the largest reduction of peak temperatures was estimated at 1.13 °C.

With reference to this parameter, the simulation results did not show a substantial impact when using the PCM panels. This can be explained by the unsuitable 1 h time step used during the simulations. Since the building type considered in this study is a heavyweight construction, 1 h time step is less time consuming and allows an adequate performance of TRNSYS software. In contrast, hourly time step prevents the implicit numerical method used by the Type 399 to detect phase change phenomena. Therefore, small time steps of around 1 min provide the best results as reported by Delcroix et al. [22] and Allerhand et al. [18]. However, 1 min time step will be limited to reduced investigation study while hourly approach will be reserved for heavyweight construction.

Generally, the incorporation of PCM has a positive impact leading to the reduction of the utilization rate of the cooling system until it reaches the lowest value for roof + walls configuration. In the same way, as shown in Fig. 8, an important reduction in discomfort hours of 25% was achieved for the roof + walls configuration.

The variation of PCM orientations did not show a large effect, which can be due to the position of the panels. As said, the installation of the panels on the interior surfaces of the envelope will allow the PCM to interact more with the indoor environment where the thermal conditions do not present large variations.

As was expected, increasing the surface of the PCM panel has resulted in a larger heat storage during summer, thereby the energy saving as well as the reduction of discomfort hours were the highest among all the configurations.

The obtained results present a good agreement with similar studies where authors reported energy savings varying from 7.30% to 15.21% depending on the number of PCM layers in semi-arid climate in Morocco [5], 17.97% for a hot climate in Biskra (Algeria) [12] and 11.7% for a Mediterranean climate in Italy [13].

Furthermore, with reference to the nature of the phase change material, although organic PCMs present the best thermal stability among the other types [23, 24], an experimental investigation is preferred to acquire better understanding of the number of thermal cycles of this PCM before the degradation of its thermo-physical properties, particularly in warm temperate climate.

In this perspective, experimental investigation will be of great importance and this is fixed for future outlook of the present study. Further studies are recommended also to explore the effect of combining other strategies such as the radiative cooling of PCM panels and also investigation of new bio-PCM materials.

6 Conclusion

The application of PCM RT26 on the internal surface of a typical classroom building walls was investigated numerically using TRNSYS 17 software. The investigation took a typical classroom building in the city of Constantine in Algeria as a case study for the warm temperate climate of the Mediterranean regions. The conclusions drawn from this paper are summarized as follows:

- Using the RT26 PCM on the roof in conjunction with the walls (roof + walls) seems to be the best configuration. This configuration has contributed to a reduction of 1.33 kWh/m^2 in cooling needs, 56 h for the utilization rate of the cooling system and 25% in discomfort hours when compared with a building without PCM. Whereas, the largest temperature reduction achieved was nearly 1.13 °C against a building with without PCM.
- The best performances obtained with the roof + walls configuration are due to the maximum PCM surface of 38 m^2 while the other configurations present a limited PCM surface of only 19 m^2.
- It is also essential to note that these results are available for PCM panels that are installed on the interior surface of the envelope. Also, they are specific to this climate as well as the building type with its occupancy schedule.

References

1. IEA – International Energy Agency. IEA. https://www.iea.org. Last accessed 28 April 2022
2. Groulx, D., Castell, A., Solé, C.: Advances in Thermal Energy Storage Systems. 2nd edn. Woodhead Publishing (2021)
3. Stritih, U., Tyagi, V.V., Stropnik, R., Paksoy, H., Haghighat, F., Joybari, M.M.: Integration of passive PCM technologies for net-zero energy buildings. Sustain. Cities Soc. **41**, 286–295 (2018)
4. Henze, G.P., Le, T.H., Florita, A.R., Felsmann, C.: Sensitivity analysis of optimal building thermal mass control. ASME. J. Sol. Energy Eng **129**(4), 473–485 (2007)

5. Salihi, M., El Fiti, M., Harmen, Y., Chhiti, Y., Chebak, A., Alaoui, F.E.M.H., ... Jama, C.: Evaluation of global energy performance of building walls integrating PCM: numerical study in semi-arid climate in Morocco. Case Stud. Constr. Mater. **16**, e00979 (2022)
6. Martelletto, F., Doretti, L., Mancin, S.: Numerical simulation through experimental validation of latent and sensible concrete thermal energy storage system. J. Energy Stor. **51**, 104567 (2022)
7. Kishore, R.A., Bianchi, M.V., Booten, C., Vidal, J., Jackson, R.: Parametric and sensitivity analysis of a PCM-integrated wall for optimal thermal load modulation in lightweight buildings. Appl. Therm. Eng. **187**, 116568 (2021)
8. Lei, J., Yang, J., Yang, E.H.: Energy performance of building envelopes integrated with phase change materials for cooling load reduction in tropical Singapore. Appl. Energy **162**, 207–217 (2016)
9. Sajjadian, S.M., Lewis, J., Sharples, S.: The potential of phase change materials to reduce domestic cooling energy loads for current and future UK climates. Energy Build. **93**, 83–89 (2015)
10. Soares, N., Gaspar, A.R., Santos, P., Costa, J.J.: Multi-dimensional optimization of the incorporation of PCM-drywalls in lightweight steel-framed residential buildings in different climates. Energy Build. **70**, 411–421 (2014)
11. Seong, Y.B., Lim, J.H.: Energy saving potentials of phase change materials applied to lightweight building envelopes. Energies **6**(10), 5219–5230 (2013)
12. Sovetova, M., Memon, S. A., Kim, J.: Thermal performance and energy efficiency of building integrated with PCMs in hot desert climate region. Solar Energy **189**, 357–371 (2019)
13. Ascione, F., Bianco, N., De Masi, R. F., Mastellone, M., Vanoli, G. P.: Phase change materials for reducing cooling energy demand and improving indoor comfort: A step-by-step retrofit of a Mediterranean educational building. Energies **12**(19), 3661 (2019)
14. Panayiotou, G. P., Kalogirou, S. A., Tassou, S. A.: Evaluation of the application of phase change materials (PCM) on the envelope of a typical dwelling in the Mediterranean region. Renew. Energy **97**, 24-32 (2016)
15. Alam, M., Jamil, H., Sanjayan, J., Wilson, J.: Energy saving potential of phase change materials in major Australian cities. Energy Build. **78**, 192-201 (2014)
16. TRNSYS 17: A Transient System Simulation Tool Homepage. http://www.trnsys.com/. Last accessed 15 June 2022
17. Dentel, A., Stephan, W.: TRNSYS TYPE 399-Phase change materials in passive and active wall constructions. 1st edn. Institute for Energy and Building, Georg Simon Ohm University of Applied Sciences, Nürnberg, Germany (2013)
18. Allerhand, J. Q., Kazanci, O. B., Olesen, B. W.: Energy and thermal comfort performance evaluation of PCM ceiling panels for cooling a renovated office room. In: E3S Web of Conferences, vol. 111, pp. 03020. EDP Sciences, Romania (2019)
19. Rubitherm R26. https://www.rubitherm.eu/en/index.php/productcategory/organische-pcm-rt, last accessed 2022/04/25
20. Document Technique Règlementaire D.T.R C 3-2. Centre National d'Etudes et de Recherches Intégrées du Bâtiment (CNERIB), Alger (2016)
21. Kuznik, F., David, D., Johannes, K., & Roux, J. J.: A review on phase change materials integrated in building walls. Renew. Sustain. Energy Rev. **15**(1), 379-391 (2011)
22. Delcroix, B., Kummert, M., Daoud, A.: Development and numerical validation of a new model for walls with phase change materials implemented in TRNSYS. J. Build. Perf. Simul. **10**(4), 422-437 (2017)

23. Rathore, P.K.S., Shukla, S.K.: Enhanced thermophysical properties of organic PCM through shape stabilization for thermal energy storage in buildings: a state of the art review. Energy Build. **236**, 110799 (2021)
24. Khan, Z., Khan, Z., Ghafoor, A.: A review of performance enhancement of PCM based latent heat storage system within the context of materials, thermal stability and compatibility. Energy Conv. Manag. **115**, 132–158 (2016)

Applications of Thermoelectricity in Buildings: From Energy Harvesting to Energy Management

Diana Enescu(✉)

Valahia University of Targoviste, Targoviste, Romania
diana.enescu@valahia.ro

Abstract. Thermoelectric modules used as electricity generators or for producing heat or cooling are being increasingly deployed in applications referring to buildings. This paper recalls the basic principles of thermoelectricity and provides an overview of the solutions used in buildings for different purposes. The applications considered include heating and cooling internal spaces, energy harvesting for power supply to sensors used in energy management systems, as well as in thermoelectric refrigerators and solutions for personal thermal management to improve the thermal comfort of the individuals inside the buildings.

Keywords: Building envelope · Building-integrated photovoltaic · Personal thermal management · Sensor · Thermoelectric generator · Thermoelectric cooling · Ventilation

1 Introduction

1.1 General Aspects of Thermoelectric Systems

Thermoelectric technology is gaining interest in the emerging context in which there is more attention to producing energy also at small scales, with low environmental impact.

Notwithstanding their relatively low performance with respect to other technologies, solutions based on thermoelectric systems are gaining interest because of various advantages over other devices for similar applications. Among the most interesting properties of thermoelectric systems, there is the possible use in any position (which allows their use also in portable devices), absence of noise and moving parts, small size and light weight, compactness and high modularity, and high reliability. Moreover, thermoelectric systems are emission-free, allowing their use an alternative to devices that need harmful working fluids (such as refrigerators) or generate pollutants.

The characteristics of the thermoelectric modules enable their deployment in different applications that need energy conversion from thermal energy to electricity, as well as from electricity to heat or cooling, in the latter case with a heat pump effect. These properties make it possible to find thermoelectric devices with a versatile role in different solutions referring to buildings, for energy harvesting and integration with other

J. Littlewood and R. J. Howlett (Eds.): SEB 2022, SIST 336, pp. 152–163, 2023.
https://doi.org/10.1007/978-981-19-8769-4_15

energy systems. In this respect, thermoelectric systems are an interesting contributor to the development of solutions aimed at promoting green energy [1].

This paper summarizes the usage of thermoelectric systems in different applications referring to buildings, mainly based on recent publications. After recalling some basic aspects of the structure and models of thermoelectric devices in the rest of this section, Sect. 2 addresses energy harvesting to provide power supply to the sensors used in energy management systems and to procure electricity and heat in building-integrated applications. Section 3 deals with the use of thermoelectric systems to provide cooling solutions for the internal spaces. Section 4 refers to solutions for personal thermal management used for improving the thermal comfort of the individuals inside the buildings. The last section contains the conclusions.

1.2 Thermoelectric Devices

A thermoelectric module is composed of single or more thermoelectric couples (or thermocouples). All thermocouples of a thermoelectric module are connected electrically in series and thermally in parallel. Standard thermoelectric modules have a minimum of three thermocouples, up to over 100 thermocouples for larger modules [2].

A single thermocouple contains two legs made of different thermoelectric materials, usually bismuth and tellurium alloyed with antimony or selenium. The two legs are physically soldered together on one end, with a small metallic interconnect (e.g., copper) to form a junction. The metallic interconnect serves as an electrical contact between the two thermoelectric legs. The contacts are arranged in such a way that all the legs are connected electrically in series (if the thermoelectric module has many thermocouples). One leg is doped to create an N-type semiconductor having an excess of free electrons e^- (the major heat carriers are the electrons). This semiconductor material has the following properties: negative Seebeck coefficient α_N, electrical resistivity ρ_N, and thermal conductivity k_N. The other leg is doped to create a P-type semiconductor having an excess of holes h^+ (the major heat carriers are the holes). This semiconductor material has the following properties: positive Seebeck coefficient α_P, electrical resistivity ρ_P and thermal conductivity k_P.

Either a single thermoelectric couple or many thermocouples are placed between two ceramic plates. The ceramic plates are used to electrically insulate the thermoelectric module from external surfaces; however, they are conductors from a thermal point of view. Furthermore, the ceramic substrates must also have good thermal conductance K to provide good heat transfer with minimal thermal resistance. The thermoelectric couples are connected in such a way that when electric current flows through the module, both N-type electrons and P-type holes move in the same direction (towards the same side of the module). When a low voltage DC power source is applied to the free ends of the thermoelectric module, the heat is transferred from one side to another side of the module through the N- and P- semiconductor legs and junctions. In this case, one side of the module is cooled, and the other side is heated.

Depending on their usage, thermoelectric modules are partitioned into two main categories [3]:

1. Thermoelectric generator (TEG): the temperature difference between the hot and cold sides is converted into an electrical voltage based on the Seebeck effect (Fig. 1a). When an electrical load is connected at the TEG terminals, an electric current flows in the electrical load. The power generated is the product of voltage and current. The power is generated in direct current (DC). If the load is in DC at a different voltage, a DC/DC converter is needed. If the load is in alternating current (AC), a DC/AC converter is needed.
2. Thermoelectric cooler (TEC): an electric voltage is applied from an external electric generator to the terminals, and an electric current flows in the thermoelectric module and produces a temperature difference between the hot side and the cold side based on the Peltier effect (Fig. 1b). If the direction of the current changes, the direction of heat transfer changes as well. Thereby, the Peltier cells can be used as heat pumps for both heating and cooling purposes. The development of thermoelectric materials is essential to improve the performance of the TEC applications, which is at present relatively limited [4].

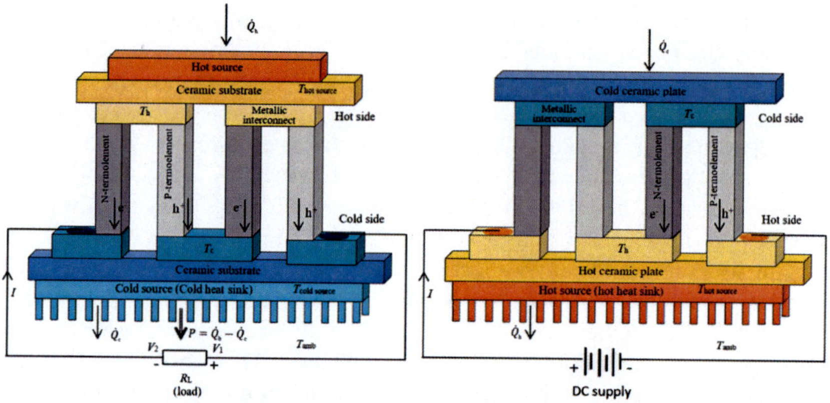

a. TEG structure b. TEC structure

Fig. 1. Sketch of TEG and TEC structures.

The parameters of the thermoelectric module considered in the sequel are:

- The Seebeck coefficient of the module α_{M}, in [V/K].
- The Thomson coefficient τ, in [V/K].
- The internal electrical resistance of the module R_{M}, in [Ω].
- The thermal conductance of the module K, in [W/K]. When needed, different representations are used for the thermal conductance of the thermoelectric legs K_{M}, and the parasitic thermal conductance of the non-ideal ceramic plates K_{C}.

The module parameters are in general functions of the area of the plates and of the thermoelectric leg length [2]. Moreover, T_{h} and T_{c} are the temperatures of the hot and cold plates of the thermoelectric module, respectively.

For a thermoelectric module, considering the input current I, the heat fluxes $\dot{Q}_h$ at the hot side and $\dot{Q}_c$ at the cold side are expressed as:

$$\dot{Q}_h = \alpha_{T_h} T_h I + \frac{1}{2} R_M I^2 - K(T_h - T_c) - \frac{1}{2} \tau I \tag{1}$$

$$\dot{Q}_c = \alpha_{T_c} T_c I - \frac{1}{2} R_M I^2 - K(T_h - T_c) + \frac{1}{2} \tau I \tag{2}$$

1.3 TEG Equivalent Circuit Modelling

The basic circuit modelling of the TEG couples the thermal model with an electrical model represented with the Thevenin equivalent circuit (i.e., open voltage source in series with an internal resistance) seen from the load terminals (Fig. 2). The parameters are determined by considering the characteristics of the thermoelectric module [5].

The circuit model corresponds to the equation:

$$V = V_{TEG,0} - R_{TEG} I \tag{3}$$

where:

$$V_{TEG,0} = \alpha_{TEG}(T_h - T_c) \tag{4}$$

$$\alpha_{TEG} = \frac{\alpha_M K_C}{2K_M + K_C} \tag{5}$$

$$R_{TEG} = R_M + \frac{\alpha_M^2 (T_h + T_c)}{2K_M + K_C} \tag{6}$$

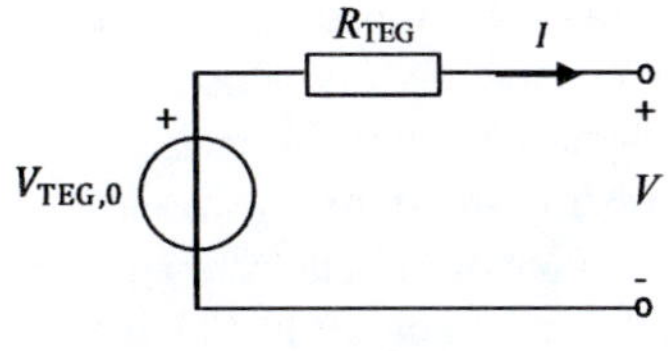

Fig. 2. TEG basic equivalent circuit.

At the TEG output terminals, in the simplest case, there is a load resistance that represents the equivalent resistance of the circuit connected. If the load resistance is equal to R_{TEG} (i.e., the load is matched to the generation system), the power transferred to the load is maximum and the voltage $V_{TEG,0}$ is double than the voltage V at the load terminals, while if the load resistance is higher than R_{TEG} the power is supplied with a load voltage closer to $V_{TEG,0}$. The maximum power reduction in the presence of internal modules in the TEG with different temperature gradients is addressed in [6] by writing the maximum power point condition for multiple modules. To determine R_{TEG}

for calculating the maximum power conditions, the contact resistance cannot be ignored [7].

To try and obtain the maximum power transfer to the load, the maximum power point tracking (MPPT) is applied by changing the impedance seen at the TEG terminals to match the TEG internal resistance (which varies with the temperature). As a control system, in a stand-alone system the MPPT is supplied by the electrical output of the TEG and is connected to the load (Fig. 3).

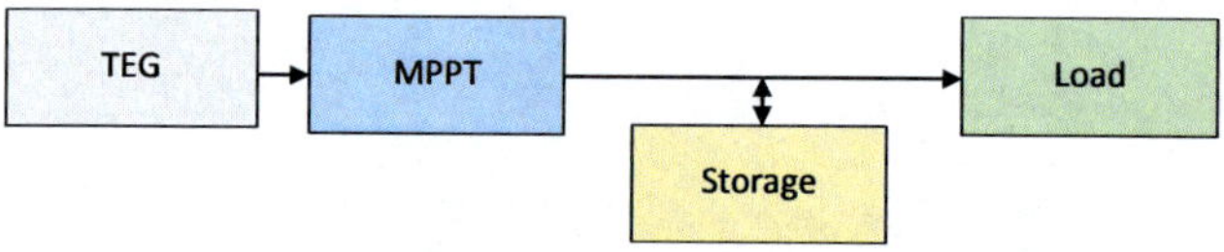

Fig. 3. TEG with MPPT and storage.

For the circuit modelling of the TEG with MPPT, combined thermal-electric coupled models can be used and analytical expressions could be searched for, borrowing the criteria used for constructing the equivalent circuits used for photovoltaic (PV) systems, based on the maximum power transfer conditions [8] in systems with MPPT [9], up to the more detailed modelling used in systems that include a local storage [10].

2 Energy Harvesting Solutions

2.1 Power Supply to Sensors

The extended usage of wireless sensors scattered in the buildings to gather many data raises the issue of how to provide an adequate long-term power supply for these sensors. Since no wired connection is needed to send the signals, the possibility of avoiding the use of power cables for providing the power supply to the sensors is highly advantageous and leads to avoiding the significant costs of installing the power cables. A typical solution is the use of batteries to supply the sensors. However, batteries need to be replaced with sufficient regularity to avoid their lack of operation, again with the related costs. For this purpose, energy harvesting solutions, in which the power supply to the sensors is guaranteed by the characteristics of the environment, are particularly useful. In the presence of batteries, the addition of energy harvesting solutions can provide further inputs, extending the service life of the batteries and guaranteeing power and energy backup to the sensors.

The energy harvesting solutions perform energy conversion from primary energy to electricity. In particular, the main energy sources for energy harvesting used in buildings, among others [11] are as follows:

- *Airflow energy*, with the conversion of the kinetic energy of an air current (e.g., available in a ventilation system) into electricity, or with the generation of vibrations by aerodynamic forces, in turn converting the vibrations into electricity by using transducers of piezoelectric or electromagnetic type.

- *Solar energy*, by exploiting the photoelectric effect to obtain the direct conversion of the energy coming from a light source into electricity.
- *Thermal energy*, by exploiting the Seebeck effect in a thermoelectric material, with output voltage depending on the temperature difference to which the thermoelectric generator is subject between the hot and cold sides.

The important limitation of the use of energy harvesting technologies comes from the power and energy needed to supply the sensor, which depends on the type of sensor. A key aspect is the source of the primary energy, which can be:

1. *Provided by the application itself during operation.* In this case, it is important to consider an internal storage system to cover the energy needed when the application is not working and at its startup. Moreover, the power needed to supply the sensors should be relatively low, to avoid significant efficiency reductions.
2. *Provided externally.* In this case, the availability of the source of energy must be guaranteed, also considering the external conditions.

For thermoelectric energy harvesting, it is important to apply high-temperature differences to the hot and cold sides of the TEG, to provide sufficient voltage for supporting the power output needed for supplying the sensors. Possible limits in the TEG dimensions due to the specific application affect the effectiveness of the voltage supply. Then, the voltage and the power depend on the characteristics of the sensor and its circuits, which form the electrical load of the TEG [12]. Operation at the maximum power point can be assisted by the presence of the MPPT. As a control system, the MPPT is supplied by the electrical output of the TEG and is connected to the sensor.

If the power supply cannot be provided with continuity, internal storage is needed to maintain sufficient energy stored for supplying the sensor at any time. To avoid the use of batteries, a supercapacitor can be used as an internal storage system. The capacitor voltage changes in the charging and discharging cycles, where charging occurs when the primary energy source is present, and discharging occurs in the absence of primary energy, or when the sensor is in operation. Moreover, the storage system has its own losses, that must be compensated by the power supply system.

In the analysis presented in [13], referring to supplying sensors in a Heating Ventilation and Air Conditioning (HVAC) system, the power requested by the wireless sensors is too high to consider an effective application of thermoelectric energy harvesting. In other applications, the temperature difference across a vacuum insulation panel for buildings is sufficient to provide self-powered pressure sensing with an ultralow-power sensor that operates at about 100 μW [14]. Energy harvesting may be based on the hot and cold pipes as the sources of the temperature difference used by the TEG for supplying power to an acoustic sensor and transmitting the monitored signals in wireless mode [15].

The TEG can be used in building envelopes in which temperature differences appear, for example, with an external surface reached by the solar radiation and an internal surface at different temperatures because of the air conditioning of the internal space. The TEG output (voltage and power) must be sufficient to supply the wireless sensors located nearby the TEG. The results shown in [16] indicate that the power that can be generated can be sufficient to supply sensors in wireless sensor networks that request power of about 12–16 mW. For providing a suitable supply to sensors, the output voltage has to be at least some volts. Thereby, more TEGs have to be connected in series, until the voltage reaches the requested value.

Passive energy harvesting can be obtained from building envelopes in which the thermal input is given by the waste heat stored into or released from a phase change material (PCM) located in the envelope. The best positions for installing the TEG-PCM blocks are in places well-heated during the day and well-cooled by the atmosphere during the night, where there are sensors to be supplied. In the experiments carried out in [17], the TEG-PCM combination has been able to provide 10 mW average electric power. More TEG-PCM blocks have to be connected in series to reach the voltage requested by the sensors. The economic profitability of the TEG-PCM solution has to be assessed in specific cases.

2.2 Building Air Heating

A thermoelectric module located between two air-to-air heat exchangers at the external ambient (cold) side and at the interior building side, respectively, forms a thermoelectric heat pump.

The electrical power needed to supply the thermoelectric heat pump is

$$P_{\mathrm{el}} = Q_{\mathrm{h}} - Q_{\mathrm{c}} = \alpha_{T_{\mathrm{h}}} T_{\mathrm{h}} I + R_{\mathrm{M}} I^2 - \tau I \tag{7}$$

Also considering the electrical power P_{fan} needed to supply the fans for air circulation [18], the coefficient of performance (*COP*) of the thermoelectric heat pump is

$$COP = \frac{Q_{\mathrm{c}}}{P_{\mathrm{el}} + P_{\mathrm{fan}}} \tag{8}$$

2.3 Building Integrated Photovoltaic Systems

The thermoelectric devices can be integrated into solutions by building integrated photovoltaic systems (BIPV) to enhance electricity production. These solutions are also denoted as active building envelopes [19].

A thermoelectric module can be integrated into a photovoltaic system to enhance the overall electrical efficiency η_{el} of the resulting system, expressed as ([20], where a thermoelectric cooler is indicated, even though the output from the thermoelectric module is electricity):

$$\eta_{\mathrm{el}} = \eta_{\mathrm{PV}} + \frac{\eta_{\mathrm{TE}} U_{\mathrm{TE}} (T_{\mathrm{TE,top}} - T_{\mathrm{TE,bottom}})}{G} \tag{9}$$

where η_{PV} and η_{TE} are the electrical efficiencies of the PV module and of the thermoelectric module, respectively, U_{TE} is the overall heat transfer coefficient from the top-end to the bottom-end of thermoelectric module in [W/m^2/K], G is the solar irradiance in [W/m^2], while $T_{\mathrm{TE,top}}$ and $T_{\mathrm{TE,bottom}}$ are the temperatures at the top-end and at the bottom-end of the thermoelectric module, respectively.

To further improve the effectiveness of BIPV-TEG systems, the combination with phase change materials (PCMs) into a PV-TEG-PCM solution (Fig. 4) offers the advantages of reducing the temperature of the solar cell and providing additional electricity generation from the TEG [21]. The benefits can be seen in the day and night periods:

- During the day, the solar radiation reaches the solar cell and generates a voltage through the photovoltaic effect, also depending on temperature and wind effects. Meanwhile, the temperature of the solar cell increases, and produces waste heat. The temperature difference between the two surfaces where the TEG is located produces further electricity by exploiting the Seebeck effect. In the PCM, when the melting temperature is reached there is a phase change from solid to liquid, maintaining the PCM temperature close to the melting temperature, thus sustaining the effectiveness of the TEG. The latent heat is absorbed in the PCM, so that the PCM operates as a cooling source and increases the electricity produced by the photovoltaic system and the TEG.
- During the night, the heat stored in the PCM is transferred through the TEG to the solar cell, with electricity generation from the TEG. Thereby, there is also continuous electricity generation during the night. The melting temperature of the PCM and the PCM layer size are key factors for properly sizing the combined system.

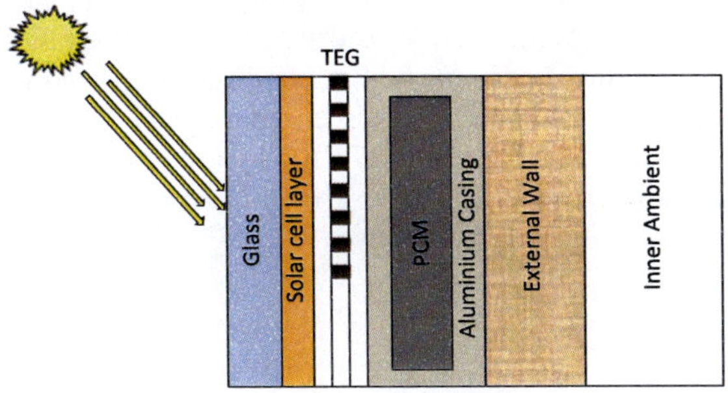

Fig. 4. PV-TEG-PCM solution.

3 Cooling Solutions

3.1 Building Envelope Cooling

The thermoelectric modules can be used in building envelopes to reduce the thermal losses or gains of the envelope. Active building walls are constructed by integrating thermoelectricity into the envelopes, in such a way that the heat flow can provide heating or cooling. A ventilated active thermoelectric envelope has been proposed in [22], with a double-skin facade formed by two opaque layers. A ventilation channel is between the layers, where air flows due to the action of fans. In general, the number and location of the thermoelectric modules are design parameters for active building walls. From the results presented in [23], better solutions are obtained by using small thermoelectric modules located close to each other.

3.2 Solar Thermoelectric Cooling

In principle, a TEC could be used to transfer heat out of the envelope by using the electrical energy provided by a photovoltaic system as the input to the TEC that operates as a heat pump [24]. In an active building wall, the photovoltaic system converts solar radiation into electricity, and the TEC converts electricity into cooling [25]. The thermoelectric devices can change the thermal heat flow based on the direction of the input current. In this way, with the same device it is possible to produce the cooling of the indoor building spaces at warm temperatures and to heat the internal spaces of the building at cold temperatures. Thermoelectric cooling can be useful when there is limited possibility of air passage on the back of the photovoltaic modules, the PV cell temperature could become relatively high [26].

On the energy efficiency point of view, thermoelectric cooling applications are of interest, as the period of time normally with the largest request of cooling is around noon, which is also the period in which there is the largest photovoltaic production. This makes a good potential match between generation and cooling demand. In addition, the TEC does not need to be reached by an electricity distribution system. A storage system (or a backup connection with the grid) is needed to cover situations in which there is no solar energy available. Some prototypes developed have been described in [19], in which however the thermoelectric module is supplied by an external source of electricity. Layout considerations concerning the use of thermoelectric modules connected in series or in parallel are illustrated in [27]. Appropriate experimental verification is needed, to establish for which temperatures and temperatures ranges the application of thermoelectric devices is practically effective [28].

3.3 Building Air Cooling

The application of thermoelectric air cooling (TEAC) devices is still limited by the low *COP* values, so that it has been developed only in laboratories [29]. The cooling performance of TEAC devices is better when the temperature differences are relatively small. This can also be obtained by improving heat dissipation, for example by adding external heat sinks. The theoretical and experimental analyses carried out in [30, 31] indicate that the development of new thermoelectric materials is needed to reach reasonable performance in comparison with vapor compression technologies.

4 Personal Thermal Management

The goal to establish satisfactory conditions for the thermal comfort of the individuals can be reached by creating a local thermal environment around the human body, as a partial alternative to heating or cooling all the space inside the building. For this purpose, the use of solutions leading to personal thermal management has the potential for reducing the energy needed in the building [32].

The solutions for personal thermal management can be categorized into:

- *External solutions*: create a comfortable environment around the human body, with personal heaters, ventilation systems and air conditioners, or thermally controlled chairs, exploiting different forms of heat transfer.
- *Wearable solutions*: consist of thermoregulated clothing made of different materials, in some cases with the integration of water circulation systems or fans. However, the size and weight of these solutions could limit their application in normal activities in buildings. Wearable solutions are also partitioned into *passive* devices, which need no energy input from external sources (e.g., phase change materials), *active* devices, which require energy input from external sources, and hybrid devices that combine the technologies applied in passive and active devices.

Wearable thermoelectric devices can be applied in heating and cooling modes. The performance can be tested by measuring the energy consumption in thermal manikins [33]. The advantages of light weight, flexibility and absence of noise of thermoelectric modules are particularly useful for wearable applications [34]. The thermoelectric modules can be applied in contact with the human body, e.g., in a cooling vest [35], or can supply power to a micro-blower that heats up or cools down the air from the ambient, sending the air to different parts of the human body through a system of small tubes [36]. Flexible and long thermoelectric fibres have been developed to enable covering any type of curves surface [37]. For wearable technologies, especially the ones in contact with the human body, an ongoing trend is to search for bio-based thermoelectric materials, which have less impact on the human body [38].

5 Conclusions

Applications of thermoelectricity have intrinsic advantages even over more efficient technologies. Thermoelectric systems are providing interesting technological solutions in buildings and for their occupants. This paper has summarized some emerging concepts, with a view on possible solutions for energy harvesting, and for heating and cooling purposes. The major advantages emerge in some building integrated solutions, also in conjunction with the use of phase change materials, and in supplying many sensors locally with no need to supply all sensors with wired connections. In other cases, such as in cooling applications, significant improvement of some properties of the thermoelectric materials should occur to make thermoelectric systems effective. A particular aspect of thermoelectric systems is their good performance with small temperature differences. Interesting solutions can be found, also improving the thermal environment close to the individuals and reducing consumption from the overall energy system.

References

1. Enescu, D. (ed.): Green Energy Advances. IntechOpen, London, UK (2019)
2. Rowe, D.M.: Handbook of Thermoelectrics. CRC Press, Boca Raton, FL (1995)
3. Bell, L.E.C.: Heating, generating power, and recovering waste heat with thermoelectric systems. Science **80**(321), 1457–1461 (2008)
4. Zhao, D., Tan, G.: A review of thermoelectric cooling: materials, modeling and applications. Appl. Therm. Eng. **66**(1–2), 15–24 (2014)
5. Kim, S.: Analysis and modeling of effective temperature differences and electrical parameters of thermoelectric generators. Appl. Energy **102**, 1458–1463 (2013)
6. Chen, M., Gao, X.: Theoretical, experimental and numerical diagnose of critical power point of thermoelectric generators. Energy **78**, 364–372 (2014)
7. Pennelli, G., Dimaggio, E., Macucci, M.: Electrical and thermal optimization of energy-conversion systems based on thermoelectric generators. Energy **240**, 122494 (2022)
8. Adak, S., Cangi, H., Yilmaz, A.S.: Thevenin equivalent of solar PV cell model and maximum power transfer. In: 2021 International Conference ICECCE. Kuala Lumpur, Malaysia (2021)
9. Turhan, M., Dai, B., Yildirim, D.: Analytical MPPT solution using Thevenin approach for solar panels. Eurocon 2013, pp. 803–808. Zagreb, Croatia (2013)
10. Bauomy, M.F., Gamal, H., Shaltout, A.A.: Dynamic modeling of DC nanogrid local branch using enhanced PV and third order battery models. In: 2016 18th European conference on power electronics and applications (EPE'16 ECCE Europe). Karlsruhe, Germany (2016)
11. Hidalgo-León, R., Urquizo, J., Macias, J., Siguenza, D., Singh, P., Wu, J., Soriano, G.: Energy harvesting technologies: analysis of their potential for supplying power to sensors in buildings. In: IEEE Third Ecuador Technical Chapters Meeting. Cuenca, Ecuador (2018)
12. Milić, D., Prijić, A., Vračar, L., Prijić, Z.: Characterization of commercial thermoelectric modules for application in energy harvesting wireless sensor nodes. Appl. Therm. Eng. **121**, 74–82 (2017)
13. Schachner, S., Sauter, T.: Comparison of energy harvesting concepts for heating, ventilation and air conditioning systems. In: IECON 2018—44th Annual Conference of the IEEE Industrial Electronics Society, pp. 6235–6240. Washington, DC, USA (2018)
14. Yun, M., Ustun, E., Nadeau, P., Chandrakasan, A.: Thermal energy harvesting for self-powered smart home sensors. IEEE MIT URTC, Cambridge, MA (2016)
15. Yajima, T., Tanaka, K., Yazawa, K.: Thermoelectric on-spot energy harvesting for diagnostics of water service pipelines. In: 17th IEEE Intersociety Conference on Thermal and Thermomechanical Phenomena in Electronic Systems, pp. 881–888. San Diego, CA (2018)
16. Al Musleh, M., Topriska, E.V., Jenkins, D., Owens, E.: Thermoelectric generator characterization at extra-low-temperature difference for building applications in extreme hot climates: experimental and numerical study. Energy Build. **225**, 110285 (2020)
17. Byon, Y.S., Jeong, J.W.: Annual energy harvesting performance of a phase change material-integrated thermoelectric power generation block in building walls. Energy Build. **228**, 110470 (2020)
18. Allouhi, A., Boharb, A., Jamil, A., Msaad, A.A., Benbassou, A., Kousksou, T.: Simulation of a thermoelectric heating system for small-size office buildings in cold climates. In: 2015 3rd international renewable and sustainable energy conference. Marrakech, Morocco (2015)
19. Xu, X., Van Dessel, S., Messac, A.: Study of the performance of thermoelectric modules for use in active building envelopes. Build. Environ. **42**, 1489–1502 (2007)
20. Dimri, N., Tiwari, A., Tiwari, G.N.: Comparative study of photovoltaic thermal (PVT) integrated thermoelectric cooler (TEC) fluid collectors. Renew. Energy **134**, 343–356 (2019)
21. Ko, J., Jeong, J.W.: Annual performance evaluation of thermoelectric generator-assisted building-integrated photovoltaic system with phase change material. Renew. Sustain. Energy Rev. **145**, 111085 (2021)

22. Ibañez-Puy, M., Martín-Gómez, C., Bermejo-Busto, J., Sacristán, J.A., Ibañez-Puy, E.: Ventilated active thermoelectric envelope (VATE): analysis of its energy performance when integrated in a building. Energy Build. **158**, 1586–1592 (2018)
23. Van Dessel, S., Foubert, B.: Active thermal insulators: finite elements modeling and parametric study of thermoelectric modules integrated into a double pane glazing system. Energy Build. **42**, 1156–1164 (2010)
24. van Sark, W.G.J.H.M.: Feasibility of photovoltaic–thermoelectric hybrid modules. Appl. Energy **88**, 2785–2790 (2011)
25. Liu, Z.B., Zhang, L., Gong, G.C., Li, H.X., Tang, G.F.: Review of solar thermoelectric cooling technologies for use in zero energy buildings. Energy Build. **102**, 207–216 (2015)
26. Choi, J.S., Ko, J.S., Kang, S.J., Jang, M.G., Back, J.W., Kim, D.K., Chung, D.K.: Development of thermoelectric cooling system for BIPV module. In: Proceedings of 31st International Telecommunications Energy Conference (INTELEC 2009) (2009)
27. Hagenkamp, M., Blanke, T., Döring, B.: Thermoelectric building temperature control: a potential assessment. Int. J. Energy Environ. Eng. **13**(1), 241–254 (2021). https://doi.org/10.1007/s40095-021-00424-x
28. Enescu, D., Spertino, F.: Applications of hybrid photovoltaic modules with thermoelectric cooling. Energy Procedia **111**, 904–913 (2017)
29. Yilmazoglu, M.Z.: Experimental and numerical investigation of a prototype thermoelectric heating and cooling unit. Energy Build. **113**, 51–60 (2016)
30. Liu, Z.B., Zhang, L., Gong, G.C., Luo, Y.Q., Meng, F.F.: Experimental study and performance analysis of a solar thermoelectric air conditioner with hot water supply. Energy Build. **86**, 619–625 (2015)
31. Duan, M., Sun, H., Lin, B., Wu, Y.: Evaluation on the applicability of thermoelectric air cooling systems for buildings with thermoelectric material optimization. Energy **221**, 119723 (2021)
32. Ma, Z., Zhao, D., She, C., Yang, Y., Yang, R.: Personal thermal management techniques for thermal comfort and building energy saving. Mater. Today Phys. **20**, 100465 (2021)
33. Zhao, D., et al.: Personal thermal management using portable thermoelectrics for potential building energy saving. Appl. Energy **218**, 282–291 (2018)
34. Ren, W., Sun, Y., Zhao, D., Aili, A., Zhang, S., Shi, C., Zhang, J., Geng, H., Zhang, J., Zhang, L., Xiao, J., Yang, R.: High-performance wearable thermoelectric generator with self-healing, recycling, and Lego-like reconfiguring capabilities. Sci. Adv. **7**(7), eabe0586 (2021)
35. Hong, Y.G.S., Seo, J.K., Wang, J., Liu, P., Meng, Y.P., Xu, S., Chen, R.: Wearable thermoelectrics for personalized thermoregulation. Sci. Adv. **5**, aaw0536 (2019)
36. Lou, L., et al.: Thermoelectric air conditioning undergarment for personal thermal management and HVAC energy saving. Energy Build. **226**, 110374 (2020)
37. Zhang, T., et al.: High-performance, flexible, and ultralong crystalline thermoelectric fibers. Nano Energy **41**, 35–42 (2017)
38. Zhang, L., Shi, X.L., Yang, Y.L., Chen, Z.G.: Flexible thermoelectric materials and devices: From materials to applications. Mater. Today **46**, 62–108 (2021)

Natural and Recycled Stabilizers for Rammed Earth Material Optimization

Giada Giuffrida(✉), Vincenzo Costanzo, Francesco Nocera, Massimo Cuomo, and Rosa Caponetto

Department of Civil Engineering and Architecture, University of Catania, 64 S. Sofia Street, 95125 Catania, Italy
giada.giuffrida@unict.it

Abstract. The interest in more environmentally sustainable building materials has led to the rediscovery of ancient and natural-based ones such as rammed earth (hereinafter RE). Nowadays, RE material is stabilized by the addition of small amount of cement in the mixture that, however, reduces its recyclability and poses concerns in terms of adverse environmental effects. According to a circular economy approach, the improvement of RE properties may be better carried out by the addiction of by-products from other production chains. In this vein, this study introduces a novel stabilization technique for RE materials that makes use of local soils, aggregates, fibers and waste materials, which allows for a significant improvement of their mechanical, physical and thermal properties.

Keywords: Rammed earth · Material characterization · Experimental analysis

1 Introduction

In the last decades small-embodied energy and low energetic costs in constructions have become the *leit motiv* of sustainable architecture, thus generating a keen interest in the rediscovery of traditional construction materials and techniques [1]. Among raw earth techniques, some of them like rammed earth (RE) are gaining momentum and undergoing strong modifications in their manufacturing process with the aim of improving their mechanical characteristics and speed up their production. In fact, traditional stabilization methods involved physical changes in the particles size distribution of natural soils quarried nearby the building sites, often integrated with the addition of finer fractions (e.g., clays) to improve the binding properties. Coarser fractions (e.g., gravels) were instead typically used to enhance the durability of the not-plastered walls and natural fibers to eventually boost the flexural properties of the composites.

Nowadays, RE is stabilized with binders like cement or lime, which are explicitly required by several laws and RE standards [2]. In this way, RE mixes easily manage to exceed minimum requirements for compressive strengths, e.g. 1.3 MPa as dictated in New Zealand [3] and 1.0 MPa as prescribed in Peru [4]. Notwithstanding these prescriptions, several authors have pointed out the ineffectiveness of cement-based stabilization from both the technical and environmental points of view [5].

J. Littlewood and R. J. Howlett (Eds.): SEB 2022, SIST 336, pp. 164–174, 2023.
https://doi.org/10.1007/978-981-19-8769-4_16

Indeed, in the last years, an intense research effort has been devoted at investigating alternative stabilization materials for raw earth mixes other than cement. Particularly, agricultural and non-agricultural waste materials have been used, as highlighted by Jannat et al. [6]. Natural by-products [7–18], rarely combined with chemical binders [7, 15, 17] or recycled synthetic materials [11, 18], have been proposed to this aim. Other works have focused on the use of industrial by-products [18–23], which are often accompanied by the use of binders like cement and lime. Only few works still maintained conventional stabilization methods based on the engineering of the soil and the addition of lime or Portland cement [24, 25].

It is remarkable that the addition of natural fibers such as wheat, barley, lavender straw, coconut coir, bagasse, date palm and banana fibers, wood aggregates, kenaf, olive waste, wool, cork, fruit shells in the earth matrix allows for a reduction of the composite's bulk densities while maintaining good mechanical performances [6–10, 13–17]. Furthermore, natural fibers enable a reduction of thermal conductivity and frequently an increase of water absorption percentages and equilibrium moisture content values [6]. On the other hand, earth materials stabilized with lime, gypsum or cement have higher densities, more brittle mechanic behavior and higher thermal conductivity [22–25], but lower water absorption percentages especially when filler materials are added [22, 23].

Based on the brief review concerning stabilization methods presented above, the present research aims at describing a novel stabilization method and testing methodology for rammed earth materials. It comprises a comprehensive investigation on rammed earth material mixes that employs several natural-based materials like volcanic sand and sisal fibers, and recycled ones (marble sawing waste namely) as stabilizers. The combined investigation of materials' mechanical, thermal and physical properties has allowed to optimize the proposed mixes and to highlight the advantages of the proposed stabilization method.

2 Material and Methods

2.1 Base Materials

The base material used for the experimental investigation comes from a quarry in Floridia, a city in the nearby of Syracuse (Sicily, Southern Italy), and from now on called "Floridia soil". The particle size distribution of Floridia Soil comprises 3% of gravel, 70% of sand, 17% of silts and 10 % of clays. The first part of the particle size curve (diameters comprised in the range 0.0075 mm $< d <$ 4 mm) is obtained through direct measurements by sieving, while the second part ($d <$ 0.0075 mm) is deduced indirectly through sedimentation analysis, in accordance with the ASTM D7928–17 methodology [26]. Floridia Soil is characterized by a comparable amount of clays ($d <$ 0.002 mm) and silts (0.002 $< d <$ 0.006 mm). The Atterberg limits [27] on the fine fraction of the soil are LL = 47.3%, PL = 30.68% and PI = 16.62%. The chosen soil thus presents an acceptable plastic index adapted to rammed earth construction [4], and is then combined with a local volcanic sand called "azolo" with diameters below 4 mm and with a filler derived from the sawing of marble. The chemical composition of azolo sand comprises 45.9% of SiO_2, 20.43% of Al_2O_3, 10.22% of CaO, 10% of Fe_2O_3, and other materials in lower percentages. Moreover, a marble sawing waste (MSW) provided by a local marble supplier is used as base material as well. This is a suspension of fine marble particles in water, generated during processing and polishing of marble slabs whose material composition—derived through a Xray diffraction analysis—confirms that the waste is mainly composed of calcite mineral ($CaCO_3$) and traces of plagioclase (Na,Ca)(Si,Al)$4O_8$. Marble sawing waste has been already proposed in combination with earthen mixes in [22, 23] as a component to reduce water absorption.

During the material testing campaign, MSW was dried and then grounded until a fine and impalpable dust was obtained. Natural fibers were also added to the mix, with the aim of improving the ductile behavior of the composite. Indeed, soils composed of finer sand particles and silts, such as Floridia soil, have more fiber-bond strength compared to coarse grained soils. Sisal fibers were then chosen; as known, they are extracted from a succulent plant belonging to the Agavaceae family, native to the Yucatán peninsula in Mexico and largely diffused in Sicily where it was traditionally used to make ropes, strings, baskets, carpets, and other handcraft items. In particular, agave filaments of average length of about 20 mm were used in the mix, whose specific weight was obtained by means of a precision balance and a standardized metal cylinder is about 612 kg/m^3.

2.2 Mix Design and Samples Manufacturing

Six RE mixes were tested, all of them containing Floridia soil with different aggregates, based on the results obtained in preliminary investigations [25]. Their composition is shown in Table 1. A control mix denoted as URE was prepared by mixing 50% of Floridia Soil and 50% of volcanic sand. In the FSRE and LSRE mixes, which had an almost equal amount of Floridia Soil and sand, sisal fibers in a percentage of 1% by weight and lime in a percentage of 5% were added, respectively. The remaining three mixes used the marble filler as partial substitution of sand: in the mix denoted as MSRE, no material was employed other than soil, sand and the marble filler. On the other hand, the mix called MLRE contained a 5% of lime while the mix denoted as MFRE included 1% of sisal fibers. A design density of 1930 kg/m^3 was used for the calculation of the mass of each mix.

Mechanical, thermal, and physical test were carried out on the six proposed mixes. To this purpose, eight cubic-shaped samples for each mix design with dimensions 150 * 150 * 150 mm were manufactured and used to perform an unconfined compressive strength test (3 samples), a capillary water absorption test (3 samples), and a thermal conductivity test (1 sample, 3 points of measure). In Fig. 1 are shown different phases of mechanical (a–b), physical (c–d) and thermal conductivity tests (e).

Sizes, manufacturing and curing processes were carried out in accordance with [3, 4]. The manufacturing protocol comprises: (1) the dry mixing of components with an electric mixer PROTTOL MXP 1000 ES; (2) the addition of the stabilizer (lime), the fiber (sisal) or the MSW; (3) the addition of an amount of water correspondent to the optimum moisture content [3] of the mix (respectively 10% and 13% of the total weight of the mix without and with fibers); (4) the weighting of the mix for manufacturing the sample; (5) the compaction of the sample in successive layers with a pneumatic tamper with air pressure 0.63 MPa, impact frequency $\geq$14 Hz and piston stroke 100 mm; (6) the flattening of the upper part of the sample; (7) immediate removal of the formwork once samples are compacted and regularized on the upper surface; (8) curing for at least 28-days (wet/dry curing for stabilized samples as [3]).

Table 1. Rammed earth mixes composition

Sample	Floridia Soil [%]	Sand [%]	Lime [%]	Sisal [%]	Marble sawing waste [%]
URE	50	50	0	0	0
FSRE	50	49	0	1	0
LSRE	50	45	5	0	0
MSRE	50	35	0	0	15
MLRE	50	35	5	0	10
MFRE	50	34	0	1	15

Note mix compositions are given as a percentage of the dry weight of the final rammed earth material

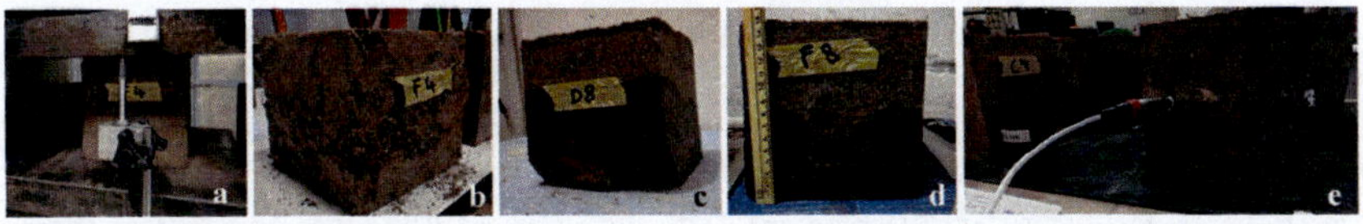

Fig. 1. Compressive strength tests (**a–b**); absorption test (**c–d**); thermal conductivity test (**e**).

2.3 Testing Methods

Samples were first visually checked to assess their compactness and the absence of any crack, while measurements were taken to ensure that no major shrinkage phenomena happened (volumetric shrinkage below 0.05% [3]).

Then, compressive strength test determined the behavior of the material with regard to compression stresses, i.e. the resistance offered by the material to vertical loads. The test was carried out on 3 samples for each mix, cured following the methodology outlined in [8] and using a mechanical press CONTROLS with a 100 KN load cell and a low initial load speed of 0.5 kN/s for samples without fibers addition. For the fiber-reinforced samples, a press with a 5MN load cell was used instead. In both cases, it was assumed that the compressive strength of the material was acceptable if it turned out to be greater than 1.3 and 1.0 MPa, respectively, according to the data found in specific raw earth standards [3, 4]. Young's Modulus was determined following the methodology indicated in [28] by testing the samples under load-unload cycles at one-third of their expected final resistance. To obtain the corresponding deformation at each stress, an electric transducer was used for measuring the strain in the center of the sample, and the Young's modulus calculated as the average of the secant slope of the unloading branch of the stress-strain curve.

An absorption test was also performed to establish the behavior of the material in relation to the phenomenon of capillary rising water. It was performed in accordance with [29], so rising heights were measured on 3 samples for each mix at predetermined time intervals over a 12 h period, placing the cubic sample on a perfectly smooth container. Then, water was poured to immerse the sample for a 3 cm a height; finally, the height reached by the liquid through capillary rise is measured at pre-established time intervals (every 15 min for the first three hours, every 30 min for the next three hours, every hour for the last six hours). At the end of the 12 h, the maximum frontier of capillary rise is measured. The test is considered passed if the height of the water does not exceed the limits provided in [29], i.e. 11.8 cm after 3 h, 13.6 cm after six hours, 14.5 cm after twelve hours. In any case, at the end of the test, the sample must be almost intact and with slight loss of material at the base.

Moisture dependent thermal conductivity and specific heat capacity of mixes were also assessed. Once samples were prepared, they were immediately removed from the formworks and allowed to dry slowly to avoid cracking. Indoor air conditions in the laboratory during the curing were 20 °C and 60% relative humidity, respectively. The samples cured for at least one month; once cured, they were put in a climatic chamber at a constant temperature of 20 °C and different percentages of relative humidity, namely 15% (dry condition), 30%, 50% and 70% and weighted periodically until the difference between two consecutive measurements 24 h apart was less than 0.1% [24]. Once removed from the climatic chamber, samples were weighted for each of the above-mentioned environmental conditions (T = 20 °C and RH = 15%, T = 20 °C and RH = 30%, T = 20 °C and RH = 50%, T = 20 °C and RH = 70%) and four set of thermal conductivity measurements per mix were realized. This procedure allowed for recording the moisture accumulation inside the sample. Thermal conductivity measurements were carried out on the samples using a ThermTest conductivity meter able to measure both thermal conductivity and thermal resistivity according to ASTM D5334 standard [30, 31]. The specific heat capacity tests were finally performed on representative crushed samples of the six RE samples [8]. Measurements were carried out with a Shimadzu DSC-60 apparatus, while enthalpy and temperature calibrations were made according to the procedure suggested by the manufacturer.

3 Results and Discussion

The results of the experimental campaign are summarized in Table 2, where it is possible to appreciate that all the resistances exceed the limits of 1.3 MPa and 1.0 prescribed by [3, 4]. Fiber-reinforced mixes (FSRE and MFRE) have the highest compressive strengths even if their densities are not the highest, while the mixes without fibers (URE, LSRE, MSRE and MLRE) show a clear increase of compressive strength values as the dry density increases. Young modulus values are found to be about 200 MPa, consistently with data found in the literature. The samples with highest compressive strength values (URE, FSRE, MSRE and MFRE) have also the highest Young's Modulus values, all in the range between 254 and 290 MPa.

Results concerning the material behavior in terms of capillary rising water revealed that mixes using marble sawing waste (MSRE) have the lowest average water absorption compared to all the other ones, followed by the fiber reinforced sample with marble sawing waste (MFRE) and the lime stabilized one (LSRE). Except for the unstabilized URE sample, all the samples can be considered acceptable with respect to the prescriptions set out in [29].

As regards thermal conductivity, at a constant temperature of 20 °C thermal conductivity increases when the relative humidity increases. The best performance in this case is obtained for the MLRE sample, followed by the LSRE sample and the URE sample. The fiber reinforced mixes, which showed more compactness, have also higher thermal conductivity values. Specific heat capacity is consistent for all the examined mixes and tendentially exceeding 1000 J/kg K, which entails good inertial properties.

Further insights on the experimental outcomes can be gathered from Fig. 2 that compares the average dry density values and average unconfined compressive strengths achieved by the samples tested in this research against those reported in previous studies regarding raw earth material optimization. As it can be seen, samples using cement or lime (corresponding to the grey hatch area of the graph) and with compressive strengths superior to 5.00 MPa are difficult to obtain without the use of those materials for stabilization purposes [32].

Table 2. Results of Rammed earth materials characterization

Properties	URE	FSRE	LSRE	MSRE	MLRE	MFRE
Dry density [kg/m^3]	2055	1944	1939	2043	1870	1989
Compr. Strength [MPa]	3.95	5.26	1.72	3.65	1.36	6.69
Young Mod. [MPa]	285	254	160	290	184	264
Avg. Water absorp. [%]	20.4	7.6	6.7	5.9	9.2	6.9
$\lambda_{dry\ (15\%RH)}$ [W/mK]	0.521	0.509	0.508	0.560	0.447	0.508
$\lambda_{30\%RH}$ [W/mK]	0.529	0.530	0.515	0.601	0.462	0.540
$\lambda_{50\%RH}$ [W/mK]	0.532	0.556	0.518	0.614	0.470	0.582
$\lambda_{70\%RH}$ [W/mK]	0.559	0.576	0.551	0.627	0.474	0.588
c [J/kg K]	1045	1760	907	690	1430	1000

On the other hand, if one considers researches that made use only of alternative stabilization materials like natural fibers, geopolymers, recycled aggregates, or by-products without using binders that interfere with the recyclability properties of the raw earth material, the outcomes reported in this study show comparable average results. As an example, if compared to the average compressive strength value (3.07 MPa) calculated from previous studies not using cement-based or lime-based stabilizations [9, 10, 12, 13, 16], the six samples analyzed in this study presented an improvement of 22.8%, thus exceeding the lower limits indicated by the abovementioned standards by 190% [3] and 277% [4], respectively. Only [9, 13], both using natural fibers to improve compressive strength, reported higher values.

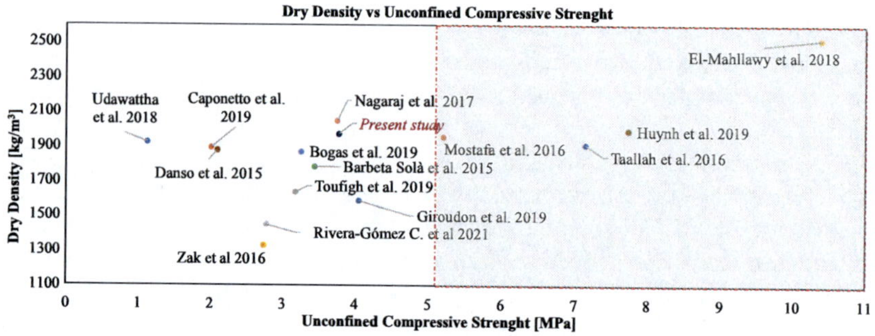

Fig. 2. Avg. Dry density vs compressive strength results from present and previous studies

As for the Young's modulus, this is a material property seldom investigated in previous studies [9, 13, 15, 18, 19]. The average Young's modulus value achieved by the six samples, equal to 282 MPa, is 14.9% lower than the values reported in these studies.

Figure 3 shows instead a comparison between the average dry thermal conductivity and the average dry density. Also in this case, the experimental results achieved in this study are located in the middle of the graph, between typical values of conventional stabilized raw and RE material thermal conductivity (ranging from 0.80 to 1.00 W/m K) and lower values (ranging from 0.30 to 0.40 W/m K) which are typical of lightweight earth materials. The average thermal conductivity values for the investigated studies [7, 9, 11, 13, 14, 16–18, 20, 24] is 0.844 W/ m K, which is 39.7% higher than the average results of the current study (0.509 W/m K). It is worth to be noted that for this comparison, we used the values of conductivity calculated at 15% of RH, being these ones the nearest values to dry thermal conductivity.

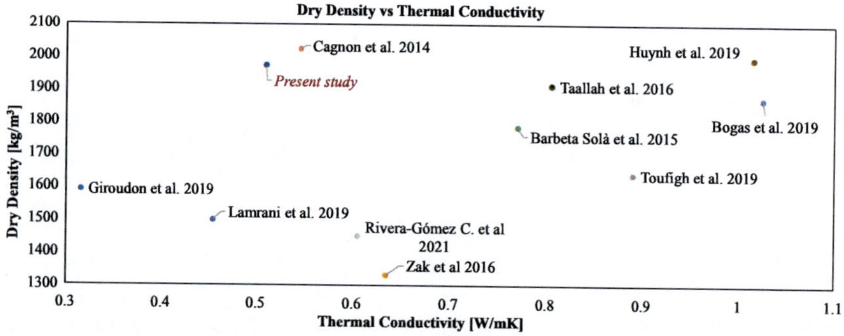

Fig. 3. Avg. Dry density vs dry thermal conductivity from present and previous studies

Also specific heat capacity is a property that is seldom assessed in the literature. The comparison of the results presented in this study with the few references available [11, 19, 24] shows really encouraging results, being the average measured value of the six samples equal to 1098 J/ kg K, near to the average specific heat capacity of 1167 J/ kg K reported in previous researches.

Finally, regarding the percentage of water absorbed for capillarity, an average value of 9.83% was calculated from previous studies [10–12, 14, 19–21, 23, 25]. Compared to this average value, this research showed a reduction of 3.89%. In this case, it is apparent the positive effect of using the marble sawing waste filler, which helped keeping this absorption relatively low. Table 3 summarises the positive effects of using combined stabilization strategies for RE materials.

Table 3. Comparison of average results from this study and previous ones

	Compressive strength [MPa]	Young modulus [MPa]	Thermal conductivity [W/mK]	Specific heat capacity [J/kgK]	Water absorption [%]
Literature avg. Value	3.069	282	0.845	1167	9.83
Present study	+22.8%	−14.9%	+39.7%	−5.93%	+3.89%

Note improvements (positive increment) and worsening (negative increment) of present study respect to previous ones

4 Conclusions

This contribution focused on the use of alternative stabilization methods for RE materials to enhance their overall mechanical and thermal properties. According to the circular economy approach, this study focused on combined stabilization methods for RE using natural, recyclable materials, and by-products from other production chains to improve different material properties. Natural sisal fibers were adopted to improve the mechanical strengths, thus enhancing the ductility of materials, while marble sawing waste was used to reduce the issue of capillary water absorption. Experimental activities proved that stabilization strategies using fibers and marble sawing waste allowed reaching the following results:

- Compressive strength up to 5.00 MPa for the FSRE and MFRE samples adopting sisal fibers in the mixes; use of marble sawing waste does not significantly alter the mechanical properties of the samples. Lime stabilization does not seem to produce any improvement.
- Young's modulus values quite uniform between the unstabilized and stabilized samples, lying in the range of 160–290 MPa.
- Lower capillary absorption water values for the LSRE lime stabilized samples and for the MSRE and MFRE samples, ranging from 5.9% to 6.9%.
- Dry thermal conductivity values in a range of 0.447 to 0.560 W/m K (at 15% RH). Moisture-dependent thermal conductivity values in the range of 0.46–0.60 W/m K at 30% RH, 0.47–0.61 W/m K at 50% RH and 0.47–0.63 W/m K at 70% RH. Samples stabilized with marble sawing waste seem to be more compact and their thermal conductivity values are the highest. Samples stabilized with natural fibers (FSRE,

MFRE) have thermal conductivity values near to the unstabilized ones but with lower densities.

- Specific heat capacity values between 690 and 1760 J/kg K, with an average value of 1098 J/kg K that is consistent with the data found in the literature.

These results are important for the upcoming generation of RE materials because they demonstrate the importance of conjugating the environmental-friendly vision of bio-based building technologies with the aim of reducing waste from construction sector, both in the manufacturing and end-of-life phases, with positive effects for the environment and local economies. Future works will focus on more in-depth material characterizations, concerning both mechanical (flexural strength, ductility properties of the material) and hygrothermal (vapor permeability, moisture buffer value) properties, and on the evolution of material performances during the lifetime of the building components (durability).

Acknowledgements. The authors acknowledge the Guglielmino Soc. Coop. For providing the base materials of the campaign and the training, Eng. Daniele Calderoni to encourage the collaboration with *Amato Marmi* company (provider of the MSW) and for the participation in the experimental campaign, Ph.D. Antonio Lo Presti for the MSW X-ray diffraction, Prof. Ignazio Blanco for the DSC measurements. The authors acknowledge the Official Material Testing laboratory and the Energetic Sustainability and Environmental control (SECA) laboratory of University of Catania.

References

1. Hall, M.R., Lindsay, R., Krayenhoff, M.: Modern earth buildings: materials, engineering, construction and applications. Woodhead Publishing Series in Energy, Philadelphia (2012)
2. Giuffrida, G., Caponetto, R., Cuomo, M.: An overview on contemporary rammed earth buildings: technological advances in production, construction and material characterization. IOP Conf. Ser. Earth Environ. Sci. **296**, 012018 (2019)
3. Standard New Zealand 4298: 2020. Materials and construction for earth buildings, Standard New Zealand, Wellington (2020)
4. NTE E 080—Diseño y Construcción con Tierra Reforzada. Perú: Ministerio de Vivienda, Construcción y Saneamiento, Lima (2017)
5. Damme, H.V., Houben, H.: Earth concrete. Stabilization revisited. Cem. Concr. Res. **114**, 90–102 (2018)
6. Jannat, N., Hussien, A., Abdullah, B., Cotgrave, A.: Application of agro and non-agro waste materials for unfired earth blocks construction: a review. Constr. Build. Mater. **254**, 119346 (2020)
7. Zak, P., Ashour, T., Korjenic, A., Korjenic, S., Wu, W.: The influence of natural reinforcement fibers, gypsum and cement on compressive strength of earth bricks materials. Constr. Build. Mater. **106**, 179–188 (2016)
8. Giuffrida, G., Detommaso, M., Nocera, F., Caponetto, R.: Design optimisation strategies for solid rammed earth walls in mediterranean climates. Energies **14**, 325 (2021)
9. Giroudon, M., Laborel-Préneron, A., Aubert, J.E., Magniont, C.: Comparison of barley and lavender straws as bioaggregates in earth bricks. Constr. Build. Mater. **202**, 254–265 (2019)

10. Danso, H., Martinson, D.B., Ali, M., Williams, J.B.: Physical, mechanical and durability properties of Soil building blocks reinforced with natural fibers. Constr. Build. Mater. **101**(1), 797–809 (2015)
11. Huynh, T.P., Nguyen, T.C., Do, N.D., Hwang, C.L., Bui, L.A.T: Strength and thermal properties of unfired four-hole hollow bricks manufactured from a mixture of cement, low-calcium fly ash and blended fine aggregates. IOP Conf. Ser.: Mater. Sci. Eng. **625**, 012010 (2019)
12. Udawattha, C., De Silva, D.E., Galkanda, H., Halwatura, R.: Performance of natural polymers for stabilizing earth blocks. Materialia **2**, 23–32 (2018)
13. Laibi, A.B., Poullain, P., Leklou, N., Gomina, M., Sohounhloué, D.K.C.: Influence of the kenaf fiber length on the mechanical and thermal properties of Compressed Earth Blocks (CEB). KSCE J. Civ. Eng. **22**(2), 785–793 (2018). https://doi.org/10.1007/s12205-017-1968-9
14. Taallah, B., Guettala, A.: The mechanical and physical properties of compressed earth block stabilized with lime and filled with untreated and alkali-treated date palm fibers. Constr. Build. Mater. **104**, 52–62 (2016)
15. Mostafa, M., Uddin, N.: Experimental analysis of Compressed Earth Block (CEB) with banana fibers resisting flexural and compression forces. Case Stud. Constr. Mater. **5**, 53–63 (2016)
16. Rivera-Gómez, C., Galán-Marín, C., López-Cabeza, V.P., Diz-Mellado, E.: Sample key features affecting mechanical, acoustic and thermal properties of a natural-stabilized earthen material. Constr. Build. Mater. **271**, 121569 (2021)
17. Barbeta Solà, G., Massó Ros, F.X.: Improved thermal capacity of rammed earth by the inclusion of natural fibers. In: Ciancio, D., Beckett, C. (eds.) Rammed Earth Construction. Taylor & Francis Group, London (2015)
18. Toufigh, V., Kianfar, E.: The effects of stabilizers on the thermal and the mechanical properties of rammed earth at various humidities and their environmental impacts. Constr. Build. Mater. **200**, 616–629 (2019)
19. Porter, H., Blake, J., Dhami, N.K., Mukherjee, A.: Rammed earth blocks with improved multifunctional performance. Cem. Concr. Compos. **92**, 36–46 (2018)
20. Bogas, J.A., Silva, M., Gomes, M.G.: Unstabilized and stabilized compressed earth blocks with partial incorporation of recycled aggregates. Int. J. Archit. Heritage **13**(4), 569–584 (2019)
21. Nagaraj, H.B., Shreyasvi, C.: Compressed stabilized earth blocks using iron mine spoil waste—an explorative Study. Procedia Eng. **180**, 1203–1212 (2017)
22. Balkis, A.P.: The effects of waste marble dust and polypropylene fiber contents on mechanical properties of gypsum stabilized earthen. Constr. Build. Mater. **134**, 556–562 (2017)
23. El-Mahllawy, M.S., Kandeel, A.M., Abdel Latif, M.L., El Nagar, A.M.: The feasibility of using marble sawing waste in a sustainable building clay industry. Recycling **3**(3), 39 (2018)
24. Cagnon, H., Aubert, J.E., Coutand, M., Magniont, C.: Hygrothermal properties of earth bricks. Energy Build. **80**, 208–217 (2014)
25. Caponetto, R., Giuffrida, G.: Innovation in rammed earth systems. In: Ingegno e costruzione nell'epoca della complessità, forma urbana e individualità architettonica. Colloqui.AT.e, Edizioni Politecnico di Torino (2019)
26. ASTM D7928—17, Standard Test Method for Particle-Size Distribution (Gradation) of Fine-Grained Soils Using the Sedimentation (Hydrometer) Analysis. ASTM International. West Conshohocken (2017)
27. ASTM D4318—17e1, Standard Test Methods for Liquid Limit, Plastic Limit, and Plasticity Index of Soils. ASTM International. West Conshohocken (2017)
28. ISO 1920–10:2010 Standard, Testing of concrete Determination of static modulus of elasticity in compression. International Organization for Standardization. Geneve (2010)
29. HBE 195–2002, The Australian Earth Building Handbook, BD-083 (Earth Building) (2002)

30. ASTM D5334–14, Standard Test Method for Determination of Thermal Conductivity of Soil and Soft Rock by Thermal Needle Probe Procedure. ASTM International. West Conshohocken (2014)
31. Giuffrida, G., Caponetto, R., Nocera, F.: Hygrothermal properties of raw earth materials: a literature review. Sustainability **11**, 5342 (2019)
32. Giuffrida, G., Caponetto, R., Nocera, F., Cuomo, M.: Prototyping of a novel rammed earth technology. Sustainability **13**(21), 11948 (2021)

Driving a Photovoltaic Panel in Manhattan with Well-Chosen Projections of the Reflections of the Mirror City

Benoit Beckers(✉), Jairo Acuña Paz y Miño, and Inès de Bort

Urban Physics Joint Laboratory, Université de Pau et des Pays de l'Adour, 2S UPPA Anglet, France
benoit.beckers@univ-pau.fr

Abstract. Since Manhattan's first glass tower, the Lever House, built about seventy years ago, entire neighborhoods of glazed facades have appeared in every major city in the world. These contemporary city centers are turning into very complex scenes, especially for radiation. Faced with the extreme difficulty of carrying out measurement campaigns in urban environments, numerical simulation, enriched with adequate representations, is the only tool available to help understanding and making informed decisions and is, therefore, the key to Urban Physics. We show how the construction of synthetic representations allows guiding the orientation of a photovoltaic panel in the complex urban scene of New York City. A public, clean and complete geometric model of Manhattan is used here. The calculations are performed on Radiance, with a thousand reflections, diffuse or specular. These calculations include the illumination of the scene and thus the luminance of its surfaces, the sky, and the sun. The input data are obtained from the Pérez model using weather information of a standard meteorological year. Such calculations generate a huge number of results, and we show, for a given time of the year, that it is then essential to master the geometric properties of the cartographic projections, to find each time the representation that synthesizes the useful information that allows making the right decision.

Keywords: Geometrical projections · Photovoltaic potential · Urban physics

1 Introduction

This year, Sciences and Arts seem willing to take another look at the cradle of the New City. *Geometry* has fine-tuned her models. Surrounded by a flock of drones and satellites, she is slowly approaching an automatic survey that would avoid the tedious task of manually tracing corners, edges, and windows. *Astronomy*, enriched by the study of climate and meteorology, promises reliable data on the Sun and clouds, by refining its frequency bands. Fostered by such boundary conditions, *Physics*, in her brand new digital finery, proposes to follow the paths of photons, wavefronts, and air currents. Only *Clio* is disappointed: she no longer recognizes the child, whose genetic heritage, since the industrial revolution, seems to be constantly undergoing random mutations, out of

J. Littlewood and R. J. Howlett (Eds.): SEB 2022, SIST 336, pp. 175–184, 2023.
https://doi.org/10.1007/978-981-19-8769-4_17

all control. How to find one's way in this mass of bricks, concrete, metal, and glass with changing reflections, full of devices that are constantly blowing hot and cold? Where has the time gone when only the breath of people and their horses, in addition to the smoke of lamps and hearths, mingled with the cold winter air, when the hot summer air could only be softened by tarpaulins stretched over the stalls, shutters on the windows, and the shelter of heavy walls offering their thermal inertia? *Architecture* further adds: worried about the energy flowing through the New City, she would at least like to draw a portrait of it, to capture some images that would allow following these incessant fluxes and, if possible, to restrain them.

In the past, this had already been done, but only for visible light, thanks to the invention of the central perspective, which allowed, according to Leonardo da Vinci, to bring the whole universe to an eye, and this eye to a point [1]. Similarly, we can capture a whole scene from a point in space, by first projecting it onto a unit sphere surrounding this point, and then, by a second projection, bringing this sphere back to the image plane. It is then possible to integrate the information of this image to quantify the energy received by the point, but we then lose all spatial information and, therefore, the understanding of the scene and the ways of its possible transformation to modify the energy under consideration at the chosen point [2].

We will show how the realization of synthetic images allows supporting the architectural project by controlling a flux—here: the solar radiation—in a complex urban scene.

2 Study Case

The chosen point is in the heart of Manhattan, one and a half meters above the low terrace of the Lever House, the famous building built by Gordon Bunshaft in 1952—seventy years ago-, considered the first glass tower in a neighborhood that now has many. The city of New York has made available to the public a clean and complete geometric model of Manhattan, which will be used here [3].

2.1 A Point in Space

To obtain a complete panoramic representation of the scene—that is, with an aperture of 4π steradians—Fig. 1 proposes, in false colors, a Mollweide projection [4, 5]: the south is at the center of the image, the north at the cutoff, on the sides, where the tower of the Lever House is located (in purple), while the ground (in magenta) is that of the low terrace. This projection has the advantage of being *equivalent*, in other words, of keeping the solid angles (that is, the surfaces on the projected unit sphere). Thus, the ratio between the number of pixels representing the sky and the total number of pixels in the image gives directly the solid angle that holds the sky (*Sky Solid Angle, SSA*) [6, 7]. This would be the ideal projection to calculate the Mean Radiant Temperature of the black globe [8]. On the other hand, it is not *conformal* (it does not respect the angles on the image). There is no such thing as an equivalent and conformal projection, but there are some that do not respect either of these two properties and that are called *aphylactic*.

Fig. 1. Mollweide projection in false colors (left). The solid angle underlying the sky (SSA) is 1.45 sr and corresponds to 11.54% of the surface of the sphere (right)

If one is interested only in the half-sphere above the horizon, one can use the equivalent Lambert projection (Fig. 2): only the sky and its obstructions can be seen. However, these two projections have a characteristic that makes them difficult to represent: they are not *geometrical*, namely, they cannot be obtained by straight projections. This is the first reason to favor the stereography (Fig. 3), which is a simple central projection of the sphere made from the nadir (the pole opposite to the zenith of the sphere, these two poles being defined as the points of intersection on the sphere of the axis perpendicular to the projection plane).

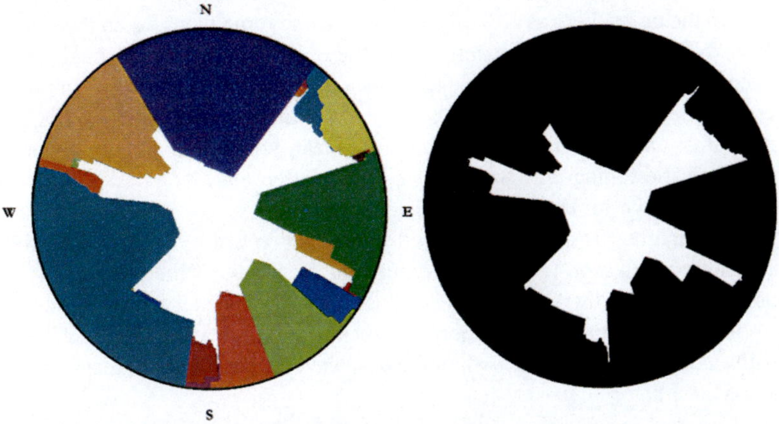

Fig. 2. Lambert equivalent projection in false colors (left), the solid angle underlying the sky (SSA) is 1.45 sr and corresponds to 23.08% of the surface of the disk (right).

The stereographic projection is conformal, which gives it an appearance all the more pleasant as, due to the position of the nadir, it moves away on the image the points close to the horizon (Fig. 3 on the right): this is a magnifying glass effect shared by the human eye (this is why the moon appears larger when it is low) and which makes sense in terrestrial scenes, where interesting events are more likely to occur near the horizon than in the middle of the sky…

However, there is one last property that belongs only to it, and which has always fascinated mathematicians: it is an *inversion*, which preserves the circles, except those

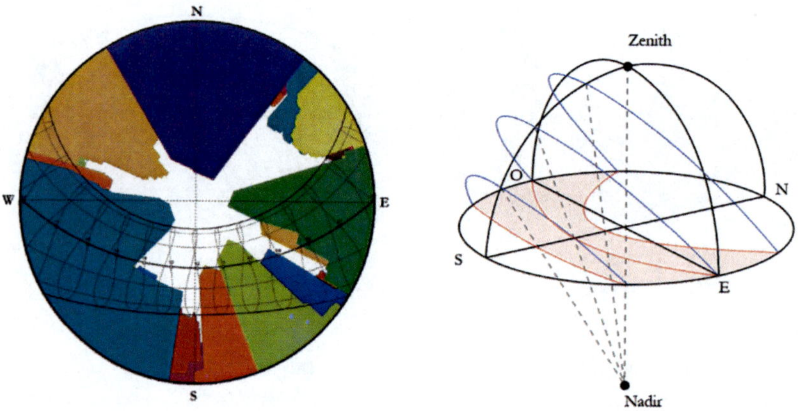

Fig. 3. Stereographic projection. Solar diagram corresponding to New York (latitude 40.73) (left). Construction of the stereographic projection (right).

that pass through the nadir, which it transforms into straight lines. Thus, the paths of the sun, which appear as arcs of a circle on the sky vault, remain circles on the projection (except for the path of the winter solstice in the tropics, whose extension passes through the nadir, and which degenerates into a straight line [9]). The straight lines of the scene are projected on the unit sphere as arcs of great circles, and remain circles on the projection, except for the verticals, whose extension on the unit sphere passes through the nadir, which degenerate into a straight line. This is why the stereography is the ideal projection for the solar diagram, which puts on the same drawing the masks of the buildings and the solar paths: only straight lines and circles appear, which was a major advantage at the time when these diagrams were drawn with a ruler and a compass and remains a significant advantage for their digital construction [10].

The solar diagram is a very condensed figure, which can be read at the same time as a spatial representation (one can recognize the buildings of the projected scene), a compass (one can see the four cardinal points on the perimeter, which represents the horizon) and a calendar (the projections are then interpreted as masks, which hide the solar paths at the precise moment when the projected buildings would mask the sun for the studied point).

The orthographic projection (Fig. 5) is a parallel projection (the projecting lines are verticals brought down to the horizontal plane). It is aphylactic, but it is of special interest. As Nusselt's analogy shows, the surfaces of the drawing directly give the view factors, in other words, the magnitude of the energy exchanged between each surface of the scene and the one (here horizontal) to which the point belongs [11]. The unmasked space gives the *Sky View Factor (SVF)*. When darkness comes, the reader unconsciously turns his book until it is parallel to the window, so that the last rays of daylight fall perpendicularly on the open page and illuminate it better. In the same way, the orthographic projection on a horizontal plane favors the zenithal directions and penalizes the grazing rays, near the horizon. In this interpretation, we no longer think of a point in space, but of a point that belongs to a plane. That is, for radiative exchanges, while the equivalent projection corresponds to a spherical sensor, the orthographic projection corresponds to a plane sensor [12].

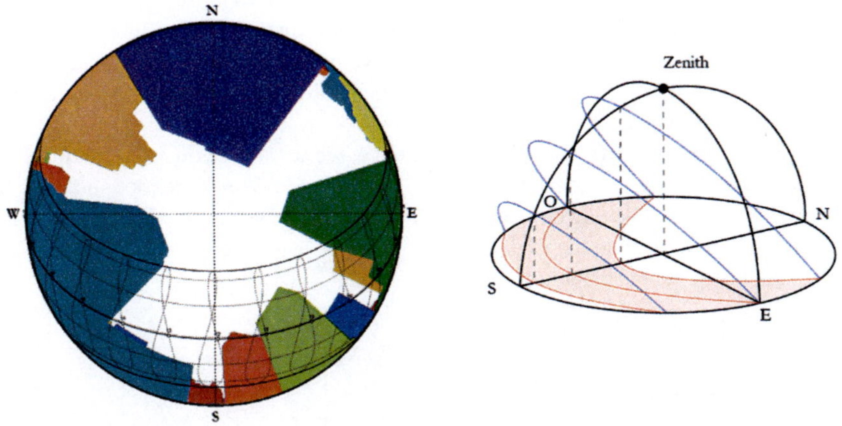

Fig. 4. Orthographic projection (left). Construction of the orthographic projection (right).

2.2 A Point on the Plane

The plane sensor, which we have placed horizontally at a height of one and a half meters above the low terrace of the Lever House, could be a photovoltaic panel mounted on a frame. The solar diagram in Fig. 3 shows us at what time of the year it will be exposed to the sun, but it does not inform us well about the two other contributions it will receive: the light from the sky and that reflected off the surfaces of the scene. As for the latter, Fig. 4 is better, because it shows us the view factors of all the elements of the scene, but these are still only masks, hence we have drawn them in false colors.

In Figs. 3 and 4, we can appreciate the solar path from the studied point. At the winter solstice, the sun appears, barely, around noon, brushing the red building. On October 28, the chosen date for the rest of this study, the sun appears briefly in the morning, between the green and the blue tower, which is visible on the stereogram only, thanks to its magnifying effect. Around noon, it is well visible, but it is necessary to wait for the equinoxes to see it for two hours in the early afternoon, and for the summer solstice to see it for four hours, in the late morning and around noon.

For a given time of the year (here, October 28 at 11:45), it is now possible to calculate the illumination of the scene and, therefore, the luminances of its surfaces, including the sky and the sun (input data, here obtained from the Pérez model [13] using weather information of a standard meteorological year [14]). The calculations are performed on Radiance [15], with a thousand reflections, diffuse or specular. All the surfaces that appear in Figs. 2, 3, and 4 are vertical, and assumed perfectly specular, with a typical reflection coefficient of 0.7. The ground of the terrace and the street, which will appear when the panel is no longer horizontal, are assumed to be diffuse, with a reflection coefficient of 0.5.

Figure 5 shows the result. It is interesting to compare this calculation to what we could do a dozen years ago, when we produced a geometrical model of the city center of Compiègne (France), composed of about twenty thousand triangles [16], to study it with the Heliodon2 software [17], using projections similar to those in Figs. 2, 3 and 4. Now, we are working on a model of one and a half million triangles, taking into account all the reflections, with calculation times that have become very reasonable (about two hours per diagram, on a 54-core computer), which allows us to color the diagram at a specific time of the year.

By integrating the different parts of the image, we obtain the direct radiation (from the sun), diffuse (from the sky), and reflected by the scene. The values are shown in the first row of Table 1. The day chosen is sunny, but humid, hence, the direct radiation is quite low; especially since the midday sun is only 35 degrees high on the horizon. The contribution of specular reflections is very important.

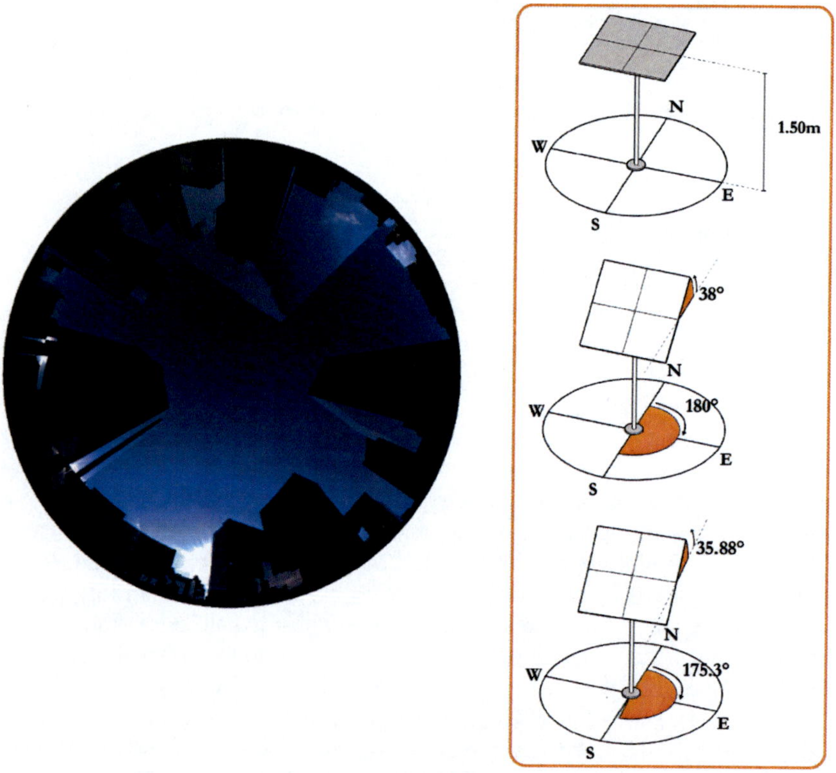

Fig. 5. Orthographic projection of the simulated scene on October 28 at 11:45. View from a horizontal panel (right). Solar panel orientations at 11:45 (left).

In New York City, to harvest the maximum amount of solar energy over the year, fixed photovoltaic panels are tilted 38 degrees upward, facing south (line 2 in Table 1). This tilt decreases the Sky View Factor, which is compensated by a better view of the sun's diffuse corona: sky radiation increases slightly, while reflections remain stable and direct radiation increases strongly. There are also motorized panels, which follow the solar path, like sunflowers, to optimize direct sunlight (line 3 of Table 1 and Fig. 6). In October, near noon, the previous configuration was already almost optimal, and the results hardly change.

Table 1. Sky view factors (SVF, in % of the sky) and irradiances (in Watts) for different PV panel orientations.

Orientation	SVF	Sun	Sky	Reflections	TOTAL
Horizontal	38.3	376	142	232.1	750.1
South facing	32	458.3	143.5	235.8	837.6
Sunflower	32.2	459	143.7	237.9	840.6
Sunflower 2	31.9	0	103.5	157.7	261.2
Reflection oriented	22.6	0	74.9	640.6	715.5

Can we predict what will happen, say, an hour later? We know of course where the sun will be then, and the sunflower panel is not mistaken and turns accordingly (Table 1 line 4). Unfortunately, the sun has disappeared behind a tower, and the balance on the panel is poor. This was still to be foreseen, on the solar diagram, but only the complete calculation shows us how the whole scene is now illuminated. Figure 6 on the right

Fig. 6. Orthographic projection of the simulated scene for a sunflower panel on October 28. At 11:45 am (left). At 12:45, the sun is hidden behind a building in the scene (right).

explains the result of the table, however, it does not allow us to reorient the panel in the best possible way, because half of the scene (the one behind the panel), is hidden from us, with its possible opportunities. To have a complete vision it is necessary to return to the projection of Mollweide (Fig. 7). We discover a specular reflection of the sun on the tower of the Lever House (north of the panel). Reoriented in the direction of this reflection (Fig. 8), the solar panel gives us the best possible results at this time (Table 1, line 5).

Fig. 7. Mollweide's projection of the simulated scene at 12:45, with the sun occluded (south), but giving rise to a specular reflection on the Lever House tower (north).

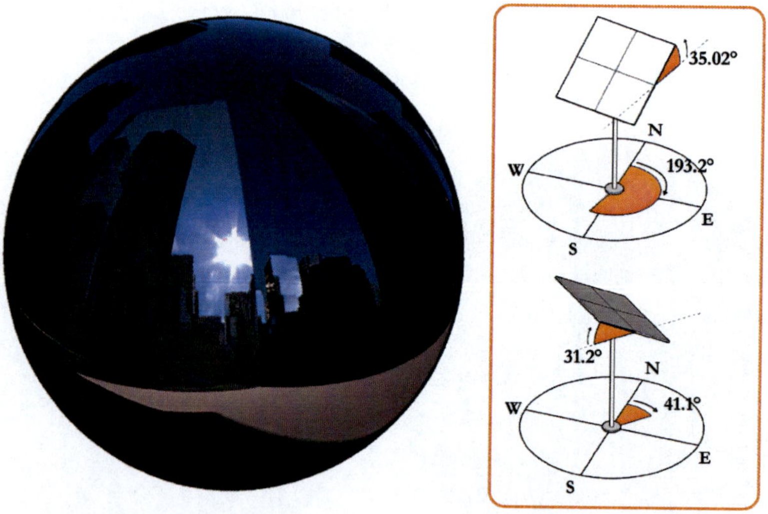

Fig. 8. Orthographic projection of the simulated scene for a sunflower panel that follows the brightest point in the scene (right). Solar panel orientations at 12:45 (left).

3 Conclusions

This example shows that the guidance of a photovoltaic panel in a complex urban environment is not intuitive. Hence, it may be necessary to look for the reflection of the sun on a glass surface in the north, while the sun itself is hidden in the south by another building... Today, good quality geometric models are available, as well as reliable and accurate meteorological data. Calculation tools based on ray tracing allow to take into account different optical properties (diffuse, specular reflection...) and to calculate a large number of reflections in a reasonable time. However, such calculations can generate a very large number of results, and we have shown on this example that it is then essential to master the geometric properties of the projections, in order to find every time the representation that clearly synthesizes the useful information and allows to take the right decision.

Due to the multiplication of glazed and metallic surfaces, contemporary city centers become very complex scenes, especially for radiation, while the variety of shapes and heights causes thermal, sound or aeraulic fields that are just as difficult to interpret, and even more difficult to correct. Given the extreme difficulty of carrying out measurement campaigns in an urban environment, digital simulation, enriched with adequate representations, is the only tool available to help understanding and deciding, and therefore the key to Urban Physics. A new science that *Architecture* and *Clio* are strongly calling for, as much at the bedside of the shaky cities of the industrial revolution, as at the cradle of a habitat composed on new patterns to better shelter the human beings of the 21st century.

References

1. Beckers, B., Garcia-Nevado, E.: Urban planning enriched by its representations, from perspective to thermography. In: Sustainable Vernacular Architecture: Innovative Renewable Energy, pp. 165–180. Springer, Cham (2019)
2. Beckers, B., Rodríguez, D.: Helping architects to design their personal daylight. WSEAS Trans. Environ. Dev. **5**(7), 467–477 (2009)
3. New York City Planning. https://www1.nyc.gov/site/planning/data-maps/open-data/dwn-nyc-3d-model-download.page. Last accessed 05May 22
4. Beckers, B., Beckers, P.: Reconciliation of Geometry and Perception in Radiation Physics. Wiley, New York (2014)
5. Lapaine, M.: Mollweide map projection. KoG **15**(15), 7–16 (2011)
6. Capeluto, I.G.: The influence of the urban environment on the availability of daylighting in office buildings in Israel. Build. Environ. **38**(5), 745–752 (2003)
7. Beckers, B.: Geometrical interpretation of sky light in architecture projects, In: CISBAT 2009, Proceedings of the International Scientific Conference on Renewables in a Changing Climate: from Nano to Urban Scale. Lausanne (2009)
8. Acuña Paz y Miño, J., Lawrence, C., Beckers, B.: Visual metering of the urban radiative environment through 4π imagery. Infrared Phys. Technol. 103463 (2020)
9. Beckers, B.: Worldwide aspects of solar radiation impact. In: Solar Energy at Urban Scale, pp. 99–118. Wiley, New York (2012)
10. Beckers, B., Masset, L., Beckers, P.: The universal projection for computing data carried on the hemisphere. Comput. Aided Des. **43**(2), 219–226 (2011)

11. Nusselt, W.: Graphische bestimmung des winkelverhaltnisses bei der wärmestrahlung. Zeitschrift des Vereines Deutscher Ingenieure **72**(20), 673 (1928)
12. Beckers, B., Beckers, P.: Fast and accurate view factor generation. In: First International Conference on Urban Physics. Quito—Galápagos, Ecuador (2016)
13. Perez, R., Seals, R., Michalsky, J.: All-weather model for sky luminance distribution—preliminary configuration and validation. Sol. Energy **50**(3), 235–245 (1993)
14. Typical Meteorological Year of New York. https://energyplus.net/weather-location/north_and_central_america_wmo_region_4/USA/NY/USA_NY_New.York.City-Central.Park.94728_TMY. Last accessed 05 May 22
15. Ward, G.J.: The radiance lighting simulation and rendering system. In: Proceedings of the 21st Annual Conference on Computer Graphics And Interactive Techniques, pp. 459–472 (1994)
16. Beckers, B., Rodríguez, D., Antaluca, E., Batoz, J.L.: About solar energy simulation in the urban framework: the model of Compiègne. In: 3rd International Congress Bauhaus SOLAR, pp. 10–11. Erfurt, Germany (2010)
17. Beckers, B., Masset, L.: Heliodon2^{TM}. http://www.heliodon.net/heliodon/index.html. Last accessed 05 May 22

One Stop Shops on Housing Energy Retrofit. European Cases, and Its Recent Implementation in Spain

Rolando Biere-Arenas(✉) and Carlos Marmolejo-Duarte

Centre for Land Policy and Valuations (CPSV), Architectural Technology Department (TA), Universitat Politècnica de Catalunya (UPC), 08028 Barcelona, Spain
rolando.biere@upc.edu

Abstract. The energy rehabilitation of buildings in the European Union would mean significant energy savings, and a 26% reduction in consumption. But, despite these benefits and the implemented programs, in Europe the rate of housing renovation is approximately 1%. In Spain, the situation is similar. The barriers (information, technical and economic) faced by homes, contractors, and finance companies are difficult to solve. So, in the international arena, promoted by European directives, appear One-Stop-Shops (OSS) as integrated management entities to promote the energy renovation of dwellings advising homeowners and users. This paper analyses the implemented OSS experiences in Europe, to identify its main elements, and propose lines of action to strengthen OSS operation in long term. To do it, regulations and documents were studied, and a cases database was elaborated. Also explores Spanish experiences through in-depth interviews, detecting barriers and problems faced and the implemented measures. The results suggested that a lack of structural funding is one reason of closing and that successful cases applied 'all inclusive' model and supported families in all process. It is highlighted the role played by EU projects in funding and knowledge; as well as the pending solution barriers for families, as obtaining resources.

Keywords: Housing policies · Renovation barriers · Retrofitting agents

1 General Framework of Housing Energy Retrofit in Europe

Buildings in Europe are responsible for 36% of greenhouse gas emissions, and 40% of energy consumption. Around 35% of the buildings are over 50 years old, almost 75% of these are energy inefficient, and only is renovated between 0.4% and 1.2% of the housing [1]. In Spain the situation is similar: 50% of buildings is prior to NBE-CT-79 rule, of minimum criteria for thermal insulation in dwellings, representing a 30% of energy consumption, mainly the residential sector.

A legislative framework to promote the rehabilitation of this obsolete building stock, and processes for improving the energy efficiency of buildings, has been developed by the European Union, with the enactment of the Energy Performance of Buildings Directive 2002/91/EC [2], and the Energy Efficiency Directive 2012/27/EU [3] that established

J. Littlewood and R. J. Howlett (Eds.): SEB 2022, SIST 336, pp. 185–196, 2023.
https://doi.org/10.1007/978-981-19-8769-4_18

that each member state must design a building renovation strategy, with specific actions, to achieve efficient, decarbonized building stock prior to 2050 [4].

Also, the EU promotes policies to create a framework for involved actors for informed investment decisions to save energy and money. The European Green Deal [5] defines energy renovation of public and private buildings as a basic measure to ensure a climate neutral Europe by 2050. Each member state decides requirements and calculation methods, implementing different models for transposing EPBD regulations [6].

Despite the implemented programs and economic efforts to promote the housing energy retrofit, the European annual rate of renovations in residential sector is only 1%, due to the barriers that families must face to perform the rehabilitate. In this context, renovations must be defined as those that reduce a building's final energy demand for heating between 50% and 80% [7], or 26% reduction in buildings energy consumption.

Therefore, it is necessary to promote the dwellings retrofit, guiding users in all steps of the process. Hence the importance of the One Stop Shops (OSS) for energy retrofit. In the energy rehabilitation sector, it represents an entity that provides, integrally, services and inputs during all the process: information on benefits and co-benefits to owners or users; diagnosis; prescription of improvements and a cost-benefit analysis; financing, managing construction permits and subsidies; search and contracting of suppliers; project management; post intervention evaluation, and maintenance. It means a unique qualified interlocutor, reducing some of the barriers to the Energy Retrofit.

In this context, the aim of this work is to analyze the OSS experiences in Europe, and deeply the recent in Spain. This analysis is aimed to understand the OSS organizational models, the services they offer, and overall the barriers it has overcome, and those that are still to be resolved. In addition, to propose strategies for future lines of action.

2 Methodology

The research develops an analysis of 31 implemented OSS European cases study (Table 1). Then an in deep qualitative analysis of 3 Spanish cases (Table 2) to generate a comparison and define the relevant topics in both contexts. The European analysis has followed the next stages: (1) identification of OSS, by official reports and studies [5, 6], research projects, papers, and institutional websites of OSS focused on housing energy renovation; (2) elaboration of a database, and variables organized by groups of contents: (a) general data (location, operation area, operational start year, leader entity, country, program/project, European, national or regional plan, website, type, (b) type of dwelling, (c) macroeconomic and environmental variables (GDP, CO2PC, RE%), (d) mass media (internet, showroom, office, visits, meetings…), (e) passive improvements (isolation, ventilation, enclosures, solar protections, water recycling), (f) active improvements (photovoltaic plates, boilers, heat pumps, heat recovery), (g) other (functional, accessibility), (h) responsibility for the works, (i) services (marketing, energy audit, project redaction, financing, grant/permission management, suppliers search, bidding for works, supervision, set up, monitoring, post evaluation), (j) information of target customers (owners, tenants), and (k) partners (providers, manufacturers, advisers, financial entities); (3) selection of cases, and (4) information and database analysis. The variables have been analyzed individually, within each group, detecting the most relevant. For Spanish case, in-depth interviews were made to promoters of operative OSS, in advanced stage.

Table 1. Analyzed European OSS cases.

OSS Name	Country	Operating	Leader	Plan/Project
RenoBooster	Austria	2019	PPP	N
Huisdokter	Belgium	2005	PPP	R
HomeGrade	Belgium	2017	Pub	N
EERSF	Bulgaria	2005	PPP	N, R
Aradippou OSS	Cyprus	2018	Pub. PPP	R
ProjectZero[1]	Denmark	2009	Pub	R
BedreBolig	Denmark	2013	Pub	E, R
PKA—Sust. Sol	Denmark	2015	PPP	–
Frederikshavn	Denmark	2017	Pub	R
Ecofurb	England (UK)	2009	Pr	E
Parity Projects[2]	England (UK)	2013	Pr., Coo	E, N
KredEx	Estonia	2009	Pub	N
Energies POSIT'IF[3]	France	2013	Pub, Coo	R
Pass Rénovation[4]	France	2013	Pub, Coo	R
ARTÉÉ	France	2015	Pub	E, N, R
OKTAVE	France	2017	Pub	N
RenoHub	Hungary	2019	PPP	E
SuperHomes	Ireland	2015	Pr	N, R
Mantova	Italy	2020	Pub	E, N, R
Leeuwarden[5]	Netherlands	2013	Pr	N, R
Woon Wijzer Winkel	Netherlands	2015	Pr	N, R
Huizenaanpak	Netherlands	2014	Pr., Coo	–
Stroomversnelling	Netherlands	2015	Pr., Coo	E
Reinmarkt	Netherlands	Pub	Pub	N
Bolig Enøk	Norway	Pr	Pr	–
Tighean Innse Gall	Scotland (UK)	Pr	Pr	R
ALIEnergy	Scotland (UK)	Pub	Pub	R
MunSEFF	Slovak Repub	Pub	Pub	N
Slovseff	Slovak Repub	Pub	Pub	N

(*continued*)

Table 1. (*continued*)

OSS Name	Country	Operating	Leader	Plan/Project
Energiesprong	UK, Fr., Germ	Pr	Pr	E, N
FinEERGo	Various *	Pub	Pub	E

Notes (a) Other name; [1]ZeroHome, [2]RetrofitWorks, [3]Île-de-France Énergies, [4]SPEE Picardie, [5]Slim Wonen Met Energies, (b) Leader entity type; PPP: Public-Private Partnership, Pub.: Public, P.: Private, Coo.: Cooperative., (c) Plan/project funding: E: European, N: National, R: Regional
Source Author's own elaboration

Table 2. Spanish OSS cases studied.

OSS Name	Leader type	Region	Type
GarrotxaDomus	Foundation	Catalonia	Coordination
OSIR	Administration	Extremadura	Coordination
OPENGELA	Administration	Basque country	Coordination

Notes In an initial implementation stage is Save the Homes (StH) in Valencian Community. It is included, in the Energy Office managed by the Valencia Climate and Energy municipal foundation. And recently, in Galicia, it has opened the housing rehabilitation office, promoted by the Galician Administration.
Source Author's own elaboration

The forms structure is: general data (leader, partners…), background, barriers, improvements and services (information, managing, financing, execution…), structure, operation, and barriers overcome. The steps have been: (1) identification, (2) collecting information, by public documents, (3) form applied to promoters, and (4) interview; barriers faced, handicaps to acting, and solutions. To establish relevant topics, an analysis, discussion, and final comments comparing European and Spanish cases has been made.

3 Brief Synthesis of the Main Barriers to Energy Retrofit

Housing renovation process is limited by barriers that affect families (economic, lack of knowledge and information, lack of capacity to implement renovations, etc.). Also, the renovation decision is affected by negative experiences of owners and lack of trust in advisors and contractors [8], also supported by the "do it yourself" culture [9] rooted in families, in the works on dwellings improvement, affecting the work of specialists.

We synthesize them and also determinant factors, as follows:

- *Barriers and market failures:* (a) informational asymmetry: actors in housing renovation do not have knowledge of energy efficiency [10], nor knowledge of the feasibility of improvements in benefits, the technical, and administrative processes; (b) economic factors: lack of funding and families do not have economic resources. Also, the process

to obtaining financing, managing financial aid, reaching consensus between homeowners, understanding the legal framework, managing the permits, and awarding the works contract. Evidence indicates a future savings penalization, and divided incentives [11]; (c) behaviorism: decision to develop renovations is influenced by personal, and external factors [12], and (d) legal framework and management: some national energy plans of some states are not adequate, and new policies are required [13].
- *Determinant factors:* (a) inconveniences in the decision process is a widespread problem for confused and asymmetric information [14], and that is not a dichotomous process, but a complex with specific problems in each stage [15]; (b) social factors, for example habits, which induce actions regardless of the context, or the reluctance to invest in residential improvements [16]; (c) understand rehabilitation as a housing adaptation process [17], or as gradual on time [18]; (d) EE lack of knowledge and fragmented supply [19], and (e) demand disaggregation [20].

In this scenario of difficulties in rehabilitation, the OSSs have gradually emerged, mainly in Europe and USA, in national, regional, and local settings, with a range of regulatory frameworks, adapted to these, with difficulties that this implies.

4 The OSS Situation in Europe

According INNOVATE project classification, based on the degree of support in the process, exist four OSS operation models: (a) Facilitation: offers an approach of the client to the benefits of energy retrofit, informing at no cost, and acts as a facilitator of the processes, (b) coordination: contacts customer with previously validated suppliers, and carry out energy renovation works, including financial entities if is needed. These are not responsible for the results, (c) all-inclusive: acts as a contractor, offering packages of services: information, coordination with suppliers, contractors and financing. It is responsible for the process and, sometimes, guarantee energy saving after works, and (d) ESCO: similar to all-inclusive, but also guarantees energy savings after the works.

OSS have appeared in Europe, provided by the EU as integral management entities to promote housing renovation. The first reference of OSS is in the work of Tommerup et al. [21]. Based on Nordic experiences, they proposed to accelerate housing rehabilitation with sustainability criteria, promoting companies to provide integrated services: consultancy, contracting, management, financing, commissioning and maintenance.

- *Jämtkraft* (Sweden) in 2006 offered an integrated service for; (a) inform the owners of homes with electric resistors the benefits to connect to a biomass system, (b) remove the resistors, install radiators with heat exchanger, (c) extend the network to connection with the home, (d) manage a credit of up to 30 years at a competitive interest rate. To add demand, pamphlets were distributed that included testimonials, informative meetings were organized with technical representatives and those of the financial entity.
- *ENRA* (Finland) begins in 2009 a pilot program to promote single-family rehabilitation. The leader specialized in rehabilitation and partners companies: manufacturers of doors, windows, ventilation and insulation systems, a supplier of heat pumps, and

an energy auditor. An initial meet with potential clients is made to explain possibilities of rehabilitation, and the environmental effects, the comfort and air quality improvement.

- *Clean Tech* (Denmark) was born in 2009 led by the energy company Dong Energy associated with a window producer, a heat pump distributor, an insulation producer and a financial company. Aimed at single-family homes heated with oil, they implemented a website with a cost-benefit calculator linked to improvement alternatives, and a single point of contact for information. A technician visits the house, prescribe improvements, and deliver budgets. The OSS handles grants, permits, and financial credits.

These pioneering commercial initiatives, due to its geographical context, were focused to replace thermal equipment in single-family homes. Mahapatra et al. [22] document later experiences in the region, highlighting those, whose OSS models provide a set of services through a single interlocutor and under the same roof.

Studies, such as Boza-Kiss and Bertoldi [23], Cicmanova et al. [8], or Krosse et al. [24], review implemented OSSs, many of its promoted by European initiatives. These enable measures to be adopted that improve dwellings' energy efficiency, offering a renovated dwelling to the owner's real needs [18]. The highlighted are:

- *BetterHome* (Denmark) offers predefined renovation packages for private homeowners. Through automated services and a web application, the potential customer informs the installers and preselects the measures. Then, the homeowner in direct contact with the technicians can adapt the package, technical and financial terms to their needs. The OSS works with trained local craftspeople, including tools to guarantee quality services. Better-Home carries out the promotion, quality control, monitoring, and customer care. In 2016, it completed over 200 projects and it is in expansion.
- *Retrofit Works* (U.K.) open in 2013 as part of the Green Deal, as a cooperative of SMEs of contractors, local suppliers, and technicians qualified in energy, also social agents and the energy consultancy Parity Projects (PP). In 2017 they have constituted as an OSS. The process includes a web tool that families can use to find out about possible improvements and necessary investment. Those interested people contact to PP, and a technical coordinator visit the home to carry out an onsite assessment.
- *OKTAVE* (France) led by town and city councils and promoted by the agency for ecological transition ADEME and the Gran Este region. It includes two financial companies (one social) and provides a service including assistance on technical, financial, and administrative aspects in a single point of contact. A financial plan is drawn up, including subsidies, tax credits, and zero interest loans for up to 15 years, and it seeks an ESCO to recover the investment with energy savings. Local contractors and suppliers, trained, and accredited by the OSS, are registered as qualified technicians.

The European OSS for retrofitting of dwellings not only induce the demand for rehabilitation, but also make visible and organize a fragmented market of providers (technicians, contractors, financiers) and claimants.

5 The Recently Implemented OSS in Spain

In Spain the OSS concept appears in 2014 ERESEE as "local rehabilitation agencies." Although the local rehabilitation offices existed, linked to rehabilitation of public housing in degraded neighborhoods. Unlike these, the OSS: (a) is not only linked to an integral rehabilitation, (b) nor is it focused at vulnerable areas, and (c) nor is in public housing.

- *GarrotxaDomus (GD).* Is the evolution of HolaDomus (HD), pilot project in the framework of the H2020 EuroPACE Project, managed by the EuroPACE Foundation, a non-profit entity with public-private participation: GNE Finance and Olot City Council, which has supported it and intends to extend it on the province (GiDomus). The model simplifies interaction between and actors, trying to solve the barriers to rehabilitation: lack of information, financing and time, management, qualified professionals, support in applying for grants and credits, and difficulties of the management in front of administrations. The process begins offering services of an energy office to advice and inform in habits to reduce consumption, and improvements. Subsidies are managed and information is provided on financing lines. IBI, ICIO bonuses, and building license are processed, and owner is assisted to request budgets, between 70 professionals.

  The OSS experts make a study to identify the ideal solutions, giving advices and information on financing lines. The subsidies are managed, license of works, IBI and ICIO discounts are processed and the owners is assisted to request budgets, helping them to choose of the most convenient. OSS has a portfolio of around sixty technically validated local professionals to require budgets. Once finished works the administrative reception is carried out, and the accompaniment is maintained during one year.

  In addition to the office, a website with all technical, and administrative information is available for costumers. Likewise, instruments are given to the network of professionals to evaluate energy efficiency measures. The implementation of the OSS has had the willingness of the Olot City Council, that has collaborated in its implementation.

  Some 430 middle-income households and single-family homes have benefited, 95 assisted in drafting energy projects; a third part executed, another third part in execution and the rest in the previous phases. The most measures applied have been: insulation, air conditioning with photovoltaic support, enclosures, etc.
- *Office of Comprehensive Services for the Energy Rehabilitation of Housing (OSIR).* It is an initiative of the Extremadura Government and Energy Agency (AGENEX), with the support of the H2020 projects: HoseEnvest and INNOVATE. It is aimed to multi-family homes. The office is in Badajoz, and is studied new openings in Mérida and Caceres. Before its opening, a diagnosis and rehabilitation potential study was made in the Interreg FINERPOL project and training was given to the construction sector through the REHABILITE project of the same European program. Then appear OSIR and the Extremadura Housing Energy Efficiency Guarantee Fund (GEEVE).

  The fund receives public resources (ERDF included), is managed by Extremadura Avante, and seeks to mitigate the risk of participating financial entities. Thus, the entities offer loans with special conditions; return terms of 15 years, reduced rates (less than 4.5% APR), without commissions (except opening). The fund accredits guarantee if renovations affect the envelope, include an active system, and improve one step in EPC energy class. These conditions can be improved in the case of healthy communities. In order for the fund to guarantee the credits, the renovations must affect the envelope and an active system, in addition to improve one step the EPC energy class.

Process begins identifying buildings with highest potential for rehabilitation, by age, lack of insulation and inefficient boilers. Then is contacted the property administration and the presidency of the community. Is made the visit of a technic to identify deficiencies and intervention opportunities. Then, with the improvements proposals, the energy savings that it represent, the economic estimate of the investment for improvement, the subsidy and financing options, the OSS summons the meeting of owners, to adopt the agreement, prioritize those cost-effective and affordable measures or reject them.

Ten months after its opening, 200 buildings were visited and 170 diagnostic and improvement reports have been issued. 3 have agreed to undertake the improvements and 2 have completed the selection of the contractor that will implement them.

- *Opengela (OG)* was born with the support of the H2020 Hiross4all project, to promote the creation of OSS in vulnerable neighborhoods of the Basque country. The first two OG offices operate in Otxarkoaga (Bilbao) and Txonta (Eibar) with particularities in its operation. GNE Finance also participates with its specific knowledge in financing for rehabilitation. OG promotes the rehabilitation of multi-family buildings, empowering their owners to become the protagonists who decide and lead the actions. Given the vulnerable nature of neighborhoods, social inclusion aspects are relevant, representing an added complexity. The rehabilitation focuses on energy efficiency (reaching "C" EPC class, improving one or two letters), including health, habit-ability, comfort (including acoustics) and accessibility improvement. Also, modification of the stairwell, installation of elevators, or fire protection measures, priority aspects in post war areas.

Approximately 50% of the rehabilitation cost is covered by subsidies. Also, owners can apply for an additional aid or credits, and the spills can be prorated (up 36 months). The OSS is developed through a technical assistance service contract that depends, (in Otxarkoaga) on Bilbao Municipal Housing. OG seeks to increase the household's confidence by a regulated, competitive and transparent action.

The process begins with the information of the problems to the owners, including improvements, based on the available aid, preferences and economic possibilities. This detailed information is presented at a Meeting and, subsequently, the improvements are chosen based on the available aid, preferences and economic possibilities. Then, an agreement is requested for the administrative contracting process that begins with the drafting of the technical conditions, continues with the public call, with the contracting table, the review of the offers and its interpretation to facilitate the understanding of the owners. Once the Board selects the most convenient offer, the OSS offers support in drafting the project, obtaining permits, monitoring the works and their certifications, management of spills, clarifying doubts, intermediation between users and professionals, and the reception of works. In the case of OG Txonta, the procedure is similar, although there is no administrative contracting. A committee to assess offers is formed, representants of owner's community, Basque Government and City Council.

In Otxarkoaga, there are 5 buildings (16 portals and 240 homes) the office also manages the remodeling of an old shopping center and the launch of entrepreneurship initiatives in empty commercial spaces. In Txonta, the action is being taken on 17 portals (221 homes) with an emphasis on energy and accessibility aspects.

6 General Analysis, Discussion, and Final Comments

In addition to initials, the operative OSS are mainly in Netherlands (5), France (4), and U.K. (4). Four of these operating since 2009 (ProjectZero in Denmark; Ecofurb in England; KredEx in Estonia; and Bolig Enøk in Norway). These are the longest, after the two that are operating since 2005 (Huisdokter in Belgium, and EERSF in Bulgaria).

Regarding nature, highlights public initiatives (51.61%). Generally public entities are town councils, supported by regional or national energy agencies. Almost half the OSS use all-inclusive model, 26% facilitating model, and 24% coordination model.

About financing for operation, most (28 cases, 90.32%) have received public financing of European programs. Many of these start as pilot test. Of these the ones that are not operating, generally shut down, if the pilot test, does not considered future funding.

Regarding offered improvements, in the passive ones, there is a tendency to more integrated. In the buildings, the main passive improvements are focused on insulation (30 cases, 96.77%), types of doors-windows (26, 83.87%), and ventilation (25, 80.65%). In active ones, there is a predominance of photovoltaic panels incorporation (26 cases, 83.87%) in multi-family housing, and heat pumps (23, 74.19%), followed by boilers.

Regarding Services, the main are: energy audit (27 cases, 87.10%), project (23 cases, 74.19%). Those frequently offered are: acceptance of works, and other improvements.

Results at European level suggest that the structural financing lack is cause of the closure in many cases, and that the most successful respond to the all-inclusive model.

The 'all-inclusive' model is used in almost half of the OSSs studied, followed by coordination. Offering an all-inclusive service requires an additional effort but enables manufacturers and service providers to meet customers' specific needs, which may range from solutions for owners of single-family to multi-family homes. But, the OSSs in Spain is 'facilitation' model, based by the leading role of the public administrations.

In Spain, OSS share the impulse of public administrations, unlike the rest of Europe, mostly private. Although the administrations responsibility in promoting energy rehabilitation is clear, its participation in the OSS (with public funds) entails difficulties.

The battery of barriers faced and solutions is different in each Spanish case (Table 3).

Table 3. Barriers and problems faced, and implemented measures of the Spanish cases.

OSS	Main Barriers and problems *Housing market*	Main implemented Measures and actions
G.D	General information lack and about affordable financing; complex management; difficulty in finding professionals; information; difficult in applying for grants and credits; difficulties of the management in front of the administrations *Single and multi-family homes*	Simplified interaction between owners and actors Providing a website with all the information Each procedure is optimized including information Offering services and advices to reduce consumption Information on habits and housing improvements Managing subsidies and information on financing lines Processing building license, IBI and ICIO bonuses Assisting owners to request budgets Including a portfolio of 70 validated professionals Financing 5.5% APR rate, 1.5% comm., 5–15 years
OSIR	Lack of specific financial products; low demand due to the absence of incentives and lack of knowledge of benefits and co-benefits; complex management and need to accompaniment of the rehabilitation process *Multi-family homes*	Offering information and contact via web Contacting administration and owner's community Technical, identifying deficiencies & opportunities Elaborating proposals for improvements Information about investment, subsidies and financing Prioritization of cost-effective and affordable measures Management services; licenses and subsidies *In study:* Implementation of a guarantee system to verifies the achievement of technical parameters, and new financing models: (a) maintaining services by subsidies; and (b) sharing costs with benefited organizations
OG	Vulnerable neighborhoods; social inclusion aspects; inherent problems to buildings of post war period; lack of information *Multi-family homes*	Neighborhood offices with operational particularities, based on the specific context Empowering owners to be protagonists of actions Health, habitability, and accessibility solutions Developing stairwell modifications, installation of elevators, fire protection, and accessibility actions Covering 50% rehabilitation cost by subsidies Apply for owners to an additional aid or credits Prorating spills up 36 months Designing a credit line for owner's communities

Source Author's own elaboration

Acknowledgments. Authors appreciate funding of the project EnerValor2, by Spanish Ministry of Science, Innovation & Universities (PID2019-104561RB-I00) and contributions of reviewers.

References

1. European Parliament (2020). Draft Report on maximizing the energy efficiency potential of the EU building stock (2020/2070(INI), Committee on Industry, Research and Energy. Available: https://www.europarl.eu/648631.pdf
2. European Parliament and Eu Council (2002). DIRECTIVE 2002/91/CE, of 16 December 2002 on the energy performance of buildings (OJEC L 1 de 65. 4.1.2003). Available: https://eur-lex.eu.eu/3A32002L0091
3. European Parliament and Eu Council (2012). DIRECTIVE 2012/27/UE, on energy efficiency, amending Directives 2009/125/EC and 2010/30/EU and repealing Directives 2004/8/EC & 2006/32/EC. Available: https://eur-lex.eu.eu/3A32012L0027
4. De Santiago, E.: La estrategia para la rehabilitación energética en el sector de la edificación residencial en España: método y principales resultados. Ciudad y Territorio. Estudios Territoriales **46**(182), 773–788 (2014). https://doi.org/10.3989/ic.63735
5. European Commission. A European Green Deal. 2019. Available: https://ec.europa.eu/info/strategy/priorities-2019-2024/european-green-deal [21 August 2021]
6. García-Hooghuis, A., Neila, J.: Modelos de transposición de las Directivas 2002/91/CE y 2010/31/UE "Energy Performance Building Directive" en los Estados miembros de la UE. Consecuencias e implicaciones. Informes de la Construcción 2013, 65, pp. 289–300
7. ENERDATA. ZEBRA2020-Data Tool. 2015. Available: https://zebra-monitoring.enerdata.net/ES/
8. Cicmanova, J., Eisermann, M., Maraquin, T.-E.C.: How to set up a One-stop-shops for integrated home energy renovation? A step-by-step guide for local authorities and other actors. INNOVATE Project (2020). Available https://e-c.eu/wp-content/INNOVATE.pdf
9. Boza-Kiss, B., Bertoldi, P.: One-stop-shops for energy renovations of buildings. Case studies. European Commission. JRC Science for Policy Report, [JRC113301] (2018). Available: https://e3p.jrc.ec.europa.eu/one-stop-shops
10. Hunkin, S., Krell, K.: Supporting energy renovation of private households through One-Stop-Shops. A Policy Brief. From the Policy Learning Platform on Low-Carbon Economy. Interreg Europe (2019). Available: https://www.interreu.eu/20190210_oss.pdf
11. Gillingham, K., Palmer, K.: Bridging the energy efficiency gap: policy insights from economic theory and empirical evidence. Rev. Environ. Econ. Policy **8**(1), 18–38 (2014). https://doi.org/10.1093/reep/ret021
12. Ebrahimigharehbagh, S., et al.: Unravelling Dutch homeowners' behaviour towards energy efficiency renovations: What drives and hinders their decision-making? Energy Policy **129**, 0301–4215 (2019). https://doi.org/10.1016/j.enpol.2019.02.046
13. Rosenow, J., Fawcett, T., Eyre, N., Oikonomou, V.: Energy efficiency and the policy mix. Build. Res. Inf. **44**, 562–574 (2016). https://doi.org/10.1080/09613218.2016.1138803
14. Arkeloff, G.A.: The market for "Lemons": quality uncertainty and the market mechanism. Q. J. Econ. **84**(3), 488–500 (1970). https://doi.org/10.2307/1879431
15. de Vries, G., Rietkerk, M., Kooger, R.: The hassle factor as a psychological barrier to a green home. J. Consum. Policy **43**(2), 345–352 (2019). https://doi.org/10.1007/s10603-019-09410-7
16. Rana, A., et al.: Evaluation of financial incentives for green buildings in Canadian landscape. Renew Sust. En. Reviews **135**, 110199 (2021). https://doi.org/10.1016/j.rser.2020.110199
17. Salom, J., Pascual, J. (ed).: Residential Retrofits at district scale. Business Models under Pub. Private Partnerships (2018)
18. Pardalis, G., et al.: Swedish house owners' intentions towards renovations: is there a market for one-stop-shop? Buildings **9**(7), 164 (2019). https://doi.org/10.3390/buildings9070164

19. Thøgersen, J.: Housing-related lifestyle and energy saving: a multi-level approach. Energy Policy **102**, 73–87 (2017). https://doi.org/10.1016/j.enpol.2016.12.015
20. Risholt, B., Berker, T.: Success for energy efficient renovation of dwellings-learning from private homeowners. Energy Policy **61**, 1022–1030 (2013). https://doi.org/10.1016/j.enpol.2013.06.011
21. Tommerup, H.M. et al. : Existing sustainable renovation concepts for single-family houses. In SB10 Finland: Sustainable Community-BuildingSMART. Helsinki, Finland (2010). Available: https://backend.orbit.dtu.dk/ws/portalfiles/portal/4673337/SB10.pdf
22. Mahapatra, K. et al.: Business models for full service energy renovation of single-family houses in Nordic countries. Appl. Energy **112** (2013). https://doi.org/10.1016/j.apenergy.2013.01.010
23. Boza-Kiss, B., Bertoldi, P.: One-stop-shops for energy renovations of buildings. Case studies. European Commission. JRC Science for Policy Report, [JRC113301] (2018). Available: https://e3p.jrc.ec.europa.eu/oss-erb
24. Krosse, L., Monclus, M., Nijrolder, A.: Building Refurbishment Initiatives and Business Models. A Global Benchmark (2021). Available: https://www.fundnaturgy.org/77630

Mitigating Multi-risks in the Historical Built Environment: A Multi-strategy Adaptive Approach

Federica Rosso[1,2(✉)], Letizia Bernabei[2], Gabriele Bernardini[3], Juan Diego Blanco Cadena[4], Martina Russo[1], Alessandro D'Amico[1], Graziano Salvalai[4], Edoardo Currà[1], Enrico Quagliarini[3], and Giovanni Mochi[2]

[1] Sapienza Università Di Roma, 00184 Rome, Italy
federica.rosso@uniroma1.it
[2] Università Degli Studi Di Perugia, 06125 Perugia, Italy
[3] Università Politecnica Delle Marche, 60121 Ancona, Italy
[4] Politecnico Di Milano, 23900 Lecco, Italy

Abstract. The built environment is subject to complex combinations of cascading disasters, sudden onset disasters (SUODs), such as earthquakes, or slow onset disasters (SLODs), such as pollution and heat islands. They can cause harm to people and destroy the built environment. Moreover, the historical built environment (HBE) possesses typical characteristics that increase the risks, for two main reasons: the HBE construction features, which are vulnerable to SUODs and SLODs; the disruption of the cultural heritage, which is part of the HBE. To preserve the HBE, suitable strategies to adapt to increasingly frequent SUODs and SLODs should be considered in a multi-risk, multi-strategy perspective, as some of them are able to mitigate more than one risk simultaneously. Therefore, the contribution of this article is to propose a multi-strategy approach for mitigating multi-risks in the HBE, by means of (i) the definition of *Built Environment Typologies (BETs)*, which are clustered to represent typical HBEs, and (ii) a critical overview of the literature and expert judgement, which is used to hypothesize strategies' combinations on peculiar BETs. Another original contribution is the focus on the open spaces portion of the HBE. Indeed, they are often overlooked, while they constitute a crucial portion of the HBE, where there is high exposure to risk, as open spaces are among the most used spaces, and are affected not only by their characteristics but also by those of their frontiers. The findings of this work contribute to frame multi-risk multi-strategies approaches for the HBE, towards adaptive and resilient HBEs.

Keywords: Multi-risk assessment · Historical built environment · Built environment typologies · SUODs · SLODs

J. Littlewood and R. J. Howlett (Eds.): SEB 2022, SIST 336, pp. 197–207, 2023.
https://doi.org/10.1007/978-981-19-8769-4_19

1 Introduction

Historical Built Environment (HBE) is particularly subject to multi-risks, such as SUdden Onset Disasters (SUODs) (e.g., earthquakes and terrorist attacks) and Slow Onset Disasters (SLODs) (e.g., urban heat island—UHI, and pollution) [1]. Suitable mitigation strategies should be studied to face these risks: while studies on the specific mitigation strategies to face each risk have been largely conducted, the simultaneous consideration of strategies to face multiple risks is less common.

The main objective of this contribution is thus to investigate the application of combined multi-strategies that are effective in mitigating multi-risk combinations in exemplificative Built Environment Typologies (BETs) of the HBE. These results could provide relevant information for tailoring effective strategies to improve resilience and adaptation to increasing risks for the HBE, in an overall, multi-risk perspective and are of interest for professionals intervening on the HBE, for policymakers and for city/towns administrations.

2 Background

The BETs were identified starting from morpho-typological and construction parameters relevant for responding to SUODs and SLODs, and for defining the typologies of historical built environment, as from [2–4].

Such relevant parameters were then numerically described, in order to have them quantitative and to define their ranges of variability, on a relevant context, the Italian one. Indeed, Italy is prone to increasing risks related to both SUODs and SLODs [5, 6], thus constituting a significant and exemplificative context.

The **parameters** were selected based on literature review and expert judgement, and their ranges were defined by means of statistical analysis by testing the parameters on a significant sample of provincial capital and small-medium towns in the entire Italian territory [3, 7]. The total sample consisted in 90 towns homogeneously distributed along the entire Italian peninsula, to represent the entire context, comprehending all of the provincial capitals of Italy. The main square of each town was selected. While Italian provincial capitals are 109, some were deleted from the sample as the main squares presented articulated shapes. To compensate for the elimination of such squares, few random small-medium towns were selected to increase the sample size. The results of the parameters' identification and description are illustrated in [7], and constitutes the basis for the identification of the BETs. Then, the **BETs** were identified by enlarging the sample of Italian squares to 1111 cases, randomly selected in the same towns of the previous analysis and in other representative towns dispersed all along the Italian territory, to allow for an advanced statistical analysis, i.e., cluster analysis [4], aimed at grouping homogeneous elements in a set of data. The cluster analysis is a multi-variate non-supervised analysis that investigates the presence of similar statistical units in a given dataset [8], and non-hierarchical k-means for the grouping around the selected variables. The previously identified parameters (P1-P9) were divided into two groups, a first group for expeditious evaluation of morphology, geometry and function (P1, P2, P4, P5, P8, P9), and a second group for detailed evaluation, comprehending construction features (the remaining P3, P6, P7). For this first general identification, the sole first group of parameters were considered as variables for the cluster analysis. Control indicators were R-squared and Pseudo-F. Moreover, given the number of cases, GIS and OpenStreetMap were employed to retrieve the characteristics of the cases in an expeditious manner. Resulting clusters are presented and briefly described in Table 1.

3 Methods

In order to hypothesize the most effective mitigation strategies for combined SUODs and SLODs, in this work we tap on the BETs, exemplificative open spaces in the HBE, illustrated in the background section. Here, in Sect. 3.1, we describe the association of multi-risk to the BETs. Then, a literature overview, supported by expert judgement, has been conducted to identify effective multi-risk strategies (Sect. 3.2). In the following subsections, these two steps are presented in greater detail.

3.1 Multi-risks Identification for BETs

For each BET, the evaluation of the possibility of being subjected to the arousal of risk combinations leading to multi-risk were evaluated. The analyses of multi-risks combinations were carried out starting from the Italian context sample of 133 provincial capital and medium-sized towns in the entire Italian territory and then associated to the parameters and the BETs [3, in D322]: each square of the sample was assigned to the corresponding BET, and thus the most frequent risks combinations were calculated. The results of this analysis are reported in the results section.

3.2 Strategies Identification: Literature Overview and Expert Judgement

The literature review was carried out on the Scopus database, by inserting the keywords "UHI" "pollution" "seismic risk" "earthquake" "terrorism" "cultural heritage" "mitigation strategies" "urban areas" "cities" "historical built environment". An extensive explanation and recall of the literature review are reported in the Deliverable 5.1.1 of the mentioned project. The strategies were then structured into categories of solutions, for different scales: morphological factors (larger scale), physical and construction factors (smaller scale), and dedicated systems and behaviour strategies (human/organizations scale). Moreover, the solutions can be applied on the entire Open Space (OS), on the OS frontier, or on the users.

The Table below (Table 2) reports the specific strategy and references to the literature review and examples from the professional field.

The combined effect of the strategies for multi-risks is considered by means of expert judgement, and the results are presented in Sect. 3.2.

Table 1. BETs clusters for the Italian context.

BET representation	General description
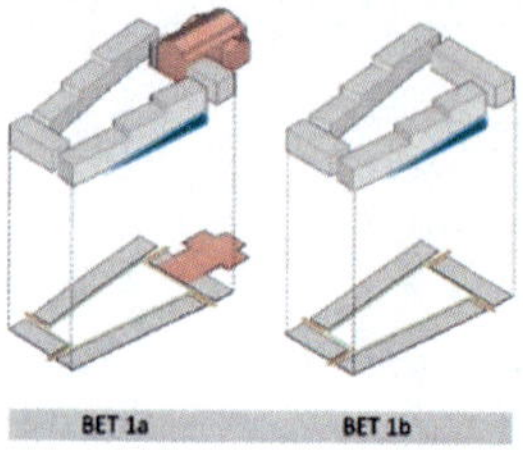	**Cluster 1** Medium compactness and regular morphology, no vulnerability to overturning of the facades of the frontiers, but critical accesses number along the perimeter; presence of a slope **Main feature**: slope
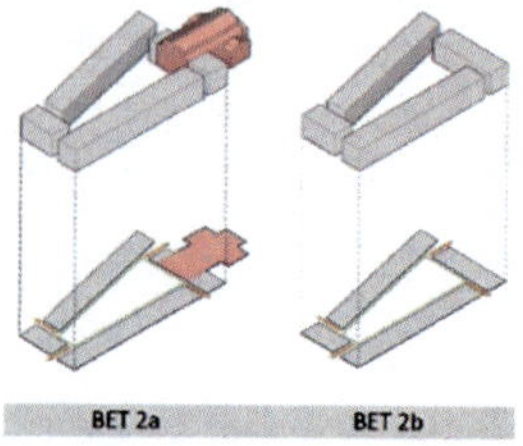	**Cluster 2** Low compactness and irregular morphology, no vulnerability to overturning of the facades of the frontiers but critical accesses number along the perimeter; no slope **Main feature:** low compactness and regularity
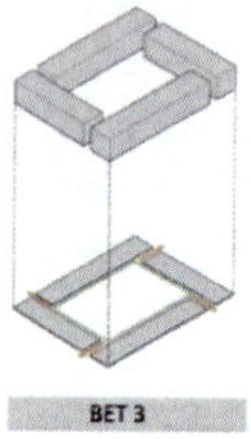	**Cluster 3** High compactness and regular morphology, no vulnerability to overturning of the facades of the frontiers but critical accesses number along the perimeter; no or low slope **Main feature**: high compactness and regularity
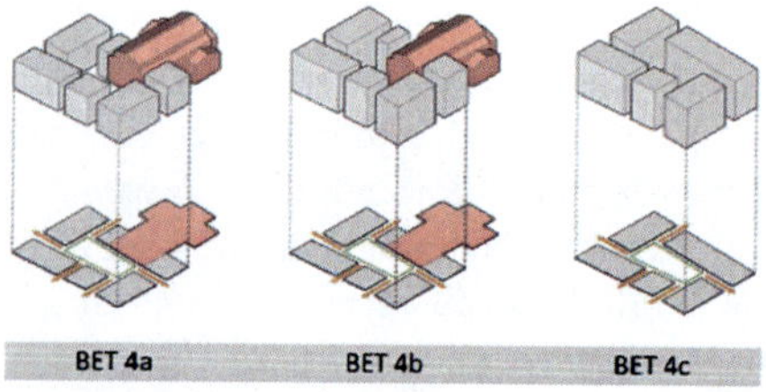	**Cluster 4** Average compactness and regular morphology, vulnerability to overturning of the facades of the frontiers but adequate accesses number along the perimeter; no or low slope **Main feature**: overturing risk of the facades, adequate accesses along the perimeter
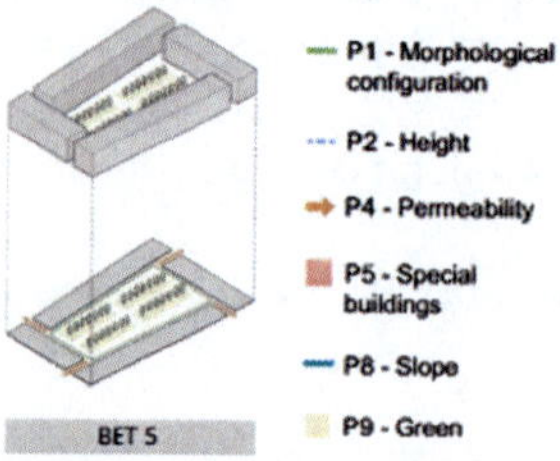	**Cluster 5** Average compactness and regular morphology, no vulnerability to overturning of the facades of the frontiers but critical accesses number along the perimeter; no or low slope; presence of green areas **Main feature**: presence of green areas

Table 2. Examples of strategies and references grouped by risk type.

Risk	Strategy	Case studies from the professional fields (where directly correlated)	References
SLOD	Improve vegetation	Hamburg, Milan	[9]
	Improve shaded areas	Expo 2015, Metrosol Parasol Seville, Umbrella sky project	[10]
	Morphological configurations of OS		[10]
	OS surface temperatures reduction	Los Angeles white painting project	[11]
		Derbyshire street	[12]
		Lunix pavers	www.ferraribk.it
	Air cleaning solutions	Green city solutions	[13]
	Building surface temperatures reduction	Santorini	[14]
		White roof project	[15]
		Vegetecture, Park lane	[16]
		Madrid city hall, Chicago city hall terrace	[16]
	Air cleaning solutions	Converse walls, Volkswagen walls UK	[17]
		Biq house, Hamburg	[18]
	Reduce vehicular traffic and emissions	EnelX	[19]
		Milan: GuidaMI, Car2go, Enjoy, E-vai, SHARE'nGo	[20]
		Carpooling Bla bla car, Uber	[21]
		Electric scooters: EM transit, Bird rides Italy, Voi technology Italia, Wind mobility, Bit mobility, Helbiz Italia and Lime technology	[22]
		Call a bike system – Stuttgart	[20]
	Increase community and individuals' awareness	nZEB, LEED certifications	[23]
		Health care education for elders—Guangzhou, China	[24]

(*continued*)

Table 2. (*continued*)

Risk	Strategy	Case studies from the professional fields (where directly correlated)	References
		PM risk self-awareness in vulnerable groups	[25]
	Waste management	Sweden example	[26]
SUOD	Morphological configurations of OS		[27]
	Permeability improvement (accesses and road network)	Nocera Umbra, Città di Castello, Gubbio (Italy)	[28]
	Reduction of BE vulnerability	Kyoto	[29]
		Washington, Monument (ha-ha barrier and water obstacle), Cardiff City Center (integrated furniture), Phoenix Police Department (trees)	[30]
		France, Germany	[31]
		Federation Square, Melbourne;	[30]
	Reduction of buildings vulnerability	Coimbra (Portugal); Sant'Antimo (Italy)	[27]
		Steel connectors; FRP and FRCM; Reticulates system	[32]
		Lorca (Spain)	[33]
		Wall Street (NYC); T-DAYS Via Rizzoli (Bologna)	[34]
	Reduce vehicular traffic		[35]
	Increase community and individuals' awareness to mitigate users' vulnerability and casualties		[27]

4 Results and Discussions

BETs are useful tools to provide preliminary assessments of the effectiveness of mitigation strategies that can be applied to HBE. In this section combinations of multi-risk scenarios for each BET are identified with the aim to preliminary defining the potential effective strategies for the SUODs/SLODs mitigation. Such identification would intrinsically consider the climate and seismic zonation of the sample considered for BETs definition.

4.1 BETS Towards Multi-risk Scenario Combinations

According to the methodological approach, each BET can be subject to all of the considered SUODs and SLODs. However, the lower or higher susceptibility to a specific risk scenario has been identified due to the specific morpho-typological characteristics according to the nine parameters that define the BETs.

A combination of characterizing risks has been defined for each BET by the statistical analysis of the frequency of the single hazard and hazards combination within the Italian context sample [8]. It is worth specifying that only the combination of a SLOD and a SUOD has been counted, while the combination of two SUODs at the same time has been excluded, given their rare simultaneous occurrence. The results of this analysis are briefly reported in Table 3. All BETs, except 4B, are more frequently prone to seismic risk, due to the high seismicity of the entire Italian territory. With respect to multi-risks, all BETs resulted to be prone to at least a combination of two risks. BET 3 represents the most critical ideal scenario to multi-risks, since it is susceptible to the four hazards. These results represent all the potential hazards to which the BETs may be subject but do not entail the possibility of a real simultaneous overlap of all of these.

Table 3. Single and multi-risk scenarios characterizing BETs, where T: Terrorism, S: seismic, P: pollution, UHI: Urban Heat Island risk

BET	Risk	BET	Risk
1A	S, S + UHI	4A	S + P
1B	S, S + P	4B	P, S + P
2A	S, S + UHI, S + P	4C	S + UHI
2B	S, S + P	5	S, P, S + P
3	S, S + UHI, S + P, T + UHI		

4.2 Potential Mitigation Strategies to Multi-risks for BETs

The first step in adopting a multi-risk mitigation approach is to identify common strategies that can be applied to each BETs for each risk. From the results of the study, illustrated in Table 4, it emerges that many strategies can mitigate both SLOD and SUOD risks. In addition, BETs may be subject to different risks at the same time. Therefore, in Table 4 mitigation measures have been considered for both the potential for each risk and for combined risks. In addition, each strategy is associated with the BETs according to the risk or multi-risk to which it is subject.

The effectiveness of the strategies is evaluated considering the general potential mitigation impact through a qualitative colour scale from green to red: very high, high, medium, low, negligible mitigation potential. The evaluation of the mitigation measures is based on the analysis of the main parameters that influence each hazard UHI-Urban Heat Island, P-Pollution, S-Seismic, and T-Terrorism risks. In particular, for P the effects on particulate matter dissipation, particulate matter reduction is considered; the effects on air temperature level reduction, solar radiation reflection, heat dissipation are considered for UHI; reduction of OS and building vulnerability for Seismic risk; reduction of OS vulnerability and hazard for Terrorism. In addition, the effects on BET users of the mentioned strategies has been addressed due to the importance of applying simulations-based approach using behavioural models (e.g. evacuation, transportation), which is considered in the BES2ECURE project. This effect is reported with a checkmark (✓) if it applies.

Comparing the effects of strategies between SUODs and SLODs reveals potential common joint-mitigation effects. Nevertheless, some of the proposed strategies could only be effective for multiple risks in determined conditions. For example, a cool plaster on the external envelope is effective on the UHI but could also be effective for the seismic risk if it is also reinforced, and against P, if it contains self-cleaning pigments; the replacement of pushing roofs is an effective measure to reduce the buildings seismic vulnerability but it could reduce the impact of solar radiation (UHI) if designed as a green or cool roof. These cases are marked with an asterisk (*). It is useful to consider these strategies when there are more risks to be mitigated at the same time.

In the last column, only those BETs in which the strategies are potentially effective for their combination of risks have been listed.

5 Conclusions

Historical built environments (HBE) are affected by critical risk scenarios given their specific vulnerability, and the high exposure due to the presence of numerous inhabitants and tourists. Mitigation strategies can produce benefits for specific SLODs or SUODs risks. However, to avoid negative effects and to consider potential joint beneficial effects, the strategies need to be analysed from a multi-risk perspective to increase the overall resilience of the HBE.

In this work, the topic is addressed in two stages: first, typical Built Environment Typologies (BETs) prone to SUODs and SLODs are defined according to morpho-typological and construction parameters; then, multi-risk mitigation strategies are collected according to the literature review and grouped depending on the scale of application. Finally, for each BET, combinations of SUODs and SLODs were defined, and potential strategies were identified to be effective simultaneously against such combinations of risks. For example, it was found that one of the most holistic strategies could be a conscious planting of trees, which are effective against UHI, P and T. These preliminary assessments could be of interest for policymakers and urban administrators towards more resilient HBE. Furthermore, BETs can be used to assess the effectiveness of the mitigation potential of each solution to multi-risks through quantitative estimation by specific simulation tools.

Table 4. Potential impact of mitigation strategies (from negligible (red) to very high (green)) for SUODs and SLODs and correlation with BETs.

Category	*Strategy*	*Examples*	*SLOD*		*SUOD*		*Users*	*BET*
			UHI	*P*	*S*	*T*		
Morphological factors	Improve vegetation and OS permeability	Trees					✓	**All; 3**
		Hedges	*					All
		Preventive evacuation plans					✓	All
		Redundancy of evacuation routes					✓	All
Physical and construction factor	OS/building surface temperature reduction	Cool pavement						All
		Permeable pavers						All
	Reduction of BE/buildings vulnerability	Cool façade			*			**All; 1A, 4C, 2A, 3**
		Protection of strategic lifelines and infrastructures						All
		Increase available free areas and define «safe» areas						All
		Masonry quality improvement						All
		Replacement of pushing roof typology	*					**All; 1A, 4C, 2A, 3**
Behavior strategies	Reduce vehicular traffic and emissions	Public transportation						All
		Shared mobility						All
	Reduce users' vulnerab.	Evacuation training					✓	All
		Support vulnerable people					✓	All

*Potentially applicable, if integrated with different strategies.
BET in bold: combined multi-risk strategy effectiveness for the BET

References

1. Fatiguso, F., Bruno, S., Cantatore, E., et al.: Built environments prone to sudden and slow onset disasters : from taxonomy towards approaches for pervasive training of users. 1, 1–15
2. Quagliarini, E., Currà, E., Fatiguso, F. et al.: Resilient and user-centered solutions for a safer built environment against sudden and slow onset disasters: The BE S2ECURe project. In: Littlewood, J., Howlett, R.J. JLC (eds) Sustainability in Energy and Buildings 2020. Smart Innovation, Systems and Technologies, vol 203. Springer, Singapore (2021)
3. PRIN (2017) BES2ECURe. Built Environment Safer in Slow and Emergency Conditions through behavioUral assessed/designed Resilient solutions. https://www.bes2ecure.net/il-progetto
4. Amico, A.D., Russo, M., Angelosanti, M., et al.: Built Environment Typologies Prone to Risk : A Cluster Analysis of Open Spaces in Italian Cities, pp. 1–32 (2021)
5. Godey, S., Bossu, R., Guilbert, J.: Improving the mediterranean seismicity picture thanks to international collaborations. Phys. Chem. Earth **63**, 3–11 (2013). https://doi.org/10.1016/j.pce.2013.04.012
6. D'Orazio, M., Quagliarini, E., Bernardini, G., Spalazzi, L.: EPES–Earthquake pedestrians′ evacuation simulator: a tool for predicting earthquake pedestrians′ evacuation in urban outdoor scenarios. Int. J. Disaster Risk Reduct. **10**, 153–177 (2014). https://doi.org/10.1016/j.ijdrr.2014.08.002
7. Rosso, F., Bernabei, L., Bernardini, G., et al.: Urban morphology parameters towards multi-risk scenarios for squares in the historical centers: analyses and definition of square typologies and application to the Italian context. J. Cult. Herit. under Revi. (2022) CULHER-D-22-00015R1
8. Halkidi, M.: On clustering validation techniques, pp. 107–145. Springer, Berlin (2001)
9. Hamburg.de Grünes Netz Majowiecki M.: Roof of the pedestrian walkway for the Milan Expo 2015 exhibition complex. In: Majowiecki website (2015)
10. Tsoka, S., Tsikaloudaki, K., Theodosiou, T.: Urban space's morphology and microclimatic analysis: a study for a typical urban district in the Mediterranean city of Thessaloniki, Greece. Energy Build. **156**, 96–108 (2017). https://doi.org/10.1016/j.enbuild.2017.09.066
11. Abdallah, C.: Los Angeles is painting the streets white (Again), and Your City Might Be Next|ArchDaily. In: ArchDaily (2018)
12. Susdrain: Case study, Derbyshire Street Pocket Park. London Borough of Tower Hamlets (2014)
13. Green City Solutions: CITYTREE—Green City Solutions (2021)
14. Akbari, H., Berdahl, P., Levinson, R., et al.: Cool color roofing materials draft final report prepared for. Energy **73** (2006)
15. Kptecki, P.: New York City Painted Over 9.2 Million Square Feet of Rooftop White. In: Insider (2018)
16. Caro, J.C., Armour, T., Fernandez, M.: Madrid + Natural: Nature-Based Climate Change Adaptation (2016)
17. Elassar, A.: Converse is sponsoring giant murals that break down air pollutants in 13 cities around the world—CNN Style. In: CNN Style (2020)
18. Loomans, T.: The World's first algae-powered building opens in Hamburg. In: inHabitat (2013)
19. Fernyhough, J.: Enel X launches bid to power Australia's electric bus revolution. In: The Driven (2021)
20. Di Bartolo, C., Bosetti, S., De Stasio, C., Malgieri, P.: Smart choices for cities Cities towards Mobility 2.0: connect, share and go! Cities towards Mobility 2.0: connect, share and go! Smart choices for cities (2020)

21. Setiffi, F., Lazzer, G.P.: Riding free-riders? A study of the phenomenon of BlaBlaCar in Italy. Contemp. Collab. Consum. 77–96 (2018). https://doi.org/10.1007/978-3-658-21346-6_5
22. Comune di Milano: Mobilità. Pubblicato il bando per contributi all'acquisto di veicoli a basso impatto ambientale—Mobilità. Pubblicato il bando per contributi all'acquisto di veicoli a basso impatto ambientale
23. U.S. Green building Council (2019) LEED—Reference guide for building design and construction
24. Ma, W.J., Lin, Q.X., Lin, H.L., et al.: Effectiveness of health education about heat wave hazard prevention in the elderly: a mixed effect model analysis. Zhonghua Liu Xing Bing Xue Za Zhi **37**, 1228–1232 (2016). https://doi.org/10.3760/cma.j.issn.0254-6450.2016.09.009
25. Guo, S.-E., Chi, M.-C., Hwang, S.-L., et al.: Effects of particulate matter education on self-care knowledge regarding air pollution, symptom changes, and indoor air quality among patients with chronic obstructive pulmonary disease. Int. J. Environ. Res. Public Health **17**, 4103 (2020). https://doi.org/10.3390/ijerph17114103
26. OECD SWEDEN conclusions and recommendations
27. Zlateski, A., Lucesoli, M., Bernardini, G., Ferreira, T.M.: Integrating human behaviour and building vulnerability for the assessment and mitigation of seismic risk in historic centres: Proposal of a holistic human-centred simulation-based approach. Int. J. Disaster Risk Reduct. **43**, 101392 (2020). https://doi.org/10.1016/j.ijdrr.2019.101392
28. Salvo, G., Di, G.M., Pellegrino, P.B.P.: Prevenzione e ricostruzione per la riduzione del rischio sismico. Planum XV Confere (2012)
29. Ahn, Y., Tsukaguchi, H., Ogawa, K., Tanaka, K.: Study on disaster risk assessment of cultural heritage and road network improvement in historical city. J. Disaster Res. **6**, 119–131 (2011). https://doi.org/10.20965/jdr.2011.p0119
30. Coaffee, J.: Beyond Concrete Barriers Innovation in Urban Furniture and Security in Public Space (2018)
31. Nationale SG de la D et de la, S.: Faire Face Ensemble. Vigilance, Prévention Et Protection Face À La Menace Terroriste (2016)
32. Corradi, M., Di Schino, A., Borri, A., Rufini, R.: A review of the use of stainless steel for masonry repair and reinforcement. Constr. Build. Mater. **181**, 335–346 (2018). https://doi.org/10.1016/j.conbuildmat.2018.06.034
33. Basset-Salom, L., Guardiola-Víllora, A.: Influence of the maintenance in seismic response of Lorca historic centre masonry residential buildings after the 11 May 2011 earthquake. In: Structural Studies, Repairs and Maintenance of Heritage Architecture XIII. WIT Transactions on the Built Environment, pp. 343–354 (2013)
34. Pedrini, L.: Bologna, le fioriere anti-terrorismo dei T-Days. Le foto (2017)
35. Quagliarini, E., Fatiguso, F., Lucesoli, M., et al.: Risk reduction strategies against terrorist acts in urban built environments: towards sustainable and human-centred challenges. Sustainability **13**, 901 (2021). https://doi.org/10.3390/su13020901

Sensorial Design—A Collaborative Approach for Architects and Engineers

P. Grant[1,2](✉), J. R. Littlewood[2], R. Pepperell[3], and F. Sanna[4]

[1] Stride Treglown Architects, Dowlais Road, Cardiff CF24 5QL, UK
pggrant@cardiffmet.ac.uk
[2] The Sustainable & Resilient Built Environment Group, Cardiff Metropolitan University, Cardiff CF5 2YB, UK
[3] Cardiff Metropolitan University, Fovography, Cardiff CF5 2YB, UK
[4] Cardiff School of Art and Design, Cardiff Metropolitan University, Cardiff CF5 2YB, UK

Abstract. STUDENT PAPER: Post occupancy evaluations (POEs), allow the design team to see how well the initial design objectives have succeeded when in use. Current POE procedures gather data relating to its sustainability. They were largely developed throughout the sixties and seventies by academics and engineers [1] resulting in a legacy for POE procedures to focus on the technical performance of the materials and components. In this paper we propose the engineer has the skills to contribute even more to the aesthetics of a building design. Architects are naturally keen to know how the fabric and components of their designs are performing technically. Quantifiable data is often sought through collaboration with the design engineer, often using electronic devices to record performance data. There is limited input from occupants. Feedback on a building's 'aesthetics' remains scant, often focussing on the visual appearance of a building design and by reviewing an occupant's visceral response; commonly known as the 'wow factor'. The first author has significant experience in the design of schools in South Wales, receiving the Eisteddfod Gold Medal Award for Architecture for a new school in South Wales on behalf of his employers in 2017. He has had many informal consultations with occupants who will reveal their sense of architectural delight by referring to their five common senses: sight, sound, smell, taste, and touch. Survival instincts have evolved these senses to respond to changes, resulting in the occupants' innate ability to sense changes in the built environment irrespective of how small the changes are. Architects and engineers can design for these changes and evaluate them post completion. A mixed methodology is recommended for gathering data and knowledge from post occupancy evaluations, making the outcomes of more appealing to more readers, including the inhabitants.

Keywords: Architecture · Phenomenology · POE · Qualitative and quantitative methodologies · Sensory response · Social value and wellbeing

J. Littlewood and R. J. Howlett (Eds.): SEB 2022, SIST 336, pp. 208–217, 2023.
https://doi.org/10.1007/978-981-19-8769-4_20

1 Introduction

This paper is the third publication by the same authors. In their previous papers [2, 3] the authors established a context for multisensory design in architecture. This paper focusses on engineering design. Whilst the architect will remain the overall design lead, there are opportunities for the engineer to weave some sensory experiences into the overall architectural and spatial experience. The objective is to increase the experiential delight of new buildings, leading to an increase in the occupants' enjoyment and to the building's greater longevity in use.

As part of his professional doctorate studies, the first author proposes a paradigm shift in the evaluation methods of his employer's Post Occupancy Evaluation (POE) procedure. The proposed change will mean that completed schemes can be evaluated more from an *architectural design concept* position. He will be looking for feedback on the quality of spaces rather than the performance of components and materials. The journey, so far, has resulted in changing his research question from looking at the final stage of a design project to looking at the changes required to the initial stages of a design: the *design brief*. His research question still proposes a change in how POE is gathered, but it now encourages evaluations to take place after sufficient time for the initial visceral response to subside. The occupants can then reveal, through informal consultations, how they have settled into their new environment.

Meaningful evaluations require changes to the brief so that equal weighting is given to the aesthetic design of spaces, including various *design interventions* for the senses. Each '*sensory intervention*' can then be reviewed throughout the design development stages, ensuring they remain intact throughout the construction period. Post occupancy evaluation will then have a meaningful set of reference points to discuss and evaluate with the occupants when the building has been in use for some time. The objective will be to gather some phenomenological feedback to reinforce the engineered performance of the building's fabric and components.

In the first author's experience, design engineers *of all disciplines* enjoy the sense of good design. He is confident that engineers welcome the opportunity to contribute more to the experiential quality (the architectural delight) of a building. Good engineering design will inevitably influence the occupant's sensory experience, and their responses will be expressed with reference to their own sensory abilities. As lead designers, architects are encouraged to whole design team to incorporate some *sensory design interventions*. Engineering designs should contribute to the overall sensory experience when the building is in use.

The aesthetics of a building are not just visual. Consciously or otherwise, a new building will prompt occupants to respond with comments that rely on to their *common senses* of sight, sound, touch smell and taste. Ricca [4] has argued the psychological benefits of good design and how it can influence the occupant's behaviour, health, and wellbeing through the reduction of stress, blood pressure, depression, and anxiety. There is a social value in this process and so there are good reasons to encourage the *whole* design team to contribute to the fine-tuning of the overall designed experience.

Nicholson [5] has argued there is a clear evolutionary basis for many psychological and physiological human responses to building designs [5]. Most human beings are effectively 'hardwired' to their survival instincts. They are always alert to sensorial change; the senses often being fatigued after constant exposure to the same stimuli. Whilst these changes can only be evaluated post completion, they are an essential part of the initial design concept, making feedback from POE an essential part of the initial design process. This paper explores an approach to building design that encourages both the engineer and architect to collaborate in design opportunities that allow for a range of sensory experiences. It sets out a methodology for gaining knowledge of how the occupants perceive and respond to the built environment. Ultimately, a multisensory design approach will inspire similar approaches in the development of landscape design and urban spaces, leading to the improvements in the inhabitants' social, cognitive, and emotional wellbeing.

2 Aesthetics

In this study the authors are referring to the wider meaning of aesthetics, derived from from the Greek word for '*perceiver*' or '*sensitive*'. In the first author's experience, the occupants will convey their feelings about spaces by describing their cognitive understanding of the way things *look and feel*. The way a space looks is both important to the designer and to the user but, as Holl [6] points out: '*the way it feels, the smell and sound of a place*' also contributes to a '*complete experience of a place*'. This is essentially the occupant's feeling of delight when using a building and they are very much a part of the 'aesthetic' design of the architecture, including the significant design input by engineers. The engineers' designs will also have an impact on the overall appreciation of a building. Good design requires more than just the minimum. All it requires is for the design team to vary from the minimum statutory and regulatory requirements by incorporating their own design flair and ingenuity.

Designers are encouraged to do more than simply satisfying an industry recognised metric that can be measured easily post-occupation. The author's change proposal requires the designers to use their imagination, ingenuity, and initiative to maximise the sensory opportunities. Below are some aesthetic design opportunities for both the architect and the engineer to consider. The intention is to provide an indication of the opportunities rather than be an exhaustive list of possibilities.

Thermal Aesthetics: This is not a new concept. Half a century ago, Lisa Heschong had been exploring the thermal delight in architecture [7]. She noted that food is as fundamental to survival as our thermal environment. Whilst it is theoretically and physiologically possible to provide for all nutritional needs with a few pills and an injection, it would lack the essential desire for eating to be a social event. There is human need for food and drink to have taste, smell texture, temperature, and colour, and there a need to allow for detectable changes in terms of temperature, colour, smell and taste. Similarly, thermal designs can be designed for variation. Lisa Heschong [ibid] has noted that comfort zones for occupants vary widely: in Britain it can be between 14 °C and 21 °C whereas in America the range is between 20 °C and 26 °C. The difference is possibly affected by the range of latitudes over which the data is drawn. A later study found that residents in the UK set their central heating to allow for internal temperatures between 17 and 23 °C which links average indoor temperature to the mild outdoor conditions of west central Kenya or the Ethiopian highlands where human life is first thought to have evolved [8]. Each assessment appears to have a range of about 6 °C and so there is a range of sensory design opportunities for engineers to vary the heating intensity by considering a variation in the way heating is transferred either by radiation, conduction, and convection or as a combination of the three.

Aesthetics of sight: Building orientation has a major influence on the access to daylight within a building and is one of the important sustainable considerations in design. Daylighting and artificial lighting will influence the visual appreciation of both the materials and the spaces created. Lighting levels are often specified in a design brief, but there are opportunities to allow for variation. Light and shade will inform the impact of what can be seen. The intensity of light can reveal the visual patterns created in colour, shape and form will influence how we respond to the visual appearance of the built environment. Internal spaces should avoid being painted white throughout, relying on signage to help guide and direct the inhabitants to various destinations. Diurnal changes can be incorporated to enhance circadian expectations. Visual impairment rarely means there is no perception of light. Designs that allow for visual impairment will also increase spatial enjoyment for all other occupants.

Aesthetics of sound: There are opportunities for the acoustician to include sensorial changes in the absorption and reflection of sound. Once the technical requirements of the building fabric and spaces have been resolved, the acoustic engineer could consider some creative changes in the design; not only to increase the delight in use, but also to provide wayfinding information and acoustic interest. Textural changes in the surface of material can be introduced in collaboration with the architect to provide both visual and acoustic changes that will enrich the occupant's sensory interest, both in the design of internal and in the design of external spaces. Delight can be achieved when the two are designed as a continuous transition from outside to inside, linking as many of the sensory aesthetics in a seamless manner.

Aesthetics of touch: Touch is the first sense to develop in human beings. In the Eyes of the Skin, Pallasmaa determined that all mammals perceive touch through physical pressure, temperature, light touch, vibration, pain etc. often in combination with another sense [9]. It is therefore important to the wellbeing of building occupants, especially for those with visual impairment [10]. Touch is often our first sense of a building, making the approach and contact with a main entrance an important feature of the architecture. The choice of materials for the handle can have a lasting impact on the impressions made in this first contact. Heating and cooling are both sensed by the skin and so the choice of heating system will influence the sensory responses of occupants.

Aesthetics of smell: Smell is omnidirectional. It is more difficult to avoid than sight, but good spatial design can help to control its movement. Every space has its own characteristic smell which can have a lasting and emotional effect. Mehta [11] noted this could possibly have a stronger and longer lasting influence on the architectural experience than sight. The way that internal air moves around internal spaces can affect the intensity and flow of smells. There is a tendency for design solutions to eradicate smells but there are opportunities to enhance smells through air displacement. The smell of baking at the supermarket entrance is a positive displacement of smell within a food store can heighten awareness of freshness. In schools there are a range of activities that will have their own intrinsic smells. These could be designed in a positive way to assist with wayfinding particularly for the visually impaired.

Aesthetics of taste: The sensation of taste is closely related to smell. Taste is more often experienced when all other senses perceive it is safe to taste. There are design opportunities in the appreciation of taste. Whilst it is not actually encouraged to taste a building, Forster [12] and Yi Hsuan [32] have shown that colours and sound can generate oral sensations related to taste. Even so, Eberhard [14] found that a restaurant can influence a customer's 'conditioned response' to the taste of the food through lighting, colour, and comfort.

A final note: Sensory designs can activate all senses to varying degrees of intensity or concentration; both in the delivery and in the reception of a sensory response. In 2014, Spence et al. revealed that the intensity of senses might be better limited to two senses in order not to confuse. This was reported in a study into 'Store Atmospherics' when it was found that there appears to be an optimal level of stimulation leading to a risk of sensory overload [15].

3 Proposed Research Methodology

The first author is seeking to support the fabric performance data with a complementary methodology that can gather feedback for the conceptual stages of building design. He is seeking a methodology that involves the occupants because he believes that feedback relating to the occupant's experience and perceptions of spaces will inform future designs and possibly lead to a more complete sensorial experience of a building. His research is seeking a methodology that will gather some of the less quantifiable responses from the users. He aims to develop a qualitative approach to evaluating buildings in use. The objective will be to capture some feedback and data based on the user's feelings and their lived experience and this can only be gathered by meeting with the occupants.

Current evaluation methodologies tend to focus on collecting the more tangible (quantitative) data that can confirm whether the design team has met the performance targets set within the original design brief. This approach to evaluation satisfies the current requirements set out in Stage M of the Royal Institute of British Architect's (RIBA) Plan of Work 2020 [16], the aim of which is to record the sustainability measures when the building is in use. These metrics are all quantifiable and so the current methodology aims to gather valuable knowledge for a Building Performance Evaluation (BPE). Comparative studies can be made with respect to industry benchmarks on performance and any disparities between the *designed* performance and the as-built performance can be assessed. The sustainability parameters sought by Part M of the Plan of Work are often achieved by meeting the quantifiable metrics currently benchmarked and assessed by the engineering industry. Data can be collected '*post occupancy*' without any input by the inhabitants, the real-time users of the spaces. The ontological position of this paper is that **Post Occupancy Evaluations** *(POEs)*, should also include an evaluation of the design with the building users. The aim is to increase the architect's *knowledge and understanding* of *why and how* things are perceived to be the way they are in the real world. Gaining knowledge in this way would be based on a *qualitative methodology* that aims to discover why the occupants are doing what they have been observed as doing.

As Bryman [17] says, a qualitative methodology can involve either '*semi-structured interviews*' to reveal views on a particular topic, or '*in-depth interviews*' to gather knowledge on the occupant's personal (*phenomenological*) experience. The first author favours the latter approach as it will allow the participants to talk freely about their experiences and it will allow the facilitator to interpret the feedback as part of the recording of data.

There is no disputing that both BPEs and POEs are appropriate methodologies for gaining knowledge of how a building is performing in use, but a combination of the data from each approach could provide a clearer picture. A combination of the two methodologies could aim to clarify and enrich the knowledge gained from each of the methodologies. As Bryman has noted, the two methodologies can address both the 'what' (quantitative and qualitative) questions and the 'how' or 'why' (qualitative) questions [ibid]. This approach to reporting on POE findings moves away from the assumption of building performance is paramount. Instead, there is the opportunity to juxtapose the 'lived experience' from the occupants with the performance data of the building fabric. There is also the opportunity to triangulate the feedback to better understand different interpretations of a view (or views) of the feedback, making the combined report far more meaningful to a reader [18].

4 The Pilot Study—About to Commence

A pilot study has been developed to capture the occupants' sensorial responses to the built environment. It focusses in on school design which is a specialism of the first author. Schools can be seen as a microcosm of the outside world in the sense that all the subjects and activities taught within a school are represented in the world outside school. Knowledge, social skills, trades, and professions are all learned through the initial years at a school.

The first author's research assumptions are based on his personal and professional experience of feedback from occupants in the schools he has designed. Irrespective of any sensory deficiencies, all occupants will 'sense' their environment; both in a conscious and in a subconscious way [19]. Whilst an occupant's feelings and perceptions will differ in many ways, often varying in intensity, most will make references to their five common senses of sight sound touch smell and taste. Some might combine more than one sense by presenting mixed modal responses.

The pilot study will involve focus groups, consisting only of the teaching staff. It will be based on a set of five questions that are aimed to start a conversation with the participants. The objective is to record, *after at least one year in use*, how the occupants *feel* about their school throughout the day. It is assumed that their responses will be less that visceral and hopefully more robust for later evaluation and assessment.

The research methodology will assume a *constructivist* approach to interviewing participants, as this will allow for a *continual review* of the participants responses within different interview settings, both in terms of time and location. Clarity in feedback may be affected by the participants mood at the time of the research study. It might also be affected by the research context and by the presence of the researcher. For these reasons, the recording of the feedback will inevitably require some degree of *interpretation*.

5 The Questionnaire

This research encourages the whole design team to introduce *experiential changes* in the built environment, deliberately designed to stimulate the occupants' senses as they move about the spaces. An objective is to see if the occupants can help to calibrate a range of experiences within the built environment.

A questionnaire has been developed to help start an informal dialogue with the school's staff. We are seeking to capture their perception and experience of the designed spaces with respect to time of the day. We are seeking feedback on how the participants feel about their designed spaces: whether they are enjoyable to use or disliked. By adding the time of day, we can evaluate the effect circadian rhythms.

Records will be grouped according to how the occupants have related their experience with respect their five common senses of sight, sound touch, taste, and smell. The interviews will be transcribed and then coded within an analysis program (ENVIVO) under the five common the senses. Future coding might include gender, age and ethnicity as these parameters might show variations in responses, but there will be a need for a further ethics application to analyse such data in this way.

As the methodology is intended for architects, there is a risk that participants may focus on what they see as an architect's remit, the *visual aesthetics*. Hopefully, participants can be steered to comment on their perception of the four other common senses, even if this could be seen as guiding the respondents. The five questions are as follows:

Q1: On entering the school grounds, please describe any feelings you may have on the route from the school gate to the point of entering the school building? Include any preferred routes across the external spaces and explain why you use a particular entrance or door. *This will record the occupant's expectations and preferred experiences upon arrival to the school. It will discuss how occupants might enter the school, which points of access are used, and their choice of route within the school ground to arrive at their first place of use.*

Q2: Turning to your morning break times, please talk about some of the preferred locations where you like to take a break from your normal workspace, noting why you prefer these locations. It will also be useful to note any places that are less preferred, including any external spaces. *This will record the occupant's capability of acclimatising to the building, identifying their preferred use of spaces. There may be preferred routes that the occupant takes to feel comfortable.*

Q3: Lunchtime is an opportunity to move away from your workplace. Please describe your favoured or usual place to meet, eat and refresh, ready for the afternoon's session. Some of the spaces you may be using, such as the dining room, may be less favourable than you would wish. Please also note about any places that are less preferred, including any external spaces. *This will aim to discover if there are any preferred areas within the building where occupants wish to take a short break away from the daily tasks, recharging their energy levels to complete the day.* We will recognize there is the socially driven aspect of people wishing to meet colleagues irrespective of whether they like the space where they agreed to meet.

Q4: Mid-afternoon is when there is often a feeling that the day is almost done. Please describe any feelings about the spaces you use in the school to complete the day. This could include classrooms, workspaces, study areas or physical activity areas, as well as any internal or external spaces that are less preferred. *This will record the difference in perception of school when the day is nearly complete. Do occupants use alternative spaces to relax in the afternoon? Will changes in daylight and artificial light and location of the sun influence their choices? Will occupants display a change in their tolerance in things they are unhappy with in the building (or vice versa), wishing they could be elsewhere.*

Q5: It is the end of the school day, and you may be looking forward to getting back home. Please could you describe your return journey away from your workspace and through the school, including leaving the site? *This might reveal how occupants feel about leaving the school buildings. Some may have a strong desire to leave and forget about the building they must occupy for significant parts of their waking day.*

6 Discussion

Human survival has evolved a arrange of sensory responses that can be incorporated into the design for buildings. Occupants are always alert to sensorial changes. Post occupancy evaluations provide the ideal opportunity for designers to discover how selected design interventions are performing in use. Good architecture requires feedback on the occupants' real-life experiences. POE cannot be based solely on the fabric's performance data as this usually collected without meeting the occupants who are the real-life users of the building. Quantitative POE data provides excellent feedback on how the building fabric is performing, but buildings are for people to use and to enjoy. Qualitative feedback is therefore equally important. Qualitative research involves direct contact with the public whose permission to participate must be sought well in advance. Data recording must be carried out with respect to potential ethical concerns that could arise over the course of the pilot study, and to any subsequent relationship with the participants [20].

A mixed-methodology approach to reporting on post occupancy evaluations is recommended so that knowledge gained from the fabric's performance can be coupled with knowledge gained from the occupants' real-life experience. Feedback can then be presented in a clear and collective manner that will appeal to more readers. The objective is to provide an opportunity for architects to fine-tune the design intent and for the inhabitants to maximise the social value of their own environment.

Research data collection requires approval from the university's Research Ethics Committee. A formal application has been approved for a pilot study and this includes precise details on how the data will be collected and from whom it will be collected. Following an initial meeting with the headteacher, a formal request to participate in the pilot study will be sent out as a letter to all staff members. This will be the first contact with the teachers and so it will help to gauge initial interest. School staff are extremely busy throughout the day and so the formal request for participation will include full details on the time required for meeting staff, the potential for sensitivity in the outcomes and the need for confidentiality. Clear assurances on data control and compliance with General Data Protection Regulations (GDPR) is essential. Data will only be used as agreed and raw data will not be shared with anyone.

The first author has significant experience of consulting with school staff and sometimes with pupils who can be supervised by a responsible member of staff. He has CRB (now known as the DBS) Enhanced Disclosure certificate that covers access to schools. Even so, the ethics committee has raised concern about *vulnerable persons*: such as children under the age of 18. This is a shame because the first author's research is specifically related to the design of school buildings where about 90% of the occupancy will be pupils. Schools will also have children with special educational needs (SEN). Some of these pupils can have a heightened sensitivity to their environment, making them more likely to say how they feel about the spaces they use. In many ways, SEN pupils would be ideal participants in a survey on sensitivity. Alas, they fall well within the category of *vulnerable persons*.

7 Conclusions

Architects desire feedback on the how the occupants feel about the spaces they have designed. This requires the designers to evaluate their designs when the occupants have had enough time to settle in after the building is completed. Connecting with the inhabitants is an essential part of the objective to gather meaningful feedback.

Engineers should be encouraged to heighten the sensory experience of all occupants. A collaborative approach to multisensory design would include a *variety* of sensory design interventions from each of the engineering disciplines, all of whom can enrich the occupant's experience and *aesthetic* appreciation of a new school design. Changes in intensity can be introduced, but possibly prioritising on the intensity in just two sensory responses at any one time [14]. Variation is a key objective.

A mixed methodology combines knowledge gained from the occupants with the data recorded in the technical evaluation. There is a triangulation data. The aim is to provide a clearer picture of what has been found, making the report more meaningful to the reader. The output should then seek to attract a broader use and understanding of the feedback process, resulting in a far more useful set of POE findings for the whole design team and for the inhabitants. It is time for a change.

References

1. Cooper, I. (2001) Post Occupancy Evaluation –Where Are You? Building Research and Information, Vol. 29 No. 2, pp. 158–63. Special Issue: Post Occupancy Evaluation
2. Grant, P., Littlewood, J., Pepperell, R.: Can architectural delight improve concept design and human sensory response in schools. Springer J 99–110 (2020)
3. Grant, P., Littlewood, J., Pepperell, R.: Article title. An exploration of the relationships between architectural delight and human senses. Springer J. 99–110 (2021)
4. Ricci, N.: The psychological impact of architectural design. CMC Senior Theses 1767 (2018). https://scholarship.claremont.edu/cmc_theses/1767
5. Nicholson, N.: How Hardwired Is Human Behavior? Harvard Business Review (1989) https://www.researchgate.net/.../13115707_How_hardwired_is_human_behavior
6. Yang, W., Ji Holl, S.: "Thin Ice." In: Juhani Pallasmaa, The Eyes of the Skin (2005)
7. Heshong, L.: Thermal Delight in Architecture, 1979 Massachusetts Institute of Technology, Cambridge, Massachusetts. ISBN: 978-0-262-58039-7. http://mitpress.mit.eduCohen; L, et al.: Functional relevance of cross-modal plasticity in blind humans. In: Nature 1997 Sep 11; 389(6647):180–3 (1997). https://doi.org/10.1038/3827PMID:9296495; https://doi.org/10.1038/38278

8. Spence, C.: Senses of place: architectural design for the multisensory mind. Cogn. Res.: Princ. Implic. **5**(1), 1–26 (2020). https://doi.org/10.1186/s41235-020-00243-4
9. Pallasmaa J.: Eyes of the Skin: Architecture and the Senses (2006)
10. Cohen, L., et al.: (1997) - Functional relevance of cross-modal plasticity in blind humans. Nature **389**(6647), 180–183 (1997). https://doi.org/10.1038/3827PMID:9296495DOI:10.1038/38278
11. Mehta, B.K.: Smell and the architectural experience. Department of Neurology University of California, Los Angeles, Assistant Professor of Neurology (2014)
12. Forster, F., Spence, C.: What Smell?" temporarily loading visual attention induces a prolonged loss of olfactory awareness. PubMed (2018). https://doi.org/10.1177/0956797618781325
13. Yi Hsuan, T., et al.: (2019) - Environmental Sounds Influence the Multisensory Perception of Chocolate Gelati, Foods Journal, Received: 12 March 2019; Accepted: 10 April 2019; Published: 15 April 2019. Foods **8**, 124 (2019). https://doi.org/10.3390/foods8040124
14. Eberhard, J.P.: A Place to Learn: How Architecture Affects Hearing and Learning, The ASHA Leader Feature (2008). https://doi.org/10.1044/leader.FTR6.13142008.26
15. Spence, et al.: Store Atmospherics: A Multisensory Perspective, Oxford Institute of Retail Management (OXIRM). Psychol. Mark. (2014). https://doi.org/10.1002/mar
16. RIBA Plan of Work: Royal Institute of British Architects Publications, London (2020)
17. Bryman, A.: Bryman's Social Research Methods. Oxford University Press (2021) ISBN 978–0–19879605–3
18. Al-Ababneh, M.: Linking Ontology, Epistemology and Research Methodology. Published in Sci. Philoso. **8**(1), 75–91 75 (2020)
19. Goldhagen, S.: "How Architecture Affects Your Brain: The Link Between Neuroscience and the Built Environment." ArchDaily (2017). https://www.archdaily.com/876465/how-architectureaffects-your-brain-the-link-between-neuroscience-and-the-built-environment
20. Saunders, M., et al.: Mark Saunders, Philip Lewis and Adrian Thornhill—'Research methods for Business Students, 5th edn. Pearson Education Limited (2009) ISBN: 978–0–273–71686–0

Hemplime Blocks: Innovative Solution for Green Buildings in Italy

Chiara Moletti[1(✉)], Patrizia Aversa[2], Bruno Daniotti[3], Giovanni Dotelli[1], Vincenza A. M. Luprano[2], Anna Marzo[4], Sergio Sabbadini[5], and Concetta Tripepi[4]

[1] Politecnico di Milano, Department of Chemistry, Materials and Chemical Engineering, "Giulio Natta", Piazza Leonardo da Vinci 32, 20133 Milan, Italy
chiara.moletti@polimi.it

[2] ENEA C.R. Brindisi - SS 706 km, 7–72100 Brindisi, Italy

[3] Politecnico di Milano, Department of Architecture, Built Environment and Construction Engineering, piazza Leonardo da Vinci 32, 20133 Milan, Italy

[4] ENEA C.R. Bologna - Martiri Monte Sole street, 4–40129 Bologna, Italy

[5] A.N Studio di Architettura Disstudio.it, a.n.a.b., Milan, Italy

Abstract. In the last years, the attention to an eco-friendly development of the building sector has increased the use of green materials, mainly because the construction sector is one of the most polluting. The hemplime block represents a valid option in this direction: it is a valuable product to improve the sustainability of the building. Like any new product, the main issues are given by the absence of specific rules, starting from the production phase until the installation. For this reason, the main objective of this work is to take a first step toward the long process of identifying possible guidelines for the production and the testing phase of the products to achieve a CE (European Commission) certification. Another essential aspect to clarify is the definition of indications for the laying phase of the prefabricated blocks. This study significantly contributes to reduce uncertainties and skepticism about this technology. With these objectives, some experimental tests have been carried out to verify the reliability of the data declared in the datasheets of hemplime blocks produced in Italy, justifying any incongruity. This study also investigates other aspects of hemplime block, such as the main pathologies that may affect this technology during its lifetime and the maintenance operations necessary to restore the product. Furthermore, the thermal performance of a wall was studied in a climatic chamber to study its behavior in conditions similar to service life.

Keywords: Hemplime · Bio-based building materials · Hemplime blocks

1 Introduction

Among bio-based building materials, one of the most promising innovative solutions is hemplime which is manufactured by mixing a mineral binder and the biomass obtained by scutching of industrial hemp plant' stem i.e., hemp shives. This production process

J. Littlewood and R. J. Howlett (Eds.): SEB 2022, SIST 336, pp. 218–227, 2023.
https://doi.org/10.1007/978-981-19-8769-4_21

reduces the waste of hemp harvest and the exploitation of virgin natural resources to obtain the aggregates. An advantage given by the utilization of hemp is that it is a resistant crop, able to resist various climatic conditions. Moreover, it is versatile, and several hemp-based products can be produced using both fibers and shives [1]. The sustainability of hemplime has been evaluated in several studies; the impacts related to the manufacturing step can be further lowered by reducing transport distances and favoring the utilization of local hemp shives [2–4]. In addition, hemplime has proved to improve the energetic efficiency of buildings; this can be crucial to reduce impacts related to operational energy consumption which causes significant environmental impacts [5]. In particular, this material is characterized by high hygrothermal regulation properties, mainly due to the presence of the vegetal aggregate [6, 7]. This work focuses on the Italian production of hemplime blocks and on the testing of such building components to determine the criticalities and identify proper experimental procedures to assess material's properties. Hemplime is employed mainly in building envelopes and, for this reason, tests for water absorption and biological resistance determination have been selected for this research. In addition, thermal transmittance has been measured to investigate the material's thermal performance and validate an experimental methodology to be applied even on-site. The aim of the work is to set the basis for further testing of hempcrete blocks to optimize their properties and increase their spread.

2 Hemplime Blocks Production in Italy

Hemplime in Italy started to be used about ten years ago, i.e. twenty years after its first use in other European countries such as France, Switzerland, and England.

Also, about ten years ago, the first company started developing industrialized hemplime products. Since then, the growing interest in hemp in all sectors has led to the development of other hemp-based building materials and other companies manufacturing blocks.

Prefabricated elements as hemplime blocks or panels can be more easily standardized than site installation techniques (casting, hand-spraying). The survey, however, immediately showed that the Italian products have such different formulations that it is difficult to determine performance or professional rules that can encompass all products at the same time. The difference lies in the components and, consequently, also in their dosages. Different binders (e.g. aerial binders, hydraulic binders, mixes of the two types of aerial binders with pozzolanic components) give different chemical reactions during hemplime maturation that, in turn, determine differences in the resulting material. Consequently, the products made with the different binders show different performances depending on their density and thermal and hygrometric behaviour [8]. In addition to the use of different binders, the biodiversity of mix designs also concerns the plant aggregates used, i.e. shives [9]. Generally, virgin shives are used, but even mineralized shives have been employed in some cases. Mineralization is a preliminary treatment of shives with lime to make them less absorbent during the mixing phase.

The binders have certified characteristics that can be traced back to standardized classifications (NHL for Natural Hydraulic Lime, CL for calcic aerial lime, DL for dolomitic or magnesian aerial lime). By contrast, the only characterization of shives refers to the French ecolabel compiled by *Construire en Chanvre* [10]. These parameters aim at guaranteeing to the users of hemp straw declared and stable characteristics; this can give constancy to the production chains. However, the considered parameters neglect the chemical interaction between binder and vegetal aggregate during the mixing, curing and evolution phase, even in the case of mixtures based on air lime that trigger carbonation processes. In particular, it was pointed out that there was a complete lack of characterization of the plant (e.g. its chemical constituents and its grain size curve). In other words, the primary information that distinguishes a plant aggregate were unknown. Besides the mentioned essential characteristics, others were also identified and contained in the working draft to characterize the hemp shives (see the Table 1).

As far as the production of blocks from Italian shives is concerned, there are currently difficulties in sourcing and finding the quality of hemp since it is not certified. These assumptions can affect the stability of the final characteristics of the product.

The agricultural development of hemp, stimulated and promoted by various Associations, Bodies and Companies of national (*Assocanapa, Federcanapa*) or regional nature (*Agricanapa, Associazione Canapa Siciliana, Canapa Sarda Onlus, Canabruzzo, Canapa Ligure, Canapamo, Canapa Puglia, Lucanapa, Romagna Canapa, Rea Canapa, Sativa Molise, Sud Canapa, Toscanapa, Versilcanapa*) is mainly aimed at the use of its fruits and leaves for food, cosmetic and therapeutic purposes. Today, the stem is still considered an agricultural by-product, and its scutching and refining processes are not yet adequate to meet production needs throughout the territory.

Table 1. Requirements for hemp shives to produce building materials [11].

Parameter	Requirement
Amount of material extracted from the hemp plant	100% from hemp cultivation
Apparent density of the final product	±15% (annual variation)
Granulometry	>95% of shives with length < ld_{max}* ±10%
Humidity content	<19%
Dust	<2% (of the fraction passing 0.25 mm sieve)

*ld is the ratio between the length and the diameter of shives

The production methods for hemplime blocks are primarily derived from the vibro-compression systems developed in the last century for concrete blocks. Each company has developed its own choice of base material quality and mix design. However, they all follow the process of dosing, mixing (usually with planetary axis mixers), filling vibrating formwork, and compressing sufficiently to obtain a homogeneous distribution of the mix throughout the formwork. The storage and curing phases follow the manufacturing. The curing step has a significant influence on the timing of production and sales of blocks; this has determined the development of many formulations that have been tested to improve this phase. Some companies have opted for natural hydraulic lime to speed up the setting. Others started with a mix of white cement and hydrated lime and then moved on to purely aerial binders with the addition of microorganisms to accelerate carbonation. At the same time, others exploit pozzolanic processes to provide hydraulic properties to the aerial lime.

The blocks produced in Italy are only used as non-load-bearing construction elements in infill, counter walls and partition walls. For this reason, their thermal, acoustic and hygrometric characteristics are particularly valued. The topic of thermal performance focuses not only on thermal insulation but also on hygro-thermal behaviour for warm climatic contexts.

At present, in Europe, only in Spain denser blocks with load-bearing capacity are used in contexts where there are no seismic risks. Although blocks with hemp are not used in Italy with load-bearing function, their mechanical behaviour was nevertheless investigated through a structural behaviour model based on preliminary tests [12]. The investigation has shown how fillings with hemplime blocks in frame structures can improve mechanical performance by contributing to the bracing system. The combination of timber structures and hemp-based materials (block fillings, casting, spraying) is particularly suitable due to the hygrometric behaviour of both materials [12]. The use of blocks does not bring moisture to the wood and limits wet building systems to bedding mortar, which, if well cared for, may not contact the wooden structures. In addition to insulation, hemplime provides the necessary specific surface mass to each climate zone.

The installation of blocks is guided by the national regulations related to the selected building systems and to the technical specifications published by the manufacturers. Each manual is specific to each block since the formulation varies even among the different products of the same company. Also, specifications on the bedding mortar to be used exist; it usually contains the same binder as the blocks with different aggregates (shives for thick bedding powder or other aggregates for mortars to be laid in low thickness). Shaped blocks require specific interlocking laying and compliance with the rules of art for laying any masonry, such as staggering the blocks at each course and clamping at the corners and in the joint walls.

Following an in-depth survey (see Fig. 1) drawn up by the ENEA-Politecnico di Milano partnership in 2018, updated in 2022, there are currently five manufacturers of hemplime blocks in Italy [13]. In addition to these companies, others are importers of foreign products. These last are mainly imported from France, either because of proximity, patent longevity or because they are subject to certification and compliance with professional rules in the country of production.

In the regulatory field, the Professional Rules drawn up by the *Construire en Chanvre* Association [10], although not dealing specifically with the subject of blocks, remain the only validated document in a European country. However, there are also Guides of Good Practice drafted by other international associations that can serve as a reference, even if not legally.

3 Materials and Methods

The experimental tests have been carried out on three types of hemplime blocks (B1, B2, and B3) produced by three different companies. Their properties are reported in Table 2.

Apparent density (ρ_a) measurements have been carried out according to the standard UNI 1602:2013 [14]. Six specimens were tested for each type of block, weighed with

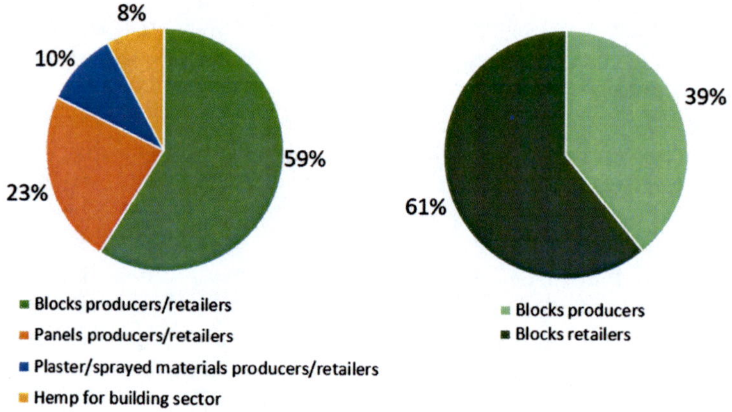

Fig. 1. Results of the survey about hemp-based building materials producers and retailers in Italy [13].

a scale with an error of 0.002 kg, and determined with a measuring tape (error of ± 0.5 cm). The tested specimens are blocks, with dimensions reported in Table 2. The apparent density has been calculated as the ratio between the mean value of the masses and the mean value of the volumes of the samples selected for each type of block.

Table 2. Properties of tested hemplime blocks declared by the producers.

Property	B1	B2	B3
Block dimensions (cm)	55 × 20 × 22	50 × 30 × 20	50 × 30 × 25
Dry density (kg/m^3)	300	330	350
Thermal conductivity (W/m K)	0.07	0.07	0.08
Thermal transmittance (W/m^{2}K)	0.33	0.22	0.25
Resistance to vapor diffusion μ	–	4.5	4.5
Specific heat (J/kg K)	–	1870	1870
Absorption coefficient (acoustic)	0.80	0.8	0.8

The values of apparent density on the day of delivery of the blocks (day 0) may be higher than the real ones depending on the conditions of the stock step of samples after their production. Indeed, absorption of water due to humidity and weather conditions can occur in this phase. For this reason, specimens have been put into a dryer machine with a controlled temperature, depending on samples conditions. Blocks B1 were left at 23 °C with a relative humidity of 50% for few days; blocks B2 were left in the dryer machine at 40 °C for 48 h, but then the samples became crumbled and too fragile to continue the test. Blocks B3 have been left in the dryer machine at 30 °C for 312 h. The apparent density measurements have been repeated during the drying phase until the difference between the expected and measured values has been lower than 10%. Then, water absorption tests have been carried out on samples with dimensions 20x20x20 cm, following the standard ISO 29767:2019 [15]. This test determines the capacity of the material to absorb water which is crucial to assess the material performance. The experimental setup is sketched in Fig. 2a: some plastic elements were used to create a mount from the bottom of the bowls. Then, some water has been poured into the bowl to achieve a level 10 ± 2 mm higher than the mounts. Specimens have been weighed and left on the mount for 24 h. After draining for 10 min, they have been weighed for the second time. Then, the absorption coefficient (W_p) has been calculated as the ratio between weight variation in 24 h and the submerged area.

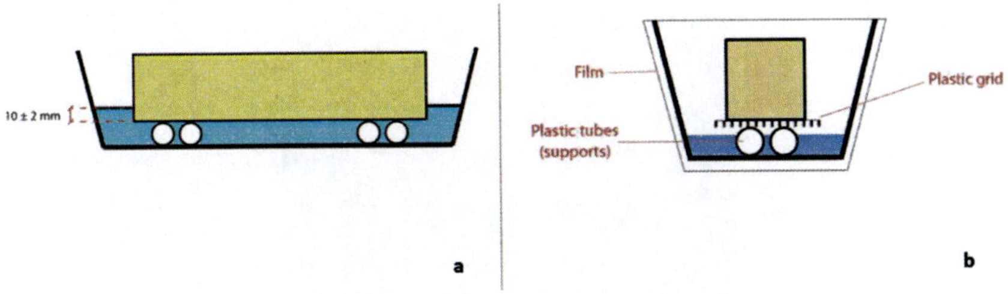

Fig. 2. Experimental setup of absorption test (**a**) and biological resistance test (**b**).

The last test carried out on blocks is a biological resistance assessment performed following the standard ÖNORM B 6010 [16]. The experimental setup is sketched in Fig. 2b, samples are not in contact with water since any capillary rise of water must be avoided. The film is necessary to keep the system's humidity at high values. The environment's temperature was 23 ± 2 °C, and the system has been left in this configuration for 4 weeks.

Eventually, some tests in climatic chamber have been carried out to evaluate the thermal transmittance value in controlled condition and validate an experimental methodology to be applied on site. The reference standard, in this case, has been the UNI 9869-1:2018 [17]. The equipment necessary to test the transmittance is a climatic chamber (able to control temperature and humidity) which can be closed with a wall having a surface equal to 1 m^2, temperature and relative humidity sensors, and flowmeters. The wall has been built with B2 hemplime blocks and plastered on both sides with a premixed hydraulic lime. The properties of the plaster are reported in Table 3.

The final thickness of the wall is 34 cm: 2 cm of plaster have been applied on each side and 30 cm thick blocks. It has been fixed into a wooden frame to close the climatic chamber (Fig. 3). The temperature and relative humidity inside the chamber were set to 10 °C and 50% RH to meet the required difference of at least 10 °C between the two surfaces of the wall (laboratory temperature was around 25 °C). The measure of the thermal transmittance is carried out after 7 days of measurements.

Table 3. Properties of premixed plaster applied on the experimental wall.

Property	Value
Density (kg/m^3)	1350
Thermal conductivity (W/m K)	0.47
Thermal resistance (m^{2}K/W)	0.03
Resistance to vapor diffusion μ	8
Specific heat (J/kg K)	1000

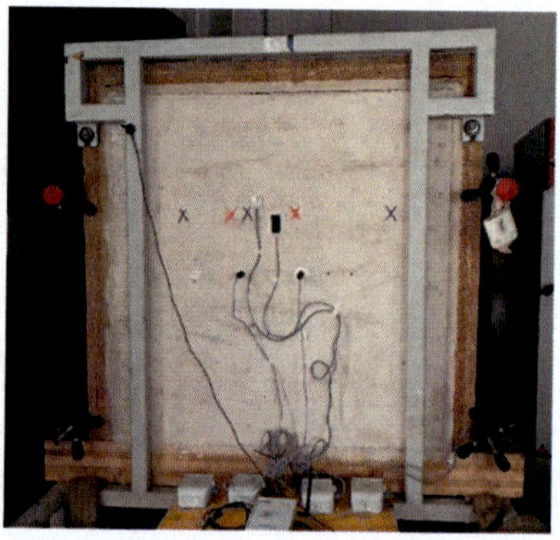

Fig. 3. Experimental wall closing the climatic chamber and sensors for T and RH measurements.

4 Results

Among the results obtained from the tests performed on blocks, the density has shown the larger difference from the values declared in the products' datasheets (Table 2). The results are reported in Fig. 4. Values on day 0 are strongly influenced by the weather conditions during the transportation phase and by the stock conditions at the manufacturing site. The other values have been measured after drying the samples in conditions selected on the basis of their integrity and humidity content at the delivery, as previously explained. For sample B3 measurements have been carried out up to day 14 due to the sizeable difference between measured and declared apparent densities recorded at days 2 and 3.

The differences highlighted in the final density values (all under or at least equal to 10%) can be attributed to different conditions of the measurements carried out in the companies, heterogeneity of the material, and experimental error related to the equipment and test procedure.

Regarding water absorption tests, the hemplime blocks' water absorption percentages are 15.8%, 22%, and 18.2% for B1, B2, and B3, respectively. Since standard bricks have a percentage of about 15–20% (of their weight) of water absorption, the results are acceptable; B2 blocks could require monitoring in high humidity conditions [18].

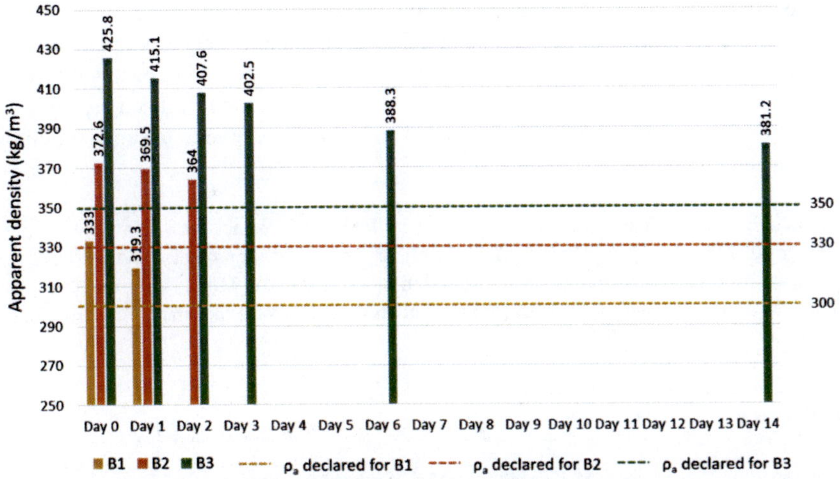

Fig. 4. Apparent density measurements.

The resistance to biological attack has been evaluated by visual inspection of samples left for 4 weeks in the closed system previously described, at high relative humidity values. All the samples were biologically resistant since, in the tested conditions, no changes were detected in the material.

Regarding the thermal transmittance measurement carried out on the experimental wall, it resulted to be 0.21 W/m^{2}K which is in accordance with the calculated theoretical value of 0.22 W/m^{2}K [18].

5 Conclusions

This work is focused on the production and determination of the properties of prefabricated hemplime blocks manufactured in Italy. Some tests have been performed to evaluate different formulations' properties and understand the criticalities of these building elements and their testing. The differences that emerged between experimental and declared bulk density values require the definition of a standard procedure to measure this property. In this way the influence of curing or transport conditions on the measurement would be excluded. The measured water absorption percentage is comparable to traditional bricks and the first evaluation of resistance to biological attack has given promising results. The tests here presented are preliminary and constitute the basis for further investigations since they are crucial to predict the behaviour of the material on site. Eventually, the measurement of the thermal transmittance performed in the climatic chamber on the basis of the standard allows to modify the experimental conditions in a controlled regime to simulate different environmental conditions. These tests can be performed to assess the behaviour of the blocks into conditions similar to the ones expected in service life, particularly when subjected to warm Mediterranean climates. Moreover, the accordance between the experimental and the theoretical values validates the experimental methodology applied in the presented measurement campaign. Consequently, the evaluation of the hygrothermal performances of a hemplime blocks wall over time is possible both ex situ and in situ. Hence, the integration of laboratory and on-site results could increase the knowledge about the behaviour of this material in real service conditions and spread its utilization as a valid alternative to traditional building material.

Acknowledgements. This work was supported by PAR2017 Project (D.2 Nearly zero energy buildings (nZEB)) funded by Italian Ministry of Economic Development. The tests have been carried out in the LPM (Laboratorio Prove Materiali) of Politecnico di Milano and the authors thank Michela Margutti and Andrea Pennati for their collaboration in the experimental measurements.

References

1. Ahmed, A.T.M.F., Islam, M.Z., Mahmud, M.S., Sarker, M.E., Islam, M.R.: Hemp as a potential raw material toward a sustainable world: a review. Heliyon **8** (2022)
2. Arrigoni, A., Pelosato, R., Melià, P., Ruggieri, G., Sabbadini, S., Dotelli, G.: Life cycle assessment of natural building materials: the role of carbonation, mixture components and transport in the environmental impacts of hempcrete blocks. J. Clean. Prod. **149**, 1051–1061 (2017)
3. Di Capua, S.E., Paolotti, L., Moretti, E., Rocchi, L., Boggia, A.: Evaluation of the environmental sustainability of hemp as a building material, through life cycle assessment. Environ. Clim. Technol. **25**(1), 1215–1228 (2021)
4. Dotelli, G., Moletti, C., Aversa, P., Sabbadini, S., Marzo, A., Tripepi, C., Lauriola, P., Luprano, V.A.M.: hempcrete buildings: environmental sustainabilityand durability of two case-studies in North and South Italy. Contribution title. In: XV International Conference Durability of Building Materials Components, pp. 1007–1014 (2020)
5. Amziane, S., Sonebi, M.: Overview on bio-based building material made with plant aggregate. RILEM Tech. Lett. **1**, 31–38 (2016)
6. Aversa, P., et al.: Hemp-lime buildings: thermo-hygrometric behaviour of two case studies in North and South Italy. Energy Build. **247**, 111147 (2021)
7. Aversa, P., Daniotti, B., Dotelli, G., Marzo, A., Tripepi, C., Sabbadini, S., Lauriola, P., Luprano, V.A.M.: Thermo-hygrometric behavior of hempcrete walls for sustainable building construction in the Mediterranean area. In: IOP Conference Series on Earth Environmental Sciences, p. 296 (2019)
8. Abdellatef, Y., Khan, M.A., Khan, A., Alam, M.I., Kavgic, M.: Mechanical, thermal, and moisture buffering properties of novel insulating hemp-lime composite building materials. Materials **13**(21), 5000 (2020)
9. Wang, L., Lenormand, E., Zmamou, H., Leblanc, N.: Effect of variability of hemp shiv on the setting of lime hemp concrete. Ind. Crops Products **171**, 113915 (2021)
10. Construire en chanvre—Homepage, https://www.construire-en-chanvre.fr/missions. Last accessed 26 Nov 2020
11. Label Granulat Chanvre Bâ1ment, Construire en chanvre, https://www.construire-en-chanvre.fr/missions. Last accessed 26 Nov 2020
12. Caruso, M., Cefis, N., Dotelli, G., Moletti, C., Sabbadini, S.: Numerical modeling of a hemp-lime blocks wall subject to horizontal in-plane loads. In: AIP Conference Proceedings. American Institute of Physics Inc. (2021)

13. Margutti, M., Pennati, A.: Definition of methods for the control of quality and for the correct laying and management of hempcrete blocks. Master thesis, Politecnico di Milano (2018)
14. BS EN 1602 :2013: Thermal insulating products for building applications—Determination of the apparent density
15. ISO 29767:2019: Thermal insulating products for building applications—Determination of short-term water absorption by partial immersion
16. ÖNORM B 6010:2015: Dämmstoffe für den Wärme- und/oder Schallschutz im Hochbau Arten und Anwendung
17. BS EN ISO 9869: 2018: Thermal insulation—Building elements—In-situ measurement of thermal resistance and thermal transmittance Part 1
18. Luprano, V.A.M., Aversa, P., Tripepi, C., Marzo, A., Daniotti, B., Dotelli, G., Sabbadini, S., Rogora, A.: Componenti edili con prestazioni energetiche ottimizzate per i climi mediterranei e basati su miscele di calcecanapulo (2018). Available online: https://www.enea.it/it/Ricerca_sviluppo/documenti/ricerca-di-sistema-elettrico/adp-mise-enea-2015-2017/edifici-nzeb/report-2017/rds-par2017-149.pdf

Have Population Growth and Economic Activity Converged Towards the Same Pattern in the Spanish Metropolises?

César Costa Costa(✉), Carlos Marmolejo Duarte, and Rolando Biere Arenas

Centre for Land Policy and Valuations (CPSV), Architectural Technology Department (TA), Universitat Politècnica de Catalunya (UPC), 08028 Barcelona, Spain
cesar.eduardo.costa@estudiantat.upc.edu

Abstract. The decentralization processes that have dominated the demographic growth of metropolitan areas in the course of recent decades have favoured in some cases the processes of dispersion and in others the growth of agglomerations that functionally orbit around the central nucleus. This paper attempts to verify whether, in the case of Spain, these patterns of population distribution within metropolitan structures have been maintained over the years, additionally, it attempts to compare the changes produced in the relocation of the population with those that have occurred in economic activity. With these objectives, based on the polycentric structure of the metropolitan areas, the percentage points of change in the population and employment of seven Spanish metropolises over the last three decades have been obtained. The results suggest that the distribution of the population has followed patterns very different from those observed in the distribution of economic activity, on the one hand, demographic decentralization, which seems to have slowed down over the last 30 years, has given rise to dispersed demographic growth in some metropolises and polycentric growth in others; while in economic activity, the loss of weight of the subcentres and their periphery seems to be the most common trait observed in the last 15 years in the metropolitan areas studied.

Keywords: Polycentrism · Monocentrism · Subcentres

1 Introduction

The process of urban decentralization that began in the last third of the 19th century with the phenomenon of suburbanization [1], would not have been possible without the improvement of rail and road transport systems that has allowed overcome spatial, topographic and orographic barriers in favour of urban expansion and generating urban structures strongly characterized by physical discontinuity [2] and an increase in the demand for mobility to access employment and other functions of the population, leading to urban areas to a vulnerability situation due to the high dependence on fossil fuels [3, 4]. Starting in the 1980s, the analysis of urban development began to show substantial

J. Littlewood and R. J. Howlett (Eds.): SEB 2022, SIST 336, pp. 228–239, 2023.
https://doi.org/10.1007/978-981-19-8769-4_22

changes in the structure of metropolitan areas that could not be explained by the traditional monocentric model [5]. These gaps were giving way to notions of poly-centric urban development associated with favouring the decentralization of employment and population, concentrated in central cities, towards other emerging ones that are linked to the former [6].

The incidence of polycentrism on the three vectors on which the sustainability paradigm is based has been widely discussed. Thus, from an environmental perspective, polycentrism could reduce land consumption and optimize mobility [7], and an improvement in transport planning in the event that the network was designed to connect the sub-centres with each other and with their hinterland [8]. From a social perspective, polycentrism is seen as an instrument to enhance cohesion in more compact and diverse social fabrics. Finally, from an economic perspective, it is considered essential in activating agglomeration economies within diverse cores, which translate into increasing returns for companies [9–11], as well as network economies between specialized nuclei [12]. Thus, in the European context, polycentrism has emerged as a sustainable alternative to the problems derived from macrocephaly and urban sprawl [13], being elevated to the level of central territorial policy in the European Spatial Development Perspective, agreed in 1999 [14]. It is therefore, on this form of urban development, on which specialized international literature has focused a good part of its attention in recent decades, emerging various empirical investigations with the aim of revealing the structure of metropolitan areas, most of them have focused on the identification of subcentres [15]. For these purposes, different methods have been used that could be grouped into the two large families of identification of subcentres: the morphological one, based on the analysis of employment and population density, and the functional one, based on residence-work flows [16].

In Spain, some fifteen years ago, polycentrism was still a phenomenon on which there was little research and even contradictory results [12]. However, the discussion of the subject in this country had been along the same lines as the international one, developing multiple research that can observed in the summary presented by Marmolejo et al. [17], standing out among them, the research that developed the functional method based on the «interaction value» that is calculated with the residence-work flows because it allows both to find the spatial borders of the metropolitan areas and to detect subcentres [18–21].

The perspective of urban resilience[1] includes the analysis of how the functional structures of modern metropolises could reduce the mobility demand and face the foreseeable scarcity and rise in the price of fossil fuels [22]. The research by Marmolejo and Tornés [23] shows that the greater the polycentric features of Spanish metropolises (polynucleation and polycentricity[2]), the lower the mobility of employed people once

[1] Resilience understood as the ability of urban systems to respond to multidimensional disturbances caused by different social, economic or ecological changes [24].

[2] Polynucleation or morphological polycentrism, which measures the number of nuclei, their relative importance as concentrations of metropolitan employment and their equipotentiality in the same terms. Polycentricity or functional polycentrism, which measures the functional interlinkage of the subcenters with the rest of the subcenters and the hinter-land that they unite [15].

differences in size and shape are weighted. Additionally, the existence of other urban and labour market factors that encourage mobility not explained by the size, shape and structure of the metropolitan area, is ratified in the aforementioned study. The latter is in line with the conclusions of recognized studies, that is, the residential decision is based on many factors in addition to the home-work separation [25, 26].

Despite the mentioned aspects, there is no updated analysis of the historical restructuring processes of the metropolises Spanish in which the internal changes that have occurred both in demographic growth and in employment can be accounted for. The objective of this paper is to analyse the way in which population and employment have been configured, over the last few years, within the seven metropolitan systems studied, to establish whether the dynamics of decentralization of the central economic nucleus, has evolved towards polycentrism or dispersion, or even towards a recentralization of the central nucleus. For the aforementioned purpose, the way in which the population and employment have been spatially distributed over the internal structure of each metropolitan system in terms of percentage points of change is analysed.

The results show that the drop in the population of the central nucleus continues to occur in all the studied metropolises, although it has been losing strength throughout the three decades analysed, especially in the last decade, drop that has given rise to two different patterns of demographic distribution, prevailing in some cases the dispersed growth and spatial expansion of the central nucleus, and in others, polycentric growth. Regarding the spatial distribution of employment, analysed for the last 15 years, it has been determined that a dynamic different from the demographic one prevails, which places the weight loss of the subcentres and their periphery as the most common element. The rest of the document is structured as follows: first, a review of the main investigations carried out in Spain around the analysis of the polycentric structure of the main metropolitan areas, followed by an explanation of the methodology used in this paper, a presentation of the results obtained that allow characterizing the evolutionary processes of the spatial distribution of the population and employment, to finally include some conclusions that put the results in perspective.

2 Review of Polycentrism in Spain

From an econometric perspective, a subcentre is considered a site with a significantly higher employment density than nearby locations, which is not fully explained by the influence of the central economic nucleus but is due to processes originating within it, and which has at the same time a significant effect on the overall employment density function [27]. However, this approach detracts from what Roca et al. [19] consider a third essential element of polycentrism: the generation of urban structure, that is, that the subcentres represent backbone elements of urban subsystems within the general structure of the metropolitan system. For this reason, this section is focused on the family of methods based on functional relationships, because it is appropriate to the diffuse city model [28] by allowing systems to be detected regardless of the physical continuity of their urbanization.

Within the family of functional methods, the research of Boix and Veneri [29] and Feria [30] could be cited, which have been carried out for the analysis of metropolitan areas on a scale of the entire Spanish territory, whose main limitation has resided in the use of thresholds in the resident employed population (REP). However, the investigations related to the so-called «interaction value» stand out, initially used in 2005 to delimit the Barcelona metropolitan area [18], later for the development of a method that allowed the detection of subcentres of the same metropolis [19], and then for a comparative analysis of the metropolitan structures of Barcelona and Madrid [20]. More complete research was developed for the analysis of the structure of the urban system of the entire Spanish territory [21]. Therefore, the functional method of «interaction value» is of great interest in the study of polycentrism because it allows simultaneously finding the spatial borders of metropolitan areas and the subcentres that structure them and because it has the following characteristics: (1) it is not based on neither relative nor absolute mobility thresholds, (2) it is capable of detecting the subtle conformation of areas of influence of peripheral subcentres, and (3) it considers mobility in a bidirectional way, as occurs in the reality of metropolitan systems complex and mature. Based on the «interaction value» method, some investigations have been developed that have delved into the analysis of the polycentric structure of the main metropolitan areas of Spain. In Roca et al. [21] it was reported that based on the information on labour flows from the 2001 Census, there were a total of 87 metropolitan systems in Spain with a population greater than 50,000 inhabitants, 20 of which corresponded to the large metropolises that exceeded half a million inhabitants, for which an analysis of the degree of complexity of the internal structures was also included through a synthetic indicator of polycentrism, based on Shannon's notion of entropy, [3] the analysis showed Barcelona, Coruna and Valencia as metropolises characterized by a high level of polycentrism, Madrid and Seville and Zaragoza as metropolises whose predominant structure is monocentric, and Bilbao and Malaga as metropolises without a clear structure.

During the period 1991–2006, the decentralization of the metropolitan population has been verified as the dominant pattern in the seven systems studied by Marmolejo et al. [17], however, the distribution of the drop in the central nucleus has occurred through two divergent urbanization models. The first case characterized by an expansion of the metropolitan centre and a structureless dispersion towards the ultra-periphery in which would locate Madrid, Seville and Zaragoza. The second model corresponds to the polycentric growth of Barcelona, Bilbao, Valencia and Malaga, characterized by a weak expansion of the central nucleus but a strong weight gain of the subcentres and their immediate peripheries. In Marmolejo et al. [31], the structure of seven metropolitan areas in Spain is studied, through the detection of potential subcentres using the analysis of labour flows, revealing as a result a marked divergence in which Barcelona, Valencia and Bilbao stand out as the systems more polycentric, while Madrid, Seville and Zaragoza are the least polycentric, and considering Malaga as an exceptional case since an important part of economic activity is concentrated in the few subcentres that make it up. On the other hand, in Marmolejo et al. [15], the measurement of the level of polycentrism of the seven main metropolises of Spain was proposed as the integrated measurement of

[3] The indicator of the level of complexity of the urban-metropolitan structure based on Shannon's notion of entropy that was built in the research of Roca et al. [21] is an approximation similar to the indicator presented by Marmolejo et al. [31], the first has used the population distribution data in the nuclei, while the second has used the employment distribution data.

the level of polynucleation (percentage of local work-places in the subcentres) and the incorporation of an indicator of functional polycentricity of the metropolitan systems that is obtained from labour mobility data [32], the results revealed a direct relationship between polynucleation and polycentricity, with the exception of Malaga, which is more polynucleated than polycentric in functional terms, it is also highlighted that the subcentre-subcentre relationships are -especially strong in the case of Barcelona and a little less Bilbao.

3 Methodology, Study Cases and Data Used

The analyses carried out in this paper have taken as a starting point the urban structure defined by Marmolejo et al. [31], which in turn was developed based on the functional method proposed by Roca et al. [19–21], described in the previous section. In order to analyse the evolution of the spatial distribution of the population and economic activity in the metropolitan areas studied in this paper, the method proposed by Marmolejo et al. [33], that is, the grouping of the municipalities into ambits or sub-areas of study: the central nucleus (central economic continuum); the periphery of the central core; nuclei detected beyond the central nucleus (subcentres); the periphery of the subcentres; and the rest of the metropolitan area; the latter also called ultra-periphery to distinguish it from the periphery of the central nucleus and the subcentres.

The central economic continuum has been defined as the set of municipalities functionally integrated through the «interaction value» to the central municipality and that also have an economic activity density greater than 700 DEP/Km^2,[4] thus, the central economic continuum can be made up of a single municipality or by a group of municipalities whose urbanized fabrics are separated by a distance of less than 200 m (conurbated municipalities). Hereinafter, the central economic continuum will be referred to only as the central core.

The evolution of the spatial distribution of demography and economic activity is analysed through the changes in concentration in the population and employment in each ambit of the metropolitan areas in terms of percentage points[5] (see Fig. 1); a comparison has been made of these changes in the concentration of the variables between different periods over the last 30 years. The case studies of this paper correspond to the set of seven metropolises that have been studied jointly with greater scope and that have been considered in some previous research as the largest metropolitan systems in Spain: Madrid, Barcelona, Valencia, Seville, Bilbao, Malaga and Zaragoza [15, 31].

The information on the population has been obtained from the 1991 and 2001 Censuses,[6] for intermediate and later years the data from the Municipal Register of the

[4] One of the GEMACA criteria was taken up again, a methodology that defines the central economic agglomeration as the set of contiguous municipalities whose density of local workplaces (DEP) is greater than 700 DEP/Km2, and which together have a minimum mass of 20,000 DEP.

[5] Percentage point are the arithmetic difference between the final and initial percentages; a positive change in one of the determined ambits represents a weight gain in the concentration of the variable analysed (population or employment). The sum of the concentration changes of all the ambits of the same metropolitan system is zero.

[6] The survey nature of the 2011 Census does not provide reliability in the data because the analyses include very small municipalities.

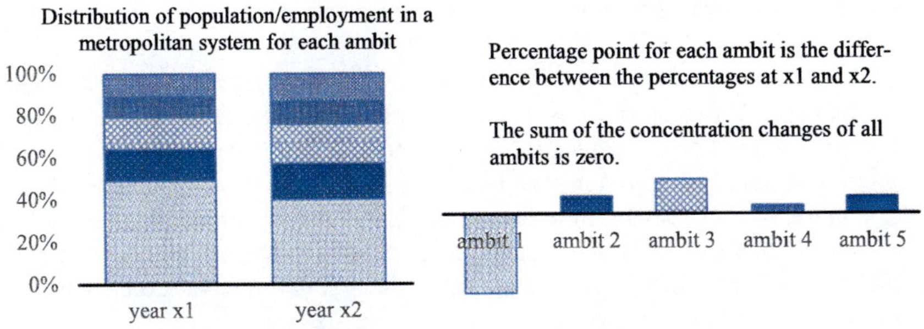

Fig. 1. Illustration of the determination of changes in population/employment concentration within the same metropolitan system

National Institute of Statistics (INE) has been used. Regarding the information related to economic activity, the number of Social Security (SS) affiliates per municipality has been used because it is the only source of labour information that is available for all the municipalities of Spain for all years within the period 2003–2020.[7] The SS data that have been used correspond to all types of workers, including the self-employed and people who work at home; however, the SS data and the results of the analysis must be read carefully, mainly due to the so-called "headquarters" effect (Table 1).[8]

Table 1. Number of municipalities in each ambit of the seven metropolitan areas

Metropolitan area	Central core	Central nucleus periphery	Subcentres	Subcentres periphery	Ultraperiphery	Total
Barcelona	12	4	23	69	76	184
Bilbao	9	12	14	40	48	123
Madrid	7	14	8	35	119	183
Malaga	1	6	3	8	14	32
Seville	1	7	7	15	22	52
Valencia	4	13	17	45	25	104
Zaragoza	1	13	7	24	43	88
Total	35	69	79	236	347	766

[7] The information on the local workplaces for all the municipalities of Spain is only available in the 2001 and 2011 Censuses, although in certain metropolitan areas there are mobility surveys that provide an idea of these values in certain years.

[8] Some companies tend to register all their affiliates in their directional headquarters, including those who work in branches located in other municipalities.

4 Evolution of the Population Spatial Distribution

The evolution of the population distribution has been analysed for the period 1991–2020; the first analysis carried out for the set of seven metropolitan areas in which the entire period has been subdivided into six five-year intervals,[9] provides an idea of the spatial distribution of population growth (see Fig. 2); initially it is clearly observed that the changes in the urban structure, in terms of population, are produced fundamentally by the relative fall of the central nucleus; another aspect that can be clearly observed is that the fall of the central nucleus has been greater in the first four intervals, that is, between the years 1991 and 2011, in subsequent years the changes are less.

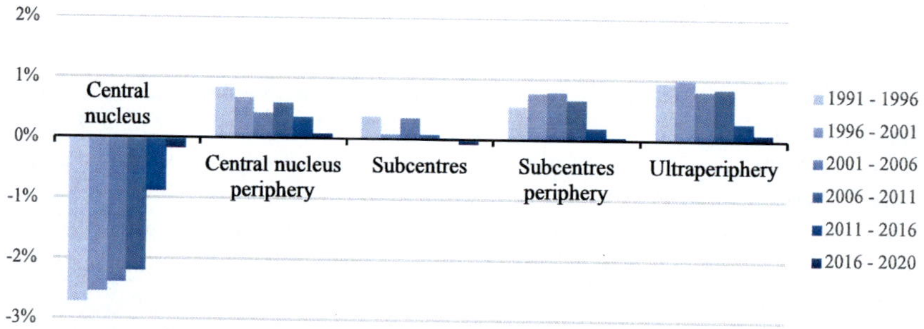

Fig. 2. Evolution of the population spatial distribution of the set of the seven metropolitan areas studied expressed in percentage points

The processes described in the previous paragraph have not occurred homogeneously among the metropolitan areas studied, so when reproducing the analysis shown in Fig. 2 for all the metropolises, it is possible to notice the marked differences in their demographic dynamics, however, in order that this article does not extend too much, these graphs have been omitted. Rather, a comparative analysis is presented between the 1991–2006 and 2006–2020 periods (see Fig. 3) that collects the information on the changes in the population concentration of the ambits included in Table 2; the size of the spheres represents the population drop of the central nucleus, the horizontal axis represents the level of polycentric growth/dispersion (subcentres + its periphery—ultraperiphery) and the vertical axis represents the spatial expansion of the central nucleus, that is, the growth of the immediate periphery of said nucleus.

The analysis allows to establish that the spatial distribution of demographic growth during the first period (1991–2006) continues to occur in a similar way in the second period (2006–2020), although with less force as indicated in previous lines. This distribution allows confirming the existence of two divergent patterns of metropolises in Spain [33]; the first corresponds to the metropolitan areas of Madrid, Seville and Zaragoza, whose predominant structure is monocentrism, and have experienced an expansion of the central nucleus towards its immediate periphery and towards the ultraperiphery; in the second group are Barcelona, Bilbao, Malaga and Valencia, characterized by a

[9] Actually, the last interval 2006–2020 covers only four years.

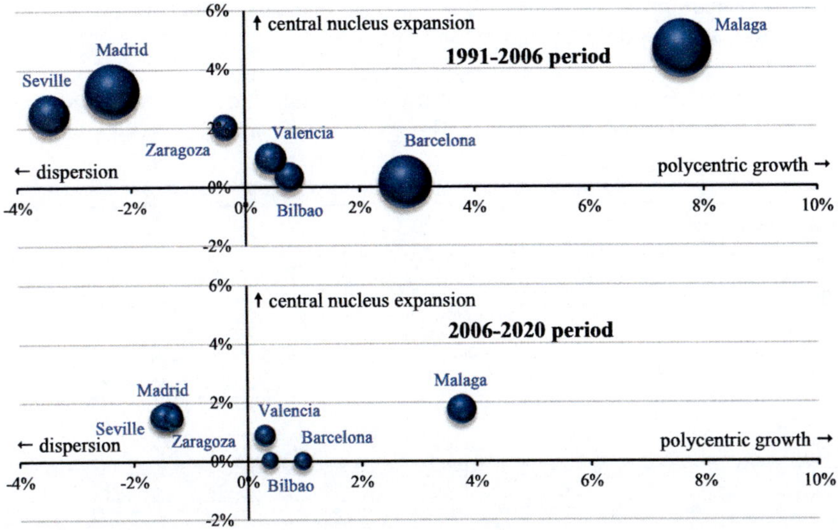

Fig. 3. Comparison of demographic dynamics between the 1991–2006 and 2006–2020 periods for each metropolitan area

higher level of polycentrism, a demographic growth of the decentralized nuclei and their immediate peripheries, and a lower expansion of the central nucleus, except Malaga, whose powerful fall of the central nucleus has favoured the expansion of its immediate periphery as well as the periphery of the subcentres.

5 Evolution of the Spatial Distribution of Economic Activity

As mentioned, the evolution of the distribution of economic activity has been analysed using the information of Social Security affiliates, in this case, the period 2006–2020 has been analysed through the same methodology used for the analysis of the demographic distribution (Table 3). Throughout the 2006–2020 period, the analysis reveals that the dynamics of the distribution of economic activity in the metropolitan areas studied have occurred in a very different way to that observed in the demographic distribution. In general, it could be said that the period is marked by the relative fall in jobs in the subcentres and their periphery in favour of recentralization and the dispersion of employment in the cases of Barcelona, Bilbao, Malaga and Seville. The dynamic of Madrid and Valencia corresponds to an employment drop in the central nucleus while it expands spatially towards its closest periphery, but also towards the ultraperiphery in the case of Valencia. Zaragoza has experienced a dynamic opposite to that observed in Valencia, the nuclei, both central and peripheral, have attracted economic activity from the periphery close to the central nucleus and from the ultraperiphery (see Fig. 4).

Another aspect to highlight in this analysis of the distribution of economic activity is the divorce with the predominant structure of the metropolises, be it monocentrism or polycentrism, showing a dispersed growth of employment in all cases except Zaragoza.

Table 2. Evolution of the population spatial distribution during 1991–2006 and 2006–2020 periods, expressed in percentage points for each metropolis

Metropolitan area	Central nucleus (%)	Central nucleus periphery (%)	Subcentres (%)	Subcentres periphery (%)	Ultraperiphery (%)
1991–2006 period					
Barcelona	−9.1	0.2	1.8	4.1	3.1
Bilbao	−2.8	0.4	1.2	0.4	0.8
Malaga	−11.1	4.7	2.5	4.5	−0.6
Valencia	−3.3	1.0	0.2	1.1	0.9
Madrid	−10.0	3.3	0.5	1.7	4.5
Seville	−5.4	2.5	0.1	−0.4	3.2
Zaragoza	−2.2	2.0	0.1	−0.2	0.3
2006–2020 period					
Barcelona	−1.9	0.0	0.2	1.2	0.5
Bilbao	−1.7	0.0	0.5	0.5	0.6
Malaga	−4.9	1.8	0.8	2.6	−0.3
Valencia	−2.6	0.9	0.2	0.8	0.7
Madrid	−4.7	1.6	−0.1	1.0	2.3
Seville	−4.1	1.5	0.3	0.3	2.1
Zaragoza	−2.3	1.5	0.1	−0.3	1.0

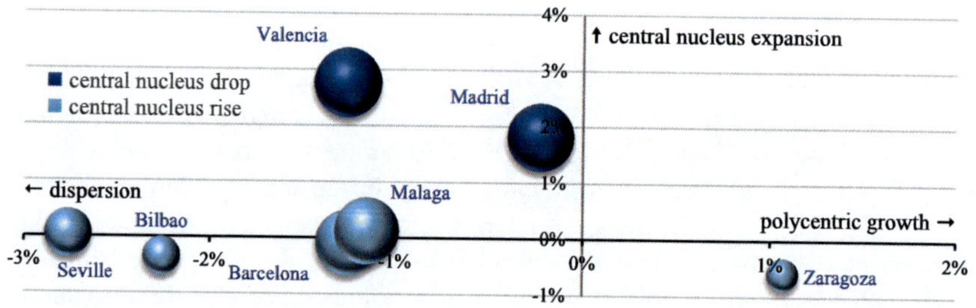

Fig. 4. Dynamics of the distribution of social security affiliates during the 2006–2020 period for each metropolitan area

The only two metropolises that had experienced a relative decline in the central nucleus, Madrid and Valencia, are in turn the only cases in which said ambit has expanded spatially towards its closest periphery in terms of economic activity.

Table 3. Evolution of the spatial distribution of Social Security affiliates during the 2006–2020 period, expressed in percentage points for each metropolis

Metropolitan area	Central nucleus (%)	Central nucleus periphery (%)	Subcentres (%)	Subcentres periphery (%)	Ultraperiphery (%)
Barcelona	1.9	−0.1	−2.1	0.5	−0.3
Bilbao	0.7	−0.3	−1.2	−0.1	1.0
Malaga	2.0	0.1	−1.7	0.0	−0.5
Valencia	−2.2	2.8	−0.2	−0.7	0.4
Madrid	−2.1	1.8	−0.3	0.4	0.2
Seville	1.0	0.1	−0.9	−1.1	0.8
Zaragoza	0.5	−0.6	0.5	0.1	−0.5

6 Discussion and Conclusions

The demographic growth rate of the set of seven metropolitan areas studied (766 municipalities in total) during the 2006–2020 period has been 10.0%, that is, it has decreased compared to the growth of the previous period 1991–2006, which was 17.4%. On the other hand, the employment growth rate (measured based on the number of workers affiliated with Social Security) in the 2006–2020 period was 8.3% for the seven metropolitan areas as a whole.

Regarding the internal demographic dynamics, the process of population deconcentration of the central nucleus (fall of the central nucleus), observed during 1991–2006 for the seven metropolitan areas studied, has continued to occur throughout the 2006–2020 period, but with less intensity. It could be said that the decentralization process of the metropolises has coincided with the period after the economic crisis in Spain produced in 2008 and with a flattening in the population growth curve of the entire country between 2012 and 2019. It would be convenient therefore to delve into a possible relationship between these economic and demographic events in future investigations, taking into account that said relationship has not been so clear in all cases. As an example of the latter is Bilbao, which for a long period has been gaining weight in the subcentres, despite the fact that demographic growth has remained practically constant. In the midst of the slowdown in the decentralization process of the population of the metropolitan areas studied, especially evidenced between 2011 and 2020, the demographic trends observed in previous years (1991–2006) have been preserved in most cases. It could be said that polycentric growth continues to be the most distinctive feature in the cases of Barcelona and Malaga, and dispersed growth in the cases of Madrid, Seville and Zaragoza, however. For the cases of Bilbao and Valencia, it is would have produced a slight shift towards dispersion, during the last five and ten years respectively.

In the distribution of economic activity there are two different patterns, in some cases towards the recentralization of employment, and in others a fall of the central nucleus in favour of its first peripheries (expansion of the central core). In all cases, except Zaragoza, there is a drop in employment in the subcentres while some ultraperipheries have gained weight, which could be suggesting the emergence of new subcentres different from those identified in 2001. In any case, under this analysis, there is no clear consensus on the patterns of the distribution of economic activity in metropolitan structures. It is necessary to have up-to-date data on labour mobility that allows establishing its relationship with the produced changes in metropolitan areas in the last two decades.

The population and economic activity data used at the date of completion of this paper correspond to the month of January of each year, in this way, the most up-to-date information corresponds to January 2020, that is, before the arrival of the COVID pandemic. However, it is likely that this global event had an impact on the dynamics of population and employment distribution, therefore, future research should include the most recent years to assess the impact of the pandemic on the population and employment structures of the metropolitan areas studied.

References

1. Hall, P., Hay, D.: Growth Centres in the European Urban System, 1st edn. Heinemann Educational Books Ltd., London (1980)
2. Roca, J.: La delimitación de la ciudad: ¿una cuestión imposible? Ciudad y Territorio: Estudios Territoriales **XXXV**(135), 17–36 (2003)
3. Dodson, J., Sipe, N.: Shocking the suburbs: Urban location, homeownership and oil vulnerability in the Australian City. Housing Stud. **XXIII**(3), 377–401 (2008). https://doi.org/10.1080/02673030802015619
4. Newman, P., Beatley, T., Boyer, H.: Resilient Cities. Responding to Peak Oil and Climate Change. Island Press, Washington, DC (2014)
5. Anas, A., Arnott, R., Small, K.: Urban spatial structure. J Econ Lit **36**(3), 1426–1464 (1998)
6. Dematteis, G.: Suburbanización y periurbanización. Ciudades anglosajonas y ciudades latinas. In F. Monclús, La ciudad dispersa, pp. 17–33. Barcelona: Centre de Cultura Contemporània de Barcelona (1998)
7. Gordon, P., Richardson, H., Wong, H.: The distribution of population and employment in a polycentric city: the case of Los Angeles. Environ. Plan. A: Econ. Space **18**(2), 161–173 (1986). https://doi.org/10.1068/a180161
8. McMillen, D.: The center restored: Chicago's residential price gradient reemerges. Econ Perspect **26**(2) (2002)
9. McMillen, D.: Employment subcenters in Chicago: past, present, and future. Econ. Perspect. **27**(2) (2003)
10. McMillen, D., Smith, S.: The number of subcenters in large urban areas. J. Urban Econ. **53**(3), 321–338 (2003)
11. McDonald, J.: Calibration of a monocentric city model with mixed land use and congestion. Reg. Sci. Urban Econ. **39**(1), 90–96 (2009)
12. Boix, R., Trullén, J.: Policentrismo y estructuración del espacio: una revisión crítica desde la perspectiva de los programas de investigación. ACE: Archit City Environ. **6**(18), 27–54 (2012). https://doi.org/10.5821/ace.v6i18.2543
13. Clark, W., Kuijpers, M.: Commuting in restructuring urban regions. Urban Stud. **31**(3), 465–483 (1994)
14. Meijers, E.: Measuring polycentricity and its promises. Eur. Plan. Stud. **16**, 1313–1323 (2008)

15. Marmolejo, C., Ruiz, N., Tornés, M.: ¿Cuán policéntricas son nuestras ciudades? Un análisis para las siete grandes áreas metropolitanas en España. Ciudad y Territorio: Estudios Territoriales **XLVII**(186), 679–700 (2015)
16. Ureña, J., Pillet, F., Marmolejo, C.: Aglomeraciones/regiones urbanas basadas en varios centros. Ciudad y Territorio: Estudios Territoriales **XLV**(176), 249–266 (2013)
17. Marmolejo, C., Chica, E., Masip, J.:. Hacia un sistema de metrópolis españolas policéntricas? Evolución de la influencia de los subcentros en la distribución de la población. (C. d. UPC, Ed.) ACE: Archit. City Environ. **VI**(18), 163–190 (2012). https://doi.org/10.5821/ace.v6i18.2548
18. Roca, J., Moix, M.: The interaction value: its scope and limits as an instrument for delimiting urban systems. Reg. Stud. **39**(3), 357–373 (2005). https://doi.org/10.1080/00343400500087372
19. Roca, J., Marmolejo, C., Moix, M.: Urban structure and polycentrism: towards a redefinition of the sub-centre concept. Urban Stud. **46**(13), 2841–2868 (2009). https://doi.org/10.1177/0042098009346329
20. Roca, J., Arellano, B., Moix, M.: Estructura urbana, policentrismo y sprawl: los ejemplos de Madrid y Barcelona. Ciudad y Territorio: Estudios Territoriales **IXLIII**(168), 299–321 (2011)
21. Roca, J., Moix, M., Arellano, B.: El sistema urbano en España. Scripta Nova. Revista Electrónica de Geografía y Ciencias Sociales **XVI**, 387–423
22. Sánchez-Toscano, G., Hernández, A.: Resilience and mobility demand. Towards a redefinition of urban polycentrism. Madrid Urban Area Analysis, 1996–2014. In: IFoU 2018: Reframing Urban Resilience Implementation: Aligning Sustainability and Resilience. Barcelona (2018).https://doi.org/10.3390/IFOU2018-05977
23. Marmolejo, C., Tornés, M.: The influence of urban structure on commuting: an analysis for the main metropolitan systems in Spain. Procedia Eng. **198**, 52–68 (2017). https://doi.org/10.1016/j.proeng.2017.07.073
24. Coaffee, J.: Risk, resilience, and environmentally sustainable cities. Energy Policy **XXXVI**(12), 4633–4638 (2008). https://doi.org/10.1016/j.enpol.2008.09.048
25. Small, K.A., Song, S.: 'Wasteful' commuting: a resolution. J. Political Econ. **100**(4), 888–898 (1992). Retrieved from https://www.jstor.org/stable/2138692
26. Giuliano, G., Small, K.A.: Is the journey to work explained by urban structure? Urban Stud **30**(9), 1485–1500 (1993). https://doi.org/10.1080/00420989320081461
27. McMillen, D.: Nonparametric employment subcenter identification. J. Urban Econ. **50**(3), 448–473 (2001). https://doi.org/10.1006/juec.2001.2228
28. Castells, M.: The Information Age: Economy, Society and Culture, vol. 1. Blackwell Publishing Ltd, Oxford (1996)
29. Boix, R., Veneri, P.: Metropolitan areas in Spain and Italy. IERMB Working Paper in Economics **9**(1) (2009)
30. Feria, J.: La delimitación y organización espacial de las áreas metropolitanas españolas: una perspectiva desde la movilidad residencia-trabajo. Ciudad y Territorio: Estudios Territoriales CyTET **XLII**(164), 189–210 (2010)
31. Marmolejo, C., Masip, J., Aguirre, C.: Policentrismo en el sistema urbano español: un análisis para siete áreas metropolitanas. Ciudad y Territorio: Estudios Territoriales 281–300 (2013)
32. Green, N.: Functional polycentricity: a formal definition in terms of social network analysis. Urban Stud. **44**(11), 2077–2103 (2007). https://doi.org/10.1080/00420980701518941
33. Marmolejo, C., Tornés, M.: Does polycentrism reduce land consumption? An analysis for the biggest metropolitan systems in Spain. In: 8th International Conference Virtual City and Territory. Río de Janeiro (2012). https://doi.org/10.5821/ctv.7834

A Community Housing Association's Strategy for the Benchmarking, Reduction and Sequestration of Carbon Towards a Resilient and Globally Responsible Wales (UK)

K. Stevens-Wood[1](✉), J. R. Littlewood[2], and F. Sanna[2]

[1] School of Education & Social Policy, The Intentional Communities Research Group, Cardiff Metropolitan University, Cardiff CF5 2YB, UK
kstevens-wood@cardiffmet.ac.uk

[2] School of Art & Design, The Sustainable & Resilient Built Environment Research Group, Cardiff Metropolitan University, Cardiff CF5 2YB, UK

Abstract. This paper explores how Cynon Taf Housing Association (CTHA hereafter)—a community-based social housing developer and landlord in Wales (UK)—has sought to respond to the challenges of climate change. In Wales, of the 1.4 million existing dwellings, circa 226,000 are owned and managed by 55 organizations such as CTHA. Housing in Wales is responsible for an estimated 21% of the UK's carbon emissions, and with ambitious carbon reduction targets set by the Welsh Government to achieve 47% of 1990 levels by 2030, there is urgent need for considering both carbon reduction as well as carbon capture and sequestration plans. Therefore, social housing developers such as CTHA, when calculating their carbon footprint, need to consider the embodied carbon of their property assets and their operational day-to-day carbon emissions, including those from their tenants' activities. CTHA's carbon sequestration plan proposes to offset and sequester carbon are discussed and illustrated, which includes an evaluation of mixed-use agricultural land for planting and coppice of native tree species for carbon sequestration. The plans discussed meet with the Welsh Government's strategies for decarbonization in Wales, by delivering a "green" recovery that will sustain Wales into the future (beyond 2022).

Keywords: Wales · UK · Embodied carbon · Operational carbon · Carbon reduction · Carbon sequestration · Agroforestry · Community-supported agriculture

1 Introduction

In 2015, the Well-being of Future Generations (Wales) (WBFG) Act, UK [1] set out an ambitious vision of a more sustainable, more resilient Wales. Also, in 2015, the 17 United Nations Sustainable Development Goals were launched with an increasing urgency for local, national, and global responses to climate change and human development [2]. This paper discusses how Cynon Taf Housing Association (CTHA), a community-based

J. Littlewood and R. J. Howlett (Eds.): SEB 2022, SIST 336, pp. 240–248, 2023.
https://doi.org/10.1007/978-981-19-8769-4_23

housing association in Wales, has sought to respond to the challenges of climate change by quantifying their organizational embodied and operational carbon, and by evaluating land-based options for carbon sequestration. This aim will be achieved through two initiatives: a SMART Partnership project [3] funded by CTHA and the Welsh Government (due to commence in September 2022), and the Down to Zero project funded by CTHA [4].

The Welsh Government has a legal duty in Part Two of the Environment (Wales) Act 2016 to reduce carbon emissions in Wales by 80% by 2050 [5]. There are approximately 1.4 million dwellings in Wales, of which circa 10% are owned and managed by housing associations [6], therefore constituting a significant holding represented by large social landlords. To put this into context, housing in Wales accounts for 21% of Welsh annual carbon emissions. When calculating their carbon footprint, housing associations need to not only consider the embodied carbon of their property assets, but also their day-to-day operational carbon emissions such as property management and maintenance, office-based activities, and tenant actions. For example, it is estimated that the average household in the UK produces approximately 8.1 tons of carbon dioxide (CO_2) per annum with a large proportion of this coming from heating and transport [7]. As such, housing associations are in a prime position to reduce housing-based carbon emissions through various layers of interventions from low-carbon, energy-efficient new build and retrofitting [8] to creating localized economic opportunities and carbon-sequestration activities.

The focus of this paper is on the latter of these activities and will seek to analyze the potential outcomes of proposed land-based carbon capture and sequestration activities, more specifically, land acquisition, tree planting, short rotation coppice and the utilization of agroforestry principles.

2 The Need for Housing Associations to Sequester Carbon

The research challenges are complex, as CTHA has an existing housing stock of circa 1,850 units, which includes different building types, from one-bedroom to eight-bedroom properties, sheltered housing, adapted homes, older-person homes and supported housing [9]. Many of CTHA's dwellings were built prior to 1919 [ibid]. Consequently, the estimation of operational and carbon emissions will be computationally considerable. The life-cycle assessment (LCA hereafter) model, indeed, will need to take into consideration key parameters that widely vary from different dwelling types, household behavior in the use of appliances, heating, ventilation, and lighting [10].

The benchmarking of embodied carbon emissions is complex, not least because the dwelling stock was constructed from pre-1919 to date and, therefore, suitable input data sets will need to be sought and utilized in the LCA model. It is likely that the carbon baseline assessment will require the development of a specific database of carbon emissions from such a wide-ranging building stock (which also includes offices). Indeed, according to Hart, D'Amico, and Pomponi [11] the assessment of life-cycle carbon in buildings is complex and needs to account for the range and composition of materials used over different periods of construction.

Formulation of a thorough carbon-reduction and carbon sequestration strategy can only be founded on a robust assessment of current emission levels. While available measures for carbon reduction and sequestration can be identified from the literature, this project looks to identify options that are sufficient to achieve the set target levels, are compatible with CTHA's ethos, opportunities, and financial constraints, and entail use of community land through the Down to Zero project.

LCA studies of this nature present an additional challenge due to the need to define and evaluate several possible land-based carbon-sequestration scenarios and to carefully assess the trade-offs associated with each of them [12]. There is some divergence on the benefits of using land for tree planting, with some suggesting that simply planting trees is a distraction from the more challenging practice of reducing fossil fuel use and protecting existing ecosystems [13]. Other studies have found that tree planting, particularly in peat rich soils, can actually have a carbon-positive effect, releasing more carbon from the soil then the trees take up over a lifetime [14], meaning that site selection and existing carbon holding is something to be considered. However, numerous studies have found that good site selection and suitable tree species selection can increase carbon uptake and sequestration [15, 16] as well as increase biodiversity. This is particularly the case when the whole life of newly planted trees is considered including the drawdown of CO_2, and also the sequestration of embodied carbon in the form of timber products.

A second strand of complexity to this is the aspiration to investigate agroforestry with a range of crops and tree species to be planted and managed for both carbon sequestration and biodiversity. Research suggests that altitude and climate can impact significantly on tree growth [17], particularly over 500 m altitude for fruit-bearing trees, and therefore site selection will need to consider which trees are planted for optimum growth and production.

The research team have developed the following step-by-step tasks to deliver the SMART Partnership project, illustrated in Fig. One.

1. Evaluate best-practice software applications for assessing organizational carbon footprint and carbon sequestration.
2. Produce a baseline carbon assessment (lifecycle assessment) of CTHA's operations across the organization using the identified software application to estimate the quantity of carbon being emitted.
3 Identify current CTHA activities that could be optimized or refined to lower embodied and operational carbon emissions; and set new target values for such activities, in the short, medium and long term (i.e., 1, 3 and 5 years, respectively).
4. Assess a range of best-practice agroforestry options to sequester carbon, through a desktop study, which could include tree planting, annual fruit and vegetable production, and crops to be utilized as natural building materials.
5. Determine the area of land, and type and quantity of agroforestry activities to meet the new target values (short, medium, and long term) for CTHA carbon footprint, within budget constraints.
6. Undertake market research of potential funding opportunities to deliver and test the land-based CTHA carbon-sequestration model (Fig. 1).

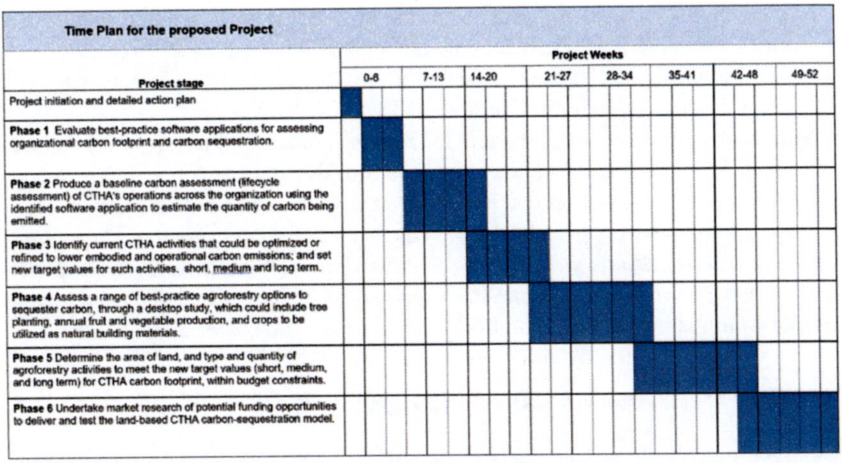

Fig. 1. Timeline for project.

3 Anticipated Outcomes

There are several anticipated outcomes to the SMART Partnership which will be delivered to support the Down to Earth project.

3.1 Baseline Assessment of Organizational Embodied and Operational Carbon

The baseline carbon assessment for any housing organization is likely to be highly complex due to the diverse range of properties, different ages, locations, and construction methods [18]. The SMART Partnership project is anticipated to deliver a comprehensive methodology to benchmark both embodied and operational carbon emissions of CTHA's 1,850 dwellings and other buildings. Forin et al. [19], identified that one of the biggest challenges to organizations undertaking carbon assessments was delineating between the different categories of activities, in particular those external to the organization such as services and purchasing.

As a result, the material variations for assessing embodied carbon associated with them may lead to the need to create unique material databases. Carbon assessment was a niche area up until the establishment of mandatory carbon reduction targets towards net zero emissions by 2050. In its 2021 Net Zero Carbon Plan, the Welsh Government set out carbon-reduction targets for every five years in budget cycles, to deliver net zero carbon by 2050 [20]. The carbon-emission reduction targets include 37%: 2020 to 2025, 58%: 2026 to 2030, 63% by 2030, 89% by 2040, and 100% by 2030 [ibid].

These carbon-emission reduction targets are a considerable challenge for organizations, as they equate to both operational carbon and embodied carbon. The built environment is responsible for more greenhouse-gas emissions (GHGs) than any other sector of the economy, more than transportation, agriculture and industry. Housing associations that own and manage existing buildings release considerable carbon emissions, including CTHA, which currently has 1,850 dwellings and other non-residential buildings.

To the authors' knowledge, CTHA are the only housing association in Wales that is seeking to benchmark its organizational embodied and operational carbon emissions, and to set annual carbon-emission reduction targets alongside delivering land-based agroforestry carbon sequestration through its Down to Zero project.

With the Welsh Government targets for carbon reduction to achieve 37% reduction in only three years by 2025 [21], the combined SMART Partnership and Down to Zero projects present a new and innovative approach within the social-housing sector and presents a commercial opportunity for CTHA and Cardiff Metropolitan University for other housing associations to follow in Wales and the rest of the UK, and potentially outside of the housing sector. To put this into context, there are 225,000 existing social housing dwellings in Wales [22] and 4.4 million across the UK as a whole [23].

3.2 Annual CTHA Carbon Reduction Targets

Once the carbon benchmarking has been undertaken, it will be possible to identify areas of high, medium and low carbon intensity. It is envisaged that there will be an emerging hierarchy of activities that CTHA will need to undertake and that these will need to be mapped out according to cost and ease of implementation. As such, some activities will bring early benefits for relatively low costs. For example, increased thermal insulation in all properties, with a longer target of replacing older gas boilers with new technology such as heat pumps [24]. Equally, carbon created through organizational business operations can be analyzed for areas of carbon intensity, for example, transport, along with a plan to move towards electric vehicles and other sustainable transport measures.

3.3 Evaluation of Best-Practice Agroforestry Carbon-Sequestration Options

It is estimated that agroforestry species appropriate for planting have the potential to sequester approximately 84 tons of carbon per hectare [25]. However, this is an average figure across the lifespan of a newly planted temperate woodland, and therefore carbon uptake in the early years can be expected to be significantly lower. A second factor to consider is the end-use of any timber extracted, as damaged or fallen trees will eventually release stored carbon as part of the natural biodegradable process [26]. Therefore, it is anticipated that alongside tree planting the use of agroforestry methods will be utilized. In addition to this, a long-term plan for the harvesting and use of timber will need to be developed.

Although initially envisaged that one land plot would be acquired by CTHA to implement agroforestry carbon sequestration options, as the Down to Zero project has progressed and a greater insight into the lack of greenfield land availability has been determined, it seems increasingly probable that the land options will include both brownfield and greenfield land in several locations in the Rhondda Cynon-Taf (RCT) region of Wales [27]. This brings with it both opportunities and challenges. As an envisaged community-led project, land which can be accessed by public transport may be more accessible to local populations, increasing involvement and adoption of the land by the proximate communities. This has the potential benefit of encouraging volunteers who in turn will ensure the sustainability of the site and associated activities.

Early carbon-mapping will enable forward planning in terms of number and types of trees as well as other types of land use that may enhance carbon reduction, such as local food production. The implementation of agroforestry techniques allows for the planting of edible crops, often in corridors between tree planting. A study by Lehmann [28] found not only a higher level of productivity (when compared to monoculture planting), but also that additional ecosystem services were provided such as greater biodiversity, soil and water retention. As such, the creation of mixed planting is anticipated to sequester carbon, but also provide additionality in the form of environmental enhancement.

3.4 Social and commercial Opportunities for a CTHA Carbon Reduction and Land-Based Carbon Sequestration Strategy

It is anticipated that these projects activities could potentially lead to remunerated employment for existing communities in RCT, volunteering opportunities in addition to local food production; all of which could result in reductions in organizational and community carbon footprints. Through the process of knowledge transfer, CTHA could become a leading social organization to meet carbon-reduction targets.

Furthermore, a potential business opportunity for CTHA and Cardiff Metropolitan University upon completion of the SMART Partnership lies in consultancy services with income generation. These services could include the implementation of the CTHA Carbon Reduction and Land-Based Carbon Sequestration Strategy for housing associations and other types of organizations with large and complex property holdings that seek to lower their carbon emissions (e.g., universities, National Trust and local authorities). To the authors' knowledge, such a strategy does not yet exist.

4 Discussion

The central focus of the CTHA project is to reduce carbon and share the knowledge from the project across other housing associations in Wales. The project has already been presented at the Community Housing Cymru (CHC) Decarbonization Conference in October 2021, and there will be ongoing involvement through a recently funded SMART Partnership. Said Partnership is an opportunity to develop and implement the lifecycle assessment of CTHA's organizational carbon emissions, and the land-based strategy for carbon reduction and sequestration. If successful, this strategy could have wider use and implications for other housing associations and organizations outside of the housing sector.

The strategy for the carbon assessment, reduction, and land-based sequestration through agroforestry would be of great value to other organizations within the sector, at a national level. This presents a significant commercial opportunity once the knowledge has been transferred to CTHA.

It is anticipated that the projects will add to the knowledge base of an emerging global priority of decarbonization and mitigation of climate-change reality.

As a project breaking new ground, the academic involvement provides critical underpinning in terms of risk management and decision making, and access to high-quality research and independent advice via academic supervision. Therefore, the project will provide a platform for knowledge sharing within and outside of the housing sector.

The SMART Partnership project is due to commence in September 2022, for 12 months.

5 Conclusions

This paper has discussed the ongoing collaboration between Cynon Taf Housing Association and Cardiff Metropolitan University around the reduction of the carbon emissions of the Association's building stock and operational activities. The project outlined is based on the application of life-cycle-assessment techniques alongside principles drawn from academic research in sustainable agroforestry to a real scenario of high complexity. Thus, the project is multi-disciplinary in nature and combines areas of knowledge that are not often associated.

The investigation has well-defined boundaries (i.e., CTHA's carbon footprint), but, thanks to its unicity in the Welsh context, its impact has potential to reach well beyond these, by setting an example for other housing associations and public organizations which intend to adopt a similar model, thus leading to wider benefits for society and environmental protection. This is the reason why this project has been conceived, from the outset, with a view to sharing the experience with other organizations founded on similar ethos and values, which take a proactive approach to mitigation of climate change and sustainable use of natural resources.

The success of the project will depend upon a detailed and carefully structured plan of action formulated in the preliminary phase, flexible enough to overcome potential difficulties or delays associated with the sourcing of data, identification of adequate computational tools, formulation of a LCA model, and so on. Upon completion of the project, the authors will share critical reflections on its implementation and outcomes, and the lessons learnt.

Acknowledgements. The authors wish to thank Cardiff Metropolitan University for providing a Get Started grant in 2021 (to enable engagement with CTHA for the development of the Down to Zero project), and a Funding Bid Development grant in 2022 (to write the 12-month SMART Partnership bid). Extensive thanks are also expressed to the Welsh Government for awarding the SMART Partnership to Cardiff Metropolitan University and CTHA in April 2022 (number 2022/02/18/SP/050). Finally, the authors are grateful to CTHA for their collaboration and co-funding of the SMART Partnership.

References

1. Welsh Government. Wellbeing of Future Generations (Wales) Act 2015. https://www.futuregenerations.wales/about-us/future-generations-act/. Accessed: 20 May 2022
2. United Nations, United Nations Sustainable Development Goals: 2015. https://sdgs.un.org/goals. Accessed: 20 May 2022
3. Welsh Government: SMART Partnerships (2022). https://businesswales.gov.wales/expertisewales/support-and-funding-businesses/SMART-partnerships. Accessed 18 May 2021

4. Down to Zero. 2022. Down to Zero. www.down-to-zero.co.uk. Accessed 18 May 2021
5. Welsh Government 2016, [online] Environment (Wales) Act 2016. https://www.legislation.gov.uk/anaw/2016/3/contents/enacted. Accessed 10 May 21
6. Stats Wales 2019: Total social housing stock by local authority area and provider type [online]. https://statswales.gov.wales/Catalogue/Housing/Social-Housing-Stock-and-Rents. Accessed 18 May 2021
7. Climate Change Committee, 2021: The fifth carbon budget how every household can help reduce the UK's carbon footprint. https://www.theccc.org.uk/wp-content/uploads/2016/07/5CB-Infographic-FINAL-.pdf. Accessed 14 May 2022
8. Lambrechts, W., Mitchell, A., Lemon, M., Mazhar, M.U., Ooms, W., van Heerde, R.: The transition of Dutch social housing corporations to sustainable business models for new buildings and retrofits. Energies **14**, 631 (2021). https://doi.org/10.3390/en14030631.Accessed18.05.21
9. Cynon Taf Housing Association: 2022. Welcome. https://www.cynon-taf.org.uk/. Accessed 18 May 21
10. Pellegrino, M. Wernert, C. Chartier, A.: Social housing net-zero energy renovations with energy performance contract: incorporating occupants' behaviour. Urban Plan 7(2) (2022); Zero Energy Renovation: How to Get Users Involved
11. Hart, J., D'Amico, B., Pomponi, F.: Whole-life embodied carbon in multistory buildings: steel, concrete and timber structures. J. Ind. Ecol. **25**(2), 403–418 (2021). https://doi.org/10.1111/jiec.13139
12. Fnais, A., et al.: The application of life cycle assessment in buildings: challenges, and directions for future research. Int. J. Life Cycle Assess. **27**, 627–654 (2022)
13. Seddon, N., et al.: Getting the message right on nature-based solutions to climate change. Glob. Change Biol. **27**(8), 1518–1546 (2021)
14. Friggens, N., Hester, A., Mitchell, R., Parker, T., Subke, J., Wookey, P.: Tree planting in organic soils does not result in net carbon sequestration on decadal timescales. Glob. Change Biol. **26**(9), 5178–5188 (2020)
15. Cunningham, S., et al.: Reforestation with native mixed-species plantings in a temperate cotinental climate effectively sequesters and stabilizes carbon within decades. Glob. Change Biol. **21**(4), 1552–1566 (2015)
16. Intergovernmental Panel on Climate Change: Climate Change and Land: Special Report on Climate Change, Desertification, Land Degradation, Sustainable Land Management, Food Security, and Greenhouse gas fluxes in Terrestrial Ecosystems (2009)
17. Körner, C., Paulsen, J.: A world-wide study of high-altitude tree line temperatures. J. Biogeogr. **31**(5), 713–732 (2004)
18. Conservation International. Carbon Footprint Calculator [online]. https://www.conservation.org/carbon-footprint-calculator/. Accessed 18 May 2022
19. Forin, S., Martínez-Blanco, J., Finkbeiner, M.: Facts and figures from road testing the guidance on organizational life cycle assessment. Int J Life Cycle Assess **24**(5), 866–880 (2018). https://doi.org/10.1007/s11367-018-1533-x
20. Welsh Government: Climate change targets and carbon budgets (2021). https://gov.wales/climate-change-targets-and-carbon-budgets. Accessed 18 May 2021
21. Welsh Government: Working together to reach net zero: all Wales plan (2022). https://gov.wales/working-together-reach-net-zero-all-wales-plan. Accessed 18.05.21
22. Stats Wales: Social housing stock and rents (2022). https://statswales.gov.wales/Catalogue/Housing/Social-Housing-Stock-and-Rents. Accessed 18 May 2021
23. Regulator of Social Housing: Annual social housing sector stock and rents statistics (2021). https://www.gov.uk/government/news/social-housing-sector-stock-and-rents-statistics-show-impact-of-pandemic#:~:text=Returns%20from%20all%20registered%20providers,of%20social%20rent%20homes%20fell. Accessed 18 May 2021

24. Duffy, A., Rogers, M., Ayompe, L.: Renewable Energy and Energy Efficiency: Assessment of Projects and Policies. Wiley Blackwell (2015)
25. Fletcher, T., Scott, C., Hall, J., Spracklen, D.: The carbon sequestration potential of Scottish native Woodland. Environ. Res. Commun. **3**(4), 41003 (2021). https://doi.org/10.1088/2515-7620/abf467. Accessed 18.05.21; IOP Publishing
26. Shannon, V.L., Vanguelova, E.I., Morison, J.I.L., Shaw, L.J., Clark, J.M.: The contribution of deadwood to soil carbon dynamics in contrasting temperate forest ecosystems. Eur. J. Forest Res. **141**(2), 241–252 (2021). https://doi.org/10.1007/s10342-021-01435-3
27. Britannica: RCT county borough, Wales, United Kingdom (2022). https://www.britannica.com/place/Rhondda-Cynon-Taff
28. Lehmann, L., Smith, J., Westaway, S., Pisanelli, A., Russo, G., Borek, R., Sandor, M., Gliga, A., Smith, L., Ghaley, B.: Productivity and economic evaluation of agroforestry systems for sustainable production of food and non-food products. Sustainability (Basel, Switzerland), MDPI AG **12**(13), 5429 (2020). https://doi.org/10.3390/su12135429. Accessed 18 May 2021

Management of Indoor Thermal Conditions in Heavy and Lightweight Buildings: An Experimental Comparison

Nicola Callegaro[1](✉), Luca Endrizzi[2], Luca Zaniboni[1], and Rossano Albatici[1]

[1] University of Trento, 38123 Trento, Italy
nicola.callegaro@unitn.it
[2] Miniera San Romedio, 38019 Ville D'Anaunia (TN), Italy

Abstract. In a Nearly Zero Energy Building (NZEB) design framework, passive design techniques are fundamental to guarantee indoor comfort of occupants, also ensuring a low energy consumption. Among these techniques, internal area heat capacity (IAHC) design still needs further field studies in full-scale constructions: a simple change in this parameter, without varying the mass of the load-bearing walls structure, can positively affect the whole building thermal behavior. Furthermore, combining this design solution with the automated control of windows shadings can help avoiding high temperature peaks due to uncontrolled solar gains. The present paper focuses on the monitoring of the indoor thermal conditions in four real scale residential buildings in the Alpine region of Northern Italy. The buildings present four different combinations of lightweight and massive load-bearing and interior cladding structure. After a first 8-days period of comparison of the pure thermal performance of the four structures, the impact of automated shadings, controlled by a basic and simplified algorithm, on lightweight cladding buildings was studied by means of a second monitoring period of eleven days. Besides confirming the remarkable impact of massive cladding on buildings' thermal performance, the results showed that the basic automated regulation of shadings allowed to mitigate the lower thermal performances of lightweight cladding, also potentially improving the indoor thermal comfort conditions.

Keywords: Building monitoring · Thermal mass · Internal areal heat capacity · Automated shading systems

1 Introduction

High thermal mass in buildings shifts and delays the reaction to sudden temperature changes. For this reason, the presence of massive walls can be beneficial for energy consumption and indoor thermal comfort in reducing the overheating phenomenon [1, 2]. Moreover, thermal inertia properties of materials can be coupled with passive buildings design techniques, e.g. the use of natural ventilation, for lowering indoor daytime temperatures [3, 4].

J. Littlewood and R. J. Howlett (Eds.): SEB 2022, SIST 336, pp. 249–260, 2023.
https://doi.org/10.1007/978-981-19-8769-4_24

This field is worth of particular interest, since thermal inertia of buildings can be varied modifying the so called internal areal heat capacity (IAHC), defined as "the modulus of the net periodic thermal conductance divided by the angular frequency" [5–8]. In literature, little research evaluating this element, with a focus on passive design, is present. Moreover, it is mainly dealing with laboratories and simulative studies. Di Perna et al. [9] studied the dynamic IAHC influence on thermal comfort in buildings with massive internal gains (classrooms) during summer. This was done by varying the insulation by inner placed panels, thus changing the thermal mass. The results highlighted how higher IAHC was the best alternative with both windows opened, closed and other scheduling for closing and opening of windows. Rossi and Rocco [7] evaluated the energy performance of eight selected massive and lightweight wall types with standard Italian transmittances and periodic thermal transmittances. Thermodynamic simulations and energy analyses showed a lack of reference values for IAHC in Italian standards and the potential of highly reducing the energy consumption only by modifying the plaster type. The study also highlighted the necessity of limiting the IAHC (and not only the periodic thermal transmittance) in Southern Europe standards. In this framework, an experimental evaluation of four types of internal linings on the South-oriented wall of a windowless test room in Central Italy (hot humid climate) was performed by Stazi et al. [8]. The results showed that design solutions with high internal thermal inertia (>30 kJ/(m^2K)) and a decrement factor that is not too low are preferable in this type of climate. The use of thermal mass can also be integrated with proper use of shadings [10–12]. In fact, large glazed surfaces can both provide daylight and useful solar gains, but may also cause the indoor environment to be overheated in summer and provoke glare issues [13–15]. During the hours with higher irradiation, solar radiation may increase indoor and envelope superficial temperature peaks [16–19]. Use of thermal mass and shading control could help dephasing and flatten these temperature spikes. For these reasons, proper design and control strategies of shadings' systems are necessary, using external devices and applying proper algorithms of home automation control [14, 20, 21]. For instance, Yun et al. [22], in their simulative study, highlighted how one of the main factors for controlling the indoor environment is the activation threshold of blinds. This parameter, together with the slat angle, depends on the climate of the area where the building is located.

However, field studies are necessary to allow a direct comparison of different solutions of shadings' control and thermal mass application. This paper describes the experimental testing of four buildings with similar shape, different load-bearing structures and different interior massive and non-massive claddings, in which the control of exterior shading is performed by a simplified home automation system. The results confirmed the role of thermal mass and showed that the automated regulation of external shadings can help improve the thermal performances of lightweight interior cladding.

2 Methodology

2.1 Aim of the Research

The study, based on an extensive monitoring of the four buildings described below and lasting 12 months, focuses specially on two different monitoring periods. The aim of the monitoring performed during the first period was to analyze the ability of different envelopes to manage the thermal heat input from solar radiation, highlighting the differences between massive and non-massive buildings. On the other hand, during the second period we investigated the role of an intelligent external shading management system to compensate for the lack of interior surface mass and to benefit indoor thermal comfort conditions.

2.2 Context and Case Studies

The study was developed in the framework of "Dhomo" project, which aimed at studying best comfort solutions in four real scale and comparable residential buildings, assessing the impact of different constructive and management solution on indoor well-being [23]. The buildings are located in Segno, a small town at an altitude of 523 m in a valley in North-East of Italy within the Dolomites—Trentino region. With 3647 degree-days, the location is classified as part of the coldest Italian climate zone. The four houses are similar in terms of shape, size, surface area, and exposure. They are two-story buildings with a total area of 160 m^2: living space, kitchen, and dining room with bathroom on the ground floor; three bedrooms and two baths on the upper floor. The main difference among the four buildings is constituted by the load-bearing and cladding structures, defined as follows (Fig. 1): Building A—wooden structure and lightweight cladding (plasterboard); Building B—concrete and bricks structure and lightweight cladding; Building C—wooden structure and massive cladding (lime plaster); Building D—concrete and bricks structure and massive cladding. Details about the stratigraphy and thermal properties of exterior walls are reported in Fig. 2 and Table 1.

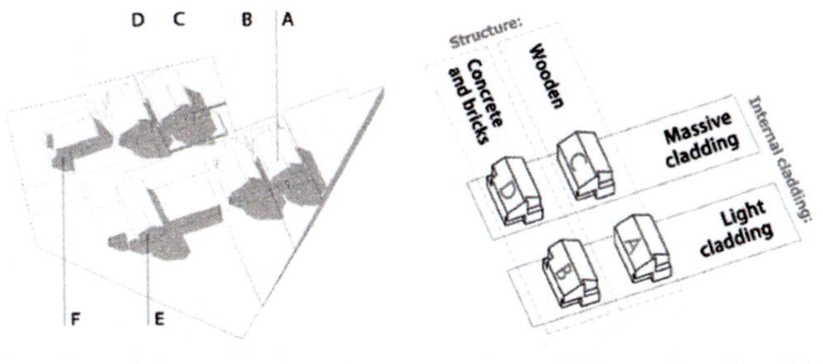

Fig. 1. Left—Layout and position of the six buildings part of "Dhomo" project. Right—Structure and cladding of the four buildings considered in the present study.

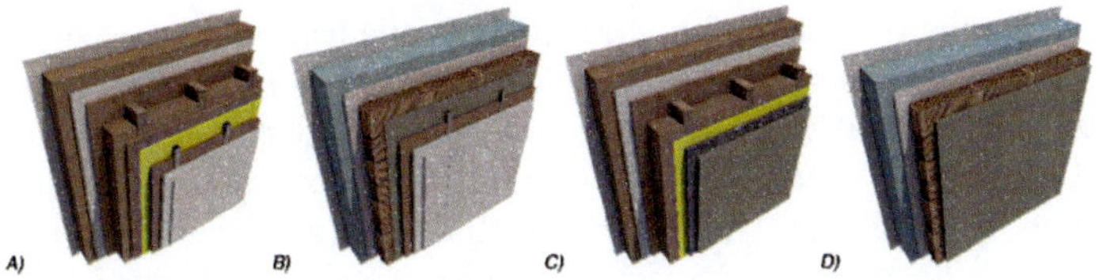

Fig. 2. Layers of the exterior walls of the four buildings

Table 1. Details about thermal propertied of exterior walls of the four buildings.

Building	Decrement factor [-]	Time lag [h]	Periodic thermal transmittance [W/m^{2}K]	Internal areal heat capacity [kJ/m^{2}K]
A	0.051	17.14	0.005	18.4
B	0.064	12.89	0.007	17.8
C	0.066	17.30	0.006	44
D	0.181	11.73	0.018	49.2

The single bedroom on the first floor (Room 1 in Fig. 3) which is identical in all four buildings, was of significant interest and the main subject of the analysis performed in this article. Limiting the investigation to a single room allowed for greater control over the environmental variables, which is not always possible in a real environment that, unlike test cells and laboratories, is uncontrolled and unsupervised.

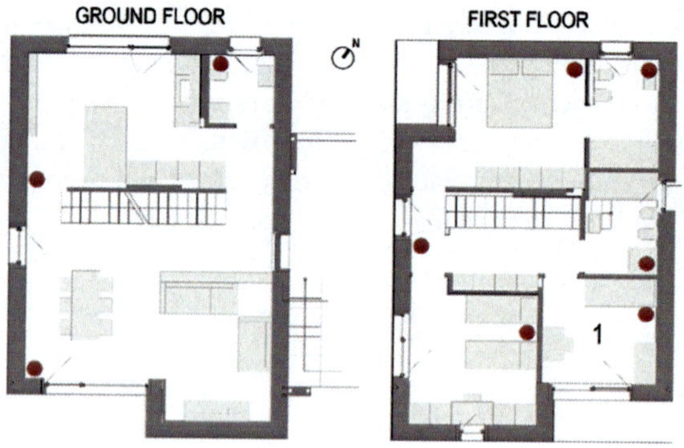

Fig. 3. Horizontal sections of Building B: positions of sensors (RED dots) and position of Room 1 (single room) on the first floor.

2.3 Tests Performed and Instrumentation

An extensive monitoring system was implemented. The system collected data on indoor and outdoor microclimatic conditions and air quality, indoor and outdoor acoustics, layer-by-layer walls temperatures and humidity, walls heat flow transfer, functioning and performance of heating and cooling, lighting, controlled mechanical ventilation and shading management systems. Figure 3 shows the positions of the temperature sensors (Agrisù—LACENTRALINA OMNIA) in Building B (note that sensor positioning is the same in all the buildings). The sensors were located at a height of 2 m from the floor level; data was collected every 15 min. A weather station was located outside of the buildings, providing data on solar irradiance on vertical planes (South-East or South-West exposed), solar irradiance on a horizontal plane, temperature, relative humidity, wind speed and direction.

The monitoring was performed with uninhabited buildings. During a first period, from the 3rd to the 11th of May 2021, the buildings were monitored in free running mode (heating and cooling systems off, no windows shadings) to investigate the different thermal behavior of the building envelope. Later, during a second lapse going from the 22nd of May to the 3rd of June 2021, a simplified home automation algorithm for controlling external blinds was applied in the two buildings with a lightweight interior cladding (A and B). Even during this second period, the buildings were nor naturally neither mechanically ventilated, heated, or cooled. Table 2 shows the boundary conditions for the two test periods.

Table 2. Test and boundary conditions during the two test periods.

	Period 1	Period 2
Period	03/05/2021–11/05/2021	22/05/2021–03/06/2021
Conditions	• Windows CLOSED • Shadings OFF • Mechanical ventilation OFF • Heating and cooling system OFF	• Windows CLOSED • Shadings OFF (C and D) • Automated shadings (A and B) • Mechanical ventilation OFF • Heating and cooling system OFF
Outdoor dry bulb temperature	• MEAN: 14.8 °C • MAX: 24.9 °C • MIN: 4.9 °C	• MEAN: 16.4 °C • MAX: 27.3 °C • MIN: 5 °C
Outdoor relative humidity	• MEAN: 61% • MAX: 96% • MIN: 20%	• MEAN: 68% • MAX: 98% • MIN: 26%
Global horizontal radiation	• MAX: 1053 W/m^2	• MAX: 1117 W/m^2
Average wind velocity	• 12 AM–06 AM: 0.38 m/s • 06AM–12PM: 0.58 m/s • 12PM–06PM: 2.77 m/s • 06PM–12 AM: 1.81 m/s	• 12 AM–06 AM: 0.34 m/s • 06AM–12PM: 0.52 m/s • 12PM–06PM: 2.09 m/s • 06PM–12 AM: 1.15 m/s

The blinds management algorithm used during the second period of investigation is depicted in Fig. 4 In this first test phase, the "shadings down" command causes both shutters and sunshades to be lowered by 100%. For the latter, the slats are tilted vertically (fully closed). The shading activation threshold was chosen based on Yun et al. [22] review. On the other hand, the reference for thermal comfort specifications was UNI EN ISO 16798-1:2019 [24], which establishes the default acceptable indoor temperatures for buildings without mechanical cooling. For this reason, the comfort temperature has been set to 20 °C with a delta of +1.5 °C.

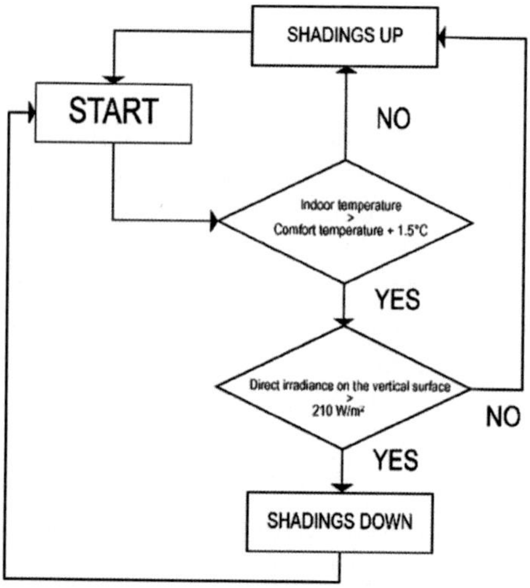

Fig. 4. Scheme of the functioning mechanism of the simplified algorithm used to control shadings systems of buildings A and B during the second test period. The specified threshold for direct irradiance on vertical surface regards the South-East facing wall.

Since the aim of the present research was to measure the efficacy of the simplified shadings' automation system, proper boundary and test conditions were necessary. For this reason, data disturbed by human presence and human windows' openings (due to occasional operations of workers inside buildings) and data referred to days with a too low solar radiation needed to be discarded. Moreover, as it will be highlighted in Results and Discussion section first days of on-field monitoring led to a change in the vertical solar radiation threshold from the initially set 250 W/m^2 to 210 W/m^2 for the South-East facing wall. This is due to the particular layout of the buildings and the fact that the surrounding context shadings caused the amount of incoming solar radiation to be significatively reduced. For the South-West wall, the threshold was kept equal to 300 W/m^2. For these reasons, as regards the second monitoring period, data referred to the last four days are particularly relevant for a full and complete comparison in the four buildings.

3 Results and Discussion

3.1 Period 1

A preliminary consideration needs to be made in order to contextualize the observations reported below. Already from the first period of analysis, it was clear that the four geometrically identical rooms (Rooms 1) were highly influenced by the different exposures of adjacent rooms. Due to the orientation of windows, they were heated up during different time periods during each day: in the morning and early afternoon for buildings A and C (South oriented), in the afternoon for buildings B and D (West oriented). This strongly influenced Room 1 temperatures in the four buildings (Fig. 5): in particular, in buildings B and D the temperature peak appears at a time when sun was not incidental on Room 1 anymore. This is due to the overheating of adjacent rooms, warm air circulation in buildings and overheating of the partition wall.

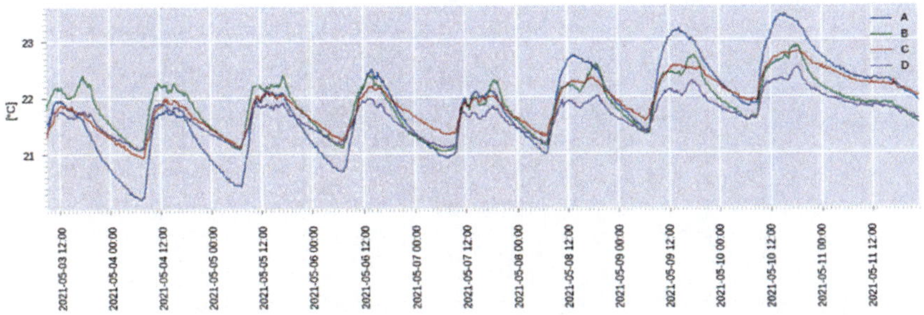

Fig. 5. Period 1—Temperature trend in Room 1 of the four different buildings.

Despite this premise, it is still possible to notice how lightweight cladding buildings present quicker temperature variations (both increasing and decreasing), if compared (Fig. 5) with the corresponding massive cladding ones (A with respect to C and B with respect to D). The ranking of the magnitudes of temperature variations is (1) Building A, (2) Building B, (3) Building C and (4) Building D. This emphasizes that, in the short run, the main influencing factor is the interior cladding, with building structure acting only secondary. Comparing Figs. 6 and 7 it is possible to state that:

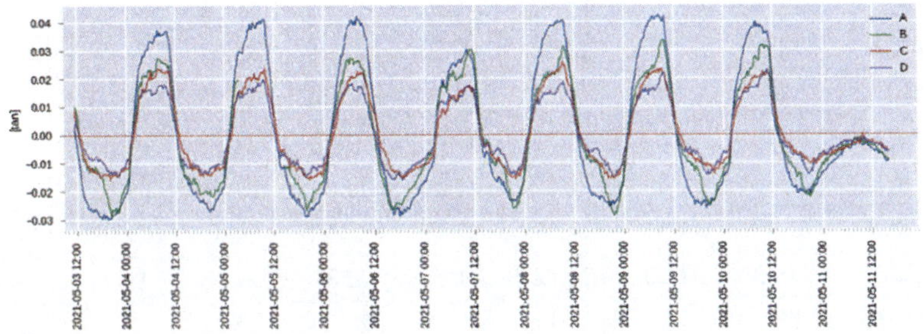

Fig. 6. Period 1—Derivative of temperature value in Room 1 of the four buildings, representing the velocity of temperature variations.

- Building A (wooden structure and lightweight cladding) has, during the day, high temperature peaks and tends to overheat over time
- Building B (concrete and bricks structure and lightweight cladding) has, during the day, high temperature peaks but manages to maintain a comfortable indoor temperature over time
- Building C (wooden structure and massive cladding) does not have high temperature peaks during the day but tends to overheat over time

- Building D (concrete and bricks structure and massive cladding) does not have high temperature peaks and is able to maintain a good indoor temperature

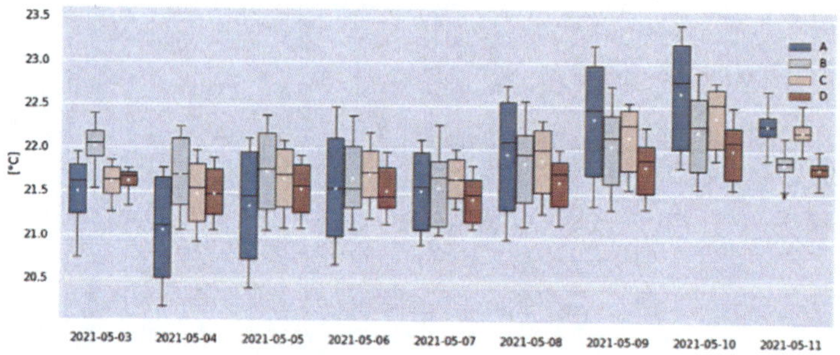

Fig. 7. Period 1—Box plots of daily temperature distributions in Room 1 of the four buildings.

Table 3 shows that internal areal heat capacity regulates daily temperature trends, but it is the inertia of the entire wall envelope that governs temperature changes over medium to long periods.

Table 3. Period 1—Initial and final daily temperatures measured in Room 1 of the four buildings.

Day	Building							
	Initial temperature [°C]—12AM				Final temperature [°C]—11 PM			
	A	B	C	D	A	B	C	D
04/05/2021	20.7	21.5	21.27	21.32	20.96	21.51	21.48	21.4
05/05/2021	20.86	21.44	21.41	21.34	21.26	21.62	21.63	21.47
06/05/2021	21.16	21.54	21.57	21.42	21.42	21.33	21.67	21.35
07/05/2021	21.32	21.28	21.62	21.3	21.5	21.61	21.63	21.46
08/05/2021	21.43	21.55	21.59	21.42	21.88	21.8	21.93	21.62
09/05/2021	21.82	21.73	21.88	21.58	22.3	21.98	22.23	21.82
10/05/2021	22.21	21.91	22.17	21.78	22.74	22.21	22.52	22.06
11/05/2021	22.67	22.15	22.49	22.0	21.91	21.52	21.98	21.58

The positive inertial effect of the interior cladding is exhausted in about a day because of its limited thickness, and it is then the inertia of the underlying layers that matters the most. The wooden structure and massive cladding building (building C) can absorb heat effectively but since it is not ventilated, it is likely to overheat, because that heat is not able to get out of the indoor environment. In a massive building (building D) that heat can be absorbed by wall structures. Over a longer period of time, even the massive building would suffer, as there would be no other building materials to release the heat to. In this case, night cooling can be an effective solution to cool down the structure and to allow heat absorption on the next day (this solution has also been tested in the "Dhomo" project and results are under publication).

3.2 Period 2

With respect to the Period 1, more similar temperature trends and variation velocities were present (Fig. 6). The introduction of a smart control for external shadings regulation changes the thermal performance ranking of the four buildings, acting as a predominant factor as concerns indoor thermal management (Fig. 8).

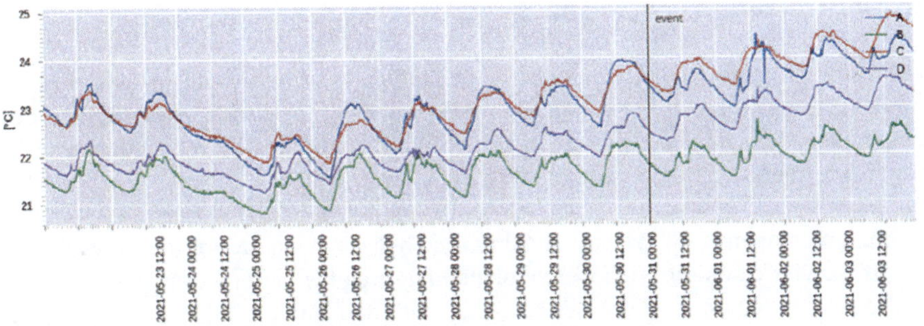

Fig. 8. Period 2—Temperature trend in Room 1 of the four buildings.

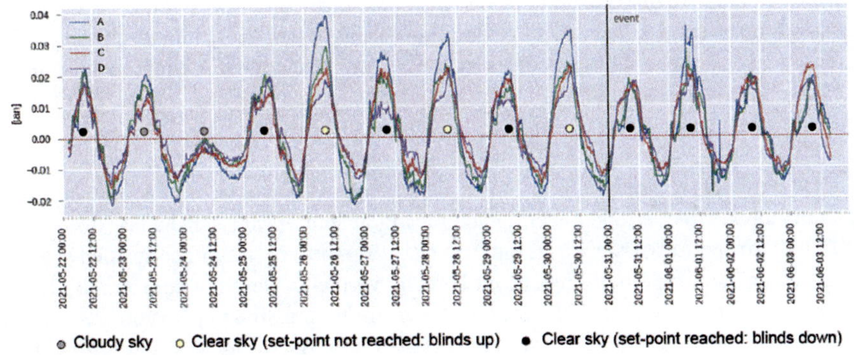

Fig. 9. Period 2—Derivative of temperature value in Room 1 of the four buildings, representing the velocity of temperature variations.

The event indicated in figures corresponds to the changing of vertical irradiance setpoint (from 250 W/m^2 to 210 W/m^2, see *Methodology* section). Before of this changing, the shadings' lowering did not happen every sunny day (see Fig. 9), causing the buildings' thermal prestation ranking to be the same as *Period 1*. After this correction, similar and comparable temperature variation velocities are visible in the four buildings.

Figure 9 shows that the ability to maintain indoor temperature is strongly influenced by home automation, especially during days when shadings were automatically lowered. During Period 1, the strongest difference was noticeable among massive structure (B and D) and lightweight structure (A and C) buildings (Fig. 6). In this case, even though that aspect remains impactive, strong differences are present between buildings with (A and B) and without (C and D) home automation.

Summarizing, home automation allows to reduce temperature peaks, aligning the behaviors of the four buildings. Variations are bigger in buildings A and B, since the two buildings suffer a higher temperature decrease during night caused by the presence of lightweight cladding.

4 Conclusion

The research conducted and described in the previous sections results in a series of recommendations for the use of automated systems in external shadings control for indoor temperature management in high energy performance buildings. Relevant considerations can also be made about the thermal mass and interior surface cladding:

- Smart management of external shadings can help compensate for the lower thermal performance of interior lightweight cladding during the warm season.
- The solar radiation set-point value for the blinds activation threshold is critical, more than the indoor temperature. This choice, therefore, requires delicate attention in its setting: the value must necessarily differ for each month, and adapt to the type of window (exposure, frame factor, possible external passive shadings).
- When blinds are opened or closed, the indoor temperature changes very quickly, which can lead to discomfort in terms of temperature perceived by the user. Cladding can mitigate sudden temperature variations in lightweight structure buildings. For these reasons, the use of appropriate massive interior cladding is considered an interesting passive indoor temperature management strategy to manage thermal indoor conditions.

The study has some limitations: (1) the shutter operation algorithm is still primitive, as it only allows for the total lowering and raising of shutters and sunshades (from 0% to 100%); (2) the period of analysis is rather short and subject to some uncontrolled external events (progress of the building site); (3) many variables were not considered, first of all the presence of the user. Despite the stated limitations, the paper represents a step forward in understanding the behavior of real buildings, providing useful indications and monitored data on passive strategies for attenuation and phase shifting of the external heat wave, such as increasing the surface thermal mass and using an automated system for the management of external shutters. Future developments will concern the field analysis of other active and passive design solutions, such as natural ventilation, with the aim of responding adequately to the comfort needs of the occupants. For this reason, within the "Dhomo" project, longer monitoring periods have been planned with buildings inhabited by users.

Acknowledgement. This work was realized within the project Dhomo funded by Covi Srl—Italy and supported by the Law n.6/1999 of the Autonomous Province of Trento "Incentives to companies".

References

1. Verbeke, S., Audenaert, A.: Thermal inertia in buildings: a review of impacts across climate and building use. Renew. Sustain. Energy Rev. **82**, 2300–2318 (2018). https://doi.org/10.1016/J.RSER.2017.08.083
2. Rodrigues, L., Sougkakis, V., Gillott, M.: Investigating the potential of adding thermal mass to mitigate overheating in a super-insulated low-energy timber house. Int. J. Low-Carbon Technol. **11**, 305–316 (2016)
3. Givoni, B.: Effectiveness of mass and night ventilation in lowering the indoor daytime temperatures. Part I: 1993 experimental periods. Energy Build. **28**, 25–32 (1998). https://doi.org/10.1016/S0378-7788(97)00056-X
4. Shaviv, E., Yezioro, A., Capeluto, I.G.: Thermal mass and night ventilation as passive cooling design strategy. Renew. Energy **24**, 445–452 (2001)
5. Esposti, R., Galbusera, G., Panzeri, A., Salani, C.: Prestazioni estive degli edifici: guida pratica per capire e progettare il comfort e il fabbisogno estivo degli edifici, 2nd edn. Collana L'isolamento termico e acustico vol.5, TEP srl, Milano (IT) (2017)
6. UNI EN ISO 13788:2012 - Hygrothermal performance of building components and building elements - Internal surface temperature to avoid criticai surface humidity and interstitial condensation - Calculation methods
7. Rossi, M., Rocco, V.M.: External walls design: The role of periodic thermal transmittance and internal areal heat capacity. Energy Build. **68**, 732–740 (2014). https://doi.org/10.1016/J.ENBUILD.2012.07.049
8. Stazi, F., Ulpiani, G., Pergolini, M., Di Perna, C.: The role of areal heat capacity and decrement factor in case of hyper insulated buildings: an experimental study. Energy Build. **176**, 310–324 (2018)
9. Di Perna, C., Stazi, F., Casalena, A.U., D'Orazio, M.: Influence of the internal inertia of the building envelope on summertime comfort in buildings with high internal heat loads. Energy Build. **43**, 200–206 (2011)
10. Kuczyński, T., Staszczuk, A., Gortych, M., Stryjski, R.: Effect of thermal mass, night ventilation and window shading on summer thermal comfort of buildings in a temperate climate. Build. Environ. **204**, 108126 (2021)
11. Figueroa-Lopez, A., Arias, A., Oregi, X., Rodríguez, I.: Evaluation of passive strategies, natural ventilation and shading systems, to reduce overheating risk in a passive house tower in the north of Spain during the warm season. J. Build. Eng. **43**, 102607 (2021)
12. Mavrogianni, A., Davies, M., Taylor, J., et al.: The impact of occupancy patterns, occupant-controlled ventilation and shading on indoor overheating risk in domestic environments. Build. Environ. **78**, 183–198 (2014)
13. Kim, G., Lim, H.S., Lim, T.S., et al.: Comparative advantage of an exterior shading device in thermal performance for residential buildings. Energy Build. **46**, 105–111 (2012). https://doi.org/10.1016/J.ENBUILD.2011.10.040
14. Nikoofard, S., Ugursal, V.I., Beausoleil-Morrison, I.: Effect of external shading on household energy requirement for heating and cooling in Canada. Energy Build. **43**, 1627–1635 (2011)
15. Callegaro, N., Pontillo, S., Albatici, R.: Influenza di modelli di gestione per il funzionamento di sistemi oscuranti interni sul consumo energetico e il comfort luminoso. EdicomEdizioni Atti di: ColloquiAT.e2020, Catania, 10th December 2020 1–10 (2020)

16. An, J., Yan, D., Guo, S., et al.: An improved method for direct incident solar radiation calculation from hourly solar insolation data in building energy simulation. Energy Build. **227**, 110425 (2020)
17. Zhao, K., Lv, G., Shen, C., Ge, J.: Investigating the effect of solar heat gain on intermittent operation characteristics of radiant cooling floor. Energy Build. **255**, 111628 (2022). https://doi.org/10.1016/J.ENBUILD.2021.111628
18. Causone, F., Corgnati, S.P., Filippi, M., Olesen, B.W.: Solar radiation and cooling load calculation for radiant systems: definition and evaluation of the direct solar load. Energy Build. **42**, 305–314 (2010)
19. De La Flor, F.J.S., Cebolla, R.O., Félix, J.L.M., Domínguez, S.Á.: Solar radiation calculation methodology for building exterior surfaces. Sol. Energy **79**, 513–522 (2005). https://doi.org/10.1016/J.SOLENER.2004.12.007
20. Zhao, J., Du, Y.: Multi-objective optimization design for windows and shading configuration considering energy consumption and thermal comfort: a case study for office building in different climatic regions of China. Sol. Energy **206**, 997–1017 (2020). https://doi.org/10.1016/j.solener.2020.05.090
21. Zaniboni, L., Pernigotto, G., Gasparella, A.: Analysis of two shading systems in a glazed-wall physiotherapy center in Bolzano, Italy. In: Proceedings of Building Simulation Application 2019, Bolzano, Italy, 19th–21st June 2019 (2019)
22. Yun, G., Park, D.Y., Kim, K.S.: Appropriate activation threshold of the external blind for visual comfort and lighting energy saving in different climate conditions. Build. Environ. **113**, 247–266 (2017)
23. Endrizzi, L.: Progetto Dhomo: la sfida del costruire per migliorare la qualità dell'abitare. Ingenio - Informazione Tecnica e Progettuale (2021)
24. UNI EN ISO 16798-1:2019 - Energy performance of buildings - Part 1: Indoor environmental input parameters for design and assessment of energy performance of buildings addressing indoor air quality, thermal environment, lighting and acoustics

Comparison of the Thermal Effect of Two Automatic Controls of Roller Shutters in an Academic Space

Marc Roca-Musach(✉), Elena Garcia-Nevado, Carlos Alonso-Montolio, Isabel Crespo Cabillo, and Helena Coch Roura

Universitat Politècnica de Catalunya, 08028 Barcelona, Spain
marc.roca.musach@upc.edu

Abstract. Roller shutters are passive systems that can be operated automatically to improve building energy efficiency based on different criteria. This paper aims to compare the effect of an occupant-based and a thermal-based control of roller shutters on heating and cooling demand in an academic space. We run dynamic energy simulations of a case study in the Mediterranean climate (Barcelona, Spain). We found that the automatic control of roller shutters can significantly decrease heating and cooling demand. Both controls here analyzed perform similarly, with a rough reduction in heating of 1/2 and in cooling of 2/3. Therefore, occupant-based controls can be good alternatives that reduce energy demand as well as preserve occupants' capacity to satisfy their needs or wills. Automated control of roller shutters could perform better if based on the energy balance of solar gains, thermal conduction and infiltration through the window. In addition to solar shading, the thermal insulation and infiltration-control provided by roller shutters are also key to improving building thermal performance in the Mediterranean climate.

Keywords: Roller shutters · Solar shading · Thermal insulation · Automation · Occupant-centric controls

1 Introduction

In 2020, the use of buildings was responsible for 30% of global energy consumption [1]. A third of energy consumption in buildings is due to the thermal conditioning of spaces [2]. The energy demand of buildings depends not only on how they are designed but also on how building systems -both passive and active- are managed.

Operable building systems can be controlled manually or automatically to increase building energy efficiency. In manually controlled systems, occupants directly operate the system based on their personal preferences and perceptions. As such, these systems require not only the presence of occupants but also an active role from them. Even if users are not satisfied, this "will to interact" is not always present. And if it is, the interaction with building systems is usually not possible because only specific users are authorized to do so [3]. This duality of users with no will and/or no possibility to manage building

J. Littlewood and R. J. Howlett (Eds.): SEB 2022, SIST 336, pp. 261–270, 2023.
https://doi.org/10.1007/978-981-19-8769-4_25

Fig. 1. Façade of the School of Architecture of Barcelona on a sunny and cloudy day in March 2022. After two weeks, only 25 out of 140 roller shutters changed position (17,8%).

systems typically occurs in public buildings, especially when it comes to windows or shadings (see example in Fig. 1), which significantly affects their thermal performance.

Automatically controlled systems, on the other hand, can be activated without the need for occupants' presence or will. Moreover, it is possible to centralize the management of multiple building systems by implementing Building Management Systems (BMS) or Building Automation Systems (BAS) that allow to operate them with no dependency on occupants' presence [4]. Because of their potential energy savings [5], the European Union is actively encouraging "the installation of active control systems such as automation, control and monitoring systems" [6].

The effectiveness and satisfaction achieved with automatically controlled systems depend on the criteria driving them. Automatic systems can be "automatons". This kind of systems are designed to do specific tasks with no connection with other systems of the building. Automatons can be based on schedules or sensors. For example, an analogic thermostat that switches on heating based on the air temperature is a simple automaton. Automatic systems can also be "cybernetic". This kind of systems are interconnected, which means that decisions can be centralized and optimized using complex criteria [7]. For example, shadings and lighting could be controlled based on an occupancy schedule and the measured indoor illuminance, from a centralized control point.

However, automated systems can lead to a lack of occupants' control or a misunderstanding of the systems [8]. Consequently, this causes dissatisfaction to occupants and, sometimes, automatic systems get manipulated or even disconnected [9]. To avoid it, occupant-centric controls (OCC) are automatisms that collect various data to adapt to different situations, and can vary depending on occupants' presence or behavior. Most of the studied OCC aim to control lighting and HVAC systems, mainly in academic and office buildings. OCC of passive thermal systems are rarely studied, as so for shading devices [10].

To contribute to filling this gap, this paper aims to compare the heating and cooling energy savings achieved with two kinds of automatic controls of roller shutters, a thermal-based and an occupant-based. Thermal-based control is scheduled with thermal optimization criteria, while occupant-based control varies according to occupants' presence and needs.

2 A Brief Background on Roller Shutters

Roller shutters are an operable passive system of buildings that can adopt multiple configurations. In this paper, we refer to roller shutters as a filtering device of windows composed of horizontal, non-orientable slats that roll around a shaft placed inside an upper box and move along vertical, fixed rails placed on each side of the window (Fig. 2). The shaft is operated by a mechanical system from the indoors, without the need of opening the window [11]. Traditionally, roller shutters were operated manually with a ribbon or a crank. Nowadays, the electrification and motorization of the shaft allow occupants to operate them just pressing a button. Motorized roller shutters could be easily automated with the implementation of a control system.

Fig. 2. Internal (left) and external (right) view of a roller shutter in the School of Architecture of Barcelona.

Roller shutters are widely used in Europe, especially in the Mediterranean region. In Spain, this system was introduced in the early 20th century mainly as solar shading devices [11]. Nevertheless, roller shutters can also serve other multiple functions: natural light control, additional thermal insulation, security barriers, protection against weather phenomena (wind, rain, snow…), ventilation and infiltration control, noise reduction, privacy control, and even aesthetics [12].

In this work, we focus on three thermal functions of roller shutters: solar shading, thermal insulation, and infiltration control. In temperate climates, the solar shading function of roller shutters is most relevant in summer, especially under direct solar radiation. The use of this kind of external shading can significantly reduce cooling needs (by 70% in buildings with window-to-wall ratios over 60% [13]). The thermal insulation function depends on the thermal conductivity and the emissivity of the slats, and the insulation effect provided by the air-cavity between the window and the roller shutters, especially if airtight [14]. Completely closed roller shutters may reduce by 33% the thermal transmittance of double windows [12], an effect especially convenient on winter nights. The function of infiltration control of roller shutters also depends on the airtightness of the cavity, influenced by the design of the rails and slats [15]. This function might be especially relevant at night in winter, and during the day in summer.

As the optimal position of roller shutters depends on multiple variables (time of the day, season, weather…), the implementation of automatic control systems can help to reduce the energy demand of buildings. Finding the criteria that maximize energy savings while preserving occupant satisfaction is then a key goal.

3 Case Study

The case study here analyzed consists of a classroom located on the fifth floor of the School of Architecture of Barcelona, Spain (Fig. 1).

Barcelona has a Mediterranean climate, with mild winters and warm to hot summers. The annual average daily temperature is 18.2 °C, with an average low temperature of 8.8 °C in January, and an average high temperature of 28.6 °C in July. In this climate, both heating and cooling systems are usually necessary in buildings.

The classroom is a single box-shaped space with just one exposed façade, oriented to the East. This façade has a 47% window-to-wall ratio and comprises eight fixed windows with operable external roller shutters. Windows have a single clear glass and aluminum frame without thermal bridge breaking. Roller shutters' slats are made of aluminum without insulation. Façade details are specified in Fig. 3.

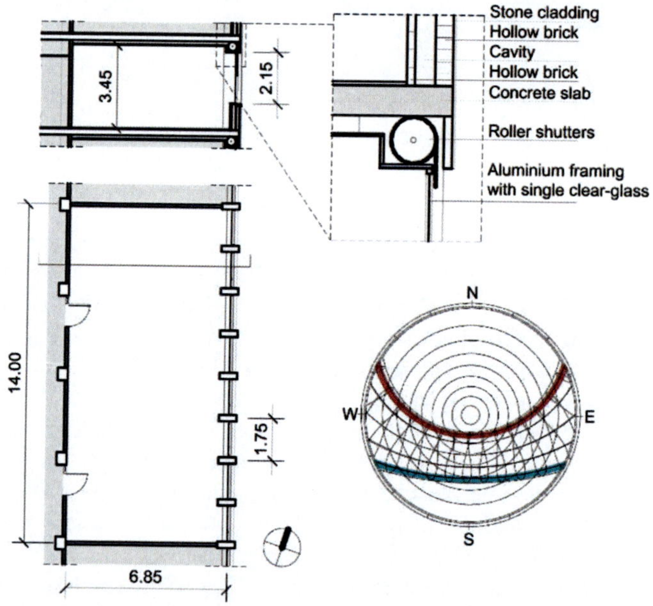

Fig. 3. Section view (top left), plan view (bottom left), and detail (top right) of the simulated space, annotations in meters. Stereographic projection of the sun path in Barcelona (bottom right).

4 Method and Tools

To study the effects of automatic control systems of roller shutters, we run dynamic energy simulations over a case study. We use the Energy Plus engine through the Honeybee interface for Grasshopper3D [16].

4.1 Simulation Model and Settings

The simulation model consists of a single box. The ceiling, the floor, and all the walls are considered adiabatic, except for one façade. Neighbor buildings and other context shadings are omitted.

We run simulation based on meteorological data from Barcelona's airport (epw file from [17]). Specifically, we focus on the extreme summer and winter weeks, with temperatures ranging between 1.7 °C and 12.8 °C (15–21 January), and between 18.3 °C and 32.2 °C (22–28 July).

Our simulations account for internal loads corresponding to an educational space. When occupied, we consider internal loads due to the people (31.5 W/m^2) and electronic equipment (10.5 W/m^2). When artificial lighting is present, we include an additional load of 4 W/m2. Heating and cooling setpoint temperatures considered in this study are specified in Table 1 [18, 19].

Table 1. Heating and cooling setpoint temperatures.

Load	Value (°C)	Schedule
Heating setpoint	21	Weekdays 7:00–20:59
	17	Weekdays 0:00–6:59 & 21:00–23:59, and weekends
Cooling setpoint	25	Weekdays 7:00–20:59

To simulate the case study, we calculate the equivalent properties of the opening system (window + roller shutters) in different states. Roller shutters can be operated in four states: completely opened (100%), half-opened (50%), nearly closed (10%), and completely closed (0%). When roller shutters are completely closed, they reduce air infiltration and create a 5 cm airtight cavity that reduces the equivalent U-value of the opening system. In the other cases, we do not consider the thermal effect of the air cavity. Table 2 shows the equivalent properties for each state of the roller shutters. The U-value of the opaque part of the facade is 1.3 W/(m^2·K).

Table 2. Equivalent properties of the opening system for different states of roller shutters.

	100% Open	50% Open	10% Open	0% Open
Visible and solar transmittance (T_{vis})	0.90	0.45	0.09	0.00
Solar heat gain coefficient (SHGC)	0.86	0.43	0.086	0.00
Equivalent U-value [W/(m^2·K)]	5.7	5.7	5.7	2.6
Air infiltration [ACH]	1.0	1.0	1.0	0.5

4.2 Control Scheduling of Roller Shutters

For this study, roller shutters are scheduled based on occupancy and solar radiation. We define three control scenarios of roller shutters (see Fig. 4):

- **No control of roller shutters (= always open).** In this scenario, roller shutters remain completely open regardless of climatic conditions or occupants' needs (such as natural light, glare, or temperature), and artificial illumination is always switched off. This scenario is taken as the baseline case for comparison.
- **Thermal-based control of roller shutters.** In this scenario, roller shutters are automatically controlled based on thermal efficiency criteria. In winter, roller shutters maximize solar gains during the day and minimize heat losses at night. In summer, roller shutters reduce solar gains during the day and maximize heat losses at night. Whenever roller shutters are completely closed, but there is occupation, artificial lights switch on. Otherwise, they are off.
- **Occupant-based control of roller shutters.** This scenario is based on occupant preferences on working days, and based on thermal efficiency at the weekends and at night. Occupant preferences are considered as a minimum natural daylighting level to carry on activities in the classroom, and the prevention of glare in case of direct solar radiation. For the simulation, roller shutters are opened a 10%, 50% or 100% based on solar radiation on the weekdays, as plotted in Fig. 4.

To generate the simulation schedule of roller shutters, we calculate the hourly sun position for each analysis week, and we determine whether there is direct solar radiation, diffuse radiation, or no radiation on the East façade. These values combined with the control criteria of each scenario, result in the different simulation schedules (Fig. 4).

5 Results and Discussion

The simulation results show that the automatic control of roller shutters reduces heating and cooling demand compared with the baseline case with no control of roller shutters (see Fig. 5).

In winter, occupant-based control decreases heating demand by 41,1%, and thermal-based control by 49,6%, compared with the baseline case. In summer, the reduction is between 65,7% with the occupant-based control and 63,3% with the thermal-based control. Therefore, both analyzed controls perform similarly.

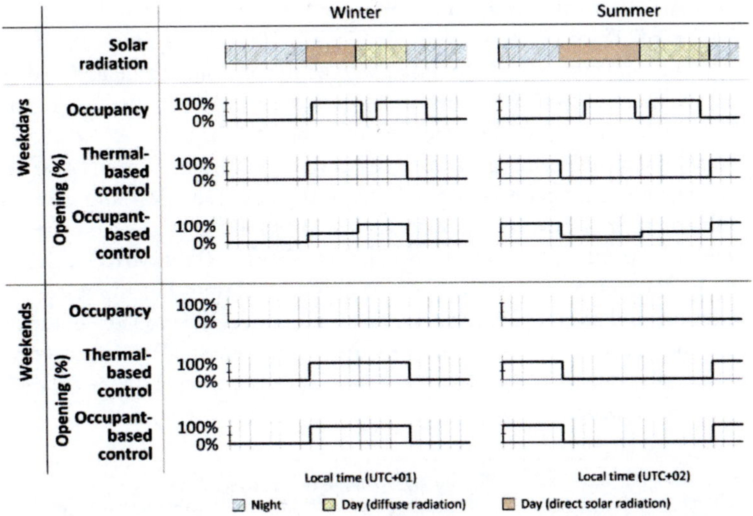

Fig. 4. Solar radiation, occupancy, and opening of roller shutters for each scenario.

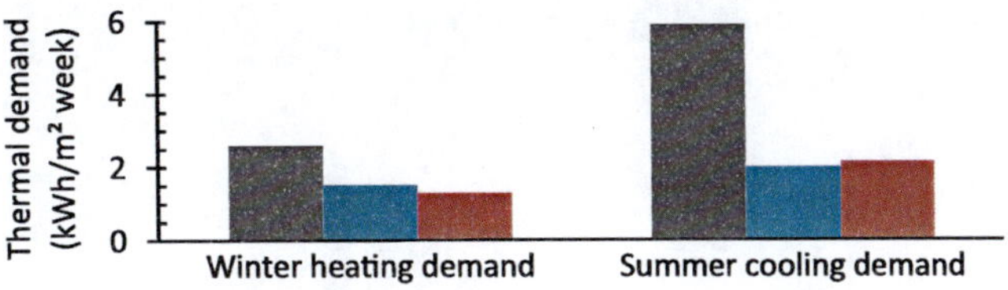

	Winter heating demand	Summer cooling demand
No control (=always open)	2.60	5.89
Occupant-based	1.53	2.02
Thermal-based	1.31	2.16

Fig. 5. Heating/cooling demand for each simulation scenario, in kWh/m^2 week.

The thermal-based scenario performs better in winter, while the occupant-based scenario performs better in summer. This is because the thermal-based control requires artificial lighting during the day in summer, which implies more internal loads, compared with the occupant-based control.

As a sample of the simulated results, Fig. 6 shows the hourly balance of a window during one day in winter for the three simulated scenarios. The baseline case with no roller shutters is plotted in Fig. 6a. Night insulation and infiltration control provided by completely closed roller shutters decreases heat losses by 52%, as shown in the gray zone of Fig. 6b and c, compared with the baseline case. Additionally, half open roller shutters for glare prevention reduce solar gains by 70%, as shown in the striped zone in the occupant-based scenario (Fig. 6c). Therefore, all analyzed functions of roller shutters have a relevant role on the energy flux through the window. However, the results depend on the properties defined in Table 2, which need further research.

Figure 6 shows that the position of roller shutters could be optimized if related to the local energy balance of the window. This means that heat gains/losses from solar gains, thermal conduction and infiltration could be dynamically evaluated to determine the optimal position each time. In the scenario with no roller shutters (Fig. 6a), from 9:00 to 11:59 (peak direct solar radiation hours in the East façade) roller shutters should be completely open since solar gains are greater than heat losses. From 12:00 to 15:59, the position of roller shutters is not obvious since solar gains are lower than heat losses; however, to completely close roller shutters would supress heat gains, so the resultant balance could be even worse. A new simulation of completely closed roller shutters could help to determine the optimal position in this situation. For the rest of the day, roller shutters should be completely closed to minimize heat losses. This timing does not correspond, though, with the one defined by the thermal-based (Fig. 6b) and occupant-based controls (Fig. 6c) for the simulations. These results show the importance of taking climate data into account.

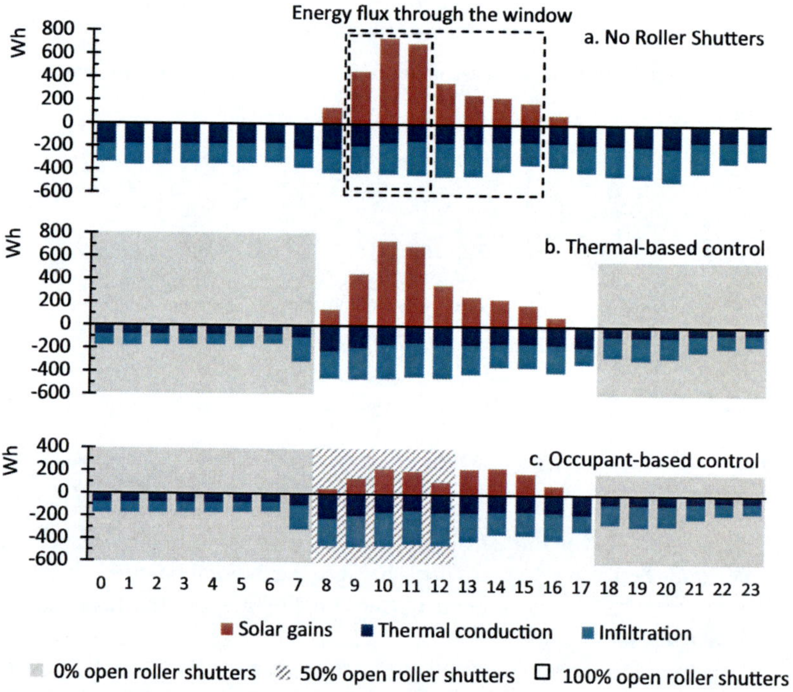

Fig. 6. Solar gains, heat losses from thermal transmission and heat losses from infiltration of a window during one day in winter for the three simulated scenarios of the case study (East façade).

In all simulated scenarios, roller shutters are controlled by a fix schedule. This means that the model does not perform optimally due to the randomness of variables that affect the results, such as weather phenomena or occupants' needs/will. Regarding weather phenomena, as an example, cloudy days would need a different schedule given that solar radiation is diffuse all day. Regarding occupants' needs/will, glare or lack of indoor illuminance would lead to subjective manual modifications: glare could be corrected with manually-activated internal shadings (blinds, curtains...), while lack of indoor illuminance would need to increase natural illumination or to switch on artificial lighting. To include weather and occupants' needs/will in the simulation, the operation of roller shutters could be linked to environmental data (e.g. solar radiation or indoor illuminance) or a dynamic occupants' discomfort evaluation.

However, the simulation of random variables is always a difficult matter. Future studies aim to implement the simulated controls in a real case study. Further research will also focus on the heat gains from diffuse solar radiation, and the reduction of heat losses/gains from thermal conduction and infiltration provided by completely closed roller shutters.

6 Conclusions

This paper studied the effect on heating and cooling demand of an occupant-based and a thermal-based scheduling of roller shutters in an academic space in Barcelona, Spain. The focus was on solar shading, thermal insulation, and infiltration-control functions of roller shutters, all them analyzed in a dynamic energy simulation.

We found that the automatic operation of roller shutters can significantly decrease heating and cooling demand in an East façade of an academic building. Both analyzed controls (occupant-based and thermal-based) perform similarly, with a reduction in heating of 1/2 and in cooling of 2/3. Therefore, occupant-based controls can be good alternatives that reduce energy demand as well as preserve occupants' capacity to satisfy their needs or wills.

The simulated results show that automated control of roller shutters could perform better if it is based on the heat gains/losses from solar gains, thermal conduction and infiltration through the window. The balance between both factors could be dynamically evaluated to determine the optimal position of roller shutters each time.

The solar shading function of roller shutters is important to control direct solar radiation from getting into the indoor space. However, the insulation and infiltration-control functions are also relevant to minimize heat losses at night in winter and reduce heat gains during the day in summer, even in the Mediterranean climate of Barcelona.

Acknowledgements. This research was developed under the project PID2020-116036RB-I100, funded by the Spanish Ministry of Science and Innovation (MCIN). This work was also carried on with the support of the Secretariat of Research and Universities of Generalitat de Catalunya and the European Social Fund.

References

1. United Nations Environmental Programme: Global Status Report for Buildings and Construction: Towards a Zero-emission, Efficient and Resilient Buildings and Construction Sector (2021). http://www.unep.org/resources/report/2021-global-status-report-buildings-and-construction

2. Zhang, T., Tan, Y., Yang, H., Zhang, X.: The application of air layers in building envelopes: a review. Appl. Energy **165**, 707–734 (2016). https://doi.org/10.1016/j.apenergy.2015.12.108
3. Pastore, L., Andersen, M.: Building energy certification versus user satisfaction with the indoor environment: findings from a multi-site post-occupancy evaluation (POE) in Switzerland. Build. Environ. **150**, 60–74 (2019). https://doi.org/10.1016/j.buildenv.2019.01.001
4. Abuimara, T., Hobson, B.W., Gunay, B., O'Brien, W., Kane, M.: Current state and future challenges in building management: practitioner interviews and a literature review. J. Build. Eng. **41**, 102803 (2021). https://doi.org/10.1016/j.jobe.2021.102803
5. Lee, D., Cheng, C.-C.: Energy savings by energy management systems: a review. Renew. Sustain. Energy Rev. **56**, 760–777 (2016). https://doi.org/10.1016/j.rser.2015.11.067
6. European Parliament and the Council: Directive 2010/31/EU of the European Parliament and of the Council of 19 May 2010 on the energy performance of buildings. EUR-Lex, 23 (2010). http://data.europa.eu/eli/dir/2010/31
7. Serra Florensa, R., Isalgué Buxeda, A., Sánchez, L.: Arquitectura i màquina : principis d'instal·lacions als edificis. Edicions UPC (1996)
8. Leaman, A., Bordass, B.: Are users more tolerant of 'green' buildings? Build. Res. Inf. **35**, 662–673 (2007). https://doi.org/10.1080/09613210701529518
9. Gilani, S., O'Brien, W.: A preliminary study of occupants' use of manual lighting controls in private offices: a case study. Energy Build. **159**, 572–586 (2018). https://doi.org/10.1016/j.enbuild.2017.11.055
10. Park, J.Y., et al.: A critical review of field implementations of occupant-centric building controls. Build. Environ. **165**, 106351 (2019). https://doi.org/10.1016/j.buildenv.2019.106351
11. Martí, N., Araujo, R.: La persiana enrollable. Revisión del sistema constructivo y sus requisitos medioambientales. Informes de la Construcción **67**, e113 (2015). https://doi.org/10.3989/ic.14.069
12. Blanco Rodríguez, M.Á.: Guía técnica de persianas y contraventanas para la mejora de la eficiencia energética en los edificios residenciales de Euskadi. https://eeecp.blogspot.com/
13. Hoffmann, S., Lee, E.: Potential energy savings with exterior shades in large office buildings and the impact of discomfort glare. Lawrence Berkeley National Laboratory, LBNL Report #: LBNL-187170 (2015). https://escholarship.org/uc/item/1rg6z0n8
14. Oleskowicz-Popiel, C., Sobczak, M.: Effect of the roller blinds on heat losses through a double-glazing window during heating season in Central Europe. Energy Build. **73**, 48–58 (2014). https://doi.org/10.1016/j.enbuild.2013.12.032
15. Alawadhi, E.M.: Effect of an incompletely closed window shutter on indoor illuminance level and heat gain. Energy Build. **110**, 112–119 (2016). https://doi.org/10.1016/j.enbuild.2015.10.040
16. Roudsar Sadeghipour, M., Pak, M.: Ladybug: a parametric environmental plugin for grasshopper to help designers create an environmentally-conscious design. In: Proceedings of the 13th International IBPSA Conference Held in Lyon, France (2013). http://www.ibpsa.org/proceedings/BS2013/p_2499.pdf
17. EnergyPlus Weather Data: https://energyplus.net/weather-location/europe_wmo_region_6/ESP/ESP_Barcelona.081810_SWEC. Accessed 20 May 2022
18. Ministerio de la Presidencia de España: Real Decreto 1027/2007, de 20 de julio, por el que se aprueba el Reglamento de Instalaciones Térmicas en los Edificios. Boletín Oficial del Estado núm. 207, BOE-A-2007-15820 35931 (2007). https://www.boe.es/eli/es/rd/2007/07/20/1027
19. Código Técnico de la Edificación: Ministerio de Transportes, Movilidad y Agenda Urbana de España (2021). https://www.codigotecnico.org/

Energy Poverty and Heatwaves. Experimental Investigation on Low-Income Households' Energy Behavior

Gianluca Maracchini(✉), Elisa Di Giuseppe, and Marco D'Orazio

Università Politecnica Delle Marche, Ancona 60131, Italia
g.maracchini@staff.univpm.it

Abstract. Global and local climate change increases the occurrence of extreme weather events such as heatwaves (HW), which in the next years will increase even more in both frequency and magnitude. This phenomenon strongly affects people's quality of life due to societal, health, and environmental repercussions, but also the building sector, increasing energy use and deteriorating the indoor thermal environment. Low-income households are the most affected by health issues during HWs, since they may not own a proper air conditioning system, or not be able to sustain the energy costs for Air Conditioning (AC) use. Several works in the literature try to investigate the impact of HW on building thermal response and energy consumption. However, these works are mainly focused on numerical simulations and do not consider the possible different AC use behavior that may occur in extreme, rare events such as HW. Moreover, none of them consider the impact of HW in an energy poverty scenario. This work presents the results of a wide experimental campaign aimed at evaluating the impact of HW on energy consumption in an energy poverty scenario. The thermal response of two Italian social housing buildings was monitored. The results indicate that, during an HW, an exponential increase in consumption for cooling is observed, which is 4 to 8 times greater than that encountered in normal summer periods. This result highlights the importance of considering a real use scenario in evaluating the impact of HW on building energy consumption.

Keywords: Heatwave · Energy poverty · Energy consumption · Social housing · Occupants' behavior

1 Introduction

Due to climate change, extreme weather events with associated environmental hazards are increasing all around the world in terms of frequency and magnitude. A heatwave (HW) is a typical extreme event associated with climate change, defined as a period of several consecutive days with very high air temperatures (higher than the seasonal average), which has become a global concern due to its several implications, and in particular those on people's health [1–4], energy consumption [5], and repercussions at the economic level [6]. According to future climate projections, HW will be twice more

J. Littlewood and R. J. Howlett (Eds.): SEB 2022, SIST 336, pp. 271–280, 2023.
https://doi.org/10.1007/978-981-19-8769-4_26

frequent in Europe between 2020 and 2049 compared to the past (occurring even 3 or 4 times per year by the end of the 21st century) [7]. In this framework, investigating the indoor and outdoor comfort conditions during HW, as well as the effect on building energy consumption, is becoming of paramount importance, and constitutes the first step to evaluating and proposing technical and feasible solutions for reducing its effects and then ensuring a more resilient and safer built environment to climate change.

Several works numerically evaluated the impact of HWs on the thermal response and energy consumption of existing and new buildings [8–10], also considering its possible synergic effects with the Urban Heat Island (UHI) phenomenon [11, 12]. In [9], for example, the authors demonstrated that new buildings can have worse performance than energy-inefficient buildings in addressing HW. In [10], the authors showed a significant increase in indoor temperatures and energy consumption during HW with respect to the rest of the summer (3.5 °C and 87%, respectively). The main limitation of these numerical works lies in the adoption of a standard deterministic approach to model the occupants' behavior and interaction with the building systems, neglecting the possible occupants' behavior changes that may occur in extreme conditions such as HW. Moreover, there are a few works that address the impact of HW on building thermal and energy performance of existing building stock through experimental investigation, thus considering a real use scenario.

Low-income households are the most affected by health issues during HWs, due to the absence of AC or the difficulties related to the high energy costs [13]. In this framework, there is a great need to pay attention to the cooling situation of low-income buildings and households in summer to find feasible technical, economic, and then policy solutions to improve their indoor thermal environment. This notwithstanding, although different works in the literature focus on energy poverty, i.e. the inability of households to afford proper indoor thermal comfort due to economic restrictions, most of them are more focused on winter issues, paying less attention to summer problems [13].

For these reasons, in this work, the impact of HW on the summer thermal response of two social-housing buildings and the related low-income households' energy behavior is evaluated through a wide experimental continuous monitoring campaign, carried out under the LIFE SUPERHERO project. The investigated buildings are two existing high-mass social housing buildings placed in the city of Reggio Emilia, Italy, in the middle of the Padan valley. The topic is very relevant since extreme weather conditions are expected to increase in the next years.

2 Materials and Methods

This work can be subdivided into three main phases. First, the monitored outdoor climatic conditions of the considered summer period are analyzed to verify and identify the occurrence of HW. Then, the thermal responses in free-floating conditions of four unoccupied apartments are analyzed to understand the thermal response of the two buildings during the HW without human interaction. Finally, the indoor thermal environment of the occupied apartments and the occupants' energy behavior (air conditioning, AC, activation) are analyzed and compared with common days to draw preliminary considerations about the influence of HW on the occupants' AC use behavior.

2.1 The Case Studies

Two existing three-story social-housing buildings built in 1980 and located in the humid subtropical climate of Reggio Emilia, Italy [14], have been selected for the monitoring campaign (Fig. 1).

Fig. 1. Monitored attic apartments of **a** Building #1 (third floor) and, **b** #2 (second and third floor). Occupied apartments are highlighted.

The buildings are characterized by poor energy quality and comfort, thus representing an appropriate object to study the effect of HW on indoor thermal comfort within an energy poverty scenario. In particular, the walls are made of prefabricated concrete panels (non-insulated for Building #1 and insulated for Building #2) with double-glazed windows. The floors are composed of reinforced concrete slabs without insulations, while the flat roofs are made of a prefabricated reinforced concrete slab ("Predalles") with a metal cladding added used as a waterproofing system covering an external airgap. For Building #1 an additional insulation layer is present within the airgap. Table 1 presents the overall estimated thermal transmittance of the components (U-value).

2.2 Experimental Setup

The monitoring campaign aims at investigating indoor and outdoor environmental conditions and the occupants' behavior in the attic apartments of the two selected buildings. A wireless communication system including several devices has been adopted to collect data with a low impact on users' daily life. A URSALINK UC1152 LoRaWAN antenna was used to send the data to an FTP server. Table 2 shows the technical specifications of the installed sensors.

2.3 Heatwave Identification

Universal methodologies to identify and quantify HWs do not currently exist [15, 16]. However, it is quite recognized that combining temperature and humidity for calculating HW indexes leads to more reliable climate projections than considering the two variables separately [17]. In this work, HW has been identified by using Thom's Discomfort Index (TDI), which combines the values assumed by the humidity and temperature parameters to describe the conditions of physiological discomfort due to humid heat [18, 19]. In particular, $TDI = 0.5T_w + 0.5T_a$ where T_a is the air temperature and T_W is the wet-bulb temperature calculated from moisture information according to [20]. The bioclimatic

Table 1. Thermal transmittance of the construction elements in $W/(m^2K)$.

Building components	Building #1	Building #2
Exterior wall	1.72	0.66
Interior partition (between dwellings)	2.66	0.66
Flat roof	0.33	1.49
Floor	1.45	1.64
Double glazed windows, wood frames	2.26	2.16

Table 2. Technical specifications of the installed sensors and monitoring system

Equipment/Sensor	Location	Measure	Accuracy
LoRaWAN Weather Station WS100LRW/SR	Roof (Building #2)	Outdoor air temperature Outdoor air humidity Solar radiation Rain gauge Wind direction Wind speed	±0.2 °C ±2% ±5% 0.2 mm or 4% of rate ±5° ±3 km/hr or 5%
ELSYS ERS CO2	Living rooms, Bedrooms	Air temperature Humidity CO2 Light	±0.2 °C ±2% RH ±50 ppm / ± 3% ±10 lx
ELSYS EMS Door	Windows	Open activity sensor	16 mg/digit (sens.)
ORNO OR-WE-514	AC systems	Energy smart-meter	class B (EN 50470–3)

distress thresholds used were identified by referring to a mortality study conducted in the urban area of the near city of Bologna for the years 1989–2003 [3]. Three different thresholds are considered, corresponding to a bioclimatic condition with low discomfort, discomfort, and severe discomfort, respectively, whose description is reported in Table 3.

Table 3. Discomfort bioclimatic conditions.

Discomfort condition	TDI	Description
Low discomfort	≥24, <25	The population feels discomfort but there are no average increases in mortality
Discomfort	≥25, <26	The weakest people can manifest health effects (headaches, dehydration, and sometimes even death). Total mortality, due to natural and cardiovascular causes, increases on average by about 15%, while mortality from respiratory causes up to 50%

(continued)

Table 3. (*continued*)

Discomfort condition	TDI	Description
Severe discomfort	26 or ≥ 25 for 3 days	The number of people that are affected by pathologies linked to heat increases. Total mortality, from natural and cardiovascular causes, increases on average by about 30%, while respiratory mortality by about 80%

3 Results and Discussion

3.1 Heatwaves Identification

In Fig. 2, the outdoor air temperature (To) and relative humidity (RHo) for the monitored period (August 2021) are reported, along with the daily average TDI values. The highest outdoor temperatures were recorded between the 11th and 16th of August with maximum values constantly higher than 36 °C (with peaks at 38 °C) and minimum values between 23 and 26 °C. In this period, TDIs are higher than 24, also exceeding the "discomfort" threshold (TDI = 25) for three consecutive days (from 13th to 15th of August), denoting the occurrence of an HW with "severe discomfort" (see Table 3), corresponding to a high-risk level for human health.

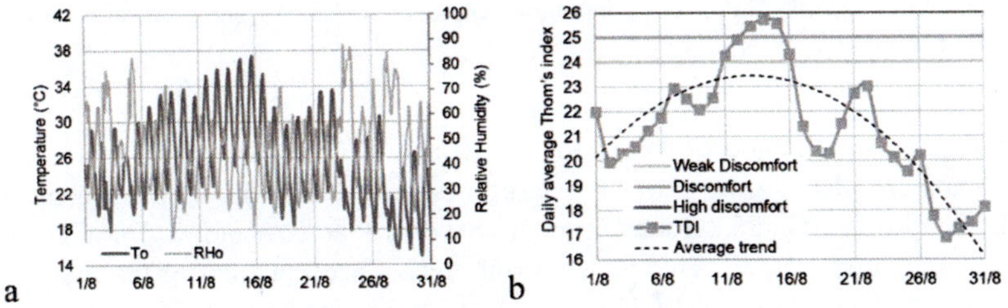

Fig. 2. **a** Outdoor air temperature (To) and relative humidity (RHo) and, **b** average daily TDI (with reference thresholds) recorded during the monitored period.

3.2 Thermal Response of Unoccupied Apartments in Free-Floating Conditions

Two unoccupied apartments for each monitored building have been considered in this section, i.e. 1A, 1B, 2C, and 2E (see Fig. 1). To investigate the impact of solar gains on the indoor conditions, all the external blinds are kept closed during the monitoring period (no solar gains), except for the windows of apartment 1B. Experimental results for apartments 1A and 1B during HW are reported in Fig. 3a,b in terms of indoor and outdoor air temperature (Ti and To) and relative humidity (RHi and RHo). All the apartments show the effect of the high thermal inertia of the construction system, i.e. the high dampening of the outdoor temperature oscillations. This effect is beneficial during the day, with indoor temperatures even 6–8 °C lower than outdoor ones, but detrimental during the night when the indoor air temperatures remain very high (higher than 28 °C). Comparing Fig. 3a and b, it is possible to observe the impact of the solar gains (through windows) on air temperature oscillations, where a marked temperature increase can be noticed during the morning due to solar beams entering through the windows.

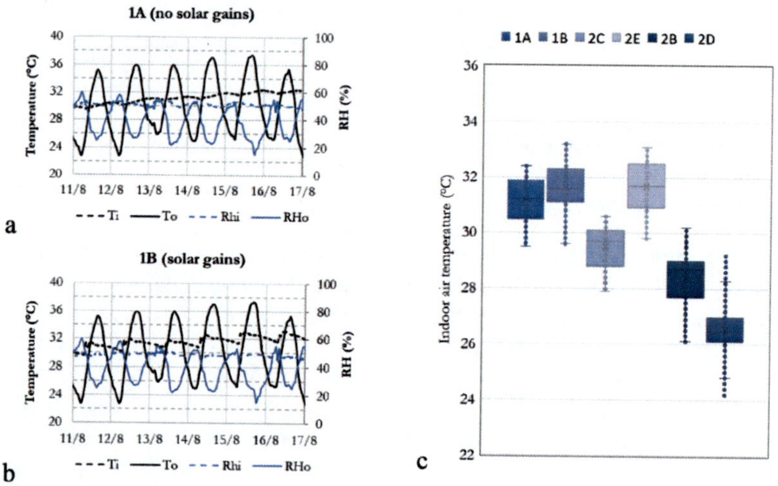

Fig. 3. Thermal response of unoccupied apartments: **a** 1A; **b** 1B. **c** Comparison between indoor air temperatures recorded in the considered unoccupied (1A, 1B, 2C, and 2E) and occupied (2B and 2D) apartments.

In Fig. 3c, the boxplots of the indoor temperatures during HW of all the unoccupied apartments are also shown. All the datasets related to unoccupied apartments show similar temperatures and interquartile ranges except for apartment 2 °C for which lower temperatures are recorded. This is due to the trees placed in front of the main south-facing façade (and in the west-facing one) that act as shading elements along with the balconies preventing the solar beam from directly hitting the wall during the day. As a result, lower solar radiations are present in this case, resulting in a lower indoor air temperature. In all cases, the highest indoor temperatures have been recorded during the last two days of HW with values equal to 32.4, 33.7, 30.6, and 33.1 °C for apartments 1A, 1B, 2C, and 2E, respectively. These are also the maximum temperatures recorded during the monitored period (August 2021).

3.3 Households' Energy Behavior During Heatwaves

In this paragraph, only the monitoring results of two occupied apartments of building #2 (2B and 2D) are analyzed for the sake of brevity since similar results have been obtained in other apartments. In particular, Fig. 4 reports the indoor air temperatures and relative humidity along with the activity of the occupants (obtained through the PIR sensors) and the AC energy consumption (EC). In both apartments, the PIR count denotes a high activity level during the HW period, thus confirming the presence of the occupants during this period. The meters data indicate the activation of the AC in both apartments during HW, with a higher cumulative period of activation recorded for apartment 2D resulting in lower indoor air temperatures. It should be noted that all the occupants express the need for closing windows during AC activation to meet the desired temperature.

In Fig. 3c, the indoor temperatures of the two occupied apartments are compared with those related to the unoccupied apartments. As expected, mainly due to AC activation and windows opening, the obtained temperatures in the two occupied apartments are lower than that in the unoccupied apartments, reaching maximum values equal to 30.2 and 29.3 °C for 2B and 2D, respectively.

To evaluate the effect of HW on occupants' AC use in the considered energy poverty scenario, Fig. 5a reports the correlation between daily EC and TDI for all the monitored days (i.e. HW and not-HW days). Despite TDIs are not computed starting from indoor bioclimatic conditions, they can be considered strictly correlated to it (also considering that indoor temperatures are quite constant during the day, thus they can be easily represented by a single daily value as TDI). As expected, a general increase of the EC during HW is recorded, characterized by a quite strong exponential relationship ($R^2 >$ 0.7) between TDI and EC. During normal periods (TDI lower than 24), the AC is rarely activated (especially for apartment 2B), despite the high indoor temperatures recorded, generally constantly higher than 26 °C. This denotes thrifty attention of the occupants to AC activation as expected in an energy poverty scenario. Conversely, during the HW, a significant change in occupants' behavior with respect to AC activation is observed. Indeed, an exponential increase in EC is observed for both apartments during HW, probably due to the higher discomfort encountered during these days and, in some cases, to a higher preoccupation concerning possible health-related consequences.

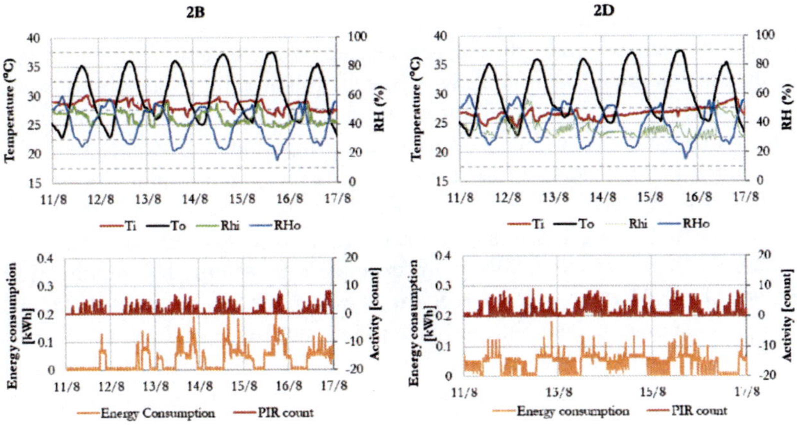

Fig. 4. Thermal response of occupied apartments during HW, PIR count, and energy consumption.

Finally, to understand the impact of HW on EC, in Fig. 5c, the box plot of the daily EC was measured during different periods, i.e. HW period, the entire monitored period (ALL), and the entire period without HW (ALL-HW). As can be seen, in terms of average value, the daily EC passed from about 1 and 2 kWh recorded in the period without HW

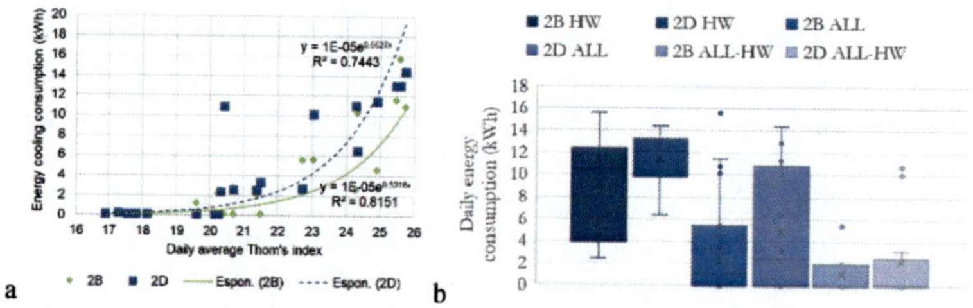

Fig. 5. **a** Correlation between TDI and energy consumption for space cooling; **b** Comparison between cooling energy during HW, the entire monitored period (ALL) and the entire monitored period without HW period (ALL-HW).

(ALL-HW) to 9 and 11 for the HW period, for apartments 2B and 2C, respectively. This indicates an increase in energy consumption of about 800 and 450%, respectively, in the considered energy poverty scenario, considerably higher than that generally found in literature through numerical simulations (a maximum of 90% is obtained in [10]). The importance of considering real use and energy behavior conditions in evaluating the HW impact on the buildings' energy performance is then highlighted.

4 Conclusions

This study analyzed and assessed the impact of HW on the thermal response and the energy occupants' behavior of two social-housing buildings placed in the city of Reggio Emilia considering a real use and energy poverty scenario, during a 6-days long heatwave in August 2021. At this aim, within the LIFE SUPERHERO project, an experimental campaign has been carried out to monitor the indoor and outdoor environments. The collected data have highlighted a general unpleasant indoor environment in both unoccupied and occupied apartments when the air conditioning system is not working. The results have also demonstrated the reluctance of the occupants to use air conditioners due to economic constraints even in case of high indoor discomfort conditions. However, during an HW, an exponential increase in energy consumption for space cooling is observed, with values 4 to 8 times greater than that observed in normal summer periods. This result strongly differs from the numerical results obtained in the literature (a maximum increase of about 90% is for example recorded in [10]) highlighting the importance of considering a real use scenario in evaluating the impact of HW on building energy consumption. The obtained increase cannot be generalized or extended to other case studies. Indeed, the main limitation of this study lies in the very small number of apartments investigated. Thus, further studies are needed to draw more general conclusions.

Acknowledgments. This work was supported by the European Union's LIFE SUPERHERO project, CUP I32F20005220006.

References

1. Ghatak, D., Zaitchik, B., Hain, C., Anderson, M.: The role of local heating in the 2015 Indian Heat Wave. Sci. Rep. **7**, 1–8 (2017). https://doi.org/10.1038/s41598-017-07956-5
2. D'Orazio, M., Maracchini, G.: An experimental investigation on the indoor hygrothermal environment of a reinforced-EPS based temporary housing solution. Energy Build. **204**, 109500 (2019). https://doi.org/10.1016/j.enbuild.2019.109500
3. Zauli Sajani, S., Scotto, F., Miglio, R., Agostini, D., Pandolfi, P., De Lisio, S., Lauriola, P.: Mortality and summer bioclimatic discomfort: an evaluation of short and medium term effects, GIMT. GIORNALE ITALIANO DELLE MALATTIE DEL TORACE (2007)
4. Maracchini, G., D'Orazio, M.: Improving the livability of lightweight emergency architectures: a numerical investigation on a novel reinforced-EPS based construction system. Build. Environ. **208**, 108601 (2022). https://doi.org/10.1016/j.buildenv.2021.108601
5. Tan, J., et al.: The urban heat island and its impact on heat waves and human health in Shanghai. Int. J. Biometeorol. **54**, 75–84 (2010). https://doi.org/10.1007/s00484-009-0256-x
6. Xia, Y., et al.: Assessment of the economic impacts of heat waves: a case study of Nanjing, China. J. Clean. Prod. **171**, 811–819 (2018). https://doi.org/10.1016/j.jclepro.2017.10.069
7. Lhotka, O., Kyselý, J., Farda, A.: Climate change scenarios of heat waves in Central Europe and their uncertainties. Theor. Appl. Climatol. **131**, 1043–1054 (2018). https://doi.org/10.1007/S00704-016-2031-3/FIGURES/9
8. Habitzreuter, L., Smith, S.T., Keeling, T.: Modelling the overheating risk in an uniform high-rise building design with a consideration of urban context and heatwaves. Indoor Built Environ. **29**, 671–688 (2020). https://doi.org/10.1177/1420326X19856400
9. Hatvani, G., Belusko, M., Pockett, J., Boland, J.: Heat stress-resistant building design in the Australian context. Energy Build. **158**, 290–299 (2018). https://doi.org/10.1016/j.enbuild.2017.10.025
10. Zinzi, M., Agnoli, S., Burattini, C., Mattoni, B.: On the thermal response of buildings under the synergic effect of heat waves and urban heat island. Sol. Energy. **211**, 1270–1282 (2020). https://doi.org/10.1016/j.solener.2020.10.050
11. Climates, U., Oke, T.R., Columbia, B., Climates, B.L., Climate, U., Service, M., Medal, P., Canadian, R., Society, G., Medal, M., Mills, G., Society, G., Climate, U., Christen, A., Program, A.S., Columbia, B., Voogt, J.A., Prediction, E., Network, C.C.: Urban Climates, n.d
12. Rasilla, D., Allende, F., Martilli, A., Fernández, F.: Heat waves and human well-being in Madrid (Spain), Atmos **10**, 288 (2019). https://doi.org/10.3390/ATMOS10050288
13. Zhang, Y., Hu, S., Yan, D., Guo, S., Li, P.: Exploring cooling pattern of low-income households in urban China based on a large-scale questionnaire survey: a case study in Beijing. Energy Build. **236**, 110783 (2021). https://doi.org/10.1016/j.enbuild.2021.110783
14. Kottek, M., Grieser, J., Beck, C., Rudolf, B., Rubel, F.: World map of the Köppen-Geiger climate classification updated. Meteorol. Zeitschrift. (2006). https://doi.org/10.1127/0941-2948/2006/0130
15. Lin, C.-Y., Chien, Y.-Y., Su, C.-J., Kueh, M.-T., Lung, S.-C.: Climate variability of heat wave and projection of warming scenario in Taiwan. Clim. Change **145**(3–4), 305–320 (2017). https://doi.org/10.1007/s10584-017-2091-0
16. Perkins, S.E.: A review on the scientific understanding of heatwaves-their measurement, driving mechanisms, and changes at the global scale. Atmos. Res. **164–165**, 242–267 (2015). https://doi.org/10.1016/j.atmosres.2015.05.014
17. Casanueva, A., et al.: Overview of existing heat-health warning systems in Europe. Int. J. Environ. Res. Public Health **16**, 2657 (2019). https://doi.org/10.3390/ijerph16152657

18. E.C. Thom, the Discomfort Index, https://doi.org/10.1080/00431672.1959.9926960; **12**, 57–61 (2010). https://doi.org/10.1080/00431672.1959.9926960
19. Epstein, Y., Moran, D.S.: Thermal comfort and the heat stress indices. Ind. Health **44**, 388–398 (2006). https://doi.org/10.2486/INDHEALTH.44.388
20. Stull, R.: Wet-bulb temperature from relative humidity and air temperature. J. Appl. Meteorol. Climatol. **50**, 2267–2269 (2011). https://doi.org/10.1175/JAMC-D-11-0143.1

Sensitivity and Uncertainty Analysis on Urban Heat Island Intensity Using the Local Climate Zone (LCZ) Schema: The Case Study of Athens

Gianluca Maracchini(✉), Fatemeh Salehipour Bavarsad, Elisa Di Giuseppe, and Marco D'Orazio

Università Politecnica Delle Marche, Ancona 60131, Italia
g.maracchini@staff.univpm.it

Abstract. The Urban Heat Island (UHI) phenomenon commonly occurs in cities worldwide and may affect the health of millions of people, especially during heatwaves. To design and propose mitigation strategies, accurate urban microclimate simulation tools are needed but retrieving information for creating reliable models is often challenging. Sensitivity (SA) and uncertainty (UA) analysis can help to achieve valuable information and to produce informed decisions even in an uncertain scenario. However, to date, that are few studies that perform SA and UA on microclimate models. In the present study, the Urban Weather Generator (UWG) software has been used to apply SA and UA to the city of Athens, Greece. The urban model is established by considering the variation of input parameters that may occur within three of the most diffused Local Climate Zones (LCZ) in city centers. Results showed that the highest average Urban Heat Island Intensity (UHII) is obtained in LCZ2 followed by LCZ_3 and LCZ_5 (2.92, 2.78 and 2.60 °C, respectively). However, due to the overlapping of UHII output values among different LCZs, it is not possible to state with absolute certainty that a "lower" LCZ corresponds to a higher UHII in a specific city. The SA confirmed that LCZ parameters such as building surface fraction, aspect ratio and mean height of buildings are the most influential ones in this case study. The percentage of buildings with an operating Air Conditioning system is also a key parameter in LCZ_2, which however is also among the most uncertain ones. Thus, this value is required to be well established in future research.

Keywords: Sensitivity analysis · Uncertainty analysis · Urban heat island · Local climate zone · Urban weather generator

1 Introduction

One of the consequences of climate change is the increased frequency, intensity, and duration of heatwaves (HWs), i.e. a period characterized by outdoor air temperatures unusually and persistently higher than the average temperatures of a specific time of the year. This condition can be potentially harmful to people's health [1–3], building energy performance [4], and economy [5] especially in urban areas, where HWs non-linearly

J. Littlewood and R. J. Howlett (Eds.): SEB 2022, SIST 336, pp. 281–290, 2023.
https://doi.org/10.1007/978-981-19-8769-4_27

interact with the so-called Urban Heat Island (UHI) phenomenon, i.e. a local warming effect caused by the complex interactions between urban canopy, buildings, and citizens [6]. In this framework, the decision process requires practical and quantitative tools to compare risks, strategies, and adaptation measures to the increasing temperatures caused by the interaction between global (HW) and local (UHI) warming.

To this aim, several Urban Microclimate Simulation (UMS) tools have been developed during the last decades. These tools mainly differ in terms of spatial scale related to the land surface represented and the resolution at which balance equations are written [7]. The mesoscale models can be considered state-of-the-art in atmospheric weather predictions. However, their application remains still limited for decision-makers due to the high computational cost and lack of boundary condition data. Conversely, parametric models, such as the Urban Weather Generator (UWG), allow having a quick estimation of the UHI effect in the urban canopy, thus being more suitable than others for decision-making processes. The UWG, in particular, has been proven to be able to produce neighborhood-specific weather files with acceptable computational cost, requiring the meteorological data obtained from a near rural weather station, and the average characteristics of the urban district under study [8].

One of the main issues in UMSs is the discrepancy, sometimes significant, between predicted and observed measured values, often resulting in poor model reliability. This may be traced back to the difficulty in characterizing a real urban scenario, due to the impossibility to retrieve the exact model input values or due to the high amount of uncertainty associated with them. In this framework, sensitivity (SA) and uncertainty (UA) analysis play an important role to achieve valuable information and increasing the model confidence, and may potentially help the building designers, facility managers, and policy-makers to produce informed decisions for the future of the cities even in an uncertain scenario. SA can be beneficial to find out the most ad less influential factors on urban microclimate also considering the possible interaction between different input values, while UA can be useful to understand the impact of input uncertainty on the outputs. However, to the authors' knowledge, there are still a few papers that perform UA and SA for estimating Urban Heat Island Intensity (UHII, i.e. the difference between rural and urban air temperatures) through UMS tools. Mao et al. [8], for example, adopt a regression technique based to understand the uncertainties brought by different inputs parameters (calibration problems) and then estimate UHII and urban energy consumption in Abu Dhabi. According to the uncertainty analysis, the authors confirmed that the UWG is a reasonably robust simulator to approximate the urban thermal behavior in Abu Dhabi for different seasons. Matinez et al. [9] performed SA by considering the Morris technique and then they characterized the local microclimate through the UA by UWG in La Rochelle (France). The result of UA identified vegetation cover, building height, and road albedo as key parameters. In another research, Kamal et al. [10] analyzed the impact of urban morphology and building energy load with uncertainty analysis via the UWG tool. The results showed that increased building footprint density could increase cooling consumption. However, an increase in greenery is not much effective in saving energy.

The urban context is often quickly classified by decision-makers through the Local Climate Zones (LCZs) approach, the universal standard for analyzing urban morphologies in terms of the UHI phenomenon. An LCZ is identified as a region of uniform surface cover and land use extending over city blocks for several kilometers that are associated with homogeneous urban environments. However, classifying the urban areas in terms of LCZ may produce inconsistent ranking in terms of UHII since the link between LCZ and UHII under uncertainty is not well investigated yet [11].

For these reasons, in this study, SA and UA techniques have been applied to numerically evaluate the impact of urban morphology and building characteristics uncertainty on UHII in the Mediterranean climate of Athens, Greece (where a strong UHI effect has been observed [12]) considering three among the most common LCZ in European city centers (LCZ_2, LCZ_3, and LCZ_5). The Sobol' sequences have been used as a quasi-random sampling technique for the Monte-Carlo UA, while the Sobol' method has been used for sensitivity analysis.

2 Phases, Materials, and Methods

2.1 Phases

The present paper can be subdivided into two main phases. First, the results of the UA considering the possible urban morphology variation in three common LCZs have been analyzed in terms of urban air temperature profiles and UHII. Then, a global SA is carried out to define the most important input parameters among urban and building characteristics.

2.2 The Urban Weather Generator

The UWG simulation tool (Python version, v5.8.0 [13]) has been used in this study to estimate the hourly urban canopy air temperature profiles. UWG has been validated in the literature in several cities such as Abu Dhabi (UAE), Toulouse (Frances), Basel (Switzerland), Singapore, Rome (Italy), and Barcelona (Spain) [8]. UWG involves four interacting sub-models, each of which is represented by one node. These are the Rural Station Model (RSM), the Vertical Diffusion Model (VDM), the Urban Boundary Layer (UBL), and the Urban Canopy and Building Energy Model (UC-BEM), which estimates heat fluxes, vertical profiles of air temperature above the rural weather station, the vertical profile of air temperature above the urban canopy layer and average urban canyon air temperature and humidity, respectively. The input data of UWG are the average urban characteristics and the weather data related to a rural area near the selected urban neighborhood. After each simulation, the UHII can be estimated as follows [14]:

$$UHII = \frac{\sum(T_{urb} - T_{rur})}{h}, T_{urb} > T_{rur} \tag{1}$$

where T_{urb} and T_{rur} are the urban and rural air temperatures, respectively, while h is the total number of hours in the simulation.

2.3 Sensitivity and Uncertainty Analysis on UHII

SA methods can be classified into local and global methods [15]. Local methods evaluate the response of the model to one local parameter. Global ones instead explore the response of the model by changing all input parameters at the same time and are considered more accurate since they provide more information about the effect of varying model inputs. In this paper, a variance-based global approach, i.e. the Sobol method, is applied to evaluate the impact of input uncertainty on UHII by using the SALib python package [16]. The Sobol' method is one of the most efficient methods for SA, able to decompose the variance of the output to the uncertainty of inputs. The main results of the Sobol' method are the variance effect indices, measuring the first order (Si) and total-effect indices (ST) of the input model parameters. Si is the measure of the main effect of the parameter, indicating how much the output variance could be reduced if a parameter is fixed. ST is the total-effect Sobol index and consider both the Si and the interaction effect with other parameters. The number of simulations required for SA are 2N(1 + D) (when both Si and ST are computed), where D is the number of input parameters with uncertainty and N is a value that should be increased until a convergence of ST values is reached. The obtained matrix of input and output from these simulations has been used for the UA.

2.4 Case Study

The Mediterranean city of Athens (Greece) has been selected as a case study where a high UHI effect has been observed in previous studies [12]. This city is characterized by the most common form of the Mediterranean climate (Csa, according to the Köppen-Geiger classification [17]), with hot and dry summers.

The rural weather file adopted for the UWG calculation is a Typical Meteorological Year (TMY) created through the Meteonorm software (version 8.0.3 [18]) referred to a weather station placed at the international airport of Athens. The simulation period goes from the 19th to the 25th of June, i.e., the hottest week in the weather file.

The input parameters needed to establish the UWG model can be categorized into three groups, i.e., meteorological parameters, urban parameters, and building parameters. Given the aim of the current study to evaluate the impact of urban and building characteristics uncertainty on UHI estimation, meteorological parameters are maintained in their default values. Conversely, urban and building parameters have been varied within uniform ranges of variation defined based on the relevant literature on urban and building characterization, as reported in Table 1. UWG parameters such as building density (blddensity), vertical-to-horizontal urban area ratio (vertohor), and grass coverage (grasscover) are not reported in Table 1 since they have been computed starting from the LCZ parameters reported in Table 2 related to the three of the most common LCZs present in the Mediterranean city centers, i.e., LCZ_2, LCZ_3, and LCZ_5 (see Fig. 1). In particular, in Athens, these LCZs cover about 3.0, 1.2, and 1.2% of the entire surface [19].

Concerning building parameters, the adopted values and range of variation are representative of the entire building stock of residential buildings in Mediterranean areas. All the unreported UWG parameters have been maintained in their default values except from internal gains schedules, which have been directly derived from [20]. About SA, convergence analysis of ST values has been carried out, obtaining a convergence bound for N = 512. Considering a total of 25 uncertain parameters, the total number of simulations is 26 624 for each LCZ.

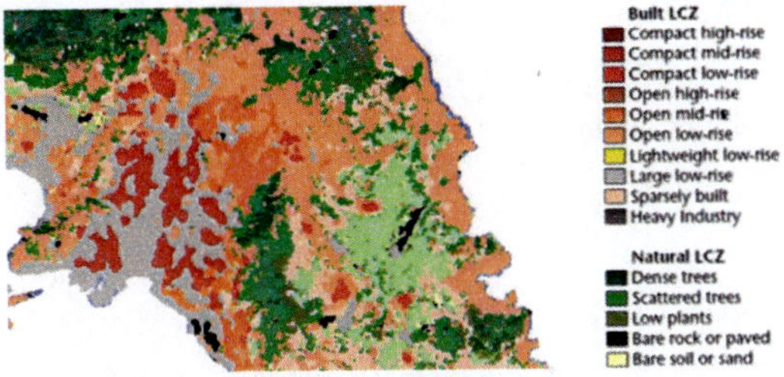

Fig. 1. LCZ map of Athens, from [19].

Table 1. Urban morphology characteristics for each LCZ.

Parameter [Unit]	Description	Variation	Refs.
Urban parameters			
Albveg [–]	Vegetation albedo	0.2–0.3	[8]
Latgrss [–]	Fraction of latent heat absorbed by urban grass	0.45–0.75	[8]
Lattree [–]	Fraction latent heat absorbed by urban trees	0.5–0.9	[8]
Albroad	Road albedo	0.08–0.24	[8]
BldcooledPerc [%]	The percentage of cooled buildings	0–45	[22]
Sensanth [W/m2]	Sensible anthropogenic heat	5–20	[23]
Treecover [–]	Number for fraction of urban ground covered in trees	0–1	–
Building parameters			
U-value [W/m^2K]	Window U-value including film coefficient	1.8–5.7	[20]
Glzr [–]	Glazing ratio	0.11–0.31	[24, 25]
SHGC [–]	Solar Heat Gain Coefficient	0.2–0.8	[8]
N_occ [m^2/person]	Occupancy density	15–25	[8]
Infil [ACH]	Infiltration rate	0.1–0.83	[20, 24]
SetpointCool [°C]	Cooling set-point	22–26	[20]
COP [–]	COP of cooling system	2.5–3	[20]
Albroof [–]	Albedo of roof	0.25–0.6	[26]
Albwall [–]	Albedo of wall	0.25–0.6	[26]

(*continued*)

Table 1. (*continued*)

Parameter [Unit]	Description	Variation	Refs.
Q-light [W/m2]	The numerical value for maximum light process load per unit area	1.6–11.5	[20]
U-value [W/m-K]	Wall thermal transmittance	0.23–2.58	[20]
U-value [W/m-K]	Roof thermal transmittance	0.33–3.44	[20]
U-value [W/m-K]	Floor thermal transmittance	0.34–2.9	[20]

Table 2. Candidate input parameters for three LCZs in the Mediterranean climate. Bld_dens: Building density; prev_sf: pervious surface fraction; H/W: canyon aspect ratio; bld_height: building height [21].

LCZ	Description	bld_sf [%]	perv_sf [%]	H/W [-]	Mean height [m]
LCZ_2	Compact mid-rise	40–70	0–20	0.75–2	10–25
LCZ_3	Compact low-rise	40–70	0–30	0.75–1.5	3–10
LCZ_5	Open mid-rise	20–40	20–40	0.3–0.75	10–25

3 Results and Discussion

3.1 Urban Microclimate Under Uncertainty Scenario

In this section, the results of the UA are reported and discussed. Figure 2 shows, for each LCZ, the results of the 26 624 simulations in terms of the envelope of urban air temperature profiles during three representative days. As expected, for all LCZ, the hourly UHII is stronger during the nighttime, mainly due to the discharge of absorbing heat from the urban structure after the sunset [27]. In particular, the nighttime maximum hourly value of UHII ranges between 4 and 8 °C. This difference is reduced during the daytime when the UHII during the hottest hours ranges between about -1 and 2.5 °C. These results are comparable with the value observed in the literature, where a maximum hourly UHII in Athens of 5.3 °C. during the nighttime and between 1.3 and 2.3 °C. during the daytime has been observed [28].

Comparing the considered LCZs, for each hour of the day, the difference obtained between the maximum and minimum hourly urban air temperature in the uncertain scenario is about 2.0, 1.5, and 1 °C for LCZ_2, LCZ_3, and LCZ_5, respectively. This denotes quite high robustness of the output regardless of the variation of the input parameters, as also seen in [8]. The air temperatures of the LCZ_2 are generally higher than the other two LCZs, probably due to the denser and higher buildings, leading to a higher heat released from Air Conditioning into the environment (see Table 2). From Fig. 2, this difference would seem more relevant for higher bounds of the envelopes (solid lines in Fig. 2) than lower bounds (dashed lines). However, it should be noted that the envelope does not take into account the different densities of the hourly temperatures within the sample.

To consider also the density of the output values, in Fig. 3a the results of UA are synthetically reported in terms of overall average UHII for each LCZ (see Eq. 1). A similar trend to that observed in Fig. 2 can be noted, with maximum, minimum, and average UHII higher for LCZ_2, followed by LCZ_3 and LCZ_5. In particular, the average UHIIs are equal to 2.92, 2.78, and 2.60 °C for LCZ_2, LCZ_3, and LCZ_5, respectively. Again, these results are in agreement with other studies in the literature. In [29], for example, the authors found an average UHII of about 2.6 °C for Hania in Greece. Concerning the variation of UHII, it can be noted that the higher the average UHII, the higher the variation of the sample. A standard deviation of 0.13, 0.08, and 0.05 °C was obtained for LCZ_2, LCZ_3, and LCZ_5, respectively, corresponding to a Coefficient of Variation equal to 4.38, 2.92, and 2.10%. From these results, it is finally interesting to note that, in general terms, UHII can be well correlated with LCZ (e.g. the "lower" LCZ, the higher the average UHII). However, from this work, it is not possible to state with absolute certainty that in a specific city a "lower" LCZ corresponds to a higher UHII, due to the overlapping of UHII values among different LCZs (which is also caused by the variation of other parameters).

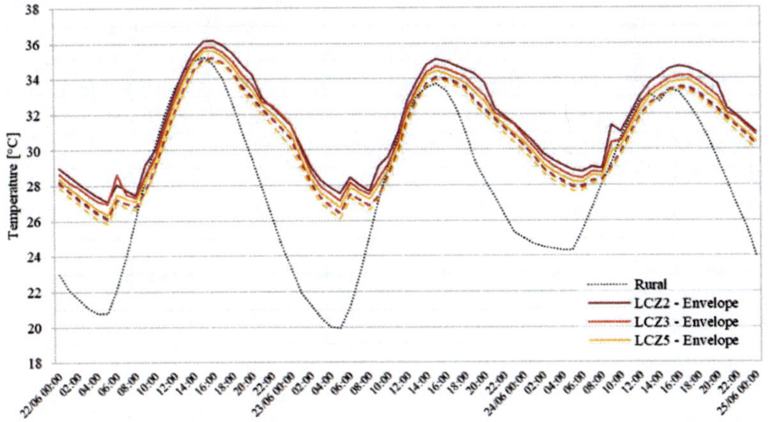

Fig. 2. Envelopes of urban air temperature profiles for different LCZ and comparison with rural temperatures in three representative days. For each LCZ, the solid line represents the higher bound of the envelope, while the dashed line the lower one.

3.2 Sensitivity Analysis

Figure 3b illustrates the ST obtained from the Sobol' SA for each LCZ. Only the input parameters with the highest value of ST are reported in the figure for the sake of brevity. Being the STs expressed in relative terms, a comparison between LCZ can be made in terms of the input ranking (from most important to less important). As expected, in all the considered LCZs, the bld_sf parameter has the highest impact on the average UHII, followed by meanHeight and aspectratio parameters (see e.g. [30]). In compact mid-rise (LCZ_2) residential building areas, the number of buildings that used an Air Conditioning

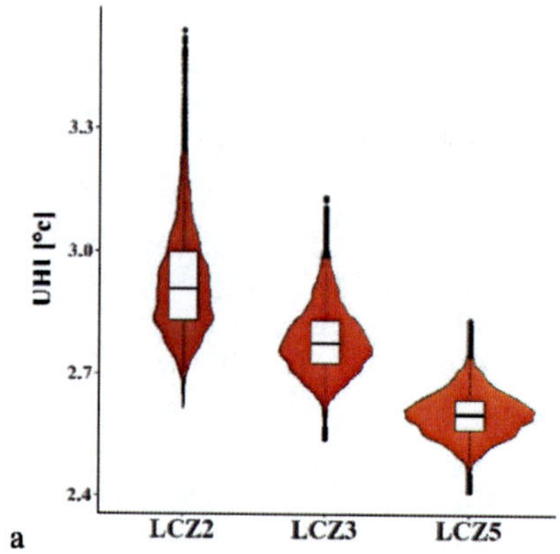

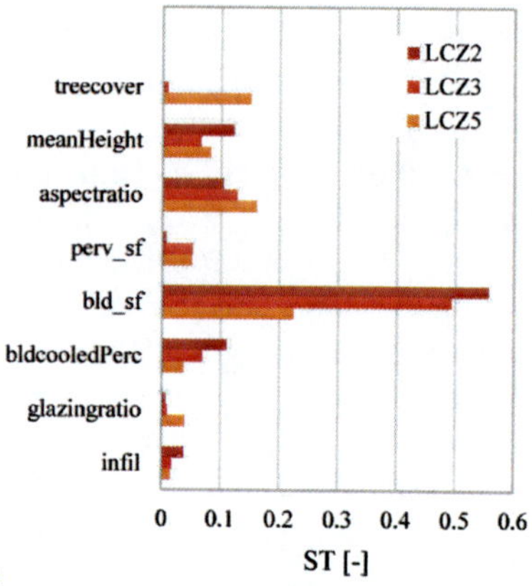

Fig. 3. **a** Violin plots with the box-whisker plots of UHII and **b** ST obtained for the most important parameters for each LCZ.

system (bldcooledPerc) is one of the key parameters, mainly due to the higher amount of building volume present in this LCZ. This is an important result since, in real applications, this parameter is also among the most difficult to estimate. Conversely, in LCZ_5, also the treecover parameter is among the most important ones. This is mainly due to the higher share of greenery in this LCZ in comparison with the other two LCZs, and then to the lower impact of building-related parameters on UHII.

Finally, despite the high uncertainty attributed to the different building parameters including all the possible residential building typologies and construction systems of the Mediterranean existing building stock, only the glazing ratio and the infiltration rate (infil) can be included among the most important parameters. However, they are always among the less important ones.

4 Conclusion

The current study applied SA and UA to UWG models to estimate UHII variation under uncertainty scenarios and to evaluate the impact of urban and building modeling uncertainty on UHII. The case study considers the Mediterranean climate of Athens, the variations in urban characteristics that may occur within three of the most diffused LCZs in Mediterranean city centers (i.e. LCZ_2, LCZ_3, and LCZ_5), and the variation in building characteristics including the entire residential building stock features.

As expected, the UA shows that the highest average UHII is obtained in LCZ_2 (2.92 °C) followed by LCZ_3 (2.78 °C), and LCZ_5 (2.60 °C). These results are in line with the results obtained in other research in the literature. However, from this work, due to the overlapping UHII values among the different LCZs, it is not possible to state with absolute certainty that in a specific city a "lower" LCZ always corresponds to a higher UHII.

The global SA confirmed as previous literature that morphological urban parameters are the most influential factor in an urban microclimate including building surface fraction, aspect ratio, and mean height of buildings. Additionally, in compact mid-rise (LCZ_2) residential building areas, the number of buildings that used an Air Conditioning system is one of the key parameters. Ironically, this parameter is the most uncertain input parameter among all the input variables. Thus, this value is required to be well established in future research. The results also suggest possible actions to reduce effectively UHII, such as increasing pervious surface fraction in LCZ_5 and LCZ_3, increasing tree coverage in LCZ_5, and reducing building heat waste in all the analyzed LCZs. This result can be potentially extended to other cities in a Mediterranean climate where these LCZs are present.

References

1. Ghatak, D., Zaitchik, B., Hain, C., Anderson, M.: The role of local heating in the 2015 Indian Heat Wave. Sci. Rep. **7**, 1–8 (2017). https://doi.org/10.1038/s41598-017-07956-5
2. D'Orazio, M., Maracchini, G.: An experimental investigation on the indoor hygrothermal environment of a reinforced-EPS based temporary housing solution. Energy Build. **204**, 109500 (2019). https://doi.org/10.1016/j.enbuild.2019.109500
3. Maracchini, G., D'Orazio, M.: Improving the livability of lightweight emergency architectures: a numerical investigation on a novel reinforced-EPS based construction system. Build. Environ. **208**, 108601 (2022). https://doi.org/10.1016/j.buildenv.2021.108601
4. Tan, J., et al.: The urban heat island and its impact on heat waves and human health in Shanghai. Int. J. Biometeorol. **54**, 75–84 (2010). https://doi.org/10.1007/s00484-009-0256-x
5. Xia, Y., et al.: Assessment of the economic impacts of heat waves: a case study of Nanjing, China. J. Clean. Prod. **171**, 811–819 (2018). https://doi.org/10.1016/j.jclepro.2017.10.069
6. Yi, C., Kwon, H.G., Yang, H.: Spatial temperature differences in local climate zones of Seoul metropolitan area during a heatwave. Urban Clim. **41**, 101012 (2022). https://doi.org/10.1016/J.UCLIM.2021.101012
7. Lauzet, N., et al.: How building energy models take the local climate into account in an urban context – a review. Renew. Sustain. Energy Rev. **116**, 109390 (2019). https://doi.org/10.1016/j.rser.2019.109390
8. Mao, J., Yang, J.H., Afshari, A., Norford, L.K.: Global sensitivity analysis of an urban microclimate system under uncertainty: design and case study. Build. Environ. **124**, 153–170 (2017). https://doi.org/10.1016/j.buildenv.2017.08.011
9. Martinez, S., Machard, A., Pellegrino, A., Touili, K., Servant, L., Bozonnet, E.: A practical approach to the evaluation of local urban overheating– a coastal city case-study. Energy Build. **253**, 111522 (2021). https://doi.org/10.1016/j.enbuild.2021.111522
10. Kamal, A., et al.: Impact of urban morphology on urban microclimate and building energy loads. Energy Build. **253**, 111499 (2021). https://doi.org/10.1016/j.enbuild.2021.111499
11. Stewart, I.D., Oke, T.R., Krayenhoff, E.S.: Evaluation of the 'local climate zone' scheme using temperature observations and model simulations. Int. J. Climatol. **34**, 1062–1080 (2014). https://doi.org/10.1002/joc.3746
12. Santamouris, M., Paraponiaris, K., Mihalakakou, G.: Estimating the ecological footprint of the heat island effect over Athens, Greece. Clim. Change **80**, 265–276 (2007). https://doi.org/10.1007/s10584-006-9128-0
13. Mackey, C., Vasanthakumar, S., Dao, A.: The Urban Weather Generator (UWG) (2020)
14. Boccalatte, A., Fossa, M., Gaillard, L., Menezo, C.: Microclimate and urban morphology effects on building energy demand in different European cities. Energy Build. **224**, 110129 (2020). https://doi.org/10.1016/j.enbuild.2020.110129

15. Saltelli, A., Ratto, M., Andres, T., Campolongo, F., Cariboni, J., Gatelli, D., Saisana, M., Tarantola, S.: Global Sensitivity Analysis. The Primer, Wiley, John, Chichester (2008). https://doi.org/10.1002/9780470725184
16. Herman, J., Usher, W.: SALib : Sensitivity analysis library in python (Numpy).Contains Sobol, SALib : an open-source Python library for sensitivity analysis. J. Open Source Softw. **2**, 97 (2017). https://doi.org/10.1016/S0010-1
17. Oliveira, A., Lopes, A., Niza, S.: Local climate zones in five southern European cities: an improved GIS-based classification method based on Copernicus data. Urban Clim. **33**, 100631 (2020). https://doi.org/10.1016/j.uclim.2020.100631
18. Meteonorm 8, Https//Meteonorm.Meteotest.Ch/En/, "Intro - Meteonorm (En)." Accessed 5 July 2021
19. Demuzere, M., Kittner, J., Bechtel, B.: LCZ Generator : a web application to create Local Climate Zone maps 9 (2020). https://doi.org/10.3389/fenvs.2021.637455
20. Carnieletto, L., et al.: Italian prototype building models for urban scale building performance simulation. Build. Environ. **192**, 107590 (2021). https://doi.org/10.1016/j.buildenv.2021.107590
21. World Urban Database and Access Portal Tools, World Urban Database (2015)
22. Eurostat, https://Ec.Europa.Eu/Eurostat
23. Manatsa, D., Chingombe, W., Matarira, C.H.: The impact of the positive Indian Ocean dipole on Zimbabwe droughts Tropical climate is understood to be dominated by. Int. J. Climatol. **2008**, 2011–2029 (2029). https://doi.org/10.1002/joc
24. Mata, É., Sasic Kalagasidis, A., Johnsson, F.: Building-stock aggregation through archetype buildings: France, Germany, Spain and the UK. Build. Environ. **81**, 270–282 (2014). https://doi.org/10.1016/j.buildenv.2014.06.013
25. Lavagna, M., Baldassarri, C., Campioli, A., Giorgi, S., Dalla Valle, A., Castellani, V., Sala, S.: Benchmarks for environmental impact of housing in Europe: definition of archetypes and LCA of the residential building stock. Build. Environ. **145**, 260–275 (2018). https://doi.org/10.1016/j.buildenv.2018.09.008
26. Maracchini, G., Latini, A., Di Giuseppe, E.: Un nuovo strumento per analisi di incertezza e sensibilità su strategie di mitigazione del fenomeno Isola di Calore Urbana, 16–19 (2021)
27. Bavarsad, F.S., Di Giuseppe, E., D'Orazio, M.: Numerical assessment of the impact of roof albedo and thermal resistance on urban overheating: a case study in Southern Italy. In: Littlewood, J.R., Howlett, R.J., Jain, L.C. (Eds.) Sustainable Energy Buildings, pp. 125–134. Springer, Singapore (2022)
28. Founda, D., Santamouris, M.: Synergies between urban heat Island and heat waves in Athens (Greece), during an extremely hot summer (2012). Sci. Rep. **7**, 1–11 (2017). https://doi.org/10.1038/s41598-017-11407-6
29. Salvati, A., Coch Roura, H., Cecere, C.: Assessing the urban heat Island and its energy impact on residential buildings in Mediterranean climate: Barcelona case study. Energy Build. **146**, 38–54 (2017). https://doi.org/10.1016/j.enbuild.2017.04.025
30. Litardo, J., et al.: Urban Heat Island intensity and buildings' energy needs in Duran, Ecuador: simulation studies and proposal of mitigation strategies. Sustain. Cities Soc. **62**, 102387 (2020). https://doi.org/10.1016/j.scs.2020.102387

Challenges in BiPV/PCM Façade System: Pathways Towards Numerical Modelling and Simulation Approaches

Jakub Čurpek[1](✉), Miroslav Čekon[1,2], Ondřej Šikula[1], and Muhammad Faisal Junaid[2]

[1] Brno University of Technology, Brno, Czechia
curpek.j@fce.vutbr.cz
[2] Slovak University of Technology, Bratislava, Slovakia

Abstract. Simulation modelling of heat and mass transfer processes is conducted in the case of a special type of a building integrated photovoltaic (BiPV) façade system with latent thermal energy storage (LTES) based on a phase change material (PCM). Experimental and simulation models are developed as part of the ventilated façade system for it to analyze and verify the adequacy of the available simulation tools. The key aspect of a BiPV/PCM façade concept is focused on reducing the peak operating temperatures of the PV modules and affecting action-reaction processes involving heat and mass transfer changes inside the façade elements. Experimental measurements were performed using an outdoor test cell to verify and validate numerical models. A comparative investigation of two façade concepts (BiPV, BiPV/PCM) is conducted using two simulation domains: the BES method (EnergyPlus) and the numerical CFD method (Ansys). The heat transfer rate through all façade elements is influenced by the high thermal inertia of the PCM differently in the diurnal/nocturnal period. The dynamic thermal response function of this façade system changes concerning the climate conditions at a small timescale (reactivity). The experimental measurements and simulation results are compared for it to provide an insight into consistency between the theoretical results and the experimental data. However, this indicates several limitations that need to be properly identified for further design.

Keywords: Building integrated photovoltaics · Phase change material · Building performance simulation · Thermal energy storage

1 Introduction

Developing and evaluating different technical solutions in façade engineering tasks is a highly relevant issue nowadays. Integrating thermal-responsive materials determines the self-adaptive characteristic of facades as per required physical stimuli. The design decision of novel façade systems requires a multi-criterion approach to analyze and recognize the potential role of uncertainties and to determine key space-time performance indicators. The outcomes from the performance-based building simulation approach can

J. Littlewood and R. J. Howlett (Eds.): SEB 2022, SIST 336, pp. 291–300, 2023.
https://doi.org/10.1007/978-981-19-8769-4_28

provide general knowledge about the investigated building element and how it could influence decision-making [1]. Generally, various simulation runs can be conducted in a simulation tool with an aggregation of observed states. The investigated system's complexity level determines the selection of a suitable simulation tool.

The holistic performance prediction of open-joint naturally ventilated façade systems is usually hard to be assessed through a BES tool due to the simplification of sub-components performance calculations [2]. Additionally, the integration of renewables (BiPV) and responsive materials (PCM) increases the complexity of the simulation campaign. The heat transfer mechanism and actual fluid dynamic behaviour of the BiPV/PCM façade is not yet sufficiently known regarding energy-saving and interior comfort conditions.

Several research studies [3–5] of the simple coupled fragment (PV/PCM) were conducted with constant values of heat flux and an ambient temperature as boundary conditions. The key intention was to decrease the PV operating temperature by passive cooling (PCM heat accumulation) and increase its electricity production. This simple system is still not economically viable with a long pay-back time. However, its direct integration into the building envelope can reveal its additional energy potential with a combination of building physics principles. Elarga et al. [6] developed a physical-mathematical model for a double-skin façade with a PV/PCM layer. They carried out numerical simulations to evaluate the integration of a PCM layer into a ventilated transparent façade. The study shows that integrating a PCM layer in a façade cavity with a semi-transparent PV layer leads to a 20 ~ to 30% reduction in monthly cooling energy demand. Kant et al. [7] performed parametric studies of forced ventilated BiPV/PCM systems to simulate the effects of different design parameters by the 1D model created in COMSOL Multiphysics. Simulation outputs obtained selected ranges of parameters with the optimum values. However, there is a lack of information regarding the real thermal performance of this type of ventilated BiPV/PCM façade that has been applied and validated in building energy simulations. Therefore, numerical investigations of BiPV/PCM façade and the verification and validation of simulation models in dynamic outdoor boundary conditions are highly relevant. Accordingly, the present study aims to provide initial insight into performance predictions of the system, where for this purpose, both experiments and simulations were employed.

2 Research Methodology

This paper is focused on preliminary simulations of heat and mass transfer processes through BiPV/PCM façade pertaining to many aspects: thermo-physical properties of PCM (temperature-enthalpy curve, latent heat capacity, etc.), heat transfer evolution in PCM, and natural ventilation regime in façade cavity during diurnal/nocturnal period of a sunny hot day. This study did not consider the effect of thermal contact between the capsulation structure of PCM and PV modules. This means that ideal case and perfect contact is considered in simulation studies. The research was performed in three phases: (i) gathering and evaluating data about input parameters taken from experimental data using outdoor test cells, and (ii) calibration of simulation models by the experimental data. Schemes of thermophysical interaction inside the façade models are depicted in

Fig. 1. An experimental campaign was conducted in Central European conditions, Brno, Czechia. Detailed information can be found in [8, 9]. Two investigated façade samples (BiPV and BiPV/PCM) were measured on the southeast-oriented part of the test cell. Organic paraffin RT27 was used as PCM. On sunny days, the measurements were taken during the late summer season, with a natural ventilation regime in the façade cavity. The data acquisition system was set for a one-minute timestep.

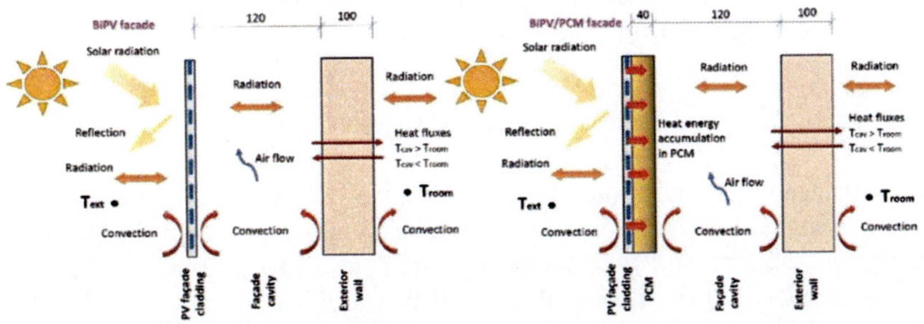

Fig. 1. Physical interaction in ventilated BiPV (left) and BiPV/PCM (right) façade element [10].

The present study was performed for one selected day (Fig. 2). The testing day is represented by the sunny day with high solar radiation (750 W/m^2) and outside air temperature (max. 26 °C, min. 12 °C). The mutual comparison within the verification and calibration process was conducted between the simulation model and experimental data, where the following physical values are contrasted: air temperature in the façade cavity (T_{ac}) and PV operating temperature (T_{pv}). The thermophysical properties of the PCM are defined by the temperature-enthalpy function (Fig. 2).

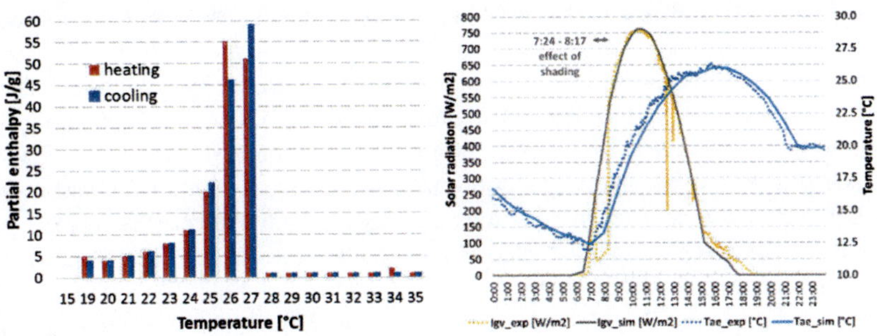

Fig. 2. Key input boundary conditions, partial temperature-enthalpy of PCM RT27 (left) and external boundary conditions (right).

The simulation campaign was performed by two different computation approaches, BES (EnergyPlus) and CFD (ANSYS). This software work on different modelling principles and mathematical theory. The computational models did not implement certain changes (modifications in the algorithms); therefore, reliability is based on implemented solutions. The effect of some of the design parameters on the behaviour of the façade, including the influence of the airflow rate in the façade cavity on the energy efficiency of the system and the effect of the fins added inside the PCM structure (uniform heat distribution) can not be studied using the BES methodology. Hence a 2D CFD model of the façade has been created by using the ANSYS software, which can provide detailed information on more specific heat transfer processes [11]. Accordingly, a more accurate building energy analysis of the façade system might be ensured by combining these two simulation principles and considering the dynamic heat transfer coefficients. This coupling method could provide an improvement that ranges between 10% and 50% for estimating the building energy requirements [12].

3 Building Energy Simulation and Modelling

The assessment of the energy performance of double-skin facades is usually difficult to perform through a BES tool. This tool is based on the assumption of fully mixed air in building zones, poor accuracy in the calculation of convective heat transfer coefficient, and infiltration/exfiltration. The simulation environment is focused on the whole building model rather than the building's subcomponents. Wildly, in this case, the BES zonal approach cannot directly extrapolate the effect of specific design parameters on the façade performance, including the influence of the airflow rate on the energy efficiency of the investigated system. The effect of the internal fins added to the structure of the PCM layer cannot be studied using this methodology (only 1D fragment structure). The model geometry of the whole outdoor test cell (with outer dimensions 3.0 m × 2.43 m × 3.09 m) was created in the simulation software DesignBuilder, a transient simulation program with an EnergyPlus calculation engine. The test cell was modelled as a multi-zone model with different internal properties of each zone. A five-zone Test-cell model was created by comprising the following independent zones: the internal space, the compensating space located behind the BiPV façade, the compensating space located behind the BiPV/PCM façade, the BiPV façade cavity and the BiPV/PCM façade cavity (Fig. 3).

The natural airflow movement in the façade cavity is ensured by two holes situated at the up and bottom sides of the façade with dimensions (i.e., 120 mm × 1200 mm) (Fig. 3 on the right). The algorithm for the convection inside/outside the façade cavity zone was adjusted to the Adaptive convection algorithm. This dynamic algorithm organizes several available convection coefficient equations and consequently selects the most suitable equation for the given surface. It is much simpler for the outside face (4 categories) than for the inside face (45 categories). The airflow through holes to the façade cavity was modelled using the Calculated Natural Ventilation option, based on the EnergyPlus Airflow Network pressure difference across the opening. Experimental measurements calibrated the coefficient of discharge Cd to a 0.5 [-] value.

The calibration of the simulation model was performed according to real climate conditions (incidence of vertical solar radiation and outdoor air temperature), however, without the consideration of the shading effect in the morning (from 7:20 to 8:20) (Fig. 2). The measurement equipment had adjusted one-minute timestep. EnergyPlus works with a one-hour timestep but carefully interpolates the weather data between data points for use at shorter timesteps. Zone Timestep was adjusted to the value of 60.

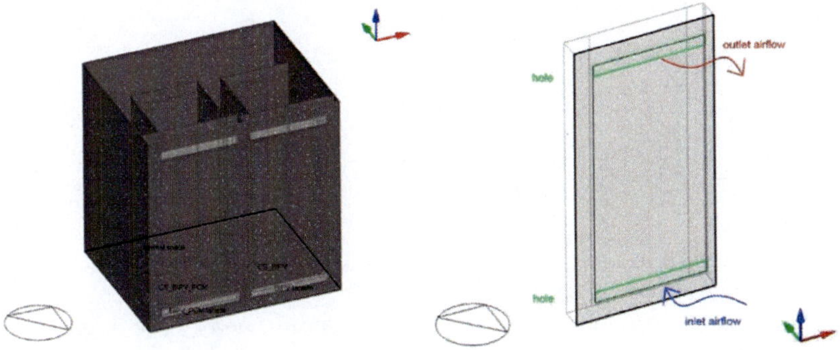

Fig. 3. The five-zone model of the outdoor test cell and the zone of the ventilated façade system.

Results presented in Fig. 4 indicate a 1.5 K difference in the case of BiPV/PCM during the solidification period (nighttime). The stored heat latent heat energy in the PCM is continuously extracted by airflow movement in the façade; however, this approach cannot extensively investigate the complex interaction between the thermal baffle effect of the PCM and natural fluid forces.

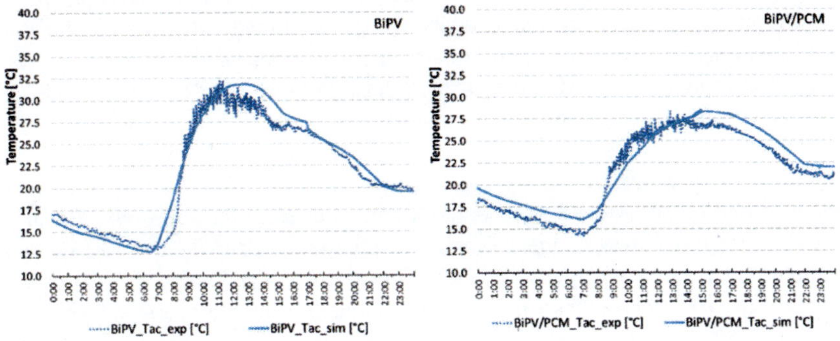

Fig. 4. Façade cavity air temperature for BiPV(left) and BiPV/PCM (right).

The one-hour shading effect significantly impacts the slope of temperature curves in the morning in both cases, where the PV layer is continuously heated up. One-hour decrease in solar radiation causes the interruption of the surface heating process. In Fig. 5. Specific differences occurred at temperature peaks in the case of BiPV/PCM. The PCM reduced the overheating of the PV by about 5 K. The temperature values of BiPV without PCM match well without any substantial deviations. However, in BiPV/PCM case, a one-hour time delay occurred during peak hours.

It should be noted that the latent heat accumulation process is sensitive to shading effects (a process of heat accumulation – melting of PCM). Additionally, the determination of the thermal field inside the PCM cannot be analyzed with internal aluminium fins. This absence slows the heat dissipation into the PCM structure and increases PV operating temperature.

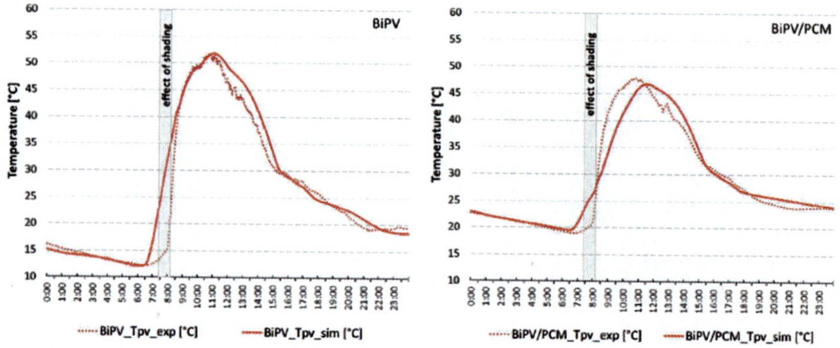

Fig. 5. PV panels' operative temperature for BiPV(left) and BiPV/PCM (right).

4 Computational Fluid Dynamics Modelling

As the drawbacks mentioned above can not be studied using the BES methodology, a 2D CFD model has been employed using the Ansys Fluent 15.07. This method can provide more specific microclimate parameters – the influence of the airflow rate on the energy efficiency of the system and the effect of the fins added inside the PCM structure. First, approximate quasi-stationary simulations were performed corresponding night and daytime. The aim was to study the independence of the solution on the computational grid, as illustrated in Fig. 6, and the approximate tuning. Further factors analyzed represent testing of thermophysical properties of materials, parameters of some boundary conditions and the initial conditions of the non-stationary simulation. Moreover, different formulations of the heat transfer coefficient can provide more accurate results [13]. After that, non-stationary simulations were performed for the test day, and all input parameters were fine-tuned. Air was considered a non-isothermal incompressible ideal gas. The turbulence model was set up as k-ω SST for convection and the Discrete Ordinates radiation model. The convergence criteria were set for the residuals of the energy and radiation Eqs. 10^{-6} and 10^{-4} for the other equations. A "double precision" solver was used.

The boundary condition of the air inlet was entered as a "pressure inlet" with the appropriate temperatures from the experiments. The turbulence was defined by the hydraulic diameter of the inlet and outlet slot and the turbulence intensity of 5%. An additional inlet pressure was tuned based on the experimentally measured outdoor wind speed, defined as $D_p = 0.2 + 0.21v$ [Pa], where v is the wind speed [m/s]. The boundary condition of the air outlet was entered as a "pressure outlet". The solar heat flux on the PV panel, corresponding to non-stationary data from a pyranometer, was modelled as an internal heat source. The absorption of the panel was set on 0.8. A non-stationary convective heat transfer coefficient to the surroundings $h_c = 4.5 + 1.5v$ [W / m^{2}K] was entered on the external surface of the PV panel. ANSYS Fluent calculated heat transfer by IR radiation from the PV panel to the surroundings for an emissivity of 0.84.

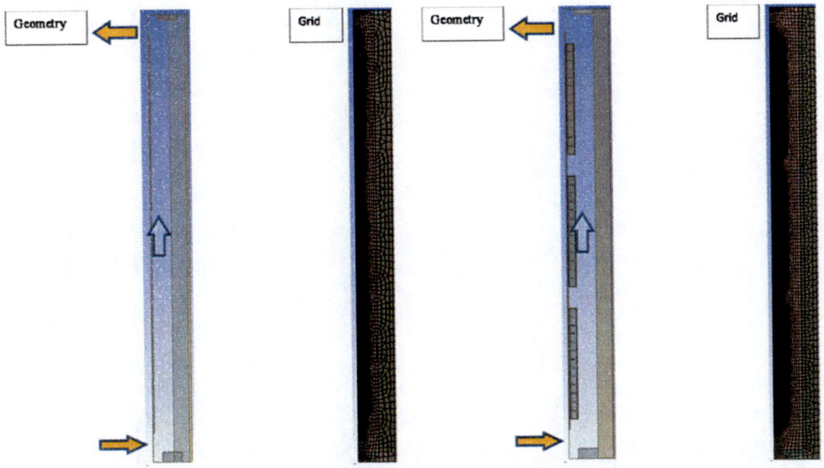

Fig. 6. The model geometry and grid of the BiPV façade (left) and BiPV/PCM façade (right).

Results presented in Fig. 7 show the non-uniform distribution of the heat transfer coefficient over the inner surfaces of the cavity over time. Further optimization of fins can improve the heat accumulation process according to the type of the PCM and façade configuration. It is necessary to consider that incomplete melting of PCM determinates lower heat transfer to the airflow in the cavity and influences the airflow regime. In natural ventilation, there are complex interactions between buoyant and wind-driven flows, which are also affected by the phenomena above. Therefore, CFD analysis can provide better insight into performance investigations.

Numerical simulations using the 2D CFD model revealed many outputs and indicated the following findings. First, the fins' effect shown in Fig. 8 significantly impacts the speed of heat transfer evolution in a PCM container during heat accumulation. Second, the heat transfer by radiation in the cavity determines the possible assessment of the impact of emissivity changes on internal surfaces in the future. Third, the higher overheating of the PV panel in the sections is out of contact with the container compared with the direct contact sections. These parts decrease the efficiency of electricity production and increase the risk of hot spots. Non-uniformity of temperatures and the cavity's height affecting the PV operating temperatures were also identified. Lastly, the intensity of the airflow in the façade cavity changes according to the state of PCM heat accumulation (day/night period).

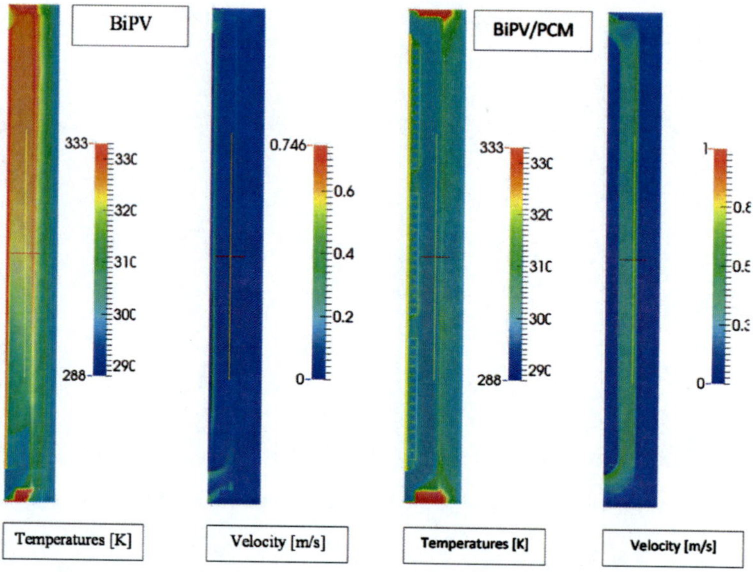

Fig. 7. Heat and mass transfers in the façade systems, BiPV (left) and BiPV/PCM (right).

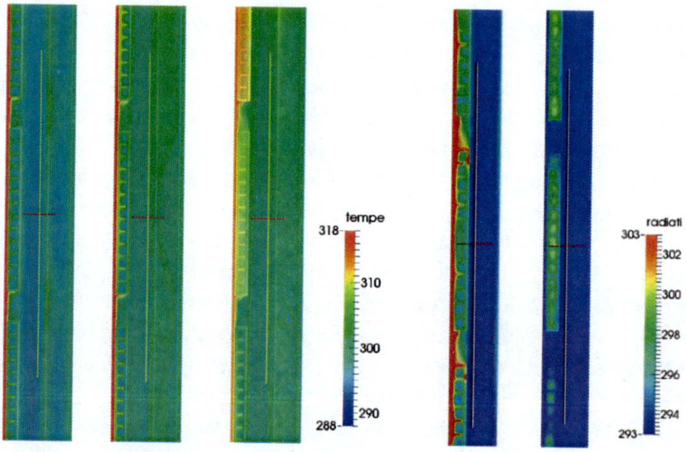

Fig. 8. The effect of internal fins in PCM structure form 10 h to 14 h, in a two-hour time step.

5 Discussions and Research Challenges

The dynamic behaviour of the ventilated BiPV/PCM façade system and its energy-saving potential is not yet sufficiently known. Its case-study performance prediction with natural airflow movement in the façade cavity is a relevant research challenge regarding particular thermophysical quantities and software performance under real boundary conditions. The experimental data collection was gathered in a real outdoor test cell and processed for validation through two main simulation approaches (BES and CFD). DSF (double-skin façade) generally represents a complex multi-functional structure for empirical validation, especially during peak solar loads. The validation study could help increase confidence and point out limitations in the use of BES/CFD tools for the performance of the façade system. No validation studies about this BiPV/PCM façade system have been performed. Likewise, this study could provide the next step for adequate façade operation regimes and control following building demands (interior temperature fluctuations, heating/cooling) in time. Increased thermal inertia of the ventilated façade caused by the integration of PCM affects overall heat and mass transfer of building elements differently under day periods. However, there are decision-making challenges for PCM-type integration in the façade LTES system concerning boundary conditions. The principles for correct estimation of phase transition temperature and thickness of the PCM are determined for ventilated regimes which lead to avoiding overheating/overcooling of the façade. An adequate identification of simulation limitations using the experimentally validated numerical model provides better insight into PCM characterization during geometrical façade configuration and the determination of the thermal delay system.

6 Conclusion

The dynamic heat and mass transfer performance of integrated BiPV/PCM façade systems is not yet sufficiently known. The objective of this work was to investigate the suitability of different simulation approaches for the performance prediction of the ventilated BiPV/PCM façade system. Two simulation runs were conducted in DesignBuilder and Ansys Fluent, where the aggregation of observed states was performed together with the calibration model. BES approach reveals that the façade system can be sufficiently stimulated. However, there are simplifications about one zone's well-mixing air temperature and the average surface temperatures for one wall area. One hour of stop heat accumulation inside the PCM due to the shading effect determining decreases at peak temperatures during noontime. However, the BES approach cannot sufficiently stimulate the investigation of airflow movement inside the façade cavity (airspeed, air temperature stratification). 2D model geometry in the CFD approach can work with more complex computational methods, which are more time-consuming than BES. However, CFD provides better insight into the nature of airflow movement inside the façade cavity (fluid flow is not uniform across the section), thermal stratification, and the effect of internal fins inside the PCM structure. The next step in this design decision investigation is the multi-criterion approach based on the coupling of BES with CFD, which poses a high potential for accurate analysis of the building energy performance of the ventilated BiPV/PCM façade system.

Acknowledgement. This research was supported by the project GA 20-00630S "Climate responsive components integrated in energy and environmentally efficient building envelope" supported by Czech Science Foundation. Further personnel support is grateful to the Scientific Grant Agency of the Ministry of Education, Science, Research and Sport of the Slovak Republic, Project No. VEGA 1/0580/20.

References

1. Bevilacqua, P., Bruno, R., Szyszka, J., Cirone, D., Rollo, A.: Summer and winter performance of an innovative concept of Trombe wall for residential buildings. Energy **258**, 124798 (2022)
2. Čurpek, J., Čekon, M.: A simple trombe wall enhanced with a phase change material: building performance study. Smart Innov. Syst. Technol. **263**, 281–291 (2022)
3. Huang, M.J.J., Eames, P.C.C., Norton, B.: Phase change materials for limiting temperature rise in building integrated photovoltaics. Sol. Energy **80**(9), 1121–1130 (2006)
4. Ma, T., Yang, Y., Zhang, L., Lu, L., Wang, X.: Using phase change materials in photovoltaic systems for thermal regulation and electrical efficiency improvement: a review and outlook. Renew. Sustain. Energy Rev. **43**, 1273–1284 (2015)
5. Hasan, T., McCormack, S.J.J., Huang, M.J.J., Norton, B.: Characterization of phase change materials for thermal control of photovoltaics using differential scanning calorimetry and temperature history method. Energy Convers. Manag. **81**, 322–329 (2014)
6. Elarga, H., Goia, F., Zarrella, A., Dal Monte, A., Benini, E.: Thermal and electrical performance of an integrated PV-PCM system in double skin façades: a numerical study. Sol. Energy **136**, 112–124 (2016)
7. Kant, K., Pitchumani, R., Shukla, A., Sharma, A.: Analysis and design of air ventilated building integrated photovoltaic (BIPV) system incorporating phase change materials. Energy Convers. Manag. **196**, 149–164 (2019)
8. Čurpek, J., Čekon, M.: Climate response of a BiPV façade system enhanced with latent PCM-based thermal energy storage. Renew. Energy **152**, 368–384 (2020)
9. Curpek, J., Cekon, M., Hraska, J.: PCM integrated in BiPV ventilated facąde concepts: experimental test cell platform and initial full-scale measurements. IOP Conf. Ser.: Earth Environ. Sci. **290**(1), 012072 (2019)
10. Curpek, J., Cekon, M.: The effect of PCM layer on the natural air flow movement in the façade cavity of BiPV system. E3S Web Conf. **172**, 19007 (2020)
11. Bruno, R., Ferraro, V., Bevilacqua, P., Arcuri, N.: On the assessment of the heat transfer coefficients on building components: a comparison between modeled and experimental data. Build. Environ. **216**, 108995 (2022)
12. Rodríguez-Vázquez, M., et al.: Coupling building energy simulation and computational fluid dynamics: an overview. J. Building Phys. **44**(2), 137–180 (2020)
13. Bevilacqua, P., Perrella, S., Bruno, R., Arcuri, N.: An accurate thermal model for the PV electric generation prediction: long-term validation in different climatic conditions. Renew. Energy **163**, 1092–1112 (2021)

Integrating Vegetation and Cities: A Review of the Applicative Solutions from Technical Component to Planning Scale

Arianna Peduzzi(✉) and Carlo Cecere

Sapienza University, 00168 Rome, Italy
{arianna.peduzzi,carlo.cecere}@uniroma1.it

Abstract. The use of greenery as an architectural component is currently among the most used strategies in urban areas aimed at mitigating climate change, reducing emissions, and restoring the evapotranspiration of natural soil, reduced nowadays because of increasing waterproofing of the built environment. This context raises a question: what can be effectively done in the urban context, to reduce the effect of the heat island and minimize the impact of emissions from buildings through the use of green integrated into the architecture? This contribution offers an updated review of the integration practices between the natural and architectural world, focusing on the possibilities of multi-scalar intervention, tracing an updated picture of the alternatives available at the planning and at the design stages, and at the technical scale.

Keywords: Bio-based technical solutions · Green strategies · Sustainable decision making · Small parks

1 Introduction

Global warming caused by greenhouse gases has increased the planet's temperature, leading to a growth of intense meteorological phenomena, connected to climate change [1]. With respect to this situation, global governments are making efforts to prevent the average temperature from rising above the critical threshold of 1.5 °C, trying to achieve carbon neutrality [1]. Within this context, it is therefore useful to frame the role of the construction sector, and its weight in this equation. As is well known, the building sector is responsible for a large part of the annual CO_2 emissions [2] as well as for a large resource consumption [3]. Currently, the main carbon dioxide absorption strategy is linked to vegetation [4] and, for this reason, the number of reforestation projects, such as the UN Trillion Trees Campaign, is growing, but what can be done in urban centres, usually densely built, to minimize the impact of building emissions and reduce the Urban Heat Island (UHI) effect? In the last two decades, the integration of architecture and nature in new buildings has become increasingly common [5] but this practice is rooted in the past. A fulgid example of this practice can be retraced in the mythical gardens of Babylon (Fig. 1a), or the garden of the Guinigi tower in

J. Littlewood and R. J. Howlett (Eds.): SEB 2022, SIST 336, pp. 301–312, 2023.
https://doi.org/10.1007/978-981-19-8769-4_29

Lucca (Fig. 1b) built in the fourteenth century. In London, during the Second Industrial Revolution, the design of multiple green spaces becomes the only solution capable of balancing the emissions of the new factories. In the nineteenth century, however, it was the oil crisis of the 1970s that reinvigorated interest in green architecture, prompting designers to rethink the sustainable development of construction. Among the different approaches adopted during that period, green roofs stand out, as in the Willis Faber and Dumas (Fig. 1c) made by Foster in 1975 in Ipswich, and in the work of Hundertwasser in Austria, which incorporated trees and plants in different complexes (Fig. 1d). Other more recent examples can be found in the Chicago green roofs (Fig. 1e), in linear parks such as the 1993 promenade planteé (Fig. 1f) in Paris, and the New York High Line built in 2009 (Fig. 2a).

Nowadays, there is a growing trend regarding the integration of medium-sized trees into the buildings' structure, among the others we can recall: the ACROS cultural centre (Fig. 2b) in Fukuoka, designed by Ambasz in 1995, the Dutch pavilion designed by MVRDV for Expo 2012 in Hanover (Fig. 2c), where the fifth floor accommodated a forest, as well as the Boeri vertical forest in Milan (Fig. 2d) inaugurated in 2015, and finally, the Maggie's Center in Leads (Fig. 2e) in the United Kingdom of 2021 and 1000 Trees (Fig. 2f) in Shanghai both designed by Heatherwick Studio and inaugurated in 2021.

Starting from the observation of current design practices, this contribution offers a review of some of the integration strategies between the plant and architectural world examining those available and those at an experimental stage.

2 Methods

The review was developed by analyzing existing technological elements, design strategies and planning practices that allow nature-architecture synergy at various intervention scales.

For doing so a bibliographic review was conducted on Web Of Science search engine using some specific keywords to compose the search string. Namely, the keywords used were: "bio-based technical solutions", "biomass building materials", "biomass construction materials", "biomass constructive materials", "bioreceptiv*", "green acupuncture" "green envelope", "green fabric", "green façad*", "green infrastructure", "green neighbourhood", "green park", "green planning", "green roof", "green residual spaces", "green spaces", "green strateg*", "green urban polic*", "green wall", "living wall", "nature-based cit*", "pocket park", "small park", "sustainable urban drainage systems", "urban green acupuncture".

This initial search yielded 80276 documents. After Web of Science's Categories filters were applied, the database was reduced to 7688 documents. Additional documents were excluded based on document type (for example: book review, correction, early access, editorial material, letter, retracted publication, retraction) deriving in a total of 7251 documents to be evaluated. After scanning the titles and the abstracts, further exclusions were made based on topical relevance resulting in 2286 contributions to be evaluated (159 were review papers). After completing the articles' impact screening, the final database was set up.

Fig. 1. **a** Gardens of Babylon, Ninive, 668–627 a.C. (credits: www.greelane.com, 2022); **b** Guinigi Tower, Lucca, XIV d.C, (credits: it.wikipedia.org, 2022); **c** Norman Foster, Willis Faber and Dumas Headquarters, Ipswich, 1975 (credits: www.archdaily.com, 2022); **d** Fridenchreich Hundertwasser, Hundertwasserhaus, Vienna, 1986 (credits: sworld.co.uk, 2022); **e** McDonough + Partners, Chicago City Hall green roofs, Chicago, 2001 (credits: www.thedifferentgroup.com, 2022); **f** Philippe Mathieux e Jacques Vergely, Promenade planteé, Parigi, 1993 (credits: www.theguardian.com, 2022).

The results were then grouped following the application/intervention scale, i.e., building, neighbourhood and planning level. The outcome of this operation was a results section articulated in four paragraphs, two addressing the building scale (paragraphs 3.1 and 3.2), one addressing the neighbourhood scale (paragraph 3.3) and finally, one addressing the planning level (paragraph 3.4).

Fig. 2. **a** Diller Scofidio + Renfro, High line, New York, 2009 (credits: www.espazium.ch, 2022); **b** Emilio Ambasz, ACROS cultural center, Fukuoka, 1995 (credits: www.stefanoboeri architetti.net, 2022); **c** MVRDV, Netherlands Pavilion at Expo 2012, Hanover, 2012 (credits: www.mvrdv.nl, 2022); **d** Stefano Boeri, Vertical forest, Milano, 2015 (credits: www.niiprogetti.it, 2022); **e** Heatherwick Studio, Maggie's Center, Leads, 2021 (credits: www.archdaily.com, 2022); **f** Heatherwick Studio, 1000 Trees, Shangai, 2021 (credits: www.isplora.com, 2022).

For each of these strategies, the possible impact, as well as the limitations, were assessed, especially in reference to applications on existing buildings or in existing cities, to trace a comprehensive picture of the design and technical alternatives available.

Finally, in paragraph 4 the proposed solutions were critically discussed, and in paragraph 5 conclusions were finally drawn.

3 Results

The bibliographic review showed an increasing interest in the research topics (Fig. 3). It is possible to observe a first phase, spanning from 1991 to 2001, where the topic emerged, followed by a second phase, from 2002 to 2011, of increasing interest and finally a third phase of exponential interest growth. The first phase account for 14 contributions whereas the second and the third account for 176 and 2087 papers respectively.

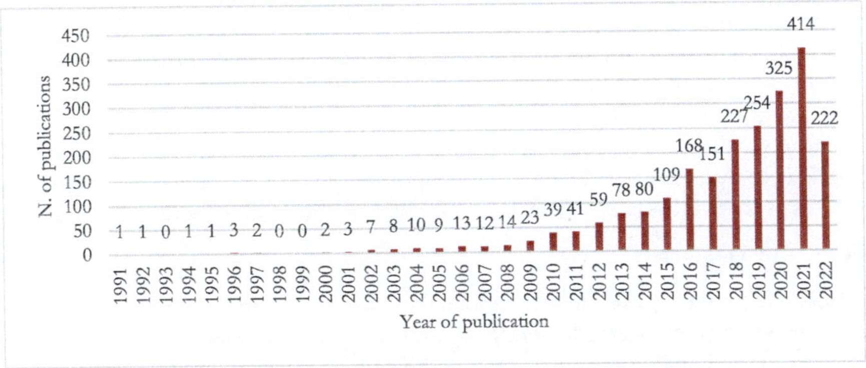

Fig. 3. Number of publications according to the publication year

3.1 Buildings and Vegetation Synergy

The solutions for integrating plants in architectural objects are divided into: (i) traditional green envelopes, with technical and system elements functional to plant growth, (ii) bio-receptive materials, which intrinsic characteristics allow the growth of organisms. In both cases, the vegetation becomes a technical component, contributing to the energy and technological performance but also to the appearance of the building.

Green envelopes

Recently, green envelopes have incorporated multiple technological possibilities, encouraging their diffusion, also promoted by the European Commission for the improvement of urban comfort [6].

The technical solutions available are divided into green roofs and green walls and facades.

Green roofs consist of different functional layers (vegetation, substrate, root barrier, filter sheet, drainage, insulation and waterproofing) [7] and are divided into: intensive, semi-intensive and extensive, listed here in descending order of: substrate thickness, weight, cost of installation and maintenance, and of the allowed variety of plants.

The vertical green systems are classified into green facades and living walls. Green facades employ climbing plants, which grow directly against the wall or through an indirect support system [8]. The living walls, modular or continuous, allow uniform growth along the surface and the application of different plant varieties, however, require frequent irrigation and are more expensive [9].

These systems allow having at the building scale: the improvement of the envelope thermal behaviour, the reduction of air conditioning cost, and the increase in the value of the property on which they are applied [7]; the reduction of the noise level [10]; the increase in the useful life of the envelope and its functional layers, reducing maintenance costs [11].

At the urban scale, the use of these solutions allows: the retention of rainwater, reducing the peak of runoff in the event of urban flooding, and absorbing the pollutants contained in the water [12]; counteracting the UHI effect thanks to the plants' evapotranspiration [7]; reduction of air pollutants, capturing particulate matter and reducing consumption for air conditioning [13]; reduction of noise pollution [14]; improvement of the perception of the environment [15]; and increase in urban fauna [16].

Thanks to the multiple benefits, the number of green envelopes continues to increase especially in Europe. Considering that the roofs occupy 40–50% of the total waterproof surfaces of urban areas [12], this makes these solutions interesting for retrofit operations.

Bioreceptive materials

In 1995, Guillitte [17] was the first to define the concept of bioreceptivity as the ability of a material to be colonized by living organisms, linking this characteristic to the biodeterioration of building materials, thus confining it to historical heritage preservation. Some scholars have instead reinterpreted this feature focusing on its potential to integrate the architectural and natural world. Most of these studies are aimed at increasing the bio-receptivity of cements, tested using different methodologies, including the reduction of pH [18]. Although the alteration of these characteristics leads to a substantial improvement in bioreceptivity in laboratory conditions, the same was not observable in urban areas [18], where the central factor for the colonization of organisms was their concentration. These results are particularly interesting, as the cement surfaces present in urban areas could be transformed into a growth medium capable of hosting fungi, lichens, and algae, all capable of sequestering pollutants, without requiring the addition of irrigation systems, the addition of weight on existing buildings, and allowing the growth of organisms even on those surfaces that do not have access to direct sunlight.

3.2 Building Materials from Biomasses

The use of biomass as construction materials is an interesting strategy to increase the sustainability of building materials, normally linked to energy-intensive production processes and resource consumption. In addition, this strategy could be help dispose of agricultural waste produced annually, equal in the European Union (EU) to 442 Mt [19]. In this case, green is therefore not intended as an organism but as an envelope component.

Agricultural waste recycling for the production of sustainable building materials is configured as an essential strategy to increase resource efficiency by finally making the economy circular, achieving the goal set by the 2050 Agenda Europe [20].

Currently, applications and experiments for the use of agro-waste regard two main topics: their use as insulation and as additions in construction mixes.

Insulations production is the main application of biomass in the building sector. These elements, available in various forms (mats, composite panels and sandwich panels) depending on the material used, are generally composed of biomass, a binder and additives added to improve fire resistance and waterproofing and as biocides.

The use of biomass for insulating purposes involves: reduction of the embodied energy thanks to the CO_2 sequestered during the life cycle of the plant and to the less energy-intensive transformative processes; revaluation of agricultural waste, and consequent creation of new economic sectors; good technical performance (high thermal capacity, low thermal conductivity, and good hygrometric behaviour); high levels of recyclability and non-harmful disposal for the environment [21].

The use of biomasses within construction mixtures has been experimented mainly in concrete [22], but also in mortars [23] and plasters [24]. The purpose of these additions is to improve some performances such as: density, thermal behaviour [24] and mechanical properties [25]; and trying to produce materials with a lower environmental impact, replacing all or part of the aggregate or binder [26]. In general, as regards the insertion of biomass into compounds, studies agree on how a great variety of organic material can be used, giving life to lightened conglomerates with good mechanical properties and with better thermal properties than those of the reference pastes [26].

3.3 Green Solutions for External Spaces

Some solutions integrate vegetation in the outdoor residual spaces between buildings. Green parks and infrastructures, even when small, can in fact mitigate the effects of climate change on a local, neighbourhood and city scale.

The benefit is directed on: the urban microclimate, reducing thermal stress in the hot season [27]; health and well-being of citizens [28]; acoustic and visual well-being [29]; resilience to floods, contributing to water management on the urban surface [30]; the value of buildings and productivity of economic activities [31]. The plants also play a crucial role in the urban ecosystem, enhancing the biodiversity in urban areas [32].

The external spaces that can be converted into green ones are many and, depending on their diffusion and size, can have a different impact with respect to the aforementioned benefits, which can however be found in all, from the smallest to the largest [33]. They can be placed in different spaces starting from parking areas, which can be made partially permeable and shaded by trees, to areas pertaining to buildings; from residual spaces between two lots to extended spaces in suburban areas.

In the compact city, rather than large parks, difficult to insert in existing urban areas, it is possible to intervene in multiple small spaces, such as residual ones. These pocket parks, once widespread in the city, contribute creating a distributed system, easily reachable by every citizen in a short time. Furthermore, they can be easily built and maintained.

Some examples of this increasingly widespread approach, which is also expressed through interventions of tactical urbanism and urban acupuncture [34], are the 100 Pocket Parks in London, and those in Amsterdam [35].

The interventions on the external spaces, therefore, have an effect not only on a local scale, on the surrounding buildings and in the external space itself, but, if put into a system, can also have an impact on neighbourhood and city scales. For this reason, it is necessary to plan their insertion using a targeted and conscious approach.

3.4 Planning Practices and Green Development

Cities are the areas most exposed to the risks associated with climate change as they are intrinsically fragile. In this regard, even within EU policies, it is requested a planning approach linked to a design approach able to consider and envisage more adaptive strategies including in the territorial development also existing buildings and urban spaces [36]. Therefore, in this paragraph, holistic and adaptive programming tools providing the inclusion of green strategies to cope with the current and complex climatic situation are referenced.

Since the second half of the twentieth century, the scientific debate has been heavily focused on overcoming mechanistic design approaches in favour of sustainable regenerative development. This paradigm aims to limit environmental damage by reconnecting man and his activities in a solid co-evolution with nature [37], providing valuable ideas at the design level to respond to the complex environmental problems to which cities are subjected today. The urban project is thus reinterpreted within a network of good practices that have a mutual benefit on environmental and social conditions [38].

Within the regenerative project, three phases are distinguished: (i) planning of subsystems, (ii) commitment of stakeholders, (iii) assessment of impacts [39]. The first group includes urban regeneration which, from a technical perspective, relies on targeted strategies inserting greenery into the built environment by addressing climatic challenges, such as reducing consumption in buildings and the use of vegetation to mitigate urban flooding and heat island effect [7, 12].

The importance of integrating natural solutions at the expense of structural interventions was also found in adaptive planning to increase resilience in the urban environment [40]. Flexible adaptation is therefore sought as a key feature to increase the system's ability to cope with climatic uncertainty [41]. This approach describes the strategies adopted by dividing them into six groups, "retain, relieve, resist, retreat, accommodate and prepare (4RAP)". Among the measures identified, blue and green infrastructures emerge to support the new paradigm that considers the inclusion of plant and water elements in the design of cities to be essential, instead of defensive measures that could lead to a false sense of security perceived by the inhabitants.

Nature-based solutions and sustainable urban drainage systems (SuDS) have been shown to be effective in mitigating the effects of floods and heavy rains [42], also providing additional benefits such as improved water quality, positive effects on urban microclimate, increasing biodiversity, improvement of urban aesthetics, and increasing the quality of life of the inhabitants [43].

4 Discussion

Despite the benefits of the systems and practices described in paragraph 3, that normally tent to be, it is also useful to highlight the limitations that could hinder their application.

For traditional green envelopes, it must be considered that: in general they are very expensive, with respect to their installation, but also during their useful life, and in their disassembling, but also for maintenance operations [9]; many of the functional layers are made using polymeric materials, while it would be interesting to find more ecological alternatives [7]; the large amount of organic material present can lead to a high fire risk [44]; the installation of these elements may be limited by the location of the building as the plants need direct sun exposure; the insertion of these elements in existing buildings could be problematic due to their weight which effect on structures originally designed without this additional load should not be underestimated.

Bioreceptive materials could allow overcoming some of the issues listed above, but further experimentation is needed. In addition, the possible colonizer organisms have a very slow growth rate, and therefore a long time is required to start to benefit from their presence.

Concerning the use of insulations from biomass, it has to be emphasized that to obtain the same thermal resistance, a greater thickness is normally required respect to the one necessary with traditional materials [45]. In addition, their availability in the market is restricted due to limited and expensive resources [21] and often the additives used are not biodegradable and can emit volatile organic compounds [46]. Finally, the use of organic insulators can lead to fire resistance problems [47].

On the other hand, as regards the biomasses use inside the pastes, good results are shown but with limitations. First of all, the reduced workability, due to the reduction of the mixing water by the organic material, limits the possibility of application in situ. The presence of organic material also raises doubts about the durability of the elements and the reduced mechanical strengths limit their application to the construction of non-structural elements. Finally, due to the variability of the characteristics of the plant material, it is difficult to maintain a standard performance.

With reference to outdoor greenery, maintenance is often critical and, if missing, it reduces its impact and effectiveness. Furthermore, especially with regard to small green spaces, the benefit is limited if they are not widespread in the city.

5 Conclusions

Climate change is a globally recognized problem and at the present day, no technology can reverse the condition reached by the planet. The only way to stop the progress of the damage is represented by a radical change in the production and industrial sectors, such as the construction one.

To reduce the impact of the building sector, as previously presented, one of the strategies lies in integrating the vegetation within the architectural objects.

The approaches and solutions adopted described in the previous intervention scales are undoubtedly beneficial, albeit with their limitations. However, if not included within an approach that considers the different factors acting in a complex system such as the built environment, there is a risk of confinement restricted to good practices or the goodwill of the individual citizen. To avoid that nature-based design strategies remain punctual actions, they require contextualization within planning policies for adaptation and resilience to climate change. Although these good practices exist, there is resistance to their use by public administrations, which means that they continue to intervene in the city without an overall picture.

References

1. Intergovernmental Panel on Climate Change.: Global warming of 1.5 °C (2019)
2. United Nations Environmental Programme.: 2020 Global Status Report for Buildings and Construction: towards a Zero-emission, Efficient and Resilient Buildings and Construction Sector, Nairobi (2020)
3. Ecorys.: Resource efficiency in the building sector. Rotterdam (2014)
4. Hill, A.C.: Vegetation: a sink for atmospheric pollutants. J. Air Pollut. Control Assoc. **21**(6), 341–346 (1971)
5. Arup.: Green Building Envelope (2016)
6. European Commission.: Green infrastructure (GI)—enhancing Europe's natural Capital (2013)
7. Shafique, M., Kim, R., Rafiq, M.: Green roof benefits, opportunities and challenges—A review. Renew. Sustain. Energy Rev. **90**(April), 757–773 (2018)
8. Vox, G., Blanco, I., Schettini, E.: Green façades to control wall surface temperature in buildings. Build. Environ.**129**(November 2017), 154–166 (2018)
9. Manso, M., Castro-Gomes, J.: Green wall systems: a review of their characteristics. Renew. Sustain. Energy Rev. **41**, 863–871 (2015)
10. Connelly, M., Hodgson, M.: Experimental investigation of the sound transmission of vegetated roofs. Appl. Acoust. **74**(10), 1136–1143 (2013)
11. Radić, M., BrkovićDodig, M., Auer, T.: Green facades and living walls—A review establishing the classification of construction types and mapping the benefits. Sustainability **11**(17), 4579 (2019)
12. Stovin, V., Vesuviano, G., Kasmin, H.: The hydrological performance of a green roof test bed under UK climatic conditions. J. Hydrol. **414–415**, 148–161 (2012)
13. Liu, H., Kong, F., Yin, H., Middel, A., Zheng, X., Huang, J., Xu, H., Wang, D., Wen, Z.: Impacts of green roofs on water, temperature, and air quality: a bibliometric review. Build. Environ.**196**(March) (2021)
14. Shafique, M., Kim, R.: Application of green blue roof to mitigate heat island phenomena and resilient to climate change in urban areas: a case study from Seoul, Korea. J. Water L. Dev. **33**(1), 165–170 (2017)
15. Cook-Patton, S.C., Bauerle, T.L.: Potential benefits of plant diversity on vegetated roofs: a literature review. J. Environ. Manag. **106**, 85–92 (2012)
16. Niu, H., Clark, C., Zhou, J., Adriaens, P.: Scaling of economic benefits from green roof implementation in Washington, DC. Environ. Sci. Technol. **44**(11), 4302–4308 (2010)
17. Guillitte, O.: Bioreceptivity: a new concept for building ecology studies. Sci. Total Environ. **167**(1–3), 215–220 (1995)
18. Manso, S., Calvo-Torras, M.Á., De Belie, N., Segura, I., Aguado, A.: Evaluation of natural colonisation of cementitious materials: effect of bioreceptivity and environmental conditions. Sci. Total Environ. **512–513**, 444–453 (2015)

19. Camia, A., Rober, N., Jonsson, R., Pilli, R., Garcia-Condado, S., Lopez-Lozano, R., van der Velde, M., Ronzon, T., Gurria, P., M'Barek, R., Tamosiunas, S., Fiore, G., Araujo, R., Hoepffner, N., Marelli, L., Giuntoli, J.: Biomass production, supply, uses and flows in the European Union. First results from an integrated assessment. Luxembourg (2018)
20. European Commission.: A Clean Planet for all. A European long-term strategic vision for a prosperous, modern, competitive and climate neutral economy, p. 114 (2018)
21. Rabbat, C., Awad, S., Villot, A., Rollet, D., Andrès, Y.: Sustainability of biomass-based insulation materials in buildings: current status in France, end-of-life projections and energy recovery potentials. Renew. Sustain. Energy Rev. **156** (2022)
22. Niyigena, C., Amziane, S., Chateauneuf, A.: Multicriteria analysis demonstrating the impact of shiv on the properties of hemp concrete. Constr. Build. Mater. **160**, 211–222 (2018)
23. Selvaranjan, K., Gamage, J.C.P.H., De Silva, G.I.P., Navaratnam, S.: Development of sustainable mortar using waste rice husk ash from rice mill plant: physical and thermal properties. J. Build. Eng. **43**(January), 102614 (2021)
24. Leiva Aguilera, M.J., del Río Merino, M.: Additived plaster with rice husk waste. In: Llinares-Millán, C., Fernández-Plazaola, I., Hidalgo-Delgado, F., Martínez-Valenzuela, M.M., Medina-Ramón, F.J., Oliver-Faubel, I., Rodríguez-Abad, I., Salandin, A., Sánchez-Grandia, R., Tort-Ausina, I. (eds.) Construction and Building Research, pp. 463–470. Springer Netherlands, Dordrecht (2014)
25. Iucolano, F., Liguori, B., Aprea, P., Caputo, D.: Evaluation of bio-degummed hemp fibers as reinforcement in gypsum plaster. Compos. Part B Eng. **138**(June 2017), 149–156 (2018)
26. Prusty, J.K., Patro, S.K.: Properties of fresh and hardened concrete using agro-waste as partial replacement of coarse aggregate—A review. Constr. Build. Mater. **82**, 101–113 (2015)
27. Chan, S.Y., Chau, C.K., Leung, T.M.: On the study of thermal comfort and perceptions of environmental features in urban parks: a structural equation modeling approach. Build. Environ. **122**, 171–183 (2017)
28. Ayala-Azcárraga, C., Diaz, D., Zambrano, L.: Characteristics of urban parks and their relation to user well-being. Landsc. Urban Plan. **189**, 27–35 (2019)
29. Rosso, F., Pioppi, B., Pisello, A.L.: Pocket parks for human-centered urban climate change resilience: microclimate field tests and multi-domain comfort analysis through portable sensing techniques and citizens' science. Energy Build. **260**, 111918 (2022)
30. Rosso, F., Mannucci, S., Morganti, M., Mariani, S., Cecere, C., Ferrero, M.: The effect of sustainable urban drainage systems on outdoor comfort and runoff. J. Phys: Conf. Ser. **1343**(1), 012023 (2019)
31. Park, J., Kim, J.: Economic impacts of a linear urban park on local businesses: the case of Gyeongui Line Forest Park in Seoul. Landsc. Urban Plan.**181**(August 2018), 139–147 (2019)
32. Connop, S., et al.: Renaturing cities using a regionally-focused biodiversity-led multifunctional benefits approach to urban green infrastructure. Environ. Sci. Policy **62**, 99–111 (2016)
33. Rosso, F., Cappa, F., Spitzmiller, R., Ferrero, M.: Pocket parks towards more sustainable cities. Architectural, environmental, managerial and legal considerations towards an integrated framework: a case study in the Mediterranean region. Environ. Challeng **7**(October 2021), 100402 (2022)
34. Herman, K., Rodgers, M.: From tactical urbanism action to institutionalised urban planning and educational tool: the evolution of park(ing) day. Land **9**(7), 217 (2020)
35. Mayor of London.: Pocket park prospectus, Mayor's foreword (2012)
36. European Commission.: Forging a climate-resilient Europe—The new EU strategy on adaptation to climate change. COM **82**, 23 (2021) (Brussels)
37. Du Plessis, C., Brandon, P.: An ecological worldview as basis for a regenerative sustainability paradigm for the built environment. J. Clean. Prod. **109**, 53–61 (2015)

38. Blanco, E., Raskin, K., Clergeau, P.: Towards regenerative neighbourhoods: an international survey on urban strategies promoting the production of ecosystem services. Sustain. Cities Soc. **80**(January), 103784 (2022)
39. Lak, A., Sharifi, A., Khazaei, M., Aghamolaei, R.: Towards a framework for driving sustainable urban regeneration with ecosystem services. Land Use Policy **111**(September), 105736 (2021)
40. Mehryar, S., Sasson, I., Surminski, S.: Supporting urban adaptation to climate change: what role can resilience measurement tools play? Urban Clim. **41**(November 2021), 101047 (2022)
41. Radhakrishnan, M., Pathirana, A., Ashley, R.M. Gersonius, B., Zevenbergen, C.: Flexible adaptation planning for water sensitive cities. Cities **78**(March 2017), 87–95 (2018)
42. Zandersen, M., Oddershede, J.S., Pedersen, A.B., Nielsen, H.Ø., Termansen, M.: Nature based solutions for climate adaptation—Paying farmers for flood control. Ecol. Econ. **179**(February 2020), 106705 (2021)
43. Griffiths, J.A.: Sustainable urban drainage. In: Encyclopedia of Sustainable Technologies, pp. 403–413. Elsevier (2017)
44. Manso, M., Teotónio, I., Silva, C.M., Cruz, C.O.: Green roof and green wall benefits and costs: a review of the quantitative evidence. Renew. Sustain. Energy Rev. **135**(August), 2021 (2020)
45. Casini, M.: Insulation Materials for the Building Sector: A Review and Comparative Analysis. Elsevier Ltd. (2020)
46. Wi, S., Kang, Y., Yang, S., Kim, Y.U., Kim, S.: Hazard evaluation of indoor environment based on long-term pollutant emission characteristics of building insulation materials: an empirical study. Environ. Pollut. **285**, 117223 (2021)
47. Kumar, D., Alam, M., Zou, P.X.W., Sanjayan, J.G., Memon, R.A.: Comparative analysis of building insulation material properties and performance. Renew. Sustain. Energy Rev. **131**(June), 110038 (2020)

RECsim—Virtual Testbed for Control Strategies Implementation in Renewable Energy Communities

Antonio Gallo[1](✉), Marco Savino Piscitelli[1], Lorenzo Fenili[2], and Alfonso Capozzoli[1]

[1] Department of Energy (DENERG), TEBE Research Group, BAEDA Lab, Politecnico Di Torino, Corso Duca Degli Abruzzi 24, 10129 Turin, Italy
antonio.gallo@polito.it

[2] ABB Smart Power Electrification, ABB SpA, Zürich, Switzerland

Abstract. In recent years, the building sector has been pointed out as critical for the energy transition, and the Energy Communities have been introduced to allow the aggregation of multiple buildings to jointly manage their energy generation and consumption. At cluster level, buildings can provide flexibility services to the energy grid by coordinating the energy consumption in order to match local energy generation. The aim of this work is to present a virtual testbed for the implementation of control strategies in Energy Communities and the assessment of their performance under different climatic conditions, generation systems, penetration of renewable energy sources or availability of energy storage. The virtual environment embeds models of the thermal building dynamics and of various systems and it is wrapped into an OpenAI gym to easily allow the implementation of advanced control algorithms such as reinforcement learning.

Keywords: Energy community · Building energy management · Virtual environment

1 Introduction

European Union has included in the Clean Energy Package several directives such as the Energy Performance of Building Directive and Renewable Energy Directive to reduce the energy intensity in the building sector. These directives push toward the improvement of building envelope and energy system performance, the liberalization of the electricity market for small producers and the penetration of Distributed Energy Resources (DER). In this context, Citizen Energy Community and Renewable Energy Community emerged as powerful instruments to involve prosumers at actively balancing the energy demand and supply at the distribution level by jointly operating with the other actors of energy grid. The shift towards a distributed approach for the management of the energy grid is however related to the increased complexity of such systems. Bi-directional energy flows, DER, Electric Vehicles (EVs), energy storage technologies together with data analytics-based and Information and Communication Technologies have opened the

J. Littlewood and R. J. Howlett (Eds.): SEB 2022, SIST 336, pp. 313–323, 2023.
https://doi.org/10.1007/978-981-19-8769-4_30

door to new possibilities for achieving a flexible energy demand giving to prosumers more influence than ever before in reaching energy efficiency and economic goals. Literature highlighted the importance of building energy flexibility also to address grid stability problems, especially if the energy fluxes are managed at a higher level than single building. In fact centralized and distributed architectures that also exploits peer-to-peer energy trading schemas may achieve better results. Advanced control strategies for the energy management of cluster of buildings are being introduced in the literature, and in many cases for their in-field implementation is required an off-line development phase that leverages digital twins of buildings and energy systems for the training of control strategies and for assessing their potential impact according to different objective functions. For this reason, the development of virtual environments for the conceptualization and testing of control strategies for energy management in Energy Communities (ECs) has become a relevant research topic.

A few tools and methodologies have been already developed for this purpose and some of them are briefly introduced and described. In [1] the virtual environment CityLearn is introduced. CityLearn is a virtual testbed for the specific implementation of Reinforcement Learning (RL)-based control algorithms for urban energy systems wrapped using OpenAI Gym library. The environment allows to consider as controllable systems Heat Pumps (HPs), an active Thermal Energy Storage (TES) and a Battery Energy Storage Systems (BESS) and photovoltaic (PV) systems as uncontrollable generation, while the thermal demand is pre-computed, thus not making it possible to exploit the building thermal mass as a flexibility source in the control problem. Pinto et al. [2], adapted the CityLearn environment introducing a data-driven approach for modelling the thermal dynamics of the buildings. A Long-Short Term Memory neural network was used to predict the indoor temperature trajectory associated to a simulated control signal, with the opportunity to feed the prediction model also by using monitored data. This approach allows to control the indoor air temperature but still relies on the availability of data or the development of a physics-based model for each building.

Also, Pigott et al. in [3] proposed an adaptation of the CityLearn environment for the evaluation of voltage levels across the distribution grid. This was done by exploiting the pandapower library from Python to simulate the Alternating Current (AC) distribution grid that connects each building to demonstrate the applicability of advanced controllers for the prevention of undervoltage and overvoltage events.

Sen et al. [4] developed a virtual environment for a real Net Zero Energy Community in Florida using the simulation tool Modelica for the control of thermostat setpoints in buildings. The environment includes both physics-based and data-driven models that are validated against measurements. Particularly, this testbed allows to consider cluster of buildings with different end-uses such as office, residential, and retail buildings, and PV systems as DER. However, the high-fidelity physics-based models need non-negligible simulation time that constraints the application to small clusters of buildings.

Modelica and EnergyPlus are combined through a Functional Mock-up Interface in Scharnhorst et al. [5], where 11 buildings with HVAC system and HPs are considered. The models were calibrated with operational data from real buildings with different end-uses. The main limitations are those related to the adoption of high-fidelity physics-based models as previously discussed. Jin et al. [6] developed a virtual environment for an urban complex to test scheduling and control strategies for the operation of HVAC systems in multiple buildings. The environment includes a detailed physical model of the HVAC system of each building and considers the thermal dynamics of the building envelope. The thermal parameters are evaluated for four typologies of buildings that are considered to be characteristic in the analysed urban complex.

Another example is proposed in [7], where a comprehensive virtual environment is developed considering models for estimating both energy demand and generation in a cluster of buildings. In particular, with physics-based models, are simulated the thermal load profiles for thermal zones, the electrical load profiles, a generation profile for PV and wind-turbine, an aggregated Plug-in Electric Vehicles (PEV) load profile, the operation of a central Energy Storage System (ESS), and the operation of an MV/LV distribution system interconnecting the buildings and the main grid. The environment allows to test control strategies for the optimization of the HVAC system operation of each building thermal zone and the power exchange scheduling between PEVs and the central ESS with the distribution grid. The models for the simulation of the building thermal dynamics adopted in [6, 7] consider a reduced number of parameters, but it is still not suitable for large cluster of buildings.

An interesting example on how to efficiently scale up the virtual environment is proposed by Wang et al. [8]. A simplified grey-box approach based on RC models was used to simulate the thermal dynamics of individual buildings for very large clusters. The proposed tool was developed as an open-source simulation-based virtual environment to train and validate control algorithms for thermostatically controlled load coordination. The key aspect of this environment is that it only needs two thermal parameters for the building envelope, that can be inferred from a real dataset and assigned to each building in a way that replicates the same distribution that they have across the building stock. However, measurements only come from residential buildings, and it is only allowed the testing of ON/OFF control strategies for HVAC systems.

From the literature review emerged that among the available virtual environments are still present issues related to the detail of simulations, computational costs and scalability. However, each of them was suitable for addressing the specific objective for which was designed. In this perspective, this study introduces the virtual environment RECsim that is specifically targeted for the simulation of energy management strategies in ECs allowing the analyst to set the control problem with the highest flexibility as possible. The main features of RECsim are related to the capability of the environment to simulate the building thermal dynamics, to scale the control problem up to large communities, to consider control actions more sophisticated than system ON/OFF and leverage simplified but accurate models of energy systems.

In particular, the simulation of the indoor air temperature is performed as in [8] in order to exploit the building thermal mass as a flexibility source to be considered in the control of HVAC system operation together with TES and BESS. In addition, RECsim allows to generate different scenarios and configurations of the EC to deploy the developed controllers in a variety of boundary conditions and assess their adaptive features.

The level of detail ensures sub-hourly evaluation of the thermal demand of each building in the EC and allows the control of HVAC systems, and active and passive energy storage. The pvlib library from Python is used to retrieve the solar position and compute the AC electricity output of the PV modules. The electrical load was disaggregated into HP load, Electric Water Heater (EHW) load and appliances load, whereas the gas load only accounts for the gas-fired boiler when available. The key features of RECsim can be then identified as follows:

- Grey-box dynamic model of the built environment validated against real measurement for sub-hourly simulation of the indoor air temperature according to [8]
- Fast implementation of a large cluster of buildings with diversified energy systems and envelopes
- The environment built on Python does not require any other software to be run and it is wrapped into the OpenAI gym library for the implementation of RL control
- The environment is composed of several sub-modules, so that it is easily accessible for future improvement to expand the existing modules or to add new ones.

The paper is structured as follows: Sect. 2. Describes the simulation environment through the various sub-modules which it is composed of. Section 3 presents the case study where a Rule-Based (RB) controller is implemented in the proposed environment and whose results are reported in Sect. 4. The last section includes the conclusions of the work and introduces the related future works.

2 Description of the Simulation Environment

The simulation environment is completely developed in Python, and it follows the OpenAI Gym framework which is a standardized interface for control-oriented simulation. The initialization is done by calling the *RECsim()* Python class and feeding the input parameters that are grouped by three categories which are: *simulation, building* and *energy system*. Before the starting of each control episode, the *reset()* function is called to reset Key Performance Indicators (KPIs), and storage SoC, and then the *step()* function implements the control action.

2.1 Definition of Building Thermal Model

A grey-box approach is adopted to evaluate the thermal demand of each building included in the environment. It has been chosen the RC model from [8], which is tuned using real data of indoor air temperature and HVAC operation collected for a very large sample of residential buildings in the United States (ECOBEE dataset). The thermal properties of each building are lumped into two parameters inferred from ECOBEE dataset that are the thermal time constant τ and the equivalent heat gain temperature $T_{\mathrm{HG,eq}}$. Each building is considered as a single-zone building where only one temperature is representative of the thermal comfort and it is indicated as T_{in}. The reduced model to compute T_{in} at the next time step is expressed by Eq. (1–4):

$$
\begin{aligned}
T_{\mathrm{in},t+1} = e^{-\Delta t/\tau} \cdot T_{\mathrm{in},t} + \left(1 - e^{-\Delta t/\tau}\right) \\
\cdot\left(T_{\mathrm{out}} + \mathrm{HG}_{\mathrm{int}} + \mathrm{HG}_{\mathrm{sol}} + R \cdot \mu + R \cdot E_{\mathrm{th,load}}\right) + \varepsilon
\end{aligned}
\tag{1}
$$

$$\mathrm{HG}_{\mathrm{int}} = T_{\mathrm{HG,eq}} \cdot \mathrm{HG}_{\mathrm{ratio}} \cdot \mathrm{HG}_{\mathrm{int,sched}} \tag{2}$$

$$\mathrm{HG}_{\mathrm{sol}} = T_{\mathrm{HG,eq}} \cdot (1 - \mathrm{HG}_{\mathrm{ratio}}) \cdot \mathrm{HG}_{\mathrm{sol,sched}} \tag{3}$$

$$R = (\tau \cdot \mathrm{RC}_{\mathrm{ratio}})^{1/2} \tag{4}$$

where Δt is the time step, T_{out} is the outdoor air temperature, $E_{\mathrm{th,load}}$ is the thermal input to the thermal zone from the HVAC system, μ is the modeling uncertainty, ε is the measurement error, $\mathrm{RC}_{\mathrm{ratio}}$ is the ratio between the thermal resistance R and the thermal capacity C and $\mathrm{HG}_{\mathrm{ratio}}$ is the ratio between the internal heat gain and the solar heat gain. $\mathrm{HG}_{\mathrm{int,sched}}$ and $\mathrm{HG}_{\mathrm{sol,sched}}$ are daily schedules for the internal and solar heat gain and are available according to the reference building developed by the U.S. Department of Energy or inferred from the ECOBEE dataset.

2.2 Estimation of Domestic Hot Water Load

The use of DHW is influenced by many factors such as location, occupant behavior and economic condition, even though, typical daily and seasonal patterns can be identified. The methodology adopted in the RECsim environment was inspired from Jordan et al. [9], that defined the probability of a DHW event as in Eq. (5):

$$p(t) = p_{\mathrm{day}}(t) \cdot p_{\mathrm{week}}(t) \cdot p_{\mathrm{season}}(t) \tag{5}$$

where $p_{\mathrm{day}}(t)$ is the probability distribution of DHW event across the day, $p_{\mathrm{week}}(t)$ is the distribution of DHW consumption among weekdays and weekend and $p_{\mathrm{season}}(t)$ is the seasonal variation across the year. In this case, holidays are not considered. Typical daily profile and weekend/weekday ratio are available in [9, 10]. The seasonal variations can be expressed as a sine function over the year.

The flowrate of each DHW event is sampled from a Gaussian distribution, and then it is associated to the energy consumption according to Eq. (6):

$$E_{\mathrm{DHW}} = \rho \cdot c_p \cdot V \cdot (T_{\mathrm{mains}} - T_{\mathrm{DHW}}) \tag{6}$$

where ρ and c_p are the density and the specific heat of the water at 15 °C, V is volume of water requested by the DHW event, T_{mains} is the water temperature from the distribution network, and T_{DHW} is the supply hot water temperature. At the current stage these temperatures are considered constant since they can be included in the seasonal variations. The average daily DHW consumption for each building is sampled according to a Gamma distribution.

2.3 Estimation of the Appliances Load

Typical daily profile for appliances electrical load is provided in [10], which is used to sample values of hourly demand. The average appliance daily electricity consumption for each building is sampled according to a Gamma distribution, while the actual daily consumption is computed by Monte Carlo estimation of the area under the daily profile and then it is spread across the whole day accordingly.

2.4 Definition of the Heat Pump Model

Each building in the environment is supposed to be equipped with an Air-to-Air HP which is sized according to the maximum thermal load. The thermal capacity is considered constant, whereas the COP is a function of temperature difference between the outdoor air temperature and the supply temperature as in [11]. The supply temperature for the cooling and heating season is an input parameter of the environment, and it is constant during each season. Equation (7) reports the COP function:

$$\text{COP} = 6.81 - 0.121\Delta T + 0.00063\Delta T^2 \tag{7}$$

2.5 Definition of Gas-Fired/Electrical Heater Model

Two options for the production of DHW are implemented. Both are considered as non-flexible loads, thus the generation matches the consumption at each simulation time-step. Each building considered in the environment could be equipped with one of those generation systems which are characterized by different energy carriers and by different efficiencies.

2.6 Definition of the PV Model

The output of the PV modules is obtained by using the pvlib library from Python. This library allows to simulate a system composed by the PV panel and the inverter that is connected to the main grid. For the sake of brevity, the reader is reminded that an extended documentation is available at [13].

2.7 Definition of Thermal and Electrical Storage Model

Let us denote E_{BESS} as the energy exchanged between the storage and the thermal/electrical system and ΔE_{BESS} the variation of the available energy in the storage. Then the SoC model is represented by Eq. (8–9):

$$\Delta E_{\text{ESS}} = E_{\text{ESS}} \cdot \eta_{\text{charge}} \text{if} E_{\text{ESS}} > 0 \tag{8}$$

$$\Delta E_{\text{ESS}} = E_{\text{ESS}} / \eta_{\text{discharge}} \text{ if} E_{\text{ESS}} < 0 \tag{9}$$

Moreover, technical constraints are considered for the SoC state and for E_{BESS} as reported in Eq. (10–11):

$$E_{\text{ESS,min}} < E_{\text{ESS}} < E_{\text{ESS,max}} \tag{10}$$

$$\text{SoC}_{\text{min}} < \text{SoC} < \text{SoC}_{\text{max}} \tag{11}$$

Eventually, the energy exchanged is limited according to the available energy for charging and to the thermal and electrical demand. TES operation is constrained by the

HP thermal capacity and the building thermal load as in Eq. (11–12) while the BESS cannot exchange electricity with the main grid, and it is expressed in Eq. (14–15).

$$|E_{TES}| < |E_{HP,max}| - |E_{th,load}| \text{if} E_{TES} > 0 \tag{12}$$

$$|E_{TES}| < |E_{th,load}| \text{if } E_{TES} < 0 \tag{13}$$

$$|E_{BESS}| < |E_{PV}| - |E_{el,oad}| \text{if } E_{BESS} > 0 \tag{14}$$

$$|E_{BESS}| < |E_{el,load}| \text{if } E_{BESS} < 0 \tag{15}$$

2.8 Action Space

The RECsim environment has been designed to allow the control of the thermal energy delivered to the building, the hot and cold TES and BESS. These variables can assume continuous values in the range $[Q_{cooling,max}, Q_{heating,max}]$, $[-0.33\ C_{TES}, 0.33\ C_{TES}]$ and $[P_{BESS,max,\ discharge}, P_{BESS,max\ charge}]$ where Q is the thermal load, C_{TES} is the TES thermal capacity and P_{BESS} is the BESS power exchange. During the simulation, the actions taken by the controller under analysis are verified according to physical constraints and then actuated.

2.9 Reward Function

The objective function is defined by the user among those terms available in the environment, which are evaluated both at building and community level. The user can specify whether to optimize a single term or a linear combination of multiple terms according to weights defined by the user itself. The available terms are the energy cost and consumption at building and community level, the comfort violations, and the shared electricity. The shared electricity is computed as the minimum between the renewable generation and community energy consumption.

2.10 Rule-Based Control

The RECsim environment is provided with three reference rule-based control logics that are independent between each other. An RB thermostatic control is implemented to keep the indoor temperature in a pre-defined comfort range. During the heating season thermal energy is delivered to the building by the HP or by the TES when the indoor temperature is below the lower acceptability range, and it is increased up to the upper limit of the comfort range. The TES control strategy aims at decreasing the peak load by shifting load toward off-peak hours. During low-price periods, the TES is charged by the HP until its complete charge, whereas during peak-price periods the TES is discharged whenever thermal energy is required by the building and the SoC is higher than 0. A simple RB control strategy for the BESS was inspired by [12]. When a PV surplus occurs, the BESS is charged otherwise it is discharged. Moreover, the constraints on charging/discharging power and on SoC have to be respected. This means that during the charging process, if the BESS cannot store all the excess electricity, the overproduction is injected in the grid. During discharging events, if the electricity from the PV and the BESS does not match the building electrical demand, the electricity is drawn from the main grid.

3 Case Study

An example of the implementation of the reference control strategies in RECsim is provided in this section. The baseline also provides a benchmark for the comparison with the tested control strategies. The simulation is set up for 50 residential buildings in Miami (FL) for 1 month during the cooling season from 01/08 to 31/08 with a simulation time-step of 5 min. Four different price schedules for buying electricity from the grid are implemented, and they are assigned to each building according to the parameters in the file "*electricity_schedule.json*". Weather data is downloaded from the pvlib module in Python where Typical Meteorological Year is available based on building location. The cost weight is not used here since the RB Control does not have an objective function to be minimized/maximized. The comfort range for the indoor air temperature is selected according to the ASHRAE Standard 55-2017 that identifies 23.95 °C and 26.85 °C as lower and upper limit of the comfort band respectively.

Thermal properties of the envelope are inferred from the ECOBEE dataset according to the selected building location. The supply water temperature of the cooling system is assumed equal to 7 °C for the considered operation mode. The gas price is set to 0.039 $/kWh. The penetration of PV, cold TES and gas-fired boiler in the cluster of 50 buildings was assumed to be 0.7, 0.7 and 0.2, respectively.

4 Results

In Fig. 1, the energy flows of the whole EC are aggregated and are plotted over one week of the simulation period. Similar daily patterns can be observed for both the electrical and thermal flows. Table 1 reports several KPIs at building and EC scale. Values range for each building from 272.4 to 1301.0 kWh for the total electricity consumption. Few buildings achieved an energy cost lower than zero due to the high number of PV modules considered together with a very low electricity demand. Self-Sufficiency (SS) and Self-Consumption (SC) are computed for buildings that are equipped with PV modules to measure the share of electricity consumption that is met by local generation and the share of local generation that is consumed on site, respectively.

When the EC is considered as a whole, values of 0.46 and 0.98 for SS and SC are obtained, while on average at single building scale SS and SC are equal to 0.48 and 0.71, respectively. SC at EC level is very high since the PV generation was generally lower than the electrical demand. The daily Peak-to-Average ratio (PAR) for the EC is equal to 1.82 which is lower than the average value of 2.9 at single building.

The Flexibility Factor (FF) may assume values between -1 and 1 and measures the amount of energy consumed during off peak price periods and has no meaning when considering the whole EC because buildings can have different peak-price periods.

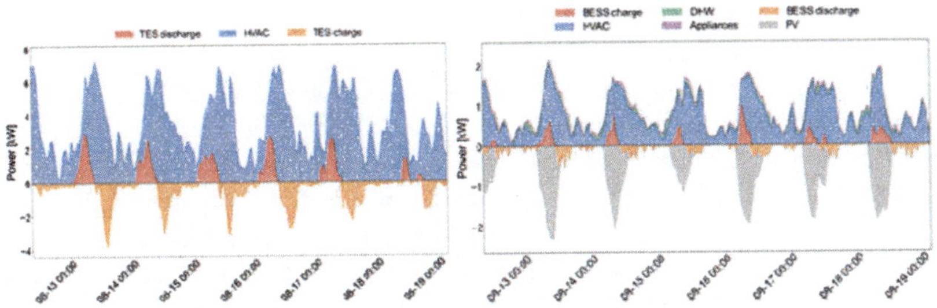

Fig. 1. Aggregated electrical and thermal balance

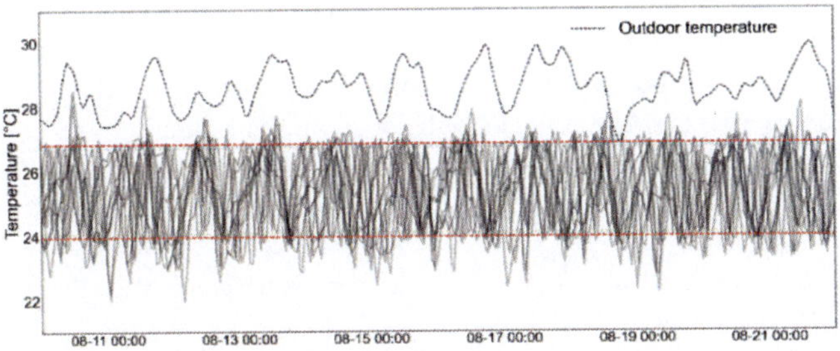

Fig. 2. Evolution of indoor temperature for each building in the EC

It shows values from 0.16 to 0.69 for the single buildings, which means that electricity is mainly consumed during off-peak periods. As shown in Fig. 2, for the considered 50 buildings, the reference RB control demonstrated its effectiveness at maintaining the indoor air temperature between the upper and lower limit of the acceptability range.

Table 1. Table KPIs for buildings and the whole Energy Community

KPI	Min	Max	Average	Community
El. Consumption [kWh]	272.4	1301.0	679.7	33985
El. Cost [$]	−15	152.5	62.2	3109.4
Shared electricity [kWh]	–	–	–	15510
SS	0.18	0.73	0.48	0.46
SC	0.21	0.91	0.71	0.98
PAR	2.2	3.8	2.9	1.82
FF	0.16	0.69	0.48	–

5 Conclusion and Future Works

RECsim is a virtual environment for the simulation of EC conceived for the implementation and comparison of advanced control strategies against a reference RB control policy. The aim of this work is to introduce this new environment by describing the various modules which is composed of, to assess pros and cons of adopting it as virtual test-bed and present a first application based on rule-based control. Future works will expand existing modules and add further energy systems. Currently the building stock is composed only by residential buildings and the same RC model is used for all of them according to [8]. In the next steps, the RC model can be improved to describe a larger variety of buildings. In addition, alternative energy modeling approach (i.e., black box modeling) will be tested. Moreover, the DHW and appliances schedule can be diversified by considering the occupant types and behavior. Other thermal and electrical generation systems will be implemented such as CHP systems, chillers, or centralized renewable energy plants. Further steps will focus on the modeling and control of EVs fleet operation and lastly, on the possibility to implement a Local Energy Market inside the EC for the negotiation of the energy flows among the members.

Acknowledgment. The work of Antonio Gallo was done in the context of a Ph.D. scholarship at Politecnico di Torino funded by ABB s.p.a.

References

1. Vázquez-Canteli, J., Dey S., Henze G., Nagy, Z.: CityLearn: standardizing research in multi-agent reinforcement learning for demand response and urban energy management. 10504 (2020). arXiv:2012
2. Pinto, G., Deltetto, D., Capozzoli, A.: Data-driven district energy management with surrogate models and deep reinforcement learning. Appl. Energy **304**, 117642 (2021)
3. Pigott, A., Crozier, C., Baker, K., Nagy, Z.: GridLearn: multiagent reinforcement learning for grid-aware building energy management (2021). arXiv:2110.06396
4. Huang, S., et al.: An open-source virtual testbed for a real net-zero energy community. Sustain. Cities Soc. **75**, 103255 (2021)
5. Scharnhorts, P., et al.: Energym: a building model library for controller benchmarking. Appl. Sci. **11**, 3518 (2021)
6. Xiaolong, J., et al.: Integrated optimal scheduling and predictive control for energy management of an urban complex considering building thermal dynamics. Int. J. Electr. Power Energy Syst. **123**, 106273 (2020)
7. Farinis, G.K., Kanellos, F.D.: Integrated energy management system for microgrids of building prosumers. Electr Power Syst Res **198**, 107357 (2021)
8. Wang, Z., Chen, B., Li, H., Hong, T.: AlphaBuilding ResCommunity: a multi-agent virtual testbed for community-level load coordination. Adv Appl Energy **4**, 100061 (2021)
9. Jordan, U., Vajen, K.: DHWcalc: program to generate domestic hot water profiles with statistical means for user defined conditions. In: Proceedings of the ISES Solar World Congress, pp. 8–12. Orlando, FL, USA (2005)
10. Hendron, B., Engebrecht, C.: Building America house simulation protocols. DOE/GO-102010-3141(2010)

11. Staffell, I., Brett, D.J.L., Brandon, N., Hawkes, A.: A review of domestic heat pumps. Energy Environ. Sci. **5**(11), 9291–9306 (2012)
12. Amato, A., Bilardo, M., Fabrizio, E., Serra, V., Spertino, F.: Energy evaluation of a PV-based test facility for assessing future self-sufficient buildings. Energies **14**(2), 329 (2021)
13. pvlib python.: https://pvlib-python.readthedocs.io/en/stable/. Accessed 22 Feb. 2022

Energy Retrofit and Fire Protection in Existing High-Rise Residential Buildings: A Case Study in Modena (Italy)

Luca Guardigli(✉) and Fausto Barbolini

Alma Mater Studiorum, University of Bologna, Bologna, Italy
luca.guardigli@unibo.it

Abstract. The deep renovation of existing buildings requires the application of thermal insulation to the façades. Unfortunately, with regard to residential buildings that are more than 24 m high, thus characterized by a high number of occupants with a long time evacuation, the presence of thermal insulation inside ventilated façades or within composite systems could fuel the fire by spreading the flames and reaching several floors in a very short time. Recent cases in Europe show that fire started in one of the apartments of the building and quickly spread along the entire façade. The paper investigates the relationship between new energy requirements and fire safety for this kind of buildings. A case study of high rise public housing owned by ACER in Modena, Italy, is then considered, introducing a retrofit project in compliance with the new 'vertical rule' V.14 of the italian Fire Prevention Code, in line with other national regulations in Europe. In the next future many insulation materials will be hardly applied in renovation projects if not certified as 'technological kits', according to new regulations.

Keywords: Energy retrofit · Thermal insulation systems · Fire protection · High-rise buildings · Ventilated façades

1 Introduction

According to the Directive 2012/27/EU on energy efficiency, deep renovations involve a modernization of buildings that reduces the energy consumption by a significant percentage compared to previous levels, leading to very high targets. In Italy, the concept of deep renovation was defined according to the law 90/2013 and the Ministerial Decree 26/06/2015 (Minimum Requirements). Deep renovations follow nearly Zero Energy Building (nZEB) objectives, which represent the ideal design in order to achieve the European goal of 2050, that is decarbonizing the residential park [1].

The actions for the purpose of energy efficiency can be applied to building enclosures, for instance insulating the perimeter walls, or to environmental services, replacing heat generators and boilers, installing thermostatic valves, photovoltaics, solar panels, etc. In this work, we focus on the enclosures of high rise residential buildings. These buildings

J. Littlewood and R. J. Howlett (Eds.): SEB 2022, SIST 336, pp. 324–336, 2023.
https://doi.org/10.1007/978-981-19-8769-4_31

are characterized by a high number of occupants with a long time evacuation, and the presence of new thermal insulation can fuel the fire reaching all the floors and producing casualties.

There is no clear definition of high-rise—or tall—buildings. By the traditional definition, high-rise buildings are those for which a fire cannot be fought using standard firefighting methods. There are significant differences to fire safety measures in high-rise buildings, depending on the countries. However, height in buildings is one of the criteria to define fire safety rules and, in fact, in Italy there are some specifications for buildings over 12 m in height and others for buildings over 24 m. The height is calculated from the ground to the top accessible level (18 m in UK, and 23 m in US, that is 75 feet). Therefore, high-rise buildings are approximately the ones with 8 storeys or more (over 24 m). Many residential buildings, especially public housing, are taller than that. A large percentage of these buildings was built in Europe between 1950 and 1980 following old regulations, and it was recently renovated without strict rules.

The paper investigates the relationship between new energy requirements and fire safety for this kind of buildings. In the first section a literature review on fire safety of high-rise residential buildings in relation to insulation systems is addressed. Then, the implication in the design of façades is investigated, applying a methodology to select the proper type of insulation system to a case study in Modena, Italy.

2 Literature Review

There are many studies on deadly fires occurred to high-rise buildings in Europe, linking the events to the characteristics of the façades [2]. The case of the Grenfell Tower in London is one of the most famous [3, 4]. The building is a 24-storey, 64 m high building erected in 1967. It consists of 4 floors for commercial use and 20 floors for apartments. In 2016 it was renovated with the aim of increasing its commercial value and it was finished with a ventilated façade, with the following elements (Fig. 1):

1. Celotex FR5000 insulation panel (PIR), 150 mm thick in polyisocyanurate. Classified as class 0 of reaction to fire, according to the British standard BS 476;
2. 50 mm ventilated cavity;
3. Outer layer composed of aluminum sandwich panel, 3 mm thick, with an internal polyethylene filling called Reynobond PE.

On 14 June 2017 a malfunction of a refrigerator on the 4th floor caused a fire that soon involved all the façades of the building. The fire spread along the insulating coating, causing the death of 70 people. A Reynobond PE sandwich panel, with poor fire resistance properties, was preferred to a more expensive version with Reynobond FR fire retardant treatment. In the United States the material was already banned for buildings taller than 12,2 m. The Celotex FR5000 PIR insulation panel is defined as reaction to fire 0, which is equivalent to class B-s3, d2 or higher, according to the European standard. During the fire, the thin and light metal material separated from the insulation causing material to fall at the base of the building. The combustible core was then exposed to the fire, favouring the very rapid spread of the flames in addition to the production of a significant amount of smoke and toxic substances. After the Grenfell Tower fire, seven BS 8414 full scale tests were conducted by the British Research Establishment (BRE) as requested by the UK Department for Communities and Local Government. These tests were planned to cover a series of combinations of Aluminum Composite Material (ACM) panels with different filler materials and insulation [5].

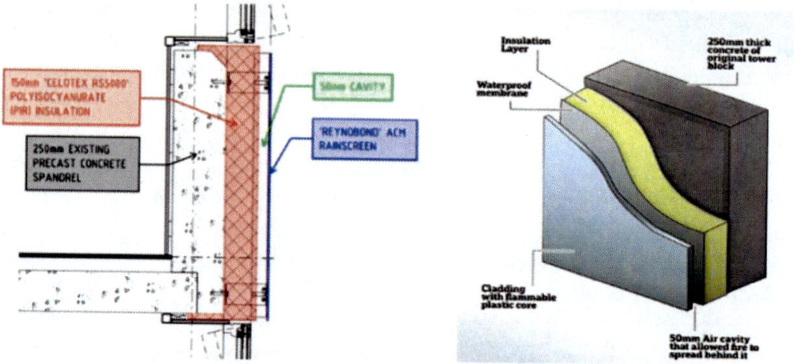

Fig. 1. Details of the Grenfell tower, with Celotex and Reynobond. Author: D'Apolito [4].

Fig. 2. Torre dei Moro in Milan with the façade 'wings', before and after the fire.

On 15 August 2009 a fire broke out in a kitchen on the 6th floor of a residential building in Miskolc, Hungary, and spread vertically through the external insulation system to the roof. The building consisted of basement, ground floor and ten floors above ground. In addition to the damage suffered by the building, there were also three victims. The building was built in 1968 and renovated in 2007. The renovation included the construction of the façade with an external polystyrene insulation, an easily combustible material. During the fire, the smoke spread quickly also along the stairs and through the plant shafts that were not adequately insulated.

Research into the causes and consequences of the fire revealed that the following factors contributed to the rapid spread of the flames through the building [6]:

- it was summer and the windows were open thus contributing to the rapid spread of the flames through the façade to the floors above, where the fire broke out;
- the polystyrene insulation was applied to the external walls without fire barriers;
- the insulating material was not installed correctly, and the polystyrene panels were not properly fixed to the walls;
- a layer of plaster with a thickness of 2–3 mm was applied against the required 5 mm.

The Torre dei Moro in Milan is a multi-functional building built in 2009, for a total of 18 floors plus two basements. It has two large asymmetrical 'wings' with an aesthetic function, but also for the ventilation of the façade (Fig. 2). A fire flared up in the building on 29 August 2020, probably generated by a cigarette butt on the 15th floor, and quickly reached the façades of the building [7]. According to the report of the Milan Fire Brigade of 21 September 2021, the building was "characterized by a geometric shape with evident aesthetic functions which, however, contributed (shape, material behavior and ventilation) to the development of the fire". In a subsequent report dated 2 November 2021, it was said that "the installation of the panels, made of material that contributed to the spread of the fire, was different from the dispositions of the test certificate and from what was approved". According to the investigators, the panels of the wings, consisting of a 3 mm polyethylene core covered with two layers of 0,5 mm of aluminum, contributed to the development of the fire. The void between the façade and the structure generated then the so-called 'stack effect'. The investigation is still ongoing. The three examples demonstrate the technological weaknesses of the insulation systems adopted in the façades, with predictable risks on a large scale, given the dimensions of the buildings. Some authors identify key variables, characterizing them in terms of risk and mitigation potential [8]: component materials, connection systems, installation techniques and geometries, occupancy type, age of application, proximity to other structures, external factors such as weather, building fire protection systems, etc.

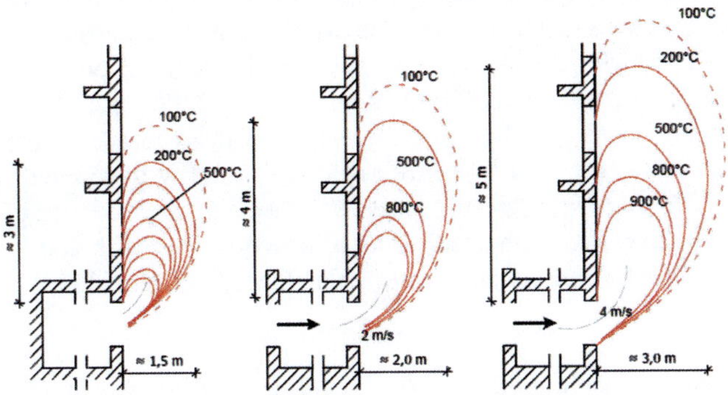

Fig. 3. Flame heights with temperatures in relation to air circulation. Author: Rigone [9].

There are two typical fire spread scenarios on the façades:

1. external fire propagation by radiation from a separate adjacent building to the combustible façade, or from fire sources located near the building itself; by radiation or by direct exposure to flames (waste on balconies, parked cars, etc.)
2. a fire inside the building, which spreads through the openings of the façade (windows, doors, etc.) on upper floors [9].

The latter is the case of the three examples above, where the factors influencing the shape of the protruding flames and therefore the possibility of propagation are quantity and nature of the fuel, dimensions of the openings, geometry of the compartments, and air velocity (Fig. 3).

However, according to BRE there is little difference in the probability of death or injury from fire as a function of building height [10]: in high-rise buildings the proportion of fire risk caused by their height is rather sensitive to the reliability and effectiveness of compartmentation and automatic suppression. The role of these systems is crucial in limiting the 'communication' of fire risk. In fact, subject to the limitations of a small sample size, the statistics show that for buildings with more than 12 storeys, there is a significant reduction in the risk of death from fire. In 2019 a research group of the American Fire Protection Research Foundation led by Michael Spearpoint with the support of Council for Tall Buildings and Urban Habitat (CTBUH) has gathered information on the fires that in recent years have involved tall buildings [11]. The results reported that 60% of deadly fires occurred in buildings that had undergone some form of refurbishment. In accidents where deaths occurred, the buildings were less than 32 storeys high and had façades with Exterior Insulation and Finish Systems (EIFS)—also known as External Thermal Insulation Composite Systems (**ETICS**). Among these systems, 14,0% employed polystyrene (EPS/XPS), 5,0% polyurethane foam, 3,0% rock wool, 2,0% polyurethane, 24% were ventilated façades with metallic composed materials (MCMs), and 8,0% not ventilated facades with MCMs; in 30% of the cases the type of combustible material engaged in the fire was unknown. According to this report, the dangers and causes of fire are internationally known but fire safety is not always integrated into the design process and not considered in every phase of the design and construction of a building. The inclusion of fire safety in the early stages of product development and in the design of the technological systems—the so called 'kits' [12, 13]—would help reduce many concerns. The fire performance of these façade systems should be taken into account as a priority, exactly as the accessibility of rescuers or other fire safety systems. Furthermore, 'green' materials and systems are expected to be designed to meet the objectives set by the society both in terms of fire risk and environmental performance [14]. As a consequence, research and legislation should move together on the following tasks:

- integration of sustainable building materials and systems into systems of fire reporting. In this way it is possible to build a statistic on the contribution of insulation and systems to the spread of flames;
- development of more robust and appropriate test methods, which produce useful data for performance evaluation of materials, components and systems (kits);

- integration of fire performance considerations into research and development of sustainable materials and technologies; fire safety should be the basis of the design and not perceived as a limitation to be inserted at a later time;
- robust methods and tools for risk and performance assessment, which are based on the extensive knowledge and experience of expert stakeholders, on available data and expert judgment where data are missing;
- better tools for holistic design and performance evaluation, leveraging BIM and other technologies that are defining the market for buildings.
- holistic and socio-technical approaches of regulation that take into account the diversity of objectives and markets for building design, construction and use; the regulation has followed fire events, and an organic vision is generally lacking on regulatory intervention in the field of fire safety.

3 Methodology

The present study has started with the identification of the fundamental indicators to guide the choice of insulating materials for energy retrofitting. With the addition of new layers to the external walls of the building, in the event of a fire, these behave differently from the existing enclosure: if made of combustible materials they can favour the spread of flames and become obstacles to the exodus of occupants or to the access of rescue teams. The main performance indicators to define the best sequence of enclosure layers, apart from fire safety, are usually the following [15]:

- thermal conductivity (λ);
- the coefficient of resistance to water vapor diffusion (μ); an insulating material with a low μ value and a small thickness is said to be 'permeable to steam';
- the density or specific weight of the material;
- the phase shift (h), that is the time it takes for the heat wave to flow from the outside to the inside through the enclosure;
- environmental impact indicators, such as material origin, embodied energy, recyclability, all identified through Life Cycle Assessment (LCA).

Thermal insulation can be applied to building enclosures in various ways but it is usually utilized as external layer to avoid condensation, with or without the presence of a ventilation cavity. The main objective of a ventilated façade is to reduce the summer thermal loads due to solar irradiation on the opaque surfaces, representing an important benefit in the Mediterranean climate; on the contrary, the benefits of a ventilated façade in winter are negligible. The efficiency in both conditions depends on the height and the section of the cavity, but also on the properties—for instance the absorbance—of the cladding systems. Specifications for ventilated façades follow ETAG 0-34 and national standards for curtain walling. Generally speaking, incombustible barriers are required to stop the fire on each floor; unfortunately, this aspect is in contrast to the principles and effectiveness of ventilation [16]; in fact, ventilation should be obtained subdividing the enclosure in different vertical sectors, which often implies using a different technological system, the so-called micro-ventilated or open joint ventilated façade.

Once having pre-selected the insulation materials, the reaction to fire class is the performance indicator to be controlled. The UNI EN 13501-1:2018 standard provides the procedure for the classification of construction products with regard to reaction to fire. Materials are classified according to Euroclasses: A1, and A2 for incombustible materials; B, C, D, E, F for the other materials for their limited or significant contribution to flash over. The European classification system favours the evaluation of heat release in function of time, considering the droplets (d0, d1, ...) and smoke production (s1, s2, ...) as accessory parameters. The European system offers a combination of performance levels that is more articulated than the old Italian system. Insulating materials can be divided into four macro categories based on their origin, with a correspondent reaction to fire:

- synthetic inorganic insulations (calcium silicate, rock wool and foam glass), all in A1 reaction to fire class, that is no contribution to a fire;
- natural inorganic insulations (clay, expanded perlite, vermiculite and pumice), all in A1 class;
- synthetic organic insulations (expanded and extruded polystyrene, polyurethane);
- natural organic insulations. The types of insulating materials that coincide with raw materials of origin are: sheep's wool, wood fiber panels, wood shavings, cork, straw (class E, significant contribution to flashover), cellulose, coconut fiber, cotton linen, hemp and wattle; B-s2-d0 reaction to fire, very limited contribution to fire growth.

In general terms, the contribution to fire is higher in natural and synthetic organic materials (e.g. EPS), while the embodied energy is higher in synthetic ones (e.g. rock wool). The first result is that combustible components and low energy natural materials ('green materials') tend to be used in small buildings, where fire certifications of the whole construction are usually not required; on the contrary, they are rarely employed in high-rise buildings, where incombustible components are required in order to meet global fire safety certifications. The certification of the insulation system is mandatory under certain circumstances, namely the size, the height, and the function of the building. The second result is that natural organic—and often recyclable—materials—for instance, wood fiber panels—can be used in large buildings only if certified within 'technological kits', that is composed building elements or systems where combustible materials are protected by incombustible layers. Thin insulation components are compatible with fire protection, especially innovative materials like mineral based ones.

A new façade system was then applied to public housing buildings in the Giardino District in Modena, between 1974 and 1979 (Fig. 4). In those years, the Istituto Autonomo Case Popolari (IACP), now ACER, implemented its interventions using industrialization and prefabrication techniques to reduce costs and construction times. Building construction was characterized by the experimentation of new technologies of French origin, such as that of the *coffrage tunnel* and *banche et tables* method: 10 cm thick reinforced concrete transversal bearing walls and 18 cm thick solid slabs cast in sequence. The prefabricated infill walls of the structure were precast on the ground and then raised and positioned in place. The chosen tower buildings have 11 floors; the dimension in plan is 23,88 × 16,08 m, the height is 33,04 m. The ground floor is used as a garage, while each floor hosts 4 apartments for a total of 36 apartments.

Fig. 4. Building in Quartiere Giardino, Modena, construction year 1970. Actual conditions.

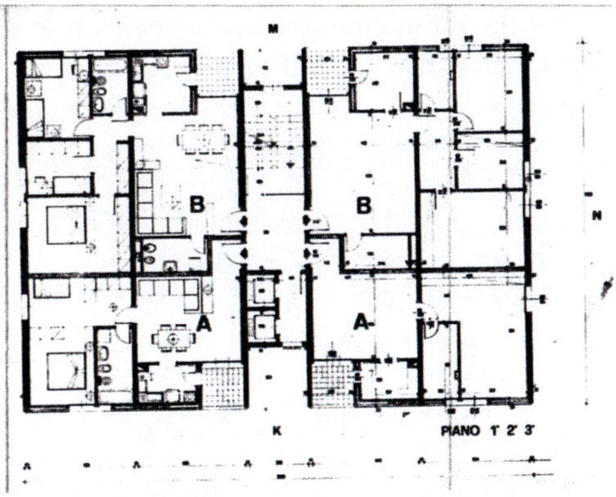

Fig. 5. Building in Quartiere Giardino, Modena, original plan. ACER archive, Modena.

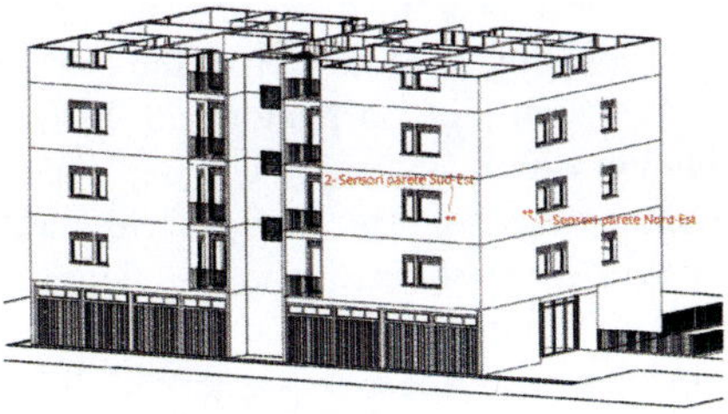

Fig. 6. BIM model of the tower; identification of the two spots for the thermofluximeter test. Author: Radighieri [17].

Archival research was needed to identify the energy characteristics of the components (Fig. 5). In addition to that, it was decided to measure the U-value of external walls in two different spots using the thermofluximeter method (Fig. 6). The measured value

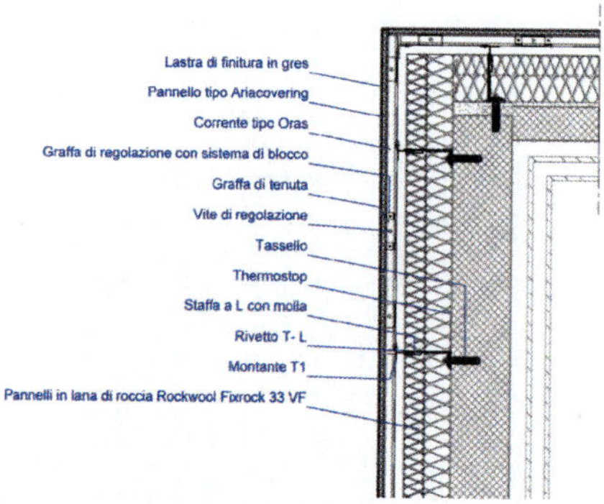

Fig. 7. New façade layers. Author: Radighieri [17].

of the first tested external wall was 1,315 W/m^2 K, very similar to the one derived from documents and calculations, and far above the minimum required. The phase shift was equal to 6,8 h, below the minimum acceptable, which is 8 h. The characteristics of other walls, partitions and floors, as well as thermal bridges, were also identified and calculated following the available documents. The building was then modeled with Edilclima EC700 software, obtaining these data (primary energy) [17]:

- Energy performance index for winter air conditioning ($EP_{H,tot}$) = 161,33 kWh/m^{2}y;
- Energy performance index for winter air conditioning, non renewable energy, ($EP_{H,nren}$) = 160,27 kWh/m^{2}y;
- Energy performance index for hot water ($EP_{W,tot}$) = 24,58 kWh/m^{2}y;
- Energy performance index for hot water, non ren. en. ($EP_{W,nren}$) = 24,42 kWh/m^{2}y;
- Global energy performance index, non ren. en. ($EP_{gl,nren}$) = 187,35 kWh/m^{2}y.

The building was built before any fire safety regulation and some mandatory adaptations were made in the last 20 years. For instance, on each floor there are a fire outlet for firefighters and new apartment doors with higher fire resistance.

The following proposals have been made to improve the thermal performance of the opaque and glazed surfaces (Fig. 7): insulation of the external walls with rock wool (14 cm) and ventilated cavity (6 cm); roof insulation with 18 cm thick rock wool panels, ground floor ceiling insulation with rock wool (6 cm); replacement of external finishes.

A new ventilated façade requires some precautions with regard to fire safety, and the choice of rock wool balances a very low reaction to fire, natural origin, recyclability and good insulation performance. The intervention falls into the category of 'second level renovations', because it involves a façade area greater than 25% of the total enclosure, without replacing the existing heating system. The minimum requirements according to DGR 967 of the Emilia-Romagna Region (2015) were verified using the same software as before (thermal zone E in Italy): absence of surface condensation and risk of mold formation, especially near thermal bridges; verification of the internal critical temperature of thermal bridges; U-value of non-air-conditioned room partitions and structures <0,80 W/m^2 K; U-value of external opaque walls <0,28 W/m^2 K; U-value of transparent structures <1,4 W/m^2 K; U-value for structures communicating with unheated rooms <0,24 W/m^2 K; global average coefficient of heat exchange by transmission (H't) < 0,65 W/m^2 K. Due to this intervention the achieved performance of the building was: $EP_{H,tot}$ = 25,26 kWh/m^{2}y; $EP_{H,nren}$ = 25,09 kWh/m^{2}y; $EP_{gl,nren}$ = 52,17 kwh/m^{2}y; $EP_{W,tot}$ and $EP_{W,nren}$ did not change.

The verification of the minimum requirements of insulation materials with respect to the class of reaction to fire, in case of buildings subject to fire prevention certificate, is currently carried out in compliance with the DM March 10, 2005 [18], with reference to the European classification system, while for other materials (curtains, upholstered furniture, etc.) is carried out in compliance with the DM June 26, 1984 [19]. Materials must also comply with the provisions of regulation EU n. 305/2011, which establishes harmonized conditions for the marketing of building products. According to article 4, construction products are placed on the market only if they follow the scope of harmonized standards or if they have been issued a European Technical Assessment, being accompanied by a Declaration of Performance (DoP).

4 Discussion

Since the current Italian legislation has no specific reference standards in force for the design of high-rise buildings over 24 m, in the case study of Modena the compliance of our solution with the Circular letter 5043 of 15 April 2013 was first analysed [20]. This is simply a set of guidelines, according to which ventilated façades that 'cannot be inspected', like the one proposed here, require insulating material with at least B-s3,d0 reaction to fire level, and at least EI30 resistance level. The use of rock wool respects the requirements, but it is possible to use less fire performing insulating materials if there are barriers in correspondence to the floors. It would have been possible also to use lower class insulating materials if the component was produced as 'kit' with a global certified (tested in laboratory) B-s3,d0 class. Therefore, according to current legislation, any type of insulation is allowed if protected and certified through laboratory tests. The use of rock wool, an incombustible material, is not necessary, but no cavity barriers are needed, according to the existing legislation, if rock wool is selected.

However, following the draft of the 'vertical rule' V.14 (Enclosures of buildings)—'vertical' meaning 'organized in relation to the parts of the building'—of the Fire Prevention Code (FPC) [21], now under approval, the rules are becoming mandatory and more restrictive. The building is more than 24 m tall, therefore it is defined as a class SC building. With regard to this class, instructions are given for materials and separation elements. In façades the following materials must meet the GM1 fire reaction requirements: thermal insulation (e.g. coats not in kit); external insulation systems in kits (e.g. ETICS, kit coats,…); gaskets, sealants and sealing materials, if they more than 10% of the entire gross area of the enclosure; other components, excluding glass components, if they occupy more than 40% of the entire gross area of the enclosure.

It is necessary to have cavity barriers in order to check the resistance to fire in correspondence of the horizontal and vertical construction elements for the compartmentalization. Barriers and any other protection must be made with materials of GM0 class of reaction to fire; on the roof, the barriers and any other protection must belong to a specific (BROOF t2, t3, or t4) or be of EI 30 fire resistance class. The closing portions

Table 1. Design strategies for the building over 24 m: comparison between current Italian guidelines and the Italian Fire Prevention Code under approval.

Topic	Current guidelines (2013)	Fire prevention code under approval (2015)
Reaction to fire inside building	None	The insulating materials in the common areas such as stairs, hallways, etc… Must be protected by materials at least of the GM2 group
Fire resistance inside building	None	Compartment materials of at least class 60
Compartments	None	Compartments for buildings taller than 12 m
Evacuation	Communication between apartments and stairwells and elevators with REI 60 doors	REI doors towards the stairs equipped with self-closing device. Security lighting in the common areas, specific indications on fire-fighting behavior to be followed and a two-way communication system on each floor
Fire control	None	Control of compliance of the existing hydrant network with the UNI 10779 standard
Safety of technological systems	None	Safety lighting with autonomous circuit capable of withstanding up to 60 min beyond the power failure
Façade	Non-mandatory indications. Installation of insulating material with fire reaction class at least Bs3, d0, or with at least EI30 fire resistance class. Alternative design solutions if the class is lower	Fire reaction class of insulating materials at least GM1, alternative design solutions with lower resistance classes. Forecast of horizontal and vertical bands of class GM0. Subdivision partitions EI60 class for each floor. Full-height class EW 60 (i → o) inner skin

of the area included in the barriers and in any other protection may have openings only if equipped with fire dampers or equivalent systems with automatic closing in case of fire, with the same fire resistance requirements of the separation, or they must be tested following EN 1364-3. If a ventilated double skin façade has an open outer skin, on each floor the cavity must be interrupted by barriers E60, the two skins must comply with the requirements of the materials for the façades and for the entire height and for all floors the internal skin must be EW 60 (i → o). The rules require the use of cavity barriers on each floor and in correspondence with the compartments with materials belonging to the group GM0, i.e. A0 or A1 class. Insulation materials inferior to B-s2, d0 class, are excluded and, if protected by a layer of higher reaction class, they must be at least D-s2, d2 class. In this way many organic insulating materials and the most common expanded polystyrene without additives are going to be excluded. Table 1 shows the comparison between the Italian guidelines and the new FPC, 'vertical rule' V.14.

5 Conclusions

At the building scale, fire safety regulations are very complex and vary significantly in different European countries. Fire regulations in Italy are not mandatory yet, but they are becoming more restrictive and the 'vertical rule' V.14 for residential buildings is under approval. Many buildings have been renovated in the past 20 years using fire safety solutions that are not acceptable today.

Façade precautions are not the only measures in fire prevention: the analysis of fire statistics shows in fact that given current fire precautions, the proportion of fire risk caused by the height of high-rise buildings is mitigated by increases in fire resistance of materials, enhanced access and facilities for the fire service, and the incorporation of sprinklers.

The use of incombustible insulation like rock wool is currently not mandatory in Italy, but it is strongly suggested; in fact, green' and natural combustible materials (class E) are still applied to renovated façades, but they will be soon accepted only if protected and certified within 'technological kits'. It is then preferable for a designer to choose tested systems that make use of incombustible materials. For innovative solutions laboratory test are necessary to certify the whole components.

Finally, the case study shows that the use of ventilated façades is possible in high-rise buildings, meeting the nZEB targets and the fire safety regulations, but some technical issues should be taken into account. In particular, if ventilation is obtained subdividing the enclosure in different vertical sectors, the efficiency of the system is reduced, as compared to full height ventilation. Therefore, in newly designed façade systems the role of micro-ventilation should be carefully considered.

References

1. Agenzia Nazionale per le nuove tecnologie, l'energia e lo sviluppo economico sostenibile ENEA, Comitato Termotecnico Italiano Energia e Ambiente CTI: Rapporto annuale 2020: La certificazione energetica degli edifici (2020)
2. Nguyen, K., Weerasinghe, P., Mendis, P., Ngo, T.: Performance of modern building façades in fire: a comprehensive review. Electron. J. Struct. Eng. **16**(1), 69–85 (2016)
3. Grenfell Tower Inquiry.: https://www.grenfelltowerinquiry.org.uk. Accessed 22 June 2022
4. D'Apolito, M.: Analisi dell'incendio avvenuto alla Grenfell Tower di Londra: simulazioni e verifiche di soluzioni progettuali. Master thesis in Building Engineering, Politecnico di Torino, supervisors: prof. R. Vancetti, Ing. E. Cereda, a.y. 2018/2019 (2019)

5. Soja, E., Wade, C., Frank, K.: Flammability tests for building facades. In: Flammability Testing of Materials Used in Construction, Transport and Mining, 2nd edn, pp. 89–126. Woodhead Publishing Series in Civil and Structural Engineering (2022)
6. Hajpál, M.: Analysis of a tragic fire case in panel building of Miskolc. Meeting COST, Malta (2012)
7. Torre dei Moro, Milano.: www.ingenio-web.it/31727-incendio-torre-dei-moro-milano-informazioni-e-prime-valutazioni
8. Lamont, S., Ingolfsson, S.: High Rise Buildings with Combustible Exterior Wall Assemblies: fire Risk Assessment Tool. National Fire Protection Association Report (2018)
9. Rigone, P.: Il comportamento al fuoco delle facciate continue. Fire Safety Academy, Milan (2019)
10. Charters, D., Cullinan, R., Warren, E.: Fire risk in high-rise and super high-rise buildings Prevention and mitigation. In: BRE, Fire Safety in Buildings. A Collection of BRE Expert Guidance of Fire Risk, Protection and Detection, DG 533, pp. 1–12, IHS BRE Press, St Albans (2016)
11. Spearpoint, M., Fu I., Frank K.: Façade fire incidents in tall buildings. CTBUH J. 34–39 (2019)
12. ETAG 034-1.: Guideline for European technical approval of kits for external wall claddings. Part I: ventilated cladding kits comprising cladding components and associated fixings (2012)
13. ETAG 034-2. Guideline for European technical approval of kits for external wall claddings. Part II: cladding kits comprising cladding components, associated fixings, subframe and possible insulation layer (2012)
14. Meacham B., McNamee, M.: fire safety challenges of 'Green' buildings and attributes. Report by Fire Protection Research Foundation for National Fire Protection Association (2020)
15. Pennacchia, E., Piras, G.: Materiali e componenti per l'efficienza energetica degli edifici, Legislazione Tecnica, Roma (2018)
16. Schabowicz, K., Zawislak. L.: Numerical comparison of thermal behaviour between ventilated facades. Studia Geotechnica et Mechanica, 1–9 (2020)
17. Radighieri G.: Isolamento termico e protezione dal fuoco negli edifici alti residenziali: un caso studio a Modena. Master thesis in Engineering of the Processes and Building Systems, University of Bologna, supervisors: prof. L. Guardigli, arch. F: Barbolini, a.y. 2018/2019 (2019)
18. DM 10/03/2005: Classi di reazione al fuoco per i prodotti da costruzione da impiegarsi nelle opere per le quali è prescritto il requisito della sicurezza in caso d'incendio (2005)
19. DM 26/06/1984: Classificazione di reazione al fuoco ed omologazione dei materiali ai fini della prevenzione incendi (1984)
20. Circ. n. 5043 del 15/03/2013: Guida tecnica su Requisiti di sicurezza antincendio delle facciate negli edifici civili. Aggiornamento (2013)
21. DM 03/08/2015 Codice di Prevenzione Incendi. Approvazione di norme tecniche di prevenzione incendi, ai sensi dell'art. 15 del DL 8 marzo 2006, n. 139, e s.m.i. Sez V, Regole tecniche verticali, cap. V.14, Edifici di civile abitazione (2015)

A Pilot Study to Evaluate Occupant Quality of Life in Optimised Retrofit Dwellings in Wales, UK

J. R. Littlewood[1] (✉), X. Zhang[1], and G. Karani[2]

[1] The Sustainable & Resilient Built Environment Research Group, Cardiff Metropolitan University, Cardiff CF5 2YB, UK
jlittlewood@cardiffmet.ac.uk

[2] Cardiff Metropolitan University, The Environmental Public Health Research Group, Cardiff CF5 2YB, UK

Abstract. This paper discusses an investigation into quality of life (QoL) as a pilot study from a sample of occupants living in existing dwellings, that have been (2021) or will be retrofitted in 2022 and 2023, funded by the Welsh Government's (WG's) Optimised Retrofit (OR) project. The pan Wales OR project aims to retrofit close to 2000 existing social housing dwellings targeting nearly-zero/zero operational energy standards, to alleviate occupant fuel poverty and reduce energy costs and carbon emissions and increase occupant comfort and QoL. The methodology presented builds on two previous research projects undertaken and completed by two of the authors in 2010 and 2020, to adapt and create a hybrid Short Form-36 (HSF36) health survey, accompanied by the established RAND analysis system. The HSF36 questionnaire survey has been further refined for the OR project and has been used to collect occupant data through face-to-face interviews and online surveys. The occupants live in dwellings manged by one of Wales's largest registered social housing landlords (RSL's) with circa 8000 dwellings. The challenges and solutions for undertaking occupant engagement for surveys during Covid19 restrictions are illustrated. Once dwelling retrofits are completed in 2022 and 2023, the intention is to conduct a second and third phase of occupant engagement.

Keywords: Occupant engagement · Quality of life · Fuel poverty · Optimised dwelling retrofit · Wales

1 Introduction

This paper discusses the WG funded OR project [1] led by Sero Life and Pobl Housing Association [2]. Researchers from Cardiff Metropolitan University are leading two sub-projects as part of the OR project, specifically to investigate occupant comfort and QoL, and a pilot study for one of these projects is discussed. Context to the need to retrofit and improve the operational energy performance of existing social housing in Wales is discussed, as is the methodology and the results of the occupant engagement assessing comfort and QoL. The research builds on two previous research projects that mapped the relationships between improvements in occupant thermal comfort and QoL and alleviation from fuel poverty from dwelling retrofit [3, 4].

J. Littlewood and R. J. Howlett (Eds.): SEB 2022, SIST 336, pp. 337–346, 2023.
https://doi.org/10.1007/978-981-19-8769-4_32

2 The Need for Optimised Dwelling Retrofit for Quality of Life

2.1 Carbon Reduction Targets for UK Housing

Mitigating and reducing global, national, and local climate change is one of the biggest challenges facing humanity for current and future generations [5]. The UK government has legislated the world-leading climate change commitment to significantly reduce carbon dioxide (CO_2) emissions by 78% by 2035 from 1990 levels, taking the UK more than three-quarters of the way to achieving net zero by 2050 [6]. The WG has also set carbon reduction targets in its Net Zero Carbon Plan for organisations, of 37% (2025), 63% (2030), 89% (2040), and net zero carbon by 2050 [7]. There are numerous challenges. The buildings and construction sector are responsible for 36% of the global final energy consumption and 37% CO_2 emissions related to energy use [8]. In the UK, the housing stock consumes a significant amount of energy accounting for 29% of anthropogenic carbon emissions. The Welsh dwellings account for 21% of the national CO_2 emissions, indicating the high level of emissions from industry in Wales, and increasing the challenge of achieving the carbon budgets [9]. Indeed, the housing stock in Wales is among the oldest and least energy efficient in Western Europe, with 32% of homes built before 1919, and just 6% built in the last 35 years, increasing energy demand for heating and reducing comfort. In 2022, the current rate of new build housing stock is 0.4%, meaning that more than 90% of the current dwellings will still be in use by 2050 [10]. So, new build housing will not have a major impact on the decarbonisation targets in a short term. In addition, despite the Welsh Government's energy efficiency initiatives, nearly a quarter of households are still facing fuel poverty, resulting in potential detrimental impacts on health and well-being of the occupants [11]. Furthermore, since the Covid-19 pandemic started in the UK in 2020 [12], more people have stayed at home or started working from home, leading to an increase in CO_2 emissions from the housing sector, with a growth rate of 1.8% in 2020 [13].

These statistics show that Welsh housing is the most important and challenging part of the UK to achieve the goals of housing decarbonisation, and dwelling retrofit plays a more significant role in reducing future energy consumption and CO_2 emissions than ever before [14].

2.2 Optimised Retrofit Programme: Pathfinder and Quality of Life

The OR Programme (ORP) was launched in 2020 by the WG, funded as part of the Innovative Housing Programme, aiming to decarbonise existing Welsh homes [15]. The aim is to develop effective, practical pathways to take the housing stocks to their lowest achievable carbon footprint and to develop a customised plan allowing for scale economies, standardisation, and to demonstrate how data analysis may optimise costs, improve performance outcomes, and provide vital lessons towards the decarbonisation for a wider programme [16]. It is a "whole house, pragmatic, approach to decarbonising existing homes", which considers the fabric or materials used in the housing stocks, along with how they are heated and how energy is stored and delivered [ibid].

Pobl Housing Group and Sero Life lead one of five OR projects known as Pathfinder, to retrofit 1768 existing homes managed by 27 registered social landlords (RSLs) in

Wales [2], with a consortium of 68 partners including Cardiff Metropolitan University. The ethos of Pathfinder has developed pathways to Zero and is designed to support incremental home upgrades over multiple steps, in a coordinated way, enabling delivery of a more comprehensive level of genuine zero carbon, with anticipated lower costs, by aligning with the decarbonisation of the UK energy grid [ibid].

There are 30 mini research projects in Sero Life's OR project, two of which are led by Cardiff Metropolitan University. The first mini project, which is the focus of this paper, is entitled "Are you comfortable Quality of Life (QoL) assessments" is investigating the impacts of dwelling retrofit on occupant QoL by capturing occupant data at both pre-retrofit and post retrofit stages on a variety of parameters. These include physical and mental health, indoor temperature, relative humidity and CO_2 levels. The second mini project entitled "Retrofit measure studies & analysis", is measuring how unwanted air leakage from/through the building fabric can impact on occupant mental wellbeing from sound transmission. Plus, also on physical wellbeing due to unwanted heat loss or gain and pathways for fire, smoke and toxic gas in the event of a building fire. This second project builds on previous research by Cardiff Met but is not the focus of this paper [17, 18].

These two mini projects will gain a better understanding of how to assess residents' health and wellbeing through a combination of occupant questionnaires and spot measurement of environmental conditions.

3 Methodology for QoL Engagement in OR Dwellings

3.1 Refinement to HSF36 Questionnaire

Since 2021, a structured questionnaire has been refined based on an adapted (hybrid) short form 36 (HSF-36) Health survey, which has been refined following a previous research project engaging with occupants [4]. It was piloted from March to May 2022, with the aim to examine the effectiveness of the occupant engagement approach and to identify the challenges and limitations that needed to be addressed.

The SF-36 questionnaire is "a set of generic, coherent, and easily administered QoL measures" [19], with indicators that rely on patient self-reporting, which are now commonly used by managed care organisations and Medicare for routine monitoring and assessment of adult patient care outcomes. The HSF-36 questionnaire enables researchers to assess eight QoL parameters of occupants living in dwellings that are part of retrofit programmes [20]. They are physical functioning and health, role limitations due to physical health, emotional health and wellbeing, role limitations due to emotional health, physical energy and fatigue, social functioning, pain, and general health. The data collected from the occupant engagement are converted into the eight dimensions listed above using a scoring algorithm. Using this scale, the respondent's health is transformed from zero to 100, where 0 indicates the worst possible health and 100 indicates the best possible health for the respondent [21].

The initial adapted QoL questionnaire combined the SF-36 questionnaire with general information about the dwelling, such as the property type, size, age, materials of the property and thermal comfort questions related to environmental conditions, such as temperature, relative humidity, ventilation, lighting, noise level and air quality, as well as questions regarding energy use at home and fuel poverty.

After several rounds of discussions, a shorter version of the QoL questionnaire was subsequently developed with property information pre-supplied by the RSLs and with reduced room options for some questions to improve occupant completion of the questionnaire. Besides, to understand occupants' perceptions of their home and energy efficiency as well as their concerns about the retrofit work, additional questions have been made by RSLs and attached to the end of the QoL questionnaire. Therefore, the final version of the QoL hybrid SF-36 questionnaire (HSF36) consists of four parts including occupant general information, indoor comfort and energy use at home, health and wellbeing (SF-36), and additional questions related to retrofit work and energy efficiency.

3.2 Case Study for Occupant QoL Engagement

Pobl Housing Group [22] and Valleys to Coast [23] are two of the 25 RSL partners in the Sero Life OR project provided case studies on two sites (Site A and Site B) totalling 40 dwellings, where most occupants are deemed to be in fuel poverty, i.e., 10% of household income is spent on heating [24]. Several meetings were held with both RSL partners in 2022 to review the HSF-36 questionnaire, leading to further refinement of the questionnaire entitled HSFv2-36 questionnaire. There are four sections covering occupant general information, indoor comfort including lighting and energy use, health and wellbeing, and impacts of retrofit work on daily living.

To commence engagement with the 40 households an information letter was sent to all dwellings by the RSL partners, introducing the OR project, the authors of this paper and the QoL survey. Due to Covid-19, a risk assessment was completed before permission was granted to commence the face-to-face occupant engagement process. The authors undertook visits to the 40 dwellings on several occasions in March and April 2022, at Site A to collate a picture of the dwellings and the occupants. During these visits occupants were offered several data collection options: structured face-to-face interviews (three agreed), postal questionnaires or email, or phone interviews. Ten occupants requested the questionnaire via email, and six by post. Site B residents were contacted directly by the RSL and arranged the face-to-face interview.

To increase the number of respondents the engagement methods were refined, including converting the questionnaire so that it could be completed online through a web-based survey software Qualtrics [25], which was emailed by both RSL partners to any remaining occupants in April 2022.

4 Results

4.1 Participants

Through the occupant engagement, a total of nine information-rich face-to-face interviews across two sites (Site A: n = 4; Site B: n = 5) were conducted between March and May 2022. However, one occupant at Site A withdrew in May 2022. Most participants were 55 years old or older. The interviews lasted between 45 min and one hour 55 min, and an average of 50 min. Two online surveys were received in April 2022 through Qualtrics. Therefore, data from ten households involved in the pilot study are presented in this paper. All the dwellings at Site A (RSL1) are pre-retrofit, and for Site B (RSL2) are post retrofit with a range of retrofit measures including external wall insulation (EWI), loft insulation, rood mounted solar photovoltaic, and door and window replacements. The heating provided at Site A is coal-fired with immersion heaters for hot water, and bottled gas, while mains natural gas central heating with electric showers were used for all properties at Site B. The participants who responded to the data collection options are listed in Table 1.

Table 1. Basic Information of QoL survey respondents

Site	Occupant	Sex	Age group	property type	Bedroom/s	Pre/Post retrofit	Survey type
A	A1	F	65+	Semi-detached house	2	Pre	Face-to-face
	A2	M	55–64	Semi-detached house	2	Pre	Face-to-face
	A3	M	65+	Semi-detached house	2	Pre	Face-to-face
	A4	F	45–54	Semi-detached house	2	pre	Online
B	B1	F	45–54	Mid terrace house	3	Post	Face-to-face
	B2	F	65+	Mid terrace house	3	Post	Face-to-face
	B3	M	65+	Flat	1	Post	Face-to-face
	B4	M	65+	Flat	1	Post	Face-to-face
	B5	M	65+	Flat	2	Post	Face-to-face
	B6	F	65+	Mid terrace house	3	Post	Online

4.2 RAND Analysis and Data Interpretation

The RAND scoring system [26] was used for the interpretation of occupant health and well-being results from the pilot study. Currently, the RAND system is considered as the most widely used survey instrument for gauging health related QoL (HRQoL) [27]. RAND provides a useful matrix for interpreting each of the eight concepts, and also to weigh individual occupants in comparison to each other. RAND has two steps. First, numeral values are recoded according to the scoring key. All items are scored from 0 to 100, with a higher score indicating a healthier state. Scores represent the percentage of the total possible score achieved. Second, the eight scale scores are calculated by averaging the items within the same scale [ibid]. Therefore, scale scores represent the average of the answers provided for all questions in the scale. For the data analysis, qualitative content analysis was applied for summaries of the interview data, combined with RAND analysis to evaluate the relationship between indoor environment, fuel poverty, energy use habits, and occupant health and well-being.

Due to the small number of participants completing the pilot surveys, the data was analysed using descriptive statistics and the results were interpreted on an individual basis. However, there are some differences in overall trends from the SF-36 results.

Figures 1 and 2 illustrate the comparisons of SF-36 subscale scores of occupants pre-retrofit (Site A) and post-retrofit (Site B).

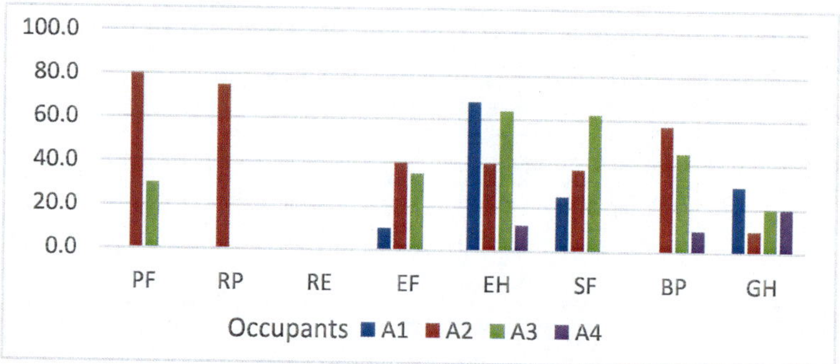

Fig. 1. Comparisons of SF-36 subscale scores of four occupants in Site A. PF = Physical Functioning, RP = Role Limitations due to physical health, RE = Role Limitations due to emotional problems, EF = Physical energy and fatigue, EH = Emotional health and well-being, SF = Social Functioning, BP = Bodily Pain, GH = General Health.

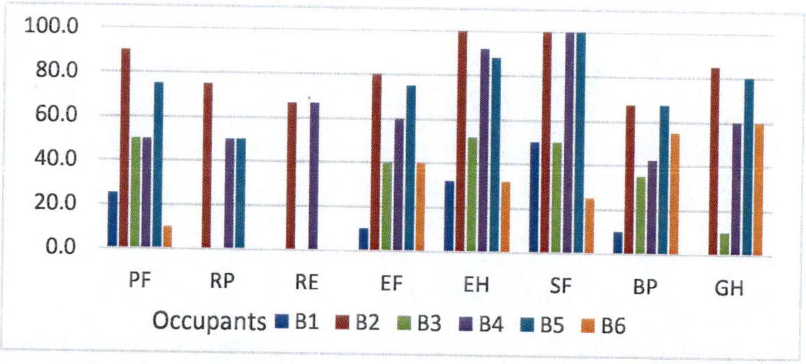

Fig. 2. Comparisons of SF-36 subscale scores of six occupants in Site B. PF = Physical Functioning, RP = Role Limitations due to physical health, RE = Role Limitations due to emotional problems, EF = Physical energy and fatigue, EH = Emotional health and well-being, SF = Social Functioning, BP = Bodily Pain, GH = General Health.

As can be seen from Figs. 1 and 2, the subscale scores of the post-retrofit occupants are generally higher than the pre-retrofit occupants at a group level. Meanwhile, on questions related to energy use, comfort and fuel poverty, post-retrofit Site B respondents generally responded more positively and scored higher on comfort satisfaction than pre-retrofit Site A respondents.

At an individual level, from the face-to-face interview data, Occupant A3 indicated they have two electric convector heaters as supplementary space heating and does not ventilate often in winter. A3's comfort satisfaction score is two for room temperature and three for RH, where the SF-36 result shows that A3 has lower subscale scores in RP, RE and GH. Occupant B3 also has lower subscale scores in RP, RE and GH, and comfort satisfaction scores across room temperature and humidity are three and two, respectively. B3 reported damp walls in the bedroom, condensation and mould in the bathroom and bedroom. Although different people's perception of temperature and RH is subjective, this finding may link poor housing temperature and RH to negative impacts on occupants' physical health.

Nearly all households surveyed were in fuel poverty [24]. The data from A4 and B6 are both collected through the online questionnaire. Their subscale scores in RE, EH and SF were lower than other occupants. For the energy use questions, they both responded by indicating they only activated the space heating when feeling cold and deactivate when leaving their homes. They also sacrificed other living options to save on heating bills. This finding may suggest that fuel poverty affects the mental health and general health of occupants to a certain extent.

For the occupants of the pre-retrofit Site A, only the dwelling occupied by A4 was installed with a smart meter. Most occupants did not understand how smart meters operate and thought it would consume more electricity than a conventional meter. In the post-retrofit Site B, all the residents who participated in the QoL survey had installed smart meters and indicated they can view and manage the energy use through it, which has also changed how they live in their homes to use less energy.

Occupants also expressed concerns about the retrofit work, for example Resident A3 pointed out that his furniture might be damaged if moved to accommodate the retrofit being undertaken. From the pilot study of occupant engagement, the online questionnaire had a lower response rate, and produced less information than the face-to-face interviews.

5 Discussion

The occupant engagement process provided an opportunity to discuss how occupants could benefit from the retrofitted home such as eradicating fuel poverty by considerably reducing energy use, improving occupant's comfort and QoL. Furthermore, occupant participation in the entire retrofit process will result in a better outcome after the retrofitted home is in use, thus improving the overall occupant QoL.

Occupant engagement is a complex and challenging process with many uncertainties. Even though the RSL's had had sent out information letters introducing the authors, the research project and the dates when face to face interviews would take place some occupants were surprised that they would be expected to have a physical face to face interview. The legacy of the Covid19 pandemic when social gatherings were prohibited could be one reason for the lack of occupant engagement. So, for the next steps in occupant engagement the authors and RSL's will also include other data collection and occupant engagement strategies, such as community meetings and welcoming occupants to empty properties that have their retrofit work completed prior to occupied dwellings as a form of 'living lab'. For the latter occupants would be invited to spend some time in the retrofitted dwellings to experience at first hand the retrofit improvements, with increased comfort and reduced energy use.

The occupant who withdrew did so as they did not want any disturbance to her living pattern during the retrofit works. So, relocating occupants to alternative accommodation during the retrofit work is under consideration.

By combining the RAND analysis of the SF-36 results with the interpretation of the face-to-face interview data, researchers will be able to understand the relationship between the physical and mental health of occupants and the internal environment of the housing as well as whether the retrofit affect occupant thermal comfort, energy use, and health and well-being.

A post-retrofit survey will be conducted upon completion of the retrofit work. In addition, environmental conditions will be monitored on site before, during and after the interview. By doing this, the environmental condition data including temperature, relative humidity, ventilation, lighting, air quality and solar radiation, etc. will be collected and compared with the interview results in order to evaluate the actual performance of buildings and their component parts, and whether they meet the predicted expectations.

6 Conclusion

This paper has discussed the context of the Optimised Retrofit project and the pilot study of the SF36 questionnaire and plan for the next stage. The gaps in knowledge, the process and the challenges of the occupant engagement have been discussed. The pilot study from two RSLs pre-retrofit and post-retrofit as well as the development of hybrid SF36 questionnaire and how the data were analysed through RAND analysis were presented. Through the pilot study, it can be concluded that the method from pilot study can be used to understand the link between the physical and mental health of occupants and the internal environment of the housing, and whether the retrofit has positive impacts on occupant thermal comfort, energy use, and health and well-being.

Acknowledgements. The research has been supported by the Welsh Government funded Optimised Retrofit (one) programme, managed by Pobl Housing Association and Sero Life. Thanks, are also given to the staff from Pobl and Valleys to Coast RSL's that manage and owns the 40 dwelling case studies, and to the occupants that engaged with the research team.

References

1. Welsh Government.: Optimised RetroFit Programme. https://gov.wales/optimised-retrofit-programme (2021). Accessed 10 June 2022
2. Sero Life. The Future – Optimised Retrofit.: https://www.optimised-retrofit.wales/ (2020). Accessed 10 June 2022

3. Atkinson, J., Littlewood, J.R., Geens, A.J., Karani, G.: Relieving fuel poverty in Wales with external wall insulation. J. Eng. Sustain. **170**(2), 93–101 (2016). https://doi.org/10.1680/jensu.15.00064
4. Jahic, D., Littlewood, J.R., Karani, G., Atkinson, J., Bolton, D.: Evaluating occupant wellbeing in retrofitted dwellings with the short form 36 questionnaire. In: Kaparaju, P., Howlett, R.J., Littlewood, J., Ekanyake, C., Vlacic, L. (eds.) KES-SEB 2018. SIST, vol. 131, pp. 359–368. Springer, Cham (2019). https://doi.org/10.1007/978-3-030-04293-6_35
5. IPCC.: IPCC Sith Assessment report—Impacts, Adaptation and Vulnerability (2022). https://www.ipcc.ch/report/ar6/wg2/
6. UK Government.: UK enshrines new target in law to slash emissions by 78% by 2035. https://www.gov.uk/government/news/uk-enshrines-new-target-in-law-to-slash-emissions-by-78-by-2035. Accessed 02 Feb. 2022
7. Welsh Government.: Net Zero Wales (2021). https://gov.wales/net-zero-wales
8. United Nations Environment Programme. 2021 Global Status Report for Buildings and Construction: Towards a Zero-emission, Efficient and Resilient Buildings and Construction Sector. Global Alliance for Buildings and Construction (2021)
9. ONS.: UK local authority carbon dioxide emissions estimates 2019. Department for Business, Energy & Industrial Strategy, p. 1, London (2021). https://assets.publishing.service.gov.uk/government/uploads/system/uploads/attachment_data/file/996056/2005-19-local-authority-co2-emissions-statistical-release.pdf
10. National Statistics. Dwelling Stock Estimates for Wales, 2019–2020. https://statswales.gov.wales/Catalogue/Housing/Dwelling-Stock-Estimates/dwellingstockestimatespercentages-by-year-tenure. Accessed 02 Feb 2022
11. Littlewood, J.R., Karani, G., Atkinson, J., Bolton, D., Geens, A.J., Jahic, D.: Introduction to a Wales project for evaluating residential retrofit measures and impacts on energy performance, occupant fuel poverty, health and thermal comfort. Energy Procedia **134**, 835–844 (2017)
12. British Foreign Policy Group. COVID-19 Timeline. https://bfpg.co.uk/2020/04/covid-19-timeline/. Accessed 10 June 2022
13. ONS.: 2020 UK greenhouse gas emissions, provisional figures. Department for Business, Energy & Industrial Strategy, London (2021). https://assets.publishing.service.gov.uk/government/uploads/system/uploads/attachment_data/file/972583/2020_Provisional_emissions_statistics_report.pdf
14. Jones, P., Li, X., Perisoglou, E., Patterson, J.: Five energy retrofit houses in South Wales. Energy and Buildings **154**, 335–342 (2017)
15. Welsh Government. Written Statement: The Optimised Retrofit Programme 2020–21. In: https://gov.wales/written-statement-optimised-retrofit-programme-2020-21. Accessed 04 Feb. 2022
16. Brown, D., Jaccarini, C., Kumar, C., Madge, C.: Financing Wales' Housing Decarbonisation (2021). https://neweconomics.org/uploads/files/Financing-Wales-Housing-Decarbonisation.pdf
17. Littlewood, J.R., Smallwood, I.: In-construction tests show rapid smoke spread across dwellings. J. Eng. Sustain. Themed Issue Sustain. Energy Build. Part 2 **170**(2), 102–112 (2017). https://doi.org/10.1680/jensu.15.00063
18. Littlewood, J.R., Alam, M., Goodhew, S.: A new methodology for the selective measurement of building performance and safety. Energy Procedia **111**, 338–346 (2017). https://www.sciencedirect.com/science/article/pii/S1876610217302254
19. Rand Health: 36-Item Short Form Survey Instrument (SF-36) (1969). https://www.rand.org/health-care/surveys_tools/mos/36-item-short-form.html
20. Bedini, S.A.: I. In: Bedini, S.A. (ed.) The Christopher Columbus Encyclopedia, pp. 313–386. Palgrave Macmillan UK, London (1992). https://doi.org/10.1007/978-1-349-12573-9_9

21. Hays, R.D., Shapiro, M.F.: An overview of generic health-related quality of life measures for HIV research. Qual. Life Res. **1**, 91–97 (1992)
22. Pobl Group.: Welcome to Pobl. https://www.poblgroup.co.uk/. Accessed 10 June 2022
23. Valleys to Coast. Welcome to Valleys to Coast. https://www.valleystocoast.wales/. Accessed 10 June 2022
24. Age Cymru.: Public Policy Statement—Fuel poverty (2017). https://www.ageuk.org.uk/globalassets/age-cymru/documents/policy/public-policy-statements/fuel-poverty-policy-statement---october-2017.pdf
25. Qualtrics XM.: https://www.qualtrics.com/uk/experience-management/. Accessed 28 Apr. 2022
26. 36-Item Short Form Survey (SF-36) Scoring Instructions (1969). https://www.rand.org/health-care/surveys_tools/mos/36-item-short-form/scoring.html. Accessed 10 June 2022
27. Ron, D.H., Leo, S.M.: The RAND-36 measure of health-related quality of life. Annal. Med. **33**(5), 350–357 (2001). https://doi.org/10.3109/07853890109002089

BIM-Based Workflow for Managing Multi-risk Factors of Open Spaces in Historical Built Environment

M. Angelosanti[1], M. Russo[1], A. D'Amico[1](✉), M. Pugnaletto[1], C. Paolini[1], E. Quagliarini[2], and E. Currà[1]

[1] La Sapienza University of Rome, Via Eudossiana 18, 00184 Rome, Italy
{marco.angelosanti,alessandro.damico}@uniroma1.it
[2] Università Politecnica Delle Marche, 60121 Ancona, Italy

Abstract. The problem of the Historical Built Environment (HBE) and risk factors representation in the Building Information Modelling (BIM)-based model is a debated topic in literature. The complex and variety of geometry and features of existing buildings and urban open spaces (OSs) raise the issue of their representation, both in terms of geometric model and information details. In the recent literature, much attention is given to the organization of workflow from the survey and data acquisition phase to the implementation phase, but most of them are focusing on single buildings or small aggregates. Moreover, there is a lack of studies that focuses on the mesoscale representation and its relationship with multi-risk assessments in BIM-based models, which include buildings, infrastructures, and Open Spaces (OSs). Considering the interoperability for the information exchange, the authors develop a specific workflow to support the setting up of integrated surveys, the related BIM models, and its risk factors parametrization to create a multi-hazard scenario. Parameters implementation in Revit BIM software and the opportunity to develop an interoperable standard are introduced and discussed.

Keywords: Built environment · Open spaces · Multi-risk scenario · Building information modelling · Narni (Italy)

1 Introduction and Research Aim

The Historical Built Environment (HBE)—composed of buildings, infrastructures, and Open Spaces (OSs)—is prone to many risks caused by disasters [1–4].

Nevertheless, most research focused mainly just on the performance of buildings and infrastructure [5], and few studies considered the OSs network that has a crucial role before, during, and after the disasters to satisfy survival needs [6]. Therefore, these OSs allow for adequate functionality and level of efficiency, both in ordinary conditions and for managing the critical phases of an emergency. Hence, the OSs can provide insights for seismic risk mitigation by considering the complex interactions between their physical components and their users [7].

J. Littlewood and R. J. Howlett (Eds.): SEB 2022, SIST 336, pp. 347–358, 2023.
https://doi.org/10.1007/978-981-19-8769-4_33

To characterize the risk level of the HBE, a variety of Built Environment Typologies (BETs) can be defined to represent typical and recurring HBEs in Italian territory [5].

In particular, a significant real-world case study is selected and deepened using digital BE models obtained from the Scan-to-BIM process.

Because of the above, the authors develop a workflow to collect BET information and implement risk parameters in BIM-based models, starting from the state of the art of the consolidated literature, codes, and standards about BIM geometry and metadata description [8–11]. Hence, the herein work evaluates BIM potential as a BET information repository, considering the information exchange needs among BIM models and simulation software. The process of BETs' representation through the BIM model includes three phases: geometric modelling, informative deepening, and their synthesis in BET parameters implementation. The approach has been tested in the case study of Narni, in Italy, affected by the earthquake and air pollution. Firstly, the Open Space of Piazza Dei Priori has been surveyed with laser scanning and photogrammetry providing an accurate point cloud, used to generate a BIM model of the content and the frontiers of the BE (Sect. 2). Successively, BE parameters have been inserted in this information three-dimensional model to be available for risk assessment procedures and risk mapping (Sect. 3). In parallel, the study proposes to structure the parameters' list according to GIS and BIM data model requirements. The digital reconstruction of the BE in BIM and the risk assessment outcomes are investigated (Sect. 4).

The present work is part of an Italian PRIN (Projects of Relevant National Interest), entitled BE S2ECURe "(make) Built Environment Safer in Slow and Emergency Conditions through behavioUral assessed/designed Resilient solutions" (Grant number: 2017LR75XK), supported by MIUR, the Italian Ministry of Education, University, and Research.

2 BIM Modelling Strategies Based on Scan-to-BIM Process

2.1 Case Study Rapid Point Cloud-Based Survey

The selected case study is Piazza dei Priori in Narni, Italy. Considering the classification proposed in D'Amico et al. [5] Piazza dei Priori belongs to the BET 2A, which presents a low level of compactness and regularity of the morphology, three special buildings, and the multi-hazard potential. Furthermore, this case study also allows evaluating the implementation of the parameters relating to the overturning of the fronts, having H_{max} greater than the width.

The expeditious survey of the BE defined the modelling strategy steps. Notably, the architectural survey is a necessarily critical operation, therefore the most appropriate instrumentation has been chosen according to the geometric-morphological needs of the selected case study. A photogrammetric and a Terrestrial Laser Scanner (TLS) survey were carried out, in order to make the methodology replicable with both instruments and obtain a point cloud for the Scan-to-BIM process [10].

The photogrammetric survey was performed with a mirrorless full-frame Sony A7. Moreover, the use of a UAV DJI Spark has allowed us to reach the upper floors and document the roofs. The shooting methodology has considered consolidated principles [5]. A frame horizontal overlap of 80% and a vertical one of 60% have been guaranteed during the acquisition of vertical and oblique photos of e.g., all the built fronts, the paving, the urban furniture. 561 photos have been acquired by an operator in a single day and the whole acquisition time has been 1 h and 20 min.

As mentioned above, the survey of the case study was also carried out with the use of the TLS. As for photogrammetry, priority was given to both accuracy and speed of survey operations. 9 scans were acquired at each access to the square, in the centre of the square and under the covered areas (e.g., porches and canopies). The TLS was set to acquire only the reflectance of the surveyed objects, without recording their colours.

2.2 BIM Modelling Strategies

Geometric and informative modelling of the Scan-to-BIM process is herein based on dual compatibility with the BIM and GIS environments. Specifically, the BIM model developed in this paper is GIS compliant for two main reasons:

- The BIM model is able to dialogue with geospatial data (i.e., OSM data) both in input and output. Indeed, the BIM reference system is corresponding with the reference cartographic system because both digital photogrammetry and TLS point clouds were geolocated in the cartographic system RDN2008/Zone 12 (N-E) (Identifier 6876) [12].
- Both the geometric entities and the list of parameters were arranged according to 2D-GIS logic and systematic modelling in BIM was then performed to allow a streamlined data transfer back to GIS as an output of the workflow.

Therefore, elements and parameters suitable for easy bi-directional conversion between 2D and 3D levels (i.e. based on point, line and surface) have been chosen [5]. As a first step, a division in High Level (HL) and Low Level (LL) elements has been made, identifying in the former the main synthetic elements that best represent the mesoscale of the OSs in the BE and that can allow an easy dialogue between GIS and BIM being a repository that hierarchically inherits and store the OS content and frontier information. Notably, the HL elements are the OS itself, the Interferent Structural Unit (SU_i) and the Interferent Structural Aggregate (SA_i), while the LL elements are all HL element components (e.g., wall, roofs, windows, road, parking, urban furniture) [13].

Nevertheless, studying SAs as a whole is a rather complex and time-consuming operation [14]. Moreover, the SU_Is composing an AS interfering with the OS being studied are often a small percentage of the total SUs due to the complexity of the historical BE, as demonstrated by the application to the case study selected in this paper. In addition, the results of structural analyses on BE performed starting from BIM modelling have highlighted that for complex structures composed of mutually interacting SUs, the seismic assessment should be performed at the scale of the building units, taking into account the effect of the adjacent masonry structures [15]. For the reasons listed above, translating the concept of LL and HL elements into LL and HL BIM families, specifically for HL families the geometric representation of SUs rather than SAs has been chosen. In this way, SAs are described by several adjacent SUs, taking into account the relationships among them.

Hence, the model elements associated with risk factors are selected based on a BE schematization for its BIM digitalization in the BIM authoring software Autodesk Revit. Specifically, the HL families are:

- **Space**: OS delimited by walls and Linear Spaces (LSs) corresponding to access between two or more US_i;
- **Generic Models (GM):** SU_i composing the frontier of the OS. GM family name is corresponding with US_i Xy, where X is the SA number and y is a letter that specifies a specific SU in the SA.

And the LL families are:

- **Wall**: frontiers walls composed of different SUs_i, specifying e.g., the core material, and finishes; windows and doors are hosted by walls;
- **Floor**: floor with different materials (i.e., grass) of the AS/LS. Floor elements do not include other LS or AS. Three different types of floors can be included in the model: uncovered (e.g., the square), covered and balcony;
- **Roof**: Four different types of SUs_i roofs can be included in the model: n/d, Flat (F), Non-Pushing (NP), Slightly Pushing (SP), Pushing (P);
- **Road**: Access family corresponding to access between two or more US_i;
- **Window and door**: openings in the OS frontiers without including decorative elements (e.g., cornices, mouldings);
- **Furniture**: Urban furniture modelled representing the footprint of e.g., fixed and temporary obstacles, monuments, fountains;
- **Parking**: parking lots in the OS.

Moreover, Table 1 shows the hierarchical relationship between LL and HL families for what concerns geometric and informative BE representation for interoperability between BIM and GIS environment. By addressing the mesoscale of OSs in BE, the analogy between Mandolesi's sub-building systems and components proposed by [10] to describe the relationship between a BIM family and its components is shifted to the relationship between Mandolesi's building system and sub-systems to describe the relationship between SUs/OSs and their composing elements.

Table 1. Hierarchical relationship between LL and HL families.

Low level (LL) family	→	High level (HL) family	BE represented
[Specific component properties]	→	[global + inherited LL family's properties]	
Wall, floor, topography, road, parking, furniture	→	Space	OS
Wall, floor, roof, window, door	→	Generic models (GM)	USi

3 Assessment of Risk Parameters in the BIM Process

3.1 Preliminary Selection of Risk Parameters

The procedure of BIM modelling is followed by risk parameters implementation. Risk parameters are preliminarily selected and categorized following previous research of BE S^2ECURe research project [1, 4] based on expert judgement and multi-hazard combination associated with the case study.

Moreover, each parameter is catalogued into two main categories, based on their BIM implementation:

- **Fillable**: parameters that need to be filled by the users;
- **Computable**: parameters that can be computable through the software.

Furthermore, each parameter is associated with a family category, either HL or LL family, according to the three implementation categories.

The first parameter implementation category includes global US/OS parameters that are fillable directly in HL families e.g., last intervention period, state of conservation, Parking area location and all environmental characteristics parameters. The second parameter implementation category includes parameters to be filled in LL families and inherited by HL families with specific synthetic operations (direct transposition, average, sum, count, minimum, maximum). Examples of such parameters are presence parameters and all specific parameters, mainly coming from the BIM environment simulations and computations (e.g., state of conservation, wall disconnection in plan, wall disconnection in elevation, masonry quality), which require an association to individual elements in order to be read synthetically and inherited into the OS/US. The last parameter implementation category includes parameters to be computed in HL families using geometric properties of LL families with specific synthetic operations (direct transposition, average, sum, count, minimum, maximum). The brackets mean that the family category is only used for its properties and the parameters are not associated with it. All the aforementioned concepts are summarised in Table 2.

3.2 BE S^2ECURe Property Set Implementation as Project Parameters Related to Specific BIM Family Categories

The parameters are inserted into the BIM model, developed in the Autodesk Revit environment, using "Shared Parameters". After creating the information structure of all the parameters necessary for the analysis of the multi-risk assessment within the BIM model, they must be filled in with the appropriate values. The integration of quantitative data derived from rapid point cloud-based survey or GIS/CAD data enrichment within the BIM model is considered. Hence, users and stakeholders can perform an analysis of the entire system on a shared virtual platform. The authors extract data from Revit elements and systematize them, so to develop a repeatable workflow for other BET multi-risk case studies.

Table 2. Reduced risk matrix for S-P multi-hazard condition. For each risk factor is specified the family category associated with according to the three main parameters implementation categories and the method for the implementation (fill: parameters that need to be filled by the users; **compute, in bold**: parameters that can be computed by the software).

BET multi-hazard parameters				
Description	Descriptor	Family category (low level)	LL to HL	Family category (high level)
Section 1: Main type				
Morpho-typology	**Canyon aspect ratio**	(Wall; floor)		Space
OS Dimension	**Width**	(Floor)		Space
H_{max} built front	**H max**	(Wall)	Max	Space
	Avg building height	(Wall)	Avg	Space
Section 2: Characteristics of geometry and space				
Frontier				
Type of aggregates	**Length of the built front**	(Wall)	Sum	Space
	Number of SU			GM
	Length of SU	(Wall)	Sum	GM
	Height of SU front	(Wall)	Avg	GM
	Number of storeys	Wall	Max	GM
Accesses	**Number**	(Road)	Count	Space
	Width	(Road)	Avg	Space
	Position/orientation (azimuth)	Road	Avg	Space
Special buildings	Presence	Wall	Direct	GM
	Length of built front	(Wall)	Sum	GM
	Height	(Wall)	Avg	GM
Porches	Presence	Floor	Direct	GM
	Linear extension	(Floor)	Sum	GM
Green area	Presence of green area	Floor	Direct	Space
	Green area position	Floor	Direct	Space
	Green area density	Floor	Direct	Space
Water	Presence of Water	Floor	Direct	Space
Quote differences/slope	Slope	Floor	Direct	Space

(*continued*)

Table 2. (*continued*)

BET multi-hazard parameters				
Description	Descriptor	Family category (low level)	LL to HL	Family category (high level)
Content				
Special buildings	Height			GM
	Lenght			GM
	Width			GM
Quote difference/slope	**Slope**	(Floor, topography)	Avg	Space
Green area	Presence of Green area	Floor	Direct	Space
	Extension (area)	Floor	Direct	Space
	Greenery adsorption capacity	Floor	Direct	Space
	Tree crown diameter	Floor	Direct	Space
Section 3: Constructive characteristics				
Frontier				
Homogeneity of built environ. Age	Last intervention period			GM
	State of conservation			GM
	Wall disconnection in plan	Wall	Direct	GM
	Wall disconnection in elevation	Wall	Direct	GM
Homogeneity of constructive techniques	Homogeneous/not homogeneous	Floor	Direct	Space
	Masonry quality	Wall	Min	GM
	Wall thickness	(Wall)	Avg	GM
	Roof types	Roof	Min	GM
	% openings	(Wall)	Avg	GM

(*continued*)

Table 2. (*continued*)

BET multi-hazard parameters				
Description	Descriptor	Family category (low level)	LL to HL	Family category (high level)
	No-structural protruding and decorative elements	Wall	Direct	GM
	Anti-seismic devices	Wall	Min	GM
	Facade finishing current roughness	Wall	Avg	GM
	Facade pollutant deposition capacity	Wall	Avg	GM
Content				
Pavement condition	Pavement finishing roughness	Floor	Direct	Space
Section 4: Characteristics of use				
Crowding	People present			Space
	Crowding potential			Space
	Exposure duration			Space
Accessibility for vehicle	Traffic intensity			Space
Parking	**Parking area location**	(Parking)	Sum	Space
Section 5: Environmental characteristics				
Seismic intensity	Ground motion severity			Space
	Seismic microzonation			Space
Climate conditions	Wind/breeze speed			Space
	Air temperature			Space
	Solar irradiation			Space
	Pollutant concentration			Space
Multi-hazard potential	Pollution sources presence Boolean			Space
Ground type	Ground roughness			Space

4 Results and Discussions

4.1 Piazza Dei Priori BIM Modelling

As for the BIM modelling, an average Level of Development (LoD) equal to 350 was chosen to appropriately represent the elements of Piazza dei Priori in Narni as it includes a more detailed definition of both geometrical and informative data from a Scan-to-BIM survey (Fig. 1). The frontier defining Piazza dei Priori has been divided into SAs and SUs. The 15 USs identified were then modelled as GM. The presence of internal courts was detected by means of a point cloud survey and GIS satellite data in Palazzo dei Priori (SU 2a), Loggia dei Priori (SU 6a) and Palazzo Calderini (SU 1a) and was modelled as extrusion voids. Starting from the faces of the GMs defining the square, 31 walls with different structural characteristics were modelled. The maximum number of walls associated with a single SU is 7 for the municipal theatre (SU 4a), while the minimum and the most recurrent case is 1 (SUs3b, 5b, 6b, 6c, 7a, 7c, 8a, 8b). The topography modelled from the point cloud shows a slight slope between SA 3 and 4. The presence of green areas is not recorded, but parking spaces and many elements of urban furniture, including a fountain, have been modelled. There are 8 accesses to the square and there is only one portico open to the square in SU 6, whose floor has been modelled. The OS frontier was completed with the various openings, whether windows or doors, for which a wide variety of typologies was recorded.

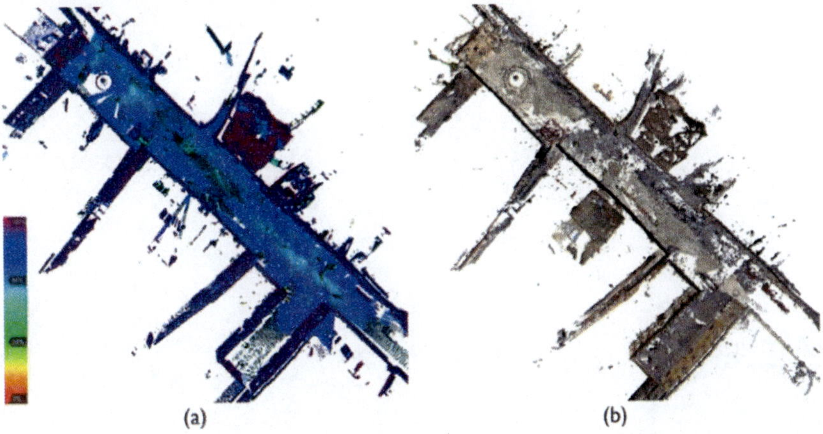

Fig. 1. Results of the TLS (**a**) and photogrammetric (**b**) point cloud survey of Piazza dei Priori, Narni. The TLS point cloud is in the reflectance intensity scale, and the photogrammetric cloud is in the RGB value scale.

Once the boundaries of the square were defined, the space was automatically generated in the BIM model. Hence, the model deepens the characterization of the building component of the OS frontier (e.g., openings, structure and finishing, types of roofs) and OS content (e.g., paving materials and finishing, topography materials and condition).

4.2 Implementation of Risk Parameters to Piazza Dei Priori BIM Model

The first parameters from Table 2 to be entered in the BIM model are the computable ones and the parameters of the HL families based on information from the geometry of the LL families can immediately be computed. This step highlights the straightforward computation automation of parameters, allowing parameters to be nested between LL elements and HL elements. Therefore, the process begins with LL element families by inserting the fields of the fillable parameters only, which are then inherited by HL families with specific synthetic operations (direct transposition, average, sum, count minimum, maximum). The last step is to fill in the HL family parameters, which are the only ones left empty in GMs and space. Figure 2 shows all the parameters entered at the end of the process in the specific space instance describing the square.

5 Conclusions and Insights

The representation of Historical Built Environment (HBE) in Building Information Modelling (BIM)-based model raises the issue of the implementation, both in terms of geometric model and information details. Considering the mesoscale nature of the OS in the BE, this paper presents a workflow to use the BIM-based models for a specific BET case study representation to collect the information for Built Environment Typologies (BETs) classification and analysis for the specific case study of Piazza dei Priori.

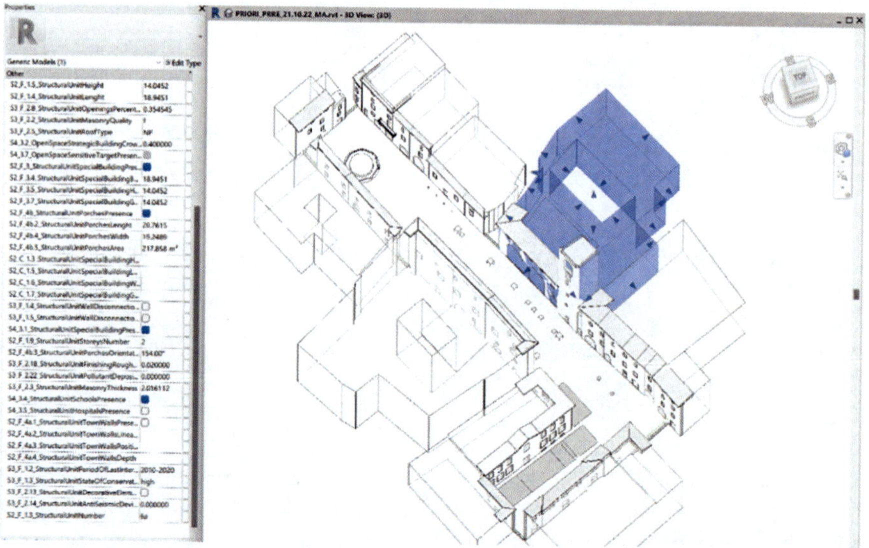

Fig. 2. BIM model of Piazza dei Priori: SUs are modelled as GM; the OS frontiers are composed of walls and roofs with windows, doors, and porticoes; the OS is modelled as floor and includes urban furniture and parking. On the right: Parameters associated with the HL GM family.

Starting from a selected risk parameters matrix for Seismic and Pollution (S-P) multi-hazard conditions, 5 categories of parameters were implemented in BIM, associated with the High Level (HL) and Low Level (LL) family categories. Specific recurrences in the parameter processing allowed to highlight 3 main parameter categories, to allow maximum automation in parameter input and inheritance, less error-prone as possible.

Moreover, to allow a more straight-lined comparison between all case studies a common ground between BIM and GIS software environment has been structured to describe the BE with interoperable geometric and informative elements, specifically the OS and the SU, which can be exported in 2D geometries to generate a GIS file with all the parameters inherited from the LL family.

Therefore, the present work presents implementation potentialities with the association of a novel BIM Property set to the GIS-based OpenStreetMap (OSM) parameters, to upload the GIS file and make it OPEN and freely available for the scientific community that can use to utilise the data collected, both at the urban tissue and building scale.

The representation of the BETs through BIM models is preparatory for the next steps of the research project for risk assessment considering human behaviour and the consequent interoperability needs for training and dissemination.

Acknowledgements. This work was supported by the MIUR (the Italian Ministry of Education, University, and Research) Project BE S^{2}ECURe—(make) Built Environment Safer in Slow and Emergency Conditions through behavioUral assessed/designed Resilient solutions (Grant number: 2017LR75XK).

References

1. Angelosanti, M., et al.: Towards a multi-risk assessment of open spaces and its users: a rapid survey form to collect and manage risk factors. Sustain. Energy Build. Proc. SEB 2021**263**, 10 (2022)
2. White, G.F., Kates, R.W., Burton, I.: Knowing better and losing even more: the use of knowledge in hazards management. Environ. Hazards **3**(3–4), 81–92 (2001)
3. Ferreira, T.M., Lourenco, P.B.: Disaster risk reduction and urban resilience: concepts, methods and applications. Resilient Struct. Infrastruct. (2019)
4. Russo, M., Angelosanti, M., Bernardini, G., Severi, L., Quagliarini, E.: Factors influencing intrinsic seismic risk of open spaces in existing built environment: a systematic review. Sustainability (preprint), 1–26 (2021)
5. D'Amico, A., et al.: Built environment typologies prone to risk: a cluster analysis of open spaces in Italian cities. Sustainability **13**(16), 9457 (2021)
6. Allan, P., Bryant, M., Wirsching, C., Garcia, D., Teresa Rodriguez, M.: The influence of urban morphology on the resilience of cities following an earthquake. J. Urban Des. **18**(2), 242–262 (2013)
7. Shrestha, S.R., Sliuzas, R., Kuffer, M.: Open spaces and risk perception in post-earthquake Kathmandu city. Appl. Geogr. **93**, 81–91 (2018)
8. CEN European Committee for Standardization, Organization and digitization of information about buildings and civil engineering works, including building information modelling (BIM)—Information management using building information modelling—Part 1: concepts and principles, p. 34 (2018)
9. Brumana, R., Banfi, F., Cantini, L., Previtali, M., Della Torre, S.: Hbim level of detail-geometry-accuracy and survey analysis for architectural preservation. ISPRS Ann. Photogr. Remote Sens. Spat. Inf. Sci. **42**(2/W11), 293–299 (2019)

10. Currà, E., D'Amico, A., Angelosanti, M.: Representation and knowledge of historic construction: HBIM for structural use in the case of villa Palma Guazzaroni in Terni. TEMA Technol. Eng. Mater. Archit. **7**(1) (2021)
11. Currà, E., D'Amico, A., Angelosanti, M.: HBIM between antiquity and industrial archaeology: former segrè papermill and sanctuary of hercules in Tivoli. Sustainable **14**(3) (2022)
12. Istituto Geografico Militare (IGM) Direzione Geodetica.: Nota per il corretto utilizzo dei sistemi geodetici di riferimento all'interno dei software GIS (2019)
13. Commissione tecnica per la microzonazione sismica, Manuale per l'analisi della Condizione Limite per l'Emergenza (CLE) dell'insediamento urbano (2014)
14. Valente, M., Milani, G., Grande, E., Formisano, A.: Historical masonry building aggregates: advanced numerical insight for an effective seismic assessment on two row housing compounds. Eng. Struct. **190**, 360–379 (2019)
15. Ponte, M., Bento, R., Vaz, S.D.: A multi-disciplinary approach to the seismic assessment of the national palace of Sintra **15**(5), 757–778. https://doi.org/10.1080/15583058.2019.1648587 (2019)

Experimental Assessment of a Preliminary Rule-Based Data-Driven Method for Fault Detection and Diagnosis of Coils, Fans and Sensors in Air-Handling Units

Mohammad El Youssef(✉), Francesco Guarino, Sergio Sibilio, and Antonio Rosato

Department of Architecture and Industrial Design, University of Campania Luigi Vanvitelli, Via San Lorenzo 4, 81031 Aversa, Italy
mohammad.elyoussef@unicampania.it

Abstract. Data-driven Automated Fault Detection and Diagnosis (AFDD) methods represent one of the most promising options for improving energy, environmental and economic performance of Air-Handling Units (AHUs). In this paper, a curated experimental faulted and unfaulted dataset associated to the field operation of a typical real AHU is firstly presented; a new rule-based data-driven AFDD method for fault detection and diagnosis of coils, fans and sensors is developed and its accuracy has been assessed in contrast with measured data.

Keywords: Air-handling unit · Data-driven fault detection and diagnosis · Rule-based method

1 Introduction

Heating Ventilation and Air-Conditioning (HVAC) systems equipped with Air-Handling Units (AHUs) are responsible of about $50 \div 60\%$ of energy demand of the building sector and $10 \div 20\%$ of the overall energy consumption [1, 2]. According to [1, 2], they are frequently operated in faulty conditions due to lack of proper maintenance, failure of components or incorrect installation. Faulty operation in AHUs could lead to significant problems, such as uncomfortable indoor environment, poor indoor air quality, larger greenhouse gas emissions and significant waste of energy and operating costs. Automated Fault Detection and Diagnosis (FDD) processes make it possible to automatically recognize fault occurrence and identify the causes and the location of faults; therefore, it could represent an effective viable solution to overcome the above-mentioned issues. Several scientific studies [1–3] underline that AFDD procedures applied to AHUs to rectify faulty operation can lead to an energy saving of about $20 \div 30\%$. AFDD methods can be knowledge-based or data-driven; the data-driven methods detect faults by analyzing the changes of selected variable patterns (derived from experimental and/or simulation data) and searching the causes according to similarities with pre-labelled anomaly libraries. Recently, the data-driven approach gained great interest thanks to the facts

J. Littlewood and R. J. Howlett (Eds.): SEB 2022, SIST 336, pp. 359–370, 2023.
https://doi.org/10.1007/978-981-19-8769-4_34

that it could be supported by various computer-aided techniques and it can be applied also when the physics-based knowledge is not wide enough [4, 5]. Although data-driven AFDD approach could be a powerful option to ensure efficient AHU operations and the related technology is maturing, it is still in the relatively early stage of adoption stock-wide. In particular, according to [1–5], there is a lack of standard datasets, methodologies and metrics for developing and assessing the accuracy of data-driven AFDD tools for AHUs. In this study, a curated experimental faulted and unfaulted testing dataset associated to the field operation of a typical AHU has been created (Sect. 2); then, a new rule-based data-driven AFDD tool has been developed based on an experimental fault free training dataset (Sect. 3) and its accuracy has been finally assessed in contrast with an experimental training database obtained under both fault free and faulty conditions (Sect. 4).

2 Fault Free and Faulty Experimental Testing Database

The operation of the single-duct dual-fan constant air volume AHU serving the 4 m × 4 m × 3.6 m test room of the SENS i-Lab of the Department of Architecture and Industrial Design of the University of Campania Luigi Vanvitelli (Aversa, Italy) has been experimentally investigated. Details about the characteristics of the AHU and related control logics, measuring sensors and served test room can be found in [6, 7]. A huge experimental campaign has been completed under fault free as well as faulty conditions under both summer and winter scenarios to obtain the testing database for assessing the performance of the proposed data-driven AFDD method. In particular, 14 daily tests (N1S, N2S, N3S, N4S, N5S, N6S, N1W, N2W, N3W, N4W, N5W, N6W, N7W, N8W) have been performed under normal (N) conditions during both the summer 2021 (S) and the winter 2021/22 (W). In addition, 22 daily faulty (F) tests (F1S, F2S, F3S, F4S, F5S, F6S, F7S, F8S, F9S, F10S, F11S, F1W, F2W, F3W, F4W, F5W, F6W, F7W, F8W, F9W, F10W, F11W) have been carried out during both the summer 2021 (S) and the winter 2021/22 (W) while manually forcing the operation of specific AHU components or sensors to artificially simulate the occurrence of 7 specific typical fault types at different severities (for a total of 11 faulty cases). Faults have been investigated one by one: (1) the fault 1 (F1) corresponds to the post-heating coil (PostHC) valve stuck at 0% (always closed); (2) the fault 2 (F2) corresponds to the PostHC valve stuck at 100% (always open); (3) the fault 3 (F3) corresponds to the cooling coil (CC) valve stuck at 0% (always closed); (4) the fault 4 (F4) corresponds to the CC valve stuck at 100% (always open); (5) the fault 5 (F5) corresponds to the humidifier (HUM) valve stuck at 0% (always closed); (6) the fault 6 (F6) corresponds to the case of a positive offset equal to +3.0 °C purposely added on the return air temperature sensor output; (7) the fault 7 (F7) corresponds to a negative offset equal to −3.0 °C artificially subtracted from the return air temperature sensor output; (8) the fault 8 (F8) corresponds to a positive offset equal to +15% purposely added on the return air relative humidity sensor output; (9) the fault 9 (F9) corresponds to a negative offset equal to −15% artificially subtracted from the return air relative humidity sensor output; (10) the fault 10 (F10) corresponds to the failure of the return air fan (RAF); (11) the fault 11 (F11) corresponds to the failure of the supply air fan (SAF). All tests have been performed from 9:00 a.m. up 6:00 p.m. while measuring all key operating parameters.

3 Rule-Based Data-Driven Fault Detection and Diagnosis Method

In this study an automated rule-based data-driven method for the AFDD of typical AHUs has been developed with the aim of recognizing the 7 fault types at different severity levels (for a total of 11 faulty cases) mentioned in Sect. 2. The method has been designed to estimate on-line the states of AHU components/sensors; the difference between measured data (assumed as inputs) and estimated values at each time step is derived and compared to threshold values in order to discover the possible presence of faults and related severity. The rules have been derived from the results of specific training experiments performed under fault free conditions and their accuracy has been assessed by contrasting their predictions with data measured under 36 normal and faulty AHU operating scenarios.

3.1 Rules for Detecting the AHU Operating Mode and the Valves' Opening Percentage

Specific rules have been developed to predict the actual AHU operating mode under fault free conditions among these 4 possible scenarios: Mode 1 (cooling + dehumidification); Mode 2 (cooling + humidification); Mode 3 (heating + dehumidification); Mode 4 (heating + humidification). This task is performed according to the AHU control logics and based on the conditions reported in the first 2 columns of Table 1 associated to the actual measured return air temperature T_{RA}, actual measured return air relative humidity RH_{RA}, selected indoor air temperature target $T_{SP,Room}$ (and related deadband DB_T) and selected indoor air relative humidity target $RH_{SP,Room}$ (and related deadband DB_{RH}). According to the predicted AHU operating mode, the opening percentages of (i) cooling coil valve $OP^{Pred}_{V\_CC}$, (ii) post-heating coil valve $OP^{Pred}_{V\_PostHC}$ and (iii) humidifier valve $OP^{Pred}_{V\_HUM}$ are predicted as specified in the last 3 columns of Table 1.

3.2 Rules for Detecting and Diagnosing the Supply/Return Air Fan Stuck

In order to identify the rules for detecting and diagnosing the supply/return air fan stuck, specific training experiments have been carried out under fault free conditions while measuring the electrical power consumption of the supply and return air fans (P^{Exp}_{SAF} and P^{Exp}_{RAF}, respectively) upon varying the experimental supply and return air fan velocities (OL^{Exp}_{SAF} and OL^{Exp}_{RAF}, respectively) from 0 to 100% with a step of 10%. The following equations, derived by interpolating the data measured during the fault free training tests, provide, respectively, the predicted electrical power consumption of the supply (P^{Pred}_{SAF}) and return (P^{Pred}_{RAF}) air fans as a function of OL^{Exp}_{SAF} and OL^{Exp}_{RAF}, respectively:

$$\begin{aligned}P^{Pred}_{SAF} =& 0.00001175 \cdot \left(OL^{Exp}_{SAF}\right)^4 - 0.00054568 \cdot \left(OL^{Exp}_{SAF}\right)^3 \\ &+0.11949019 \cdot \left(OL^{Exp}_{SAF}\right)^2 + \\ &-0.89476545 \cdot \left(OL^{Exp}_{SAF}\right) + 89.71257596 \end{aligned} \quad (1)$$

Table 1. Rules for predicting AHU operating mode and valves' opening percentages.

AHU operating mode	Conditions for mode ON	Conditions for mode OFF	$OP^{Pred}_{V\_CC}$ (%)	$OP^{Pred}_{V\_PostHC}$ (%)	$OP^{Pred}_{V\_HUM}$ (%)
Mode 1	$T_{RA} \geq (T_{SP,Room} + DB_T)$ AND $RH_{RA} \geq (RH_{SP,Room} + DB_{RH})$	$T_{RA} \leq (T_{SP,Room} - DB_T)$ AND $RH_{RA} \leq (RH_{SP,Room} - DB_{RH})$	100	0	0
Mode 2	$T_{RA} \geq (T_{SP,Room} + DB_T)$ AND $RH_{RA} \leq (RH_{SP,Room} - DB_{RH})$	$T_{RA} \leq (T_{SP,Room} - DB_T)$ AND $RH_{RA} \geq (RH_{SP,Room} + DB_{RH})$	100	0	100
Mode 3	$T_{RA} \leq (T_{SP,Room} - DB_T)$ AND $RH_{RA} \geq (RH_{SP,Room} + DB_{RH})$	$T_{RA} \leq (T_{SP,Room} - DB_T)$ AND $RH_{RA} \leq (RH_{SP,Room} - DB_{RH})$	100	100	0
Mode 4	$T_{RA} \leq (T_{SP,Room} - DB_T)$ AND $RH_{RA} \leq (RH_{SP,Room} - DB_{RH})$	$T_{RA} \leq (T_{SP,Room} - DB_T)$ AND $RH_{RA} \geq (RH_{SP,Room} + DB_{RH})$	0	100	100

$$P_{RAF}^{Pred} = -0.00000014 \cdot \left(OL_{RAF}^{Exp}\right)^5 + 0.00002882 \cdot \left(OL_{RAF}^{Exp}\right)^4 \\ -0.00137993 \cdot \left(OL_{RAF}^{Exp}\right)^3 + \\ +0.04013849 \cdot \left(OL_{RAF}^{Exp}\right)^2 \\ -0.15186884 \cdot \left(OL_{RAF}^{Exp}\right) + 17.10118159 \quad (2)$$

The maximum absolute percentage difference between the experimental electrical power consumption of the supply air fan and the corresponding prediction (Eq. 1) is 6.4%, while the maximum absolute percentage difference between the measured electrical power consumed by the return air fan and the predicted value (Eq. 2) is 3.5%. The following Eqs. 3 and 4 (derived by interpolating the data measured during the fault free training tests) provide the predicted supply and return air fan velocities (OL_{SAF}^{Pred} and OL_{RAF}^{Pred}, respectively) as a function of P_{SAF}^{Exp} and P_{RAF}^{Exp}, respectively:

$$OL_{SAF}^{Pred} = 30.45438 \cdot \ln\left(P_{SAF}^{Exp}\right) - 128.89480 \quad (3)$$

$$OL_{RAF}^{Pred} = 26.61240 \cdot \ln\left(P_{RAF}^{Exp}\right) - 66.64871 \quad (4)$$

The comparison between the experimental supply air fan velocity and the predicted value (Eq. 3) indicated a maximum absolute difference of 7.2%, together with a maximum absolute difference between the experimental return air fan velocity and the predicted value (Eq. 4) equal to 8.5%. According to the actual values of the electrical power consumption of both the supply and return air fans (assumed to be measured and provided as inputs), the proposed AFDD algorithm calculates the following parameters every time-step: (i) the arithmetic mean of the electrical power consumption of both the supply and return air fans measured in the previous hour ($P_{SAF,AVG,h}^{Exp}$ and $P_{RAF,AVG,h}^{Exp}$), and (ii) the arithmetic mean of both the supply and return air fan velocities (predicted by Eqs. 3 and 4, respectively) in the previous hour ($OL_{SAF,AVG,h}^{Exp}$ and $OL_{RAF,AVG,h}^{Exp}$). In addition, based on the actual values of both the supply and return air fan velocities (assumed to be measured and provided as inputs), the proposed AFDD algorithm calculates the following parameters every time-step: (i) the arithmetic mean of the predicted electrical power consumption of the supply air fan (Eq. 1) in the previous hour $P_{SAF,AVG,h}^{Pred}$ and (ii) the arithmetic mean of the predicted electrical power consumption of the return air fan (Eq. 2) in the previous hour $P_{RAF,AVG,h}^{Pred}$. Table 2 reports the rules developed for detecting and diagnosing the supply/return air fan stuck as well as its severity in the 36 fault free and faulty tests of the testing dataset (Sect. 2), where $OL_{SAF,AVG,h}^{Pred}$ and $OL_{RAF,AVG,h}^{Pred}$ are the hourly averages of the predicted supply and return air fan velocities, respectively. The threshold parameters ε_{SAF} and ε_{RAF} have been assumed equal to 37 W and 9 W, respectively, according to (i) the maximum error of the sensors used for measuring the electrical power consumption and (ii) the maximum difference between experimental and predicted values detected during the 14 fault free tests of the testing dataset (Sect. 2).

Table 2. Detection and diagnosis rules for the supply/return air fans stuck.

Condition to be satisfied	Fault diagnosis	Fault severity
$\left\lvert P_{SAF,AVG,h}^{Exp} - P_{SAF,AVG,h}^{Pred}\right\rvert > \varepsilon_{SAF}$	Supply air fan stuck	$OL_{SAF,AVG,h}^{Pred}$
$\left\lvert P_{SAF,AVG,h}^{Exp} - P_{SAF,AVG,h}^{Pred}\right\rvert \le \varepsilon_{SAF}$	Normal operation of supply air fan	Not applicable
$\left\lvert P_{RAF,AVG,h}^{Exp} - P_{RAF,AVG,h}^{Pred}\right\rvert > \varepsilon_{RAF}$	Return air fan stuck	$OL_{RAF,AVG,h}^{Pred}$
$\left\lvert P_{RAF,AVG,h}^{Exp} - P_{RAF,AVG,h}^{Pred}\right\rvert \le \varepsilon_{RAF}$	Normal operation of return air fan	Not applicable

3.3 Rules for Detecting and Diagnosing the Cooling/Post-heating Coil Valve Stuck

In order to identify the cooling and post-heating coil valve stuck, additional specific fault free training tests have been carried out to measure the heat carrier fluid volumetric flowrate flowing through the cooling and post-heating coil ($\dot{V}_{CC}^{Exp}$ and $\dot{V}_{PostHC}^{Exp}$, respectively) upon varying the measured opening percentage of the cooling and post-heating coil valves ($OP_{V\_CC}^{Exp}$ and $OP_{V\_PostHC}^{Exp}$, respectively), manually regulated from 0 to 100% with a step of 10%. The following equations, providing the predicted flowrate through the cooling coil $\dot{V}_{CC}^{Pred}$ (Eq. 5) and post-heating coil $\dot{V}_{PostHC}^{Pred}$ (Eq. 6) as a function of $OP_{V\_CC}^{Exp}$ and $OP_{V\_PostHC}^{Exp}$, have been developed by interpolating the data measured during the fault free training tests:

$$\begin{aligned}\dot{V}_{CC}^{Pred} = & 8.7634 \cdot 10^{-15} \cdot \left(OP_{V\_CC}^{Exp}\right)^6 - 2.7851 \cdot 10^{-12} \cdot \left(OP_{V\_CC}^{Exp}\right)^5 \\ & +3.5266 \cdot 10^{-10} \cdot \left(OP_{V\_CC}^{Exp}\right)^4 + \\ & -2.2881 \cdot 10^{-8} \cdot \left(OP_{V\_CC}^{Exp}\right)^3 + 7.2459 \cdot 10^{-7} \cdot \left(OP_{V\_CC}^{Exp}\right)^2 \\ & -1.1845 \cdot 10^{-7} \cdot \left(OP_{V\_CC}^{Exp}\right)\end{aligned} \tag{5}$$

$$\begin{aligned}\dot{V}_{PostHC}^{Pred} = & 8.8760 \cdot 10^{-15} \cdot \left(OP_{V\_PostHC}^{Exp}\right)^6 - 2.8725 \cdot 10^{-12} \cdot \left(OP_{V\_PostHC}^{Exp}\right)^5 \\ & +3.6481 \cdot 10^{-10} \cdot \left(OP_{V\_PostHC}^{Exp}\right)^4 + \\ & -2.3006 \cdot 10^{-8} \cdot \left(OP_{V\_PostHC}^{Exp}\right)^3 + 6.7718 \cdot 10^{-7} \cdot \left(OP_{V\_PostHC}^{Exp}\right)^2 \\ & +8.1955 \cdot 10^{-7} \cdot \left(OP_{V\_PostHC}^{Exp}\right)\end{aligned} \tag{6}$$

The comparison between the experimental values of $\dot{V}_{CC}^{Exp}$ and the corresponding predicted values (Eq. 5) highlighted a maximum absolute difference of 3.8%, together with a maximum absolute difference between the measured values of $\dot{V}_{PostHC}^{Exp}$ and the predicted data (Eq. 6) equal to 3.7%. According to the actual volumetric flowrates entering the cooling and post-heating coils (assumed to be measured and provided as inputs), every time-step the proposed AFDD method calculates the hourly average of the measured volumetric flowrates entering the cooling coil and the post-heating coil ($\dot{V}^{Exp}_{CC,AVG,h}$ and $\dot{V}^{Exp}_{PostHC,AVG,h}$, respectively). In addition, based on the actual opening percentages of the cooling and post-heating coil valves (assumed to be measured and provided as inputs), every time-step the proposed AFDD tool also calculates: (i) the hourly average of the predicted volumetric flowrates entering the cooling coil $\dot{V}^{Pred}_{CC,AVG,h}$ (Eq. 5) and the post-heating coil $\dot{V}^{Pred}_{PostHC,AVG,h}$ (Eq. 6), (ii) the hourly average of the predicted (based on Table 1) opening percentages of the cooling coil valve $OP^{Pred}_{V\_CC,AVG,h}$ and the post-heating coil valve $OP^{Pred}_{V\_PostHC,AVG,h}$. Table 3 reports the rules for detecting and diagnosing the cooling/post-heating coil valve stuck and its severity in the 36 fault free and faulty tests of the testing dataset (Sect. 2). The threshold parameters ε_{CC} and ε_{PostHC} have been assumed equal to 10% of the $\dot{V}^{Exp}_{CC,AVG,h}$ and $\dot{V}^{Exp}_{PostHC,AVG,h}$, respectively, based on the accuracy of flowrate measurements (2% of reading) and the maximum difference between experimental and predicted values detected in the 14 fault free testing tests (Sect. 2).

Table 3. Detection and diagnosis rules for the cooling/post-heating coil valve stuck.

Condition to be satisfied	Fault diagnosis	Fault severity
$\left\|\dot{V}^{Exp}_{CC,AVG,h} - \dot{V}^{Pred}_{CC,AVG,h}\right\| > \varepsilon_{CC}$	Cooling coil valve stuck	$OP^{Pred}_{V\_CC,AVG,h}$
$\left\|\dot{V}^{Exp}_{CC,AVG,h} - \dot{V}^{Pred}_{CC,AVG,h}\right\| \leq \varepsilon_{CC}$	Normal operation of the cooling coil valve	Not applicable
$\left\|\dot{V}^{Exp}_{PostHC,AVG,h} - \dot{V}^{Pred}_{PostHC,AVG,h}\right\| > \varepsilon_{PostHC}$	Post-heating coil valve stuck	$OP^{Pred}_{V\_PostHC,AVG,h}$
$\left\|\dot{V}^{Exp}_{PostHC,AVG,h} - \dot{V}^{Pred}_{PostHC,AVG,h}\right\| \leq \varepsilon_{PostHC}$	Normal operation of the post-heating coil valve	Not applicable

3.4 Rules for Detecting and Diagnosing the Humidifier Valve Stuck

The detection and diagnosis of the humidifier valve fault is based on the calculation (every time-step) of the hourly average of the measured ($OP^{Exp}_{V\_HUM,AVG,h}$) and predicted ($OP^{Pred}_{V\_HUM,AVG,h}$) opening percentages of the humidifier valve (assumed to be measured and provided as inputs). Table 4 reports the rules for detecting and diagnosing the humidifier valve stuck (in this case the fault severity cannot be defined) in the 36 fault free and faulty tests of the testing dataset, where $\varepsilon_{HUM} = 10\%$.

Table 4. Detection and diagnosis rules for the humidifier valve stuck.

Condition to be satisfied	Fault diagnosis
$\left\|OP^{Exp}_{V\_HUM,AVG,h} - OP^{Pred}_{V\_HUM,AVG,h}\right\| > \varepsilon_{HUM}$	Humidifier valve stuck
$\left\|OP^{Exp}_{V\_HUM,AVG,h} - OP^{Pred}_{V\_HUM,AVG,h}\right\| \leq \varepsilon_{HUM}$	Normal operation of the humidifier valve

3.5 Rules for Detecting and Diagnosing the Return Air Temperature/Relative Humidity Sensor Offset

Under the hypotheses of perfect mixing of the air in the test room, the measured return air temperature/relative humidity and the measured room air temperature/relative humidity should be equal; however, this condition usually fails in real case studies. The experimental data associated to the 14 fault free tests of the testing database (Sect. 2) indicate that (i) the difference between the measured room air temperature and the measured return air temperature ranges from 0.30 °C up to 1.50 °C, while (ii) the difference between the measured room air relative humidity and the measured return air relative humidity is almost negligible. In order to detect and diagnose the return air temperature/relative humidity sensor offset, the AFDD tool developed in this study firstly calculates every time-step the hourly averages of (i) the measured room air temperature $T_{Room,AVG,h}$, (ii) the measured return air temperature $T_{RA,AVG,h}$, (iii) the measured room air relative humidity $RH_{Room,AVG,h}$ and (iv) the measured return air relative humidity $RH_{RA,AVG,h}$. Then, the following conditions (Eqs. 7 and 8) are checked:

$$\left|T_{RA,AVG,h} - T_{Room,AVG,h}\right| > \varepsilon_T \tag{7}$$

$$\left|RH_{RA,AVG,h} - RH_{Room,AVG,h}\right| > \varepsilon_{RH} \tag{8}$$

where the temperature threshold ε_T has been assumed equal to 1.50 °C (based on the experimental results) and the relative humidity threshold ε_{RH} has been considered equal to 2% (based on the patterns associated to the 14 fault free tests of the testing dataset described in Sect. 2). In the case of Eq. 7 is verified, a temperature offset is detected by the AFDD tool; similarly, in the case of Eq. 8 is verified, a relative humidity offset is detected by the AFDD tool; however, Eqs. 7 and 8 can't identify which sensor, between the room air sensor and the return air sensor, is faulty. Therefore, in order to understand which sensor is faulty, a second diagnosis phase is carried out by applying the first law of thermodynamics to the mixing chamber of the AHU (Eq. 9) as follows:

$$\dot{m}_{RIC,AVG,h} \cdot h^{theoretical}_{Room,AVG,h} = \dot{m}_{MA,AVG,h} \cdot h_{MA,AVG,h} - \dot{m}_{OA,AVG,h} \cdot h_{OA,AVG,h} \tag{9}$$

where $\dot{m}_{RIC,AVG,h}$ and $h^{theoretical}_{Room,AVG,h}$ are, respectively, the hourly averages of (i) the recirculation air mass flowrate and (ii) the theoretical room air specific enthalpy. Equation 9 allows to obtain the parameter $\dot{m}_{RIC,AVG,h} \cdot h^{theoretical}_{Room,AVG,h}$ based on the hourly averages of (i) the measured mixed air mass flowrate $\dot{m}_{MA,AVG,h}$, (ii) the mixed air specific enthalpy $h_{MA,AVG,h}$ (calculated based on the measured mixed air temperature/relative humidity),

(iii) the measured outside air mass flowrate $\dot{m}_{OA,AVG,h}$ and (iv) the outside air specific enthalpy $h_{OA,AVG,h}$ (calculated based on the measured outside air temperature/relative humidity); then, the parameters $\dot{m}_{RIC,AVG,h} \cdot h_{Room,AVG,h}$ and $\dot{m}_{RIC,AVG,h} \cdot h_{RA,AVG,h}$ are calculated every time-step by assuming $\dot{m}_{RIC,AVG,h}$, $h_{Room,AVG,h}$ and $h_{RA,AVG,h}$, respectively, equal to the hourly average of (i) the measured recirculation air mass flowrate, (ii) the room air specific enthalpy (calculated based on the measured room air relative humidity and the difference between the measured room air temperature and ε_T) and (iii) the return air specific enthalpy (calculated based on the measured return air temperature and relative humidity). Table 5 reports the rules used for detecting and diagnosing the offsets of the return air temperature/relative humidity sensors (together with their severity) in the 36 faulted and unfaulted tests of the testing dataset (Sect. 2).

4 Results and Discussion

The performance assessment of the proposed AFDD method consisted of 3 main steps: (1) curate the set of input samples drawn from the data measured during the 36 tests of the testing dataset (Sect. 2); (2) assign ground truth information to each input, e.g. faulted or unfaulted, and, if faulted, specify fault cause and its severity; (3) execute the AFDD tool for each input of the testing database calculating the following metrics [3]:

$$\text{Correct Diagnosis Rate (CDR)} = \frac{\text{\#of input samples with correct fault type diagnosis}}{\text{\#of input samples}} \tag{10}$$

$$\text{Correct Severity Rate (CSR)} = \frac{\text{\#of input samples with correct fault severity diagnosis}}{\text{\#of input samples}} \tag{11}$$

$$\text{True Positive Rate (TPR)} = \frac{\text{\#of input samples with true positive fault type}}{\text{\#of faulted input samples}} \tag{12}$$

$$\text{True Negative Rate (TNR)} = \frac{\text{\#of input samples with true negative fault type}}{\text{\#of unfaulted input samples}} \tag{13}$$

$$\text{False Positive Rate (FPR)} = \frac{\text{\#of input samples with false positive fault type}}{\text{\#of unfaulted input samples}} \tag{14}$$

$$\text{False Negative Rate (FNR)} = \frac{\text{\#of input samples with false negative fault type}}{\text{\#of faulted input samples}} \tag{15}$$

Figure 1a–f report, respectively, the values of these metrics as a function of the tests.

CDR has been calculated for all the 36 normal/faulty experiments of the testing dataset (described in Sect. 2); CSR, TPR and FNR have been calculated for the 22 faulty tests only of the testing dataset; TNR and FPR have been calculated for the 14 normal tests only of the testing dataset. The data in these figures indicate that: (1) CDR is equal to 100% with reference to the return/supply air fan fault (tests F10S, F10W, F11S, F11W), whatever the experimental test is; (2) CDR and CSR are always equal to 100% with reference to the tests F3S and F3W (cooling coil valve kept closed); (3) CSR is equal

Table 5. Detection and diagnosis rules for air temperature/relative humidity sensors' offset.

Condition 1 to be satisfied	Condition 2 to be satisfied	Fault diagnosis	Fault severity
$\left\|T_{RA,AVG,h} - T_{Room,AVG,h}\right\| \leq \varepsilon_T$	Not applicable	Normal operation	Not applicable
$\left\|T_{RA,AVG,h} - T_{Room,AVG,h}\right\| > \varepsilon_T$	$\left\|\dot{m}_{RIC,AVG,h} \cdot h^{theoretical}_{Room,AVG,h} - \dot{m}_{RIC,AVG,h} \cdot h_{RA,AVG,h}\right\| > \left\|\dot{m}_{RIC,AVG,h} \cdot h^{theoretical}_{Room,AVG,h} - \dot{m}_{RIC,AVG,h} \cdot h_{Room,AVG,h}\right\|$	Offset of T_{RA} sensor	$T_{RA,AVG,h} - T_{Room,AVG,h}$
$\left\|T_{RA,AVG,h} - T_{Room,AVG,h}\right\| > \varepsilon_T$	$\left\|\dot{m}_{RIC,AVG,h} \cdot h^{theoretical}_{Room,AVG,h} - \dot{m}_{RIC,AVG,h} \cdot h_{RA,AVG,h}\right\| \leq \left\|\dot{m}_{RIC,AVG,h} \cdot h^{theoretical}_{Room,AVG,h} - \dot{m}_{RIC,AVG,h} \cdot h_{Room,AVG,h}\right\|$	Offset of T_{Room} sensor	$T_{Room,AVG,h} - T_{RA,AVG,h}$
$\left\|RH_{RA,AVG,h} - RH_{Room,AVG,h}\right\| \leq \varepsilon_{RH}$	Not applicable	Normal operation	Not applicable
$\left\|RH_{RA,AVG,h} - RH_{Room,AVG,h}\right\| \geq \varepsilon_{RH}$	$\left\|\dot{m}_{RIC,AVG,h} \cdot h^{theoretical}_{Room,AVG,h} - \dot{m}_{RIC,AVG,h} \cdot h_{RA,AVG,h}\right\| > \left\|\dot{m}_{RIC,AVG,h} \cdot h^{theoretical}_{Room,AVG,h} - \dot{m}_{RIC,AVG,h} \cdot h_{Room,AVG,h}\right\|$	Offset of RH_{RA} sensor	$RH_{RA,AVG,h} - RH_{Room,AVG,h}$
$\left\|RH_{RA,AVG,h} - RH_{Room,AVG,h}\right\| \geq \varepsilon_{RH}$	$\left\|\dot{m}_{RIC,AVG,h} \cdot h^{theoretical}_{Room,AVG,h} - \dot{m}_{RIC,AVG,h} \cdot h_{RA,AVG,h}\right\| \leq \left\|\dot{m}_{RIC,AVG,h} \cdot h^{theoretical}_{Room,AVG,h} - \dot{m}_{RIC,AVG,h} \cdot h_{Room,AVG,h}\right\|$	Offset of RH_{Room} sensor	$RH_{Room,AVG,h} - RH_{RA,AVG,h}$

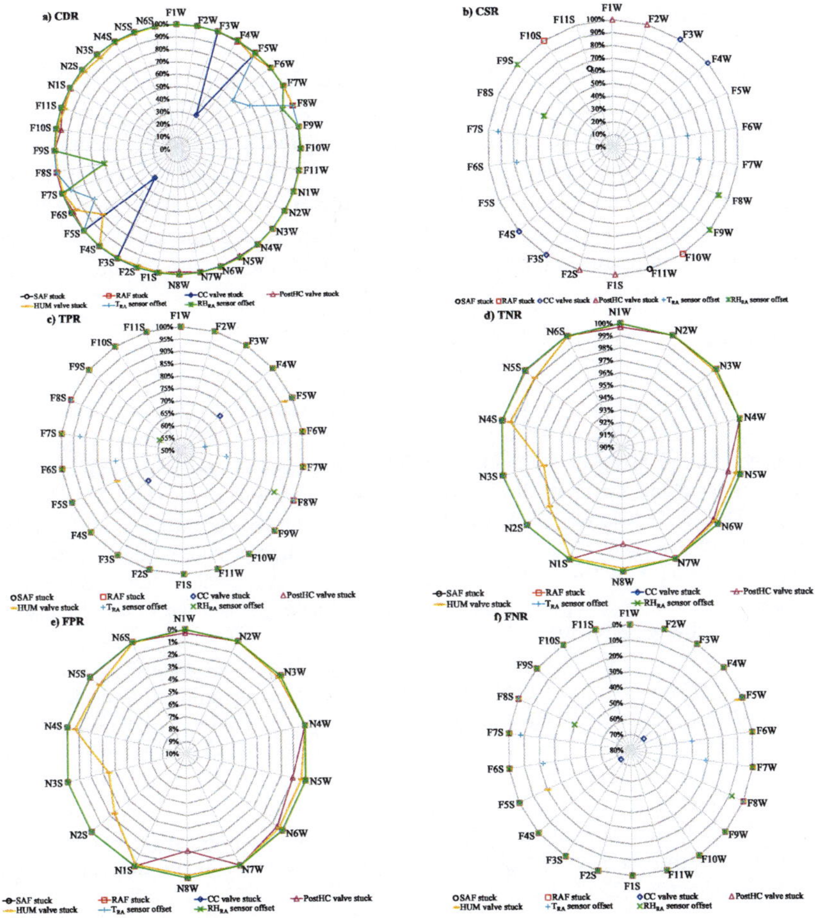

Fig. 1. CDR (**a**), CSR (**b**), TPR (**c**), TNR (**d**), FPR (**e**) and FNR (**f**) as a function of the tests.

to 100% also in the cases of tests F4W (CDR = 31.4%) and F4S (CDR = 29.0%); (4) proper predictions are obtained for the post-heating coil valve fault (tests F1S, F1W, F2S, F2W), with both CDR and CSR equal to 100%; (5) satisfactory values of CDR (from 79.7% to 96.5%) are achieved for the humidifier valve fault (tests F5S, F5W); (6) return air temperature sensor offset (tests F6S, F6W, F7S, F7W) is quite well predicted, with both CDR and CSR in the range 59.7% ÷ 92.2%; (7) TPR ranges from 79.7% up to 100% (highlighting a good predictive performance) with reference to the supply/return air fan stuck (F10S, F11S, F10W, F11W) and the cooling/post-heating/humidifier valve stuck (tests F1S, F2S, F3S, F5S, F1W, F2W, F3W, F5W), while TPR is 29.0% and 31.4%, respectively, for tests F4S and F4W; (8) TNR is between 96.5 and 100%, denoting excellent results; (9) maximum FPR (3.5%) is very small (meaning that AFDD tool is effective), while FNR is unsatisfactory in some cases (≈70% for F4S and F4W).

5 Conclusions

A new rule-based data-driven AFDD method has been developed and assessed in contrast with the experimental faulty and normal operation of a typical AHU. The performance metrics denoted a high correct diagnosis rate with results consistent with those available in the literature [3]. The AFDD method proved to be functional for detecting and diagnosing typical faults of coils, fans and sensors of AHUs in a relatively wide range of fault types/severities and operating conditions. In the future, additional faulty scenarios will be experimentally investigated and contrasted with the proposed tool.

References

1. Yun, W.S., Hong, W.H., Seo, H.: A data-driven fault detection and diagnosis scheme for air handling units in building HVAC systems considering undefined states. J. Build. Eng. **35** (2021)
2. Mirnaghi, M.S., Haghighat, F.: Fault detection and diagnosis of large-scale HVAC systems in buildings using data-driven methods: a comprehensive review. Energy Build. **229** (2020)
3. Lin, G., Kramer, H., Granderson, J.: Building fault detection and diagnostics: achieved savings, and methods to evaluate algorithm performance. Build. Environ. **168**, 106505 (2020)
4. Zhao, Y., Zhang, X., Zhang, C.: Artificial intelligence-based fault detection and diagnosis methods for building energy systems: advantages, challenges and the future. Renew. Sust. Energ. Rev. **109**, 85–101 (2019)
5. Huang, J., Wen, J., Yoon, H., Pradhan, O., Wu, T., O'Neill, Z., Candan, K. S.: Real versus simulated: questions on the capability of simulated datasets on building fault detection for energy efficiency from a data-driven perspective. Energy Build. **259**, 111872 (2022)
6. Rosato, A., Guarino, F., Sibilio, S., Entchev, E., Masullo, M., Maffei, L.: Healthy and faulty experimental performance of a typical HVAC system under Italian climatic conditions: artificial neural network-based model and fault impact assessment. Energies **14**, 1–41 (2021)
7. Rosato, A., Guarino, F., Filomena, V., Sibilio, S., Maffei, L.: Experimental calibration and validation of a simulation model for fault detection of HVAC systems and application to a case study. Energies **13**, 1–27 (2020)

Bridging the Flexibility Concepts in the Buildings and Multi-energy Domains

Gianfranco Chicco[1](✉), Diana Enescu[2], and Andrea Mazza[1]

[1] Politecnico di Torino, Turin, Italy
{gianfranco.chicco,andrea.mazza}@polito.it
[2] Valahia University of Targoviste, Targoviste, Romania
diana.enescu@valahia.ro

Abstract. This paper aims to stimulate a discussion on how to create a bridge between the concept of flexibility used in power and energy systems and the flexibility that buildings can offer for providing services to the electrical system. The paper recalls the main concepts and approaches considered in the power systems and multi-energy systems, and summarises some aspects of flexibility in buildings. The overview shows that there is room to strengthen the contacts among the scientists operating in these fields. The common aim is to identify the complementary aspects and provide inputs to enhance the methodologies and models to enable and support an effective energy and ecologic transition.

Keywords: Ancillary services · Buildings · Flexibility · Grid services · Minkowski sum · Multi-energy · Operation · Thermal comfort

1 Introduction

The concept of *flexibility* has been used in the engineering literature in the last decades, sometimes with different meanings. A contribution given in the Eighties defines *flexibility* as "*the capability of a system to maintain feasible operation over a range of uncertain/random conditions*" [1]. This definition contains three main aspects used to characterise flexibility, namely:

- Flexibility concerns the system *operation*, and as such depends on the *variable conditions in which the system undertakes its mission* during time, both internal (e.g., depending on the system constraints) and external (i.e., the variability of the environment outside the system).
- Flexibility is addressed after knowing the *feasibility* of the operational conditions. Feasibility can be determined by *establishing the operating regions of the system and their dependence on internal and external parameters*, variable in time. For this purpose, it is essential to define the operating point inside the operating region. When the analysis considers variations in time, the time series that contains the operating points becomes the operational *baseline* [2].

J. Littlewood and R. J. Howlett (Eds.): SEB 2022, SIST 336, pp. 371–385, 2023.
https://doi.org/10.1007/978-981-19-8769-4_35

- Flexibility refers to the usage of the system in a short-term time frame in the future, and what will happen is subject to *uncertainty*. Thereby, *predictions and estimations* of the uncertainty referring to the future operation *have to be included in the flexibility assessment*. In this respect, the operational baseline and its uncertainty bands must be determined by resorting to predictions and progressively updated to provide a presumably effective expected reference.

In recent years, flexibility has become a keyword in various sectors. Each sector has developed some definitions, representations, and analysis tools almost independently. The number of publications referring to flexibility is already huge. Bridging the formulations and techniques of analysis becomes of interest, particularly in the sectors in which the interactions are becoming more intense. This is for example the case of the interactions between the buildings and the energy networks. There is an increasing interest in assessing the potential of the buildings for providing services to the energy networks (e.g., for electricity, gas, and district heating/cooling), also based on recent legislation and regulatory provisions that enable the demand side to play a more active role in the energy systems and the emerging energy communities.

On the side of the buildings, many aspects are addressed by the Energy in Buildings and Community Programme, with several documents available from the EBC Annex 67 [3], to show the energy flexibility that the buildings can provide to the energy networks. In the EBC Annex 67 framework, "*the Energy Flexibility of a building is the ability to manage its demand and generation according to local climate conditions, user needs, and energy network requirements*" [4].

On the electrical side, many definitions of flexibility contain elaborated descriptions, while others are shorter. Some examples are as follows:

- International Smart Grid Action Network (ISGAN) [5]: "*Power system flexibility relates to the ability of the power system to manage changes*"
- Electric Power Research Institute (EPRI), Flexible power operation [6]: "*Flexible power operation (FPO) is any mode of operation that is not baseload*"
- Council of European Energy Regulators (CEER) – Conclusion paper [7]: "*the capacity of the electricity system to respond to changes that may affect the balance of supply and demand at all times*"
- International Renewable Energy Agency (IRENA) [8]: "*the capability of a power system to cope with the variability and uncertainty that VRE (variable renewable energy) generation introduces into the system in different time scales, from the very short to the long term, avoiding curtailment of VRE and reliably supplying all the demanded energy to customers*"
- European Smart Grids Task Force Expert Group 3 [9]: "*the ability of a customer (prosumer) to deviate from its normal electricity consumption (production) profile, in response to price signals or market incentives*"
- International Energy Agency (IEA) [10]: "*the ability of a power system to reliably and cost-effectively manage the variability and uncertainty of demand and supply across all relevant timescales, from ensuring instantaneous stability of the power system to supporting long-term security of supply*".

The above definitions add further aspects to the flexibility concept, namely:

- The *balance between generation and demand*, including the time-coupling effect of storage, that has to be addressed within different timescales, from operation stability to security of supply [11].
- The *presence of economic incentives for changing the demand*, which enables the involvement of the consumers or prosumers by formulating specific incentives and considering the users' needs, especially regarding user preferences and thermal comfort depending on climate conditions.
- The *role of variable renewable energy sources* (RES), including the energy exchanges within energy communities and the possible curtailment in case of excessive production with respect to the grid limits [12].

This paper addresses the specific case of the interactions between the buildings and the electrical network (also indicated as the *grid*). The focus is on the connection point of the buildings with the grid. A discussion on energy management aspects inside the buildings [13] is outside the scope of this paper. The gas and district heating/cooling networks [14] are considered here, especially when heating/cooling are driven by electricity [15].

In particular, the interactions between buildings and the grid will be considered under two frameworks—demand response [16] and ancillary services [17] (or grid services). Participation of buildings to demand response programmes is still very limited [18], even though there is a remarkable potential [19]. The number of buildings is so high that even a small individual contribution by a relatively high number of buildings would be useful. Conversely, a response from most buildings could become excessive, so the careful design of the incentives included in the programme is crucial. From the grid side, the interaction is managed by the distribution system operator (DSO) for what concerns the technical aspects, the retailer with reference to the contract aspects of the electricity supply, and in some cases by an aggregator, which manages a set of assets together to reach an overall size sufficient for being part of the power and energy management at the grid level. The grid could also be a microgrid, with the corresponding management and flexibility provision [20]. Aggregators manage a portfolio of end-users, activating possible smart contract options based on the definition of time-variable flexibility bands in which the aggregate flexible demand should vary [21].

The specific contributions of this paper are:

(a) To provide information aimed at stimulating closer discussions among scientists working in the buildings and electrical systems domains.
(b) To highlight the benefits of exploiting flexibility in multi-energy systems with grid-connected buildings.
(c) To summarise the approaches and tools used for determining the contribution and profitability of multi-energy systems that provide grid services.

The next sections of this paper are organised as follows. Section 2 recalls the conceptual flexibility framework used in power and energy systems and summarises some aspects of flexibility in buildings. Section 3 outlines some tools used for flexibility analysis in the electrical energy systems and multi-energy fields. The last section contains the conclusions.

2 Concepts and Operational Aspects of Flexibility

2.1 Flexibility in Power and Energy Systems

For power and energy systems, flexibility has been defined in different ways, referring to the generation side, the grid side, and the demand side [22]. The analysis tools have been formulated accordingly.

On the *generation* side, flexibility has been addressed by exploiting approaches taken from traditional problems such as *unit commitment* (i.e., the minimisation of the total cost of power generation for serving the demand in a given period through the appropriate scheduling of the generation units in pre-defined sub-periods) and *economic dispatch* (i.e., the determination of the power outputs of each scheduled generation unit in each sub-period). Optimisation techniques for stochastic systems are needed to incorporate the effects of uncertainty. The relevant indicators are based on the maximum power (capacity), minimum stable generation output, and the up/down ramp rates of conventional generators.

On the *grid* side, flexibility has been defined as the ability of a power network to deploy its flexible resources to cope with volatile changes in the power system state during operation [23]. The increased uncertain renewable generation may cause higher operational risks of *congestion* (i.e., electric lines or transformers that exceed their technical limits). Congestions further increase operating costs and establish stricter limits on using the available flexibility resources from generation and/or demand.

On the *demand* side, flexibility can be achieved by using components able to adapt their operation and *shift* their consumption to different time intervals. Specific contributions come from individual and aggregate residential demand, thermostat-controlled loads, thermal energy systems, multi-energy systems, storage systems, electric vehicles, and other loads.

2.2 Flexibility in Power System Operation and Grid Services

In a power system, the possible flexibility services that can be offered to the grid are determined by the needs of the power system itself in terms of power and energy during operation. Furthermore, sufficient reserves must be activated in case one or more power system components are unavailable. These services are also known as ancillary services by using the traditional nomenclature of adopted in the power system and in the electricity markets. In the traditional view of the power systems after the restructuring of the electricity business (at the end of the Nineties), the ancillary services were provided by generators only, in particular, were achieved from large generators. Successively, it has been recognised that some demand-side resources could be able to provide faster and more flexible responses than large generators [24].

For the electrical system, the relevant quantity at a given node is the *net demand*, that is, the difference between local demand and local generation. Hence, a demand reduction of a given amount and type in a given node is equivalent to a local generation increase of the same amount and type.

A summary of the main ancillary services is as follows:

- *Scheduling and dispatch*: the power produced in the system has to balance at any instant the power demand. Scheduling is carried out by preparing the units to use, while dispatch is used close to the real-time. Storage systems (if any) play either as generators or additional demand during time, with the main limitation of the storage capacity available at a reasonable cost [25].
- *Frequency control* (or *regulation*) and *reserves*: aim at maintaining the grid frequency close to the nominal frequency also responding to large fluctuations mainly due to RES, and at providing additional power in case of a lack of production, through generation systems that start their service at different times.
- *Reactive power and voltage support*: the electrical system node and its components are designed to operate under the nominal voltage. Any deviation from the nominal voltage should be compensated [26].

In a context with available resources and reserves, *operational flexibility* has been defined as the "*technical ability of a power system unit to modulate electrical power feed-in to the grid and/or power out-feed from the grid over time*" [27]. The operational flexibility in power systems is quantified by using four metrics [28] that consider the power provision capacity π (MW), the power ramp-rate capacity ρ (MW/min), the energy provision capacity ε (MWh), and the ramp duration δ (min). On these bases, the maximum available flexibility is defined by the limits on π, ρ and ε and can be visualised inside a *flexibility cube* [27]. However, the maximum available flexibility cannot be used at any time, mainly because of time-related constraints. Thereby, the available operational flexibility that can be deployed depends on the time-variable constraints.

2.3 Flexibility in Multi-energy Systems

Flexibility in Multi-Energy Systems (MES) is defined as the *technical ability of a system to regulate multi-energy supply, demand and power flows* subject to steady-state and dynamic constraints, while operating within predefined/desired boundary regions for certain energy vectors [29]. In a MES, there are different energy carriers and connections to the energy networks. Within the MES, the feasible energy flows depend on the specific constraints on the equipment (e.g., ranges of temperatures at which each equipment works). The constraints in the energy networks affect the available flexibility that can be obtained by adjusting the operation of the different equipment inside the MES. Once the initial operating point is defined, it is possible to calculate:

- the reduction of the electricity input from the grid, depending on local generation increase or local demand reduction), shortly denoted as *upward flexibility*; and,
- the increase of the electricity input from the grid, depending on local generation reduction or local demand increase), shortly denoted as *downward flexibility*.

In the presence of multiple energy carriers, the flexibility can be defined in different directions for each one of the energy carriers. For example, if the heat demand is kept unchanged (e.g., to maintain the same comfort level for the users), the electrical flexibility changes depending on whether to curtail or increase the electrical demand [30]. Moreover, changes in the user's electrical demand and the corresponding flexibility regions can be addressed by considering the timing at which the resources become available after being requested. In addition, it is important to consider that the feasibility ranges of some components can be limited, also with minimum technical limits, which may impact the construction of the overall feasibility regions of the MES [31].

The MES operational flexibility is strictly linked to the energy shifting capability that appears inside the MES, which in turn depends on the existing energy needs with different equipment fed with different energy carriers (e.g., considering electricity and gas as substitutable resources [32]). This dependency does not guarantee perfect interchangeability among the components, because of different possible timings in the use of these components, for which the requested service cannot be provided in the same way. In general, the thermal dynamics are slower than the dynamics of the electrical systems. However, when the service is requested in a given time, depending on external aspects (such as the user's lifestyle), some thermal systems (e.g., a gas boiler for heating water) could be faster than their electrical counterpart (an electric boiler) to provide the requested service starting from the instant of request.

2.4 Flexibility in Buildings Applications

The IEA EBC Annex 67 [3] indicates among the main aims the increase of the RES exploitation and the mitigation of the CO_2 emissions. It considers residential and non-residential buildings, addressing both new constructions and renovated buildings. In the latter case, the building renovation can include energy flexibility. Single buildings could have an energy demand too low for providing flexibility services to the grid. For this purpose, *clusters* of buildings are considered, either physically connected or commercially aggregated. The commercially aggregated buildings are owned by the same entity and are not connected to the same point of the electrical distribution network. This situation, already existing in commercial activities, is also occurring with the energy communities. The aggregation of buildings is helpful to reach a sufficient impact to enter the market or provide grid services. The interaction of clusters of buildings with the grid is an open research topic, especially when one or more buildings contain prosumers. The literature on multi-energy communities is making remarkable steps in this direction [30].

The IEA EBC Annex 67 defines the so-called *penalty signals*, as variable boundary conditions partitioned into high-frequency signals (e.g., ambient temperature, energy prices, user behaviour, and indications depending on the energy mix such as the amount of RES in the grid) and low-frequency signals (e.g., climate change, technology improvement, macro-economic factors and buildings use) [33]. The above notion of high and low frequency refers to the variations of the boundary conditions in the time scale of the energy usage [34], not to the notions of frequency used in the electrical systems. In fact, most grid services (e.g., for frequency and voltage control) require remarkably higher frequencies than those indicated before. Thereby, the thermal inertia of the buildings makes the buildings unable to provide such fast services, because of the slow temperature variation and the energy payback effect that follows demand response actions [35]. Services with a longer time, such as load shifting and reserves, could be provided to some extent. Exploiting the thermal inertia of the buildings could provide an additional flexibility gain, provided that appropriate incentives are available [36]. However, the contribution of the thermal inertia heavily depends on the thermal characteristics of the buildings and the specific weather conditions [37], also considering the impact on the comfort level [38]. Assessing flexibility helps define future strategies, so uncertainty has to be considered explicitly in the studies [39].

In the review presented in [40], most of the studies on energy flexibility refer to energy shifting. However, different aspects of energy shifting are considered, referring to energy price, electricity, energy infrastructures, and interactions with thermal comfort and external energy systems.

The determination of the available energy flexibility of clusters of buildings depends on the technologies and their control and the external conditions (climate, energy networks, markets) and the interactions with the occupants of the buildings. The flexibility of buildings is based on the definition of *flexibility functions* [41] by considering demand variations in response to a stepwise increase in the energy price [42]. Six flexibility characteristics have been defined in [43]. These characteristics include three time-based characteristics (i.e., the delay from the energy price increase and the appearance of initial effects on decreasing the energy demand, the time elapsed from initial demand decrease to the point of minimum demand, and the duration from the initial demand decrease to when the demand reaches again the level before the price change). Two further characteristics are defined in energy terms, by considering the final demand level at the equilibrium point after the energy price variation, determining the total energy corresponding to the demand lower than the demand at the equilibrium point, and the total energy demand corresponding to the demand higher than the demand at the equilibrium point. The last characteristic is the maximum demand reduction with respect to the equilibrium point after the energy price variation. The use of the flexibility characteristics in a control system framework enables the exploitation of energy flexibility resources and the increase of the amount of RES that can be deployed [43]. These aspects are useful for DSOs and aggregators.

Absolute and relative grid support coefficients have been introduced in [44] and applied in [45] to four flexibility and storage cases in buildings (i.e., batteries, water tanks, thermal building mass, and fuel switch). Electricity consumption is assessed to identify when it occurs above or below the average electricity demand. Grid support coefficients lower than unity identify grid-supportive buildings. The results indicate that batteries are the most viable alternative for providing grid services, but also that, under the present electricity price variations, there is not enough reward to make these grid services solutions attractive for the user.

In a MES context, the potential of buildings to contribute developing grid services is based on the existence in the buildings of electrical devices or electrically-driven thermal components and systems, e.g., for electric heating, ventilation and air conditioning (HVAC), or electric cooking. In particular, the availability of electric heat pumps is a major asset for performing energy shifting to provide services to the grid [35], especially if there is sufficient redundancy on the thermal side to avoid reductions in the thermal comfort of the occupants quantified in different ways [46]. Depending on the season, interactions with thermal energy storage could provide further inputs to increase flexibility [47]. To provide scheduling and dispatch ancillary services, possible strategies for curtailing HVAC demand are summarised in [48]. Dynamic models to address load shifting in HVAC that consider electricity consumption, variations in electricity prices, and indoor air temperature have been formulated in [49, 50].

The provision of energy flexibility from buildings can be limited by (i) the age of the buildings (needing renovation), (ii) the limited revenues that can be given for providing grid services, and (iii) the need for installing or updating energy management systems for dealing with the communication requested to execute demand response programmes or to manage the interactions with the grid. These reasons partially explain why the building managers could consider that energy efficiency (handled with internal energy management systems, including RES and, in prospect, the charging of electric vehicles) is more important than providing flexible services to the grid. Further insights on understanding the inefficiency of buildings in providing ancillary services are discussed in [51]. However, energy flexibility in buildings aimed at providing grid services may also be used to enable wider usage of RES [52], particularly photovoltaic systems [3]. Concerning RES, the main objectives refer to cost-effectiveness and the improvement of self-consumption (i.e., the percentage of the local RES generation used to cover the local demand) and self-sufficiency (i.e., the percentage of the local demand covered by the local RES generation).

2.5 Virtual Energy Storage

Virtual Energy Storage (VES) represents one of the most interesting bridge applications able to couple the built environment with the network infrastructure. In fact, any building presents a certain degree of thermal inertia: its exploitation allows to apply demand response actions by acting on thermal loads. It is worth noting that the model must include not only parameters of the built environment, but also comfort functions related to the occupants.

The American Society of Heating, Refrigeration, and Air Conditioning Engineers (ASHRAE) classifies space thermal modelling into forward approaches and data-driven approaches [53]. The first family (also called white-box) of models has been classically developed for design optimisation. It is necessary to know the natural phenomena that may affect the system behaviour, as well as the interaction magnitude. However, these approaches do not require that the building is ready, hence can be used for preliminary design. On the other hand, data driven approaches include validated simulation models (based on approaches similar to the ones of the forward methods, but calibrated using real data), black-box models (i.e., empirical approaches, obtained by fitting historical data gathered from the system) and grey-box models (which include the physical description of the system and the parameters calculated through system identification methods).

Many past studies included VES as a flexibility asset in the grid, for example, by introducing an equivalent thermal model based on temperatures (internal and external) and heat source (both artificial and natural) [54]. This model was used to develop an optimal load scheduling by considering the electricity prices and the energy balance.

The building thermal mass can be exploited to operate a grid or a microgrid [55], introducing proper pre-heating/precooling schemes [40] acting as thermal load shifting for the benefit of the prosumer (comparable with or better than battery-based energy storage [56]). Another VES model, included into a grid simulation framework, is based on the distinction between the thermal dynamics of the wall and the thermal dynamics of the space air, leading to a multi-resistance and multi-capacitance model [57].

3 Tools for Flexibility Analysis of Electrical Systems and MES

3.1 The Minkowski Sum

A key contribution to flexibility studies comes from the work of Hermann Minkowski on geometrical methods. In particular, the Minkowski sum (or Minkowski addition) has been used for adding two sets by taking one set and moving it along the borders of the other set. The Minkowski addition can be carried out progressively by summing up a new set to the result of the previous additions. In this way, it is possible to determine the boundaries of the aggregated feasibility regions by using Minkowski sums on polytopes [27]. The Minkowski addition is a powerful basic tool for constructing the MES feasibility regions starting from the feasibility regions of the individual MES components [29]. If one or more components have limited feasibility ranges, the result of their incorporation could lead to non-convex or even non-connected feasibility regions [31]. Possible limitations of using the Minkowski addition for calculating flexibility are discussed in [58], where a flexibility gap is determined by comparing the results of the Minkowski sum and of individual power profile summation.

3.2 MES Feasibility, Flexibility and Profitability Regions

Energy shifting is carried out by reducing (or increasing) the electricity exchanged with the grid, covering the shifted demand from fuel-based sources. In the provision of grid services from a MES, determining the possible reduction (or increase) of the electricity exchanged with the grid is the first step for understanding to what extent the MES is able to modify its internal energy flows. The maximum reduction (or increase) of the electricity exchanged with the grid has been indicated as *electricity shifting potential* [59], which is also the technical limit for providing flexibility. On the economic side, the changes in the energy flows needed to provide energy shifting are associated with extra operational costs for moving away from the baseline. If the baseline is optimal, the extra costs are always positive. Otherwise, some variations could lead to cost reductions with respect to the initial solution. The comparison between the extra operational costs and the possible revenues, carried out through profitability maps, identifies the profitability regions for which energy shifting may be convenient (considering the profits = revenues – costs), also finding out the most convenient solution [60]. If appropriate revenues do not compensate for the extra costs, there is no convenience in carrying out energy shifting.

3.3 The Energy Hub Model

An effective way to model a MES is to use the matrix representation formulated in the energy hub model [61]. This representation is based on an input-output model. All components are modelled by means of their efficiency matrix. The whole system is modelled through a coupling matrix that accounts for the individual efficiency matrices and the internal MES topology. A remarkable aspect is that the coupling matrix can be constructed directly by visual inspection of the MES topology and of the efficiency matrices, or by exploiting automatic procedures for analyses with non-linear [62] or linearised [63] models. Moreover, the shares of a given output that go in different directions can be used as decision variables in an optimisation problem.

3.4 The Virtual Battery Model

In the flexibility studies, loads with particular dynamics have been represented by using a virtual battery model [64]. This model considers as parameters the energy capacity, the charge/discharge rates (which set up the power limits, starting from baseline conditions), and the self-discharging rate. The model has been applied to estimate the aggregate flexibility from thermostat-controlled loads (TCLs) [65] and has been enhanced in [66] by handling the effect of coupling constraints referring to the grid to which the individual TCLs are connected. The model has been used to formulate multi-period optimal scheduling to exploit the aggregate flexibility from heterogeneous TCLs for providing multiple grid services and ancillary services [67].

In applications referring to buildings, the virtual battery model has been adopted for different purposes, e.g., to address the control of HVAC systems using a detailed building model [68] and to set up a unified approach for addressing flexibility of building loads and energy storage [69].

3.5 Power Node and Multi-energy Node

The power node model [70] has been formulated as a unified approach for representing different units and their dynamics during the operation of a power system. The power node model enables the incorporation of energy storage and intermittent RES. The power node equation has been established with contributions that come from a baseline scheduling model, a model for schedule updates, and a model that addresses real-time control. Starting from the power node, a multi-energy node has been formulated in [29], which incorporates the basic energy hub model and adds the components that represent curtailments of demand or excess of production from multi-energy sources.

4 Concluding Remarks

The synthetic overview presented in this paper has highlighted that different definitions, approaches and tools are used to assess the potential of individual or aggregate buildings to provide services to the grid in the specific domains. Some steps can be done to prepare a common field for discussions, for example, following the points outlined below. It could be helpful to check whether the effective tools and specific analysis techniques applied in different domains can be integrated for carrying out multi-domain assessments.

A network-based analysis is quite common in the electrical field, using the corresponding methodologies based on interconnected networks. The energy hub has been applied as a tool for hosting multi-energy modelling, again topology-based. Moreover, power system analysis is based on the network representation suitable to study fast dynamics close to real-time. Conversely, the traditional electrical modelling is less based on temperatures and details of the thermal modelling, such as phase changes and non-linearities due to the behaviour of the fluids. In this respect, even though the basic models have to remain closer to the specific domain, there are complementary contents to share, with synthetic but effective models to be exchanged among the scientists operating in the different disciplines.

In all domains, there are new aspects to integrate, concerning the incorporation of social and ecologic contents in the models, with the adaptation of the analysis techniques for considering, besides the uncertainty, the indeterminacy of the users' decisions. The social aspects are becoming more and more essential to address when dealing with the solutions for managing the interactions with the grid in the context of the energy communities, in which the energy management strategies may be affected by the personal decisions of final users who could not be willing to cooperate with the system. Moreover, the availability of huge amounts of data could lead to the increasing diffusion of data-driven approaches supported by machine learning. However, machine learning alone cannot solve all situations. The role of the domain experts remains crucial for checking the data to use and interpreting the results.

Other new contents to include depend on the expected diffusion of electric and hybrid vehicles, further increasing the demand in the distribution grids. The challenge is not only the management of the battery charge (and discharge, in case of active vehicle-to-grid options that could also lead to shaving the peak of the electrical demand). The key point is to integrate the presence of electric or hybrid vehicles within the multi-energy management inside the buildings.

On top of the previous aspects, the economics of the interactions between buildings and the grid will have to be developed by formulating smart contracts with sufficient revenues and moderate risks to become attractive for the prosumers. In this evolving scenario, the scientific communities are expected to follow the common aim of identifying the complementary aspects and provide inputs to enhance the methodologies and models used for creating a common language and improving their algorithmic tools. Flexibility assessment is a solid common ground for developing consistent experiences, and will represent one effective way to guarantee the implementation of multi-sector policies to support the energy and ecological transition.

References

1. Swaney, R.E., Grossmann, I.E.: An index for operational flexibility in chemical process design. Part I: Formul. Theory. AIChE **31**(4), 621–630 (1985)
2. Lin, Y., Barooah, P., Mathieu, J.L.: Ancillary services through demand scheduling and control of commercial buildings. IEEE Trans. Power Syst. **32**(1), 186–197 (2017)
3. Energy in Buildings and Community Programme, EBC Annex 67. http://annex67.org

4. Østergaard Jensen, S., et al.: IEA EBC Annex 67 energy flexible buildings. Energy Build. **155**, 25–34 (2017)
5. Hillberg, E., et al.: Flexibility needs in the future power system—Discussion paper. ISGAN Annex 6 Power T&D Systems (2019)
6. EPRI: Flexible Operations Program (2021). https://www.epri.com/portfolio/programs/106194
7. CEER Distribution Systems Working Group: Flexibility use at distribution level—A CEER conclusions paper. CEER C18-DS-42-04 (2018)
8. IRENA: power system flexibility for the energy transition: part 1, Overview for policy makers. Int. Renew. Energy Agency, Abu Dhabi (2018)
9. European Smart Grids Task Force—Expert Group 3: Demand Side Flexibility—Perceived barriers and proposed recommendations, Final Report (2019)
10. IEA: Status of power system transformation 2019: power system flexibility. International Energy Agency (2019)
11. Zhou, Y., Cao, S.: Quantification of energy flexibility of residential net-zero-energy buildings involved with dynamic operations of hybrid energy storages and diversified energy conversion strategies. Sustain. Energy Grids Netw. **21**, 100304 (2020)
12. Li, H., Wang, Z., Hong, T., Piette, M.A.: Energy flexibility of residential buildings: a systematic review of characterization and quantification methods and applications. Adv. Appl. Energy **3**, 100054 (2021)
13. Li, R., You, S.: Exploring potential of energy flexibility in buildings for energy system services. CSEE J. Power Energy Syst. **4**(4), 434–443 (2018)
14. Ghilardi, L.M.P., Castelli, A.F., Moretti, L., Morini, M., Martelli, E.: Co-optimization of multi-energy system operation, district heating/cooling network and thermal comfort management for buildings. Appl. Energy **302**, 117480 (2021)
15. Arteconi, A., Mugnini, A., Polonara, F.: Energy flexible buildings: a methodology for rating the flexibility performance of buildings with electric heating and cooling systems. Appl. Energy **251**, 113387 (2019)
16. Albadi, M.H., El-Saadany, E.F.: A summary of demand response in electricity markets. Electr Power Syst Res **78**, 1989–1996 (2008)
17. Heffner, G., Goldman, C., Kirby, B., Kintner-Meyer, M.: Loads providing ancillary services: review of international experience. Ernest Orlando Lawrence Berkeley National Laboratory, LBNL-62701, ORNL/TM-2007/060, PNNL-16618 (2007)
18. Amin, A., Kem, O., Gallegos, P., Chervet, P., Ksontini, F., Mourshed, M.: Demand response in buildings: unlocking energy flexibility through district-level electro-thermal simulation. Appl. Energy **305**, 117836 (2022)
19. Avramidis, I.A., Capitanescu, F., Evangelopoulos, V.A., Georgilakis, P.S., Deconinck, G.: In pursuit of new real-time ancillary services providers: hidden opportunities in low voltage networks and sustainable buildings. IEEE Trans. Smart Grid **13**(1), 429–442 (2022)
20. Kazemi-Razi, S.M., Askarian Abyaneh, H., Nafisi, H., Ali, Z., Marzband, M.: Enhancement of flexibility in multi-energy microgrids considering voltage and congestion improvement: robust thermal comfort against reserve calls. Sustain. Cities Soc. **74**, 103160 (2021)
21. Saavedra, A., Negrete-Pincetic, M., Rodríguez, R., Salgado, M., Lorca, Á.: Flexible load management using flexibility bands. Appl. Energy **317**, 119077 (2022)
22. Degefa, M.Z., Bakken Sperstad, I., Sæle, H.: Comprehensive classifications and characterizations of power system flexibility resources. Electric Power Syst. Res. **194**, 107022 (2021)
23. Li, J., Liu, F., Li, Z., Shao, C., Liu, X.: Grid-side flexibility of power systems in integrating large-scale renewable generations: a critical review on concepts, formulations and solution approaches. Renew. Sustain. Energy Rev. **93**, 272–284 (2018)

24. Ma, O., et al.: Demand response for ancillary services. IEEE Trans. Smart Grid **4**(4), 1988–1995 (2013)
25. D'Ettorre, F., De Rosa, M., Conti, P., Testi, D., Finn, D.: Mapping the energy flexibility potential of single buildings equipped with optimally-controlled heat pump, gas boilers and thermal storage. Sustain. Cities Soc. **50**, 101689 (2019)
26. Kim, Y., Blum, D.H., Xu, N., Su, L., Norford, L.K.: Technologies and magnitude of ancillary services provided by commercial buildings. Proc. IEEE **104**(4), 758–779 (2016)
27. Ulbig, A., Andersson, G.: Analyzing operational flexibility of electric power systems. Electr. Power Energy Syst. **72**, 155–164 (2015)
28. Makarov, Y., Loutan, C., Ma, J., de Mello, P.: Operational impacts of wind generation on California power systems. IEEE Trans. Power Syst. **24**(2), 1039–1050 (2009)
29. Chicco, G., Riaz, S., Mazza, A., Mancarella, P.: Flexibility from distributed multienergy systems. Proc. IEEE **108**(9), 1496–1517 (2020)
30. Good, N., Mancarella, P.: Flexibility in multi-energy communities with electrical and thermal storage: a stochastic, robust approach for multi-service demand response. IEEE Trans. Smart Grid **10**(1), 503–513 (2019)
31. Hinker, J., Knappe, H., Myrzik, J.M.A.: Precise assessment of technically feasible power vector interactions for arbitrary controllable multi-energy systems. IEEE Trans. Smart Grid **10**(1), 1146–1155 (2019)
32. Neyestani, N., Yazdani-Damavandi, M., Shafie-khah, M., Chicco, G., Catalão, J.P.S.: Stochastic modeling of multienergy carriers dependencies in smart local networks with distributed energy resources. IEEE Trans. Smart Grid **6**(4), 1748–1762 (2015)
33. Vigna, I., Pernetti, R., Pasut, W., Lollini, R.: New domain for promoting energy efficiency: energy flexible building cluster. Sustain. Cities Soc. **38**, 526–533 (2018)
34. Tang, H., Wang, S.: Energy flexibility quantification of grid-responsive buildings: energy flexibility index and assessment of their effectiveness for applications. Energy **221**, 119756 (2021)
35. Zhang, L., Good, N., Mancarella, P.: Building-to-grid flexibility: modelling and assessment metrics for residential demand response from heat pump aggregations. Appl. Energy **233–234**, 709–723 (2019)
36. Arkhangelski, J., Abdou-Tankari, M., Lefebvre, G.: Ancillary services for distribution grid: demand response of building thermal inertia case. In: Proceedings of the 2020 International Conference on Computational Intelligence for Smart Power System and Sustainable Energy (CISPSSE), Keonjhar, India (2020)
37. Gao, Q., Demoulin, M., Wang, H., Riaz, S., Mancarella, P.: Flexibility characterisation from thermal inertia of buildings at city level: a bottom-up approach. In: Proceedings of the 2020 55th International Universities Power Engineering Conference (UPEC) (2020)
38. Good, N., Karangelos, E., Navarro-Espinosa, A., Mancarella, P.: Optimization under uncertainty of thermal storage based flexible demand response with quantification of residential users' discomfort. IEEE Trans. Smart Grid **6**(5), 2333–2342 (2015)
39. Amadeh, A., Lee, Z.E., Zhang, K.M.: Quantifying demand flexibility of building energy systems under uncertainty. Energy **246**, 123291 (2022)
40. Reynders, G., Amaral Lopes, R., Marszal-Pomianowska, A., Aelenei, D., Martins, J., Saelens, D.: Energy flexible buildings: An evaluation of definitions and quantification methodologies applied to thermal storage. Energy Build. **166**, 372–390 (2018)
41. Grønborg Junker, R., et al.: Characterizing the energy flexibility of buildings and districts. Appl. Energy **225**, 175–182 (2018)
42. Vigna, I., Lollini, R., Pernetti, R.: Assessing the energy flexibility of building clusters under different forcing factors. J. Build. Eng. **44**, 102888 (2021)
43. Grønborg Junker, R., Relan, R., Madsen, H.: Designing individual penalty signals for improved energy flexibility utilisation. IFAC-PapersOnLine **52**(4), 123–128 (2019)

44. Klein, K., Langner, R., Kalz, D., Herkel, S., Henning, H.M.: Grid support coefficients for electricity-based heating and cooling and field data analysis of present-day installations in Germany. Appl. Energy **162**, 853–867 (2016)
45. Klein, K., Herkel, S., Henning, H.M., Felsmann, C.: Load shifting using the heating and cooling system of an office building: quantitative potential evaluation for different flexibility and storage options. Appl. Energy **203**, 917–937 (2017)
46. Enescu, D.: A review of thermal comfort models and indicators for indoor environments. Renew. Sustain. Energy Rev. **79**, 1353–1379 (2017)
47. Stinner, S., Huchtemann, K., Müller, D.: Quantifying the operational flexibility of building energy systems with thermal energy storages. Appl. Energy **181**, 140–154 (2016)
48. Motegi, N., Piette, M.A., Watson, D.S.,Kiliccote, S., Xu, P.: Introduction to commercial building control strategies and techniques for demand response. Lawrence Berkeley Nat. Lab., Tech. Rep. (LBNL-59975) (2007)
49. Nguyen, D.T., Le, L.B.: Joint optimization of electric vehicle and home energy scheduling considering user comfort preference. IEEE Trans. Smart Grid **5**(1), 188–199 (2014)
50. Papadaskalopoulos, D., Strbac, G., Mancarella, P., Aunedi, M., Stanojevic, V.: Decentralized participation of flexible demand in electricity markets—Part II: application with electric vehicles and heat pump systems. IEEE Trans. Power Syst. **28**(4), 3667–3674 (2013)
51. Lin, Y., Mathieu, J.L., Johnson, J.X., Hiskens, I.A., Backhaus, S.: Explaining inefficiencies in commercial buildings providing power system ancillary services. Energy Build. **152**, 216–226 (2017)
52. Chen, Y., Xu, P., Gu, J., Schmidt, F., Li, W.: Measures to improve energy demand flexibility in buildings for demand response (DR): a review. Energy Build. **177**, 125–139 (2018)
53. ASHRAE: Energy estimating and modeling methods. ASHRAE Handbook, Chapter 19. https://handbook.ashrae.org/Handbooks/F17/SI/f17_ch19/f17_ch19_si.aspx
54. Jin, X., Wang, X., Mu, Y., Jia, H., Xu, X., Qi, Y., Yu, X., Qi, F.: Optimal scheduling approach for a combined cooling, heating and power building microgrid considering virtual storage system. In: Proceedings of the 2016 IEEE Power and Energy Society General Meeting (PESGM), Tianjin, China (2016)
55. Sikder, O., Jansson, P.M.: Thermal inertia of a building as virtual energy storage: a sustainable solution for smart grids. In: Proceedings of the 2018 53rd International Universities Power Engineering Conference (UPEC), Glasgow, Scotland (2018)
56. Asare, P., Ononuju, C., Jansson, P.M.: Preliminary quantitative evaluation of residential virtual energy storage using power sensing. In: Proceedings of the 2017 IEEE Sensors Applications Symposium (SAS), Glassboro, NJ, USA (2017)
57. Fambri, G., Badami, M., Tsagkrasoulis, D., Katsiki, V., Giannakis, G., Papanikolaou, A.: Demand flexibility enabled by virtual energy storage to improve renewable energy penetration. Energies **13**, 5128 (2020)
58. Gusain, D., Cvetković, M., Palensky, P.: Quantification of operational flexibility from a portfolio of flexible energy resources. Int. J. Electr. Power Energy Syst. **141**, 107466 (2022)
59. Mancarella, P., Chicco, G.: Real-time demand response from energy shifting in distributed multi-generation. IEEE Trans. Smart Grid **4**(4), 1928–1938 (2013)
60. Mancarella, P., Chicco, G., Capuder, T.: Arbitrage opportunities for distributed multi-energy systems in providing power system ancillary services. Energy **161**, 381–395 (2018)
61. Geidl, M., Koeppel, G., Favre-Perrod, P., Klöckl, B., Andersson, G., Fröhlich, K.: Energy hubs for the future. IEEE Power Energy Mag. **5**(1), 25–30 (2007)
62. Chicco, G., Mancarella, P.: Matrix modelling of small-scale trigeneration systems and application to operational optimization. Energy **34**(3), 261–273 (2009)
63. Wang, Y., Cheng, J., Zhang, N., Kang, C.: Automatic and linearized modeling of energy hub and its flexibility analysis. Appl. Energy **211**, 705–714 (2018)

64. Hao, H., Sanandaji, B.M., Poolla, K., Vincent, T.L.: Aggregate flexibility of thermostatically controlled loads. IEEE Trans. Power Syst. **30**(1), 189–198 (2015)
65. Zhao, L., Zhang, W., Hao, H., Kalsi, K.: A geometric approach to aggregate flexibility modeling of thermostatically controlled loads. IEEE Trans. Power Syst. **32**(6), 4721–4731 (2017)
66. Wang, G., Li, Z., Wang, F.: Enhanced sufficient battery model for aggregate flexibility of thermostatically controlled loads considering coupling constraints. IEEE Trans. Sustain. Energy **12**(4), 2493–2496 (2021)
67. Wu, D., Wang, P., Ma, X., Kalsi, K.: Scheduling and control of flexible building loads for grid services based on a virtual battery model. IFAC-PapersOnLine **53**(2), 13333–13338 (2020)
68. Hughes, J.T., Domínguez-García, A.D., Poolla, K.: Identification of virtual battery models for flexible loads. IEEE Trans. Power Syst. **31**(6), 4660–4669 (2016)
69. Hao, H., Wu, D., Lian, J., Yang, T.: Optimal coordination of building loads and energy storage for power grid and end user services. IEEE Trans. Smart Grid **9**(5), 4335–4345 (2018)
70. Heussen, K., Koch, S., Ulbig, A., Andersson, G.: Unified system-level modeling of intermittent renewable energy sources and energy storage for power system operation. IEEE Syst. J. **6**(1), 140–151 (2012)

Offsite Manufacturing of Timber-Frame Woodfibre Insulated Construction Systems for Nearly-to-Zero Carbon Dwellings in Wales, UK

J. R. Littlewood[1](✉), R. J. M. Hawkins[1], N. I. Evans[2], and C. Hale[2]

[1] The Sustainable and Resilient Built Environment Research Group, Cardiff Metropolitan University, Cardiff CF5 2YB, UK
jlittlewood@cardiffmet.ac.uk
[2] Sevenoaks Modular Ltd, Neath SA11 1NJ, UK

Abstract. This paper discusses some of the findings of a knowledge transfer partnership (KTP) research project that has refined the materials and processes of offsite manufactured timber-frame wood fibre insulated construction systems for nearly-to-zero embodied and operational carbon dwellings in Wales, UK. Context to the need for reducing carbon emissions is given from a UK and Wales perspective and targets for 60% of housing to be constructed from timber in Wales. The methodology presented outlines a step-by-step process from material supply to fully manufactured two-dimensional (2D) closed panel systems for walls, roofs, and floors, using equipment that only three manufacturers use in the UK for fibre-based insulation. The outcomes of the methodological implementation have led to full scale production of the 2D prefabricated panels systems for Gwynfaen, Wales' largest zero-carbon social housing development. The next steps are discussed which include performance testing of the panel systems in manufacture and during construction.

Keywords: Nearly-zero operational and embodied carbon · Offsite manufacturing of timber-frame modern methods of construction · Wood fibre insulation · Circular economy

1 Introduction

This paper discusses how an Offsite Manufacturer (OSM) in Wales, UK of Modern Methods of Construction (MMC) has introduced new processes for timber-frame building fabric closed panels, to deliver nearly-to-zero embodied and operational carbon new housing [1]. The changes to the material supply, design for manufacture, manufacturing and construction site erection processes developed as part of a knowledge transfer partnership research project are discussed and illustrated [2]. Choice of materials have the capacity to help significantly improve carbon reduction through performance enhancement, and the refinements includes the use of blown wood fibre. Aspirations of achieving a circular economy will directly inform material choices, where it will be expected that

J. Littlewood and R. J. Howlett (Eds.): SEB 2022, SIST 336, pp. 386–396, 2023.
https://doi.org/10.1007/978-981-19-8769-4_36

materials used are recycled, or they are recyclable so that they can be reintroduced into the supply chains at the end of their working life through the process of recycling, repurposing and natural biodegrading. A case study will be drawn upon as an example of the implementation of the research findings. This will examine the use of natural insulation sources and innovative methods in which they are applied in the manufacturing process. The case study discussed is the Gwynfaen housing project, Wales's largest near to net-zero carbon housing project at the time of this paper's submission [3–5] and is funded by the third tranche of the Welsh Government's (WG) Innovative Housing Programme (IHP3) [6]. The aim of IHP3 is to trial the use of innovative techniques and MMC to increase the scale and pace of social housing delivery.

2 Carbon Reduction Targets for New UK Dwellings and OSM

2.1 UK Carbon Reduction Targets

In June 2019 the UK was one of the first major world economies to make legal obligations to end its global warming contribution by 2050. This requires all greenhouse gas (GHG) emissions to be brought to net zero by 2050, replacing the previous target of an 80% reduction from emission levels in 1990 [7]. The European Commission estimates that the construction sector is responsible for 40% of all energy consumption and 36% of all GHG emissions in Europe [8]. More recent proposed changes to building regulations would require new homes in the UK to produce 30% less operational carbon [9]. Such regulations are set to come into force in June 2022.

Throughout Wales, one of four principalities in the UK, several housing associations have pledged to meet net zero targets from as early as 2025, whilst the WG require 60% of new homes to be constructed using timber as of 2018 [10]. This includes new built and retrofit projects [11]. However, the current housing stock in the UK has a significant impact on energy consumption and carbon emissions, where 30% of energy consumption and 25% of carbon emissions can be linked to housing. Of that energy use, 78% is used for heating purposes [12]. More recent estimations attribute 40% of the UK's total energy consumption to the heating and powering of buildings [13]. This evidences a need to reduce heating demand through better building fabric performance. This is having a negative environmental impact through excessive energy consumption, and a negative economic impact for its occupants. Furthermore, poorly performing building fabric traps moisture, can encourage mould and risks affecting the health and quality of life of occupants through poor air quality [14, 15].

The Warm Homes and Energy Conservation Act 2000 outlines the statutory requirement of the WG to eradicate fuel poverty, defined as a household that spends more than 10% of its income on energy costs [16]. Switching to renewable or regenerative material sources helps to make a transition to a more circular economy and move away from a reliance on petrochemical based materials, and the turbulent oil markets that dictate their pricing.

2.2 OSM of MMC Using Natural and Renewable Materials

OSM helps to significantly reduce embodied carbon through the control and standardisation of processes in a weather-resistant environment [17]. This results in a reduction in carbon emissions from transportation, waste material and time. Furthermore, the level precision and repeatability that MMC facilitate allow for improved building performance and reduced operational carbon [18]. OSM involves the production of a range of building component sub-assemblies and modules within the factory environment. The primary intent is to produce as many components as possible in a controlled environment [19]. This is in contrast to the traditional methods of construction in the UK, where materials are assembled on a building site in the location where a building is erected [20]. This can often limit the materials and processes used to construct a building and introduce environmental influences that risk delaying the building process. Furthermore, on-site construction lacks the precision that can be achieved in a factory setting with the use of bespoke tooling, process standardisation and automation [17]. As a result, there can be a negative effect on the building performance, generation of waste and transport-related carbon emissions.

OSM offers complete process control in which tolerances can be significantly reduced through standardisation, and waste can be monitored, controlled and recycled. The resulting building is significantly more likely to have superior acoustic, thermal and fire performance, with lower carbon footprint and shorter lead time when compared with a traditional onsite build process [21].

Sevenoaks Modular (the company hereafter) have been manufacturing timber-framed housing systems for over 30 years. The company specialise in the design and manufacture of internal and external wall panels, floor and roof cassettes and roof trusses [1]. All these components can be supplied to a high precision, with factory fitted insulation and certified thermal, structural, acoustic and fire ratings. The OSM process allows components to be digitally designed before being sent to automated cutting and assembly systems. This ensures the highest precision and consistency and removes the risk of human error.

Once foundations are completed on the building site, panels are delivered in an assembly-ready state, and are often erected in a matter of days. More recently the company have expanded their portfolio to include volumetric building modules that are delivered to site and bolted together to produce a complete building. This MMC has the advantage of including all the utilities and services completely integrated before arriving onsite. Once delivered to a site and bolted together, utilities simply need connecting to an external mains supply before being ready for occupation. The company have a strong, long-term partnership with Hale Construction who are responsible for the on-site building process [22]. This long-term collaboration has resulted in a narrowing performance gap between design, through manufacture and into construction. The continuous collaborative approach to knowledge transfer has resulted in improved quality control processes and ultimately facilitated a streamlined process of construction.

The company's continuous support of the KTP project has funded a dedicated researcher to help introduce a more detached perspective on the design, manufacture, and collaboration process. This paper demonstrates how this collaboration has enabled the provision and implementation of insights, suggestions and improvements from other industrial experiences and academic sources to further improve the OSM process and the capacity to deliver nearly net-zero carbon housing construction.

3 Methodology

The assembly of a timber frame closed panel begins with the cutting of all the constituent timber stud lengths (Fig. 1a). Using a state-of-the-art automated cutting station, each job can be digitally loaded into the system directly from the design office. The intelligent system software optimises timber yield for each job and has a waste timber conveyor section to ensure controlled collection of any waste material generated. Sawdust is also extracted, filtered and collected for reuse as a source of insulation where possible. Upon completion (Fig. 1b) the cut studs are assembled in an automated framing station where they are fixed into a frame to make up a single panel (Fig. 1c). An automated nailing bridge applies an OSB sheathing to one side of the panel, followed by the manual application of a vapour barrier. At this point the part-open frame is inverted to expose the cavities between the studs which are to be filled with blown wood fibre insulation (Fig. 1d).

Fig. 1. Process stages of cutting and framing a modular panel

The advantage with the blown insulation system is that the timber-frame closed panels are manufactured on a gantry above a conveyor, allowing ease of integration into a linear, production-line-based workflow (Fig. 2a). A central control unit houses the blown insulation material where it is fed into delivery head that it mounted to this gantry. Within this unit is a Human-Machine Interface (HMI) in which settings can be adjusted or pre-set parameters can be selected. The software-controlled nature of the system helps to deliver a consistent density and fill level of insulation for a guaranteed performance level. Post-filling (Fig. 2b), OSB is applied to close the filled panel. At this stage the filled panel can have an external breather membrane fixed in position on the timber-frame closed panel, in addition to any fenestration and cladding battens necessary before the weather protection material finish is fixed in position on the external surface of the timber-frame closed panel. In some cases, where brick slips are required for example, the panel is not completely clad as spacing is left to accommodate the panel joining process on site. These are then filled post-installation. Designs are generated to allow for as much operations as possible to be conducted in the factory environment, to limit operation time during construction and retain as much quality control of the manufacturing process as possible.

Fig. 2. Examples of blown insulation production stages and a wood fibre filled panel

3.1 Gwynfaen Farm Case Study

The first project for the use of the new timber-frame closed panels with blown wood fibre insulation is located at Gwynfaen Farm in South Wales [11]. To date, Gwynfaen Farm will be the largest nearly-to-zero embodied and operational carbon housing project in Wales. This project represents a milestone for Wales and the objectives of the WG to achieve overall net-zero carbon emissions.

The project's architects, Stride Treglown, supported the use of OSM for this project due to the thermal performance of the building fabric as well as the improved speed, precision, waste reduction and health and safety of the manufacturing and construction process [ibid]. The precision achievable with OSM was of value to the reduction in operational carbon emissions due to the consistent level of airtightness achievable [23]. Through the KTP, specialist airtightness testing training was undertaken, which enabled the transfer of knowledge on best practices to be specifically transferred to this project.

Furthermore, the KTP provided the resources for in-depth materials and equipment research to help deliver the net-zero carbon targets from the concept design stages. The company purchased and installed new insulation filling equipment to facilitate a lower carbon approach. Furthermore, their ongoing close relationship with Hale Construction was key to ensuring a fluid transfer of low carbon aspirations, and the transfer of expertise to ensure this translated into the onsite construction phase of the project.

3.2 Refinement to OSM Processes for Insulating MMC Panel Systems

Successful utilisation of the applied research developed during the KTP project allowed for in-depth investigation into the existing OSM process within the company factory. Having dedicated personnel to drive continuous improvement helped lead to the refinement of production processes and integration of new equipment into the factory. The existing Trisowarm closed panel wall product involved the in-filling of a closed panel with a polyisocyanurate (PIR) expanding foam material [24]. Whilst the use of PIR as an insulation material yielded a good thermal performance and limited thermal bridges, it relied upon the use of unsustainable, environmentally damaging, non-renewable and non-reusable petrochemical-based insulation products.

With the support of the company and Cardiff Metropolitan University an evaluation of low carbon insulation alternatives highlighted the potential benefits of using blown insulation forms using specialised filling equipment from X-Floc [25]. Blown insulation is more commonly associated with insulation retrofitting rather than at the point of manufacture, and cellulose—which utilises wastepaper pulp—is a typical example of this and has been used for several years [26].

There are several other forms of blown and natural insulation, including sheep wool, rockwool glass fibre or glass beads [27]. Through the KTP, wood fibre insulation was identified as a material that could help propel the Gwynfaen project towards its net zero carbon target. The use of shredded and processed timber as an insulative material added value to a waste product currently generated by the company. This promoted a circular approach to material use and helped cut embodied carbon associated with material sourcing. Furthermore, the use of blown wood fibre over alternative sources of insulation helped to limit the range of materials used in the prefabricated wall panels. The use of wood fibre was a key decision taken during the early design phase of the wall panels. Throughout the design process the whole lifecycle of the building was considered, including consideration of the viability of selected materials for recycling, re-purposing and downcycling during the deconstruction phase. This detailed research process led to the design of the prototype system shown below in Fig. 3.

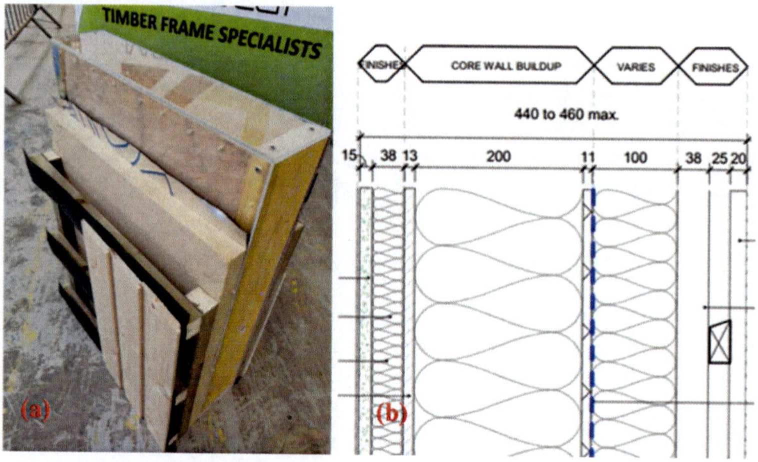

Fig. 3. Gwynfaen project exterior wall panels. **a** Prototype construction detail and **b** drawing

4 Results: In-Factory Outcomes of Refinement to OSM Processes for Woodfibre Insulated MMC Panel Systems

Funding from the KTP enabled training on lean manufacturing principals and six-sigma accreditation [28]. In line with these principals, the transition to a lower waste, higher efficiency manufacturing process was broken down into two main stages: firstly, the removal of waste from production through a Value Stream Mapping study [28]. This helps to cut the production of physical waste as well as wasted time and energy that all contribute to the embodied carbon of a product. Secondly, following the upfront removal of production waste, a process improvement project was undertaken to increase the consistency in production and yield the highest possible performance from the products in manufacture. This second stage was done in conjunction with the refinement of product design and its constituent materials. The aim is to ensure the expected design performance matched the real-world performance through the minimisation of inefficiencies in the wall makeup.

The economic outcomes for both the company and the end customer were easily quantifiable. The Gwynfaen project enabled a complete return on investment for the X-Floc filling station and resulted in a saving of over 300% on materials relative to the initial investment. Time and Motion studies were conducted to evaluate the real-word costs and embodied carbon of the production process [29, 30]. These activities uncovered that the manufacturing cost was lower than previously estimated. As a result, quotations were more competitive and the company's low-carbon solutions were more accessible to customers, such as the housing associates involved in the Gwynfaen housing project.

With respect to the performance of wood fibre, as well as its low thermal conductivity (0.038 W/m·K), it has a lower risk of waning performance overtime. Alternative blown insulation mediums such as cellulose risk vertical slumping resulting from gravitational forces. This results in the formation of voids within the insulative layer and leads to thermal bridging. The wood fibre holds to its original form within the wall cavity to ensure uniform thermal performance across the panel. The absence of any synthetic chemical adhesives also removes any Volatile Organic Compounds (VOC's) from the building makeup. This approach was mirrored with façade finishing, which uses locally sourced, low VOC materials and treatments [23]. This ensures the building fabric should not contaminate internal air quality and should provide a safer living environment for the occupants. Furthermore, the use of natural materials helps to maintain a circular approach to material use by retaining a high level of recyclability for end-of-life processing.

Manufacturing consistency and resulting thermal performance can be ensured with programmable flow volume and fill density parameters inputted on the X-Floc machine [31]. The flow head has configurable outputs that can be turned on and off depending on the geometry of the panel, as well as positional and fill level sensors to ensure consistency from the operator. As an additional quality control process, samples of the filled insulation are intermittently removed and weighed to reference against the density of that programmed into the system controller.

The semi-automated nature of the X-Floc system removes the need for a separate cutting and shaping process [31]. This also removes any human error associated with panel and batt-based insulation. The results are a more efficient use of personnel and equipment resources; it reduces the travel of materials across the production floor and utilises the space on the production line through which a panel under construction would typically pass. The result is a leaner process a more linear workflow, a lower embodied carbon, lower cost of manufacture and reduction of waste materials. Figure 4 illustrates a timber-frame closed panel which will form part of a roof prior to insulation installation (Fig. 4a) and after installation using the X-Floc equipment (Fig. 4b).

Fig. 4. Roof panel, **a** prior to insulation. **b** Injected with shredded wood fibre insulation.

5 Discussion

Through the application of tried-and-tested timber framed OSM techniques it was possible to support the delivery of a building project that delivered a lower carbon alternative to traditional on-site building methods. Through the utilisation of the KTP scheme and its resources, it was possible to further enhance the low-carbon offerings from the company using novel production methods and placing equal weighting on the performance, carbon footprint and applicability to a circular economy when selecting materials. This was achieved through desk-based research, primary data collection, technical training, and constructive engagement with industry bodies and events, and competitor companies.

The next phase of the project will involve airtightness and thermography testing during manufacture and the construction phases of the Gwynfaen project, to compare the real-world performance data with the as-designed thermal performance. A comparative analysis between the two will highlight the extent to which research has been applied successfully during the production and construction processes. Any discrepancies will likely demonstrate areas for improvement and further refinement. Initial data will give a strong indication of the extent to which carbon negating targets have been achieved, and the scale of the challenge faced to successfully meet the existing and upcoming legislative requirements for operational and embodied carbon of buildings in Wales and the rest of the UK. A post-completion evaluation of waste generation and material consumption can also be undertaken in line with the onsite waste management plan. This will evaluate further opportunities for waste reduction through better practice or and expansion of application of a circular economy in the building makeup. In addition, the feasibility of implementing an automated 'smart factory' software infrastructure across the company's manufacturing operations are also part of the next steps. There is the potential to explore real-time workflow tracking and environmental monitoring and management. Systems already in use in the automotive, aerospace and medical manufacturing sectors for legislative and productivity monitoring use could be adopted [32]. For example, providing a unique ID to products from the point of delivery bring the opportunity to create a cloud-based database of materials, panel assemblies and their constituent components. This could help to provide a live, paperless, configurable, and futureproof digital infrastructure to support continuous improvement across the entire product lifecycle and provide tangible data for project evaluation.

6 Conclusion

This paper has examined the significance of OSM in the decarbonisation of the construction industry and how the advantages brought forward by OSM can be capitalised upon to further improve the process of decarbonisation in the quest for achieving net-zero. Through the application of Value Stream Mapping, Time and Motion studies and embodied carbon evaluation of timber framed panel makeup, it was possible to reduce the embodied carbon amounting from material selection, production processes and waste when compared with the existing inhouse Trisowarm system. The process of investigation and successful outcomes from the project have given rise to further areas that can be subjected to a similar level of analysis and scrutiny in order to further reduce the operational and embodied carbon of building products used in MMC.

Acknowledgements. The research has been funded and supported by SoModular, the WG and InnovateUK. Thanks, also to J G Hale Construction, Coastal and Pobl Housing Associations.

References

1. SoModular: What we do. https://somodular.co.uk/ (2022). Last accessed 11 May 2022
2. InnovateUK: Knowledge Transfer Partnership—details, Ref Knowledge Transfer Partnership—details. https://info.ktponline.org.uk/action/details/partnership.aspx?id=11178 (2022). Last accessed 30 May 2022
3. Design Commission for Wales: IHP Design Review Report Gwynfaen, Swansea. http://dcfw.org/wp-content/uploads/2020/02/DR-Report-Gwynfaen-Swansea-June-2019.pdf (2019). Last accessed 30 May 2022
4. Stride Treglown: Gywnfaen Innovative, ultra-low carbon housing development with great social value. https://stridetreglown.com/projects/gywnfaen/ (2020). Last accessed 30 May 2022
5. Coastal Housing Group: Gwynfaen - Penyrheol, Swansea. https://www.coastalha.co.uk/our-developments/ (2022). Last accessed 30 May 2022
6. WG: Innovative housing programme: guidance. https://gov.wales/innovative-housing-programme-guidance (2020). Last accessed 30 May 2022
7. Now is the Time for Timber: https://timefortimber.org/wp-content/uploads/2021/05/Time-for-Timber-insurance-industry-white-paper-FINAL.pdf (2021). Last accessed 30 May 2022
8. European Commission: 2020 Report on the State of the Energy Union Pursuant to Regulation (EU) 2018/1999 on Governance of the Energy Union and Climate Action. European Commission, Brussels, Belgium (2020)
9. https://governmentbusiness.co.uk/news/15122021/new-homes-produce-third-less-carbon
10. Green, E., Forster, W.: More|Better: An Evaluation of the Potential of Alternative Approaches to Inform Housing Delivery in Wales. Cardiff University, Cardiff, UK (2017)
11. WG: WG43508 Working together to reach net zero: all Wales plan 2021-25. https://gov.wales/working-together-reach-net-zero-all-wales-plan (2021)
12. Wright, A.: What is the relationship between built form and energy use in dwellings? Energy Policy **36**, 4544–4547 (2008)
13. UK Government: New homes to produce nearly a third less carbon. http://www.gov.uk/government/news/new-homes-to-produce-nearly-a-third-less-carbon (2021). Last accessed 06 June 2022
14. Houses of Parliament Post Note: UK Indoor Air Quality. http://www.parliament.uk/globalassets/documents/post/postpn366_indoor_air_quality.pdf (2010). Last accessed 06 June 2022
15. Jahic, D., Littlewood, J.R., Karani, G.: Piloting a management and evaluation protocol for occupant quality of life in Welsh dwelling retrofits. In: Littlewood, J.R., Howlett, R.J., Lakhmi, C. (eds.) Sustainability in Energy and Buildings 2020. Howlett, R.J., Lakhmi, C. (series eds.) Smart Innovation, Systems and Technologies, vol. 203, chap. 39, pp. 467–478. Springer, Singapore (2021). https://doi.org/10.1007/978-981-15-8783-2_39
16. Future Generations Commissioner for Wales: The Future Generations Report 2020. http://www.futuregenerations.wales/wp-content/uploads/2020/06/Chap-5-Housing.pdf (2020). Last accessed 06 June 2020
17. Freeman, H., Christie, L.: Reducing the whole life carbon impact of buildings. https://researchbriefings.files.parliament.uk/documents/POST-PB-0044/POST-PB-0044.pdf (2021). Last accessed 30 May 2022
18. Bertram, N., Fuchs, S., Mischke, J., Palter, R., Strube, G., Woetzel, J.: Modular construction: from projects to products. https://ivvd.nl (2019). Last accessed 30 May 2022
19. Arif, M., Egbu, C.: Making a case for offsite construction in China. Eng. Constr. Archit. Manag. **17**(6), 536–548 (2010). https://doi.org/10.1108/09699981011090170
20. Pan, W., Gibb, A.G.F., Dainty, A.R.J.: Leading UK housebuilders' utilization of offsite construction methods. Build. Res. Inf. **36**(1), 56–67 (2008)

21. Hsieh, T.-Y.: The economic implications of subcontracting practice on building prefabrication. Autom. Constr. **6**(3), 163–174 (1997). https://doi.org/10.1016/S0926-5805(97)00001-0. Stride Treglown: Gwynfaen. http://stridetreglown.com/projects/gwynfaen (2022)
22. Hale Construction: Home. http://www.haleconstruction.co.uk (2022). Last accessed 07 June 2022
23. Stride Treglown: Gwynfaen. http://stridetreglown.com/projects/gwynfaen (2022). Last accessed 06 June 2022
24. Sevenoaks Modular: Trisowarm. http://somodular.co.uk/products/timber%20frame/trisowarm (2022). Last accessed 07 June 2022
25. X-Floc: Products. http://www.x-floc.com/en/products/ (2022)
26. Technology Strategy Board: Retrofit for the future. https://assets.publishing.service.gov.uk/government/uploads/system/uploads/attachment_data/file/669113/Retrofit_for_the_future_-_A_guide_to_making_retrofit_work_-_2014.pdf (2014). Last accessed 07 June 2022
27. US Department of Energy: Types of insulation. http://www.energy.gov/energysaver/types-insulation (2022). Last accessed 07 June 2022
28. The Council for Six Sigma Certification: Lean Six Sigma Yellow Belt Certification. http://www.sixsigmacouncil.org/lean-six-sigma-yellow-belt-certification (2022). Last accessed 07 June 2022
29. Moorhouse, V.L., Littlewood, J.R., Hale, E.: A pilot study evaluating offsite manufacturing of timber frame panels using lean manufacturing principles for dwellings. In: Sustainability in Energy and Buildings 2020. Howlett, R.J., Jain, L.C. (series eds.) Smart Innovation, Systems and Technologies, vol. 203. Springer, Singapore (2021). https://doi.org/10.1007/978-981-15-8783-2_42. Last accessed 11 May 2022
30. Moorhouse, V.L., Littlewood, J.R., Hale, E.: Optimising offsite manufacturing of timber-frame roof trusses for UK housing. In: Vol 1: Emerging Research in Sustainable Energy and Buildings for a Low-Carbon Future. Advances in Sustainability Science and Technology, chap. 21, pp. 341–364 (2021). https://link.springer.com/chapter/10.1007/978-981-15-8775-7_21. Last accessed 11 May 2022
31. X-Floc: X-Floc Pneumatic Insulation Technology. https://www.x-floc.com/en/start/ (2022)
32. Gartner, P., Benfer, M., Kuhnle, A., Lanza, G.: Potentials of traceability systems—a cross-industry perspective. In: 54th CIRP Conference on Manufacturing Systems. Procedia CIRP **104**, 987–992 (2021)

Energy Communities: The Concept of Waste to Energy-CHP Based District Heating System for an Italian Residential District

L. Pompei(✉), F. Nardecchia, V. Lanza, L. M. Pastore, and L. de Santoli

Department of Astronautical Electrical and Energy Engineering, University of Rome La Sapienza, Rome, Italy
laura.pompei@uniroma1.it

Abstract. Carbon dioxide, the main greenhouse gas responsible for global warming, comes primarily from fossil fuel-based energy production, which is still prevalent today. Therefore, European countries are adopting several directives and guidelines, as well as promoting many pilot projects. In this context, the district heating powered by renewable sources is still considered a sustainable future energy infrastructure for cities to face environmental changes and pollution. A suitable energy generator for district heating systems is still combined heat and power plants, wherein the use of the waste as an energy source is still a promise. Several works in literature investigated the advantages of Waste-to-Energy to supply the energy demands of a city, underlining their impact on energy savings and environmental sustainability. However, districts and cities still require many efforts to shift from fuel-based energy systems to renewable source ones. In this framework, this study aims to develop a renewable and efficient energy system applied to an Italian neighbourhood. A Waste-to-Energy-Combined heat and power-based district heating is analyzed coupled with photovoltaic systems installation. Furthermore, hydrogen production and storage are also involved, to maximize the super-plus of energy obtained by the cogenerator plant. Results underline the positive impact of these strategies, in terms of energy savings and independence from fuel sources.

Keywords: Renewable energy systems · Residential district · District heating · Waste-to-energy

1 Introduction

During the last decades, the effects of the global warming are visible worldwide [1]. European countries are facing this urgent issue, promoting several directives and guidelines to reduce the fuel emissions, energy consumptions and levels of pollutions [2]. Carbon dioxide, the main greenhouse gas responsible for global warming, comes primarily from fossil fuel-based energy production, which is still prevalent today [3]. To limit global warming, the Paris Agreement signed at COP 21 in December 2015 was a major step forward. In the Paris Agreement, the goal is to limit global temperature rise to less than 2 °C over pre-industrial levels, and to pursue efforts to reduce temperature increases to 1.5 °C above pre-industrial levels by 2050 [4].

J. Littlewood and R. J. Howlett (Eds.): SEB 2022, SIST 336, pp. 397–406, 2023.
https://doi.org/10.1007/978-981-19-8769-4_37

In this framework, several works aimed to exploit the available renewable energy sources (RES) combined with efficient energy systems for providing green and clean energy [5, 6]. Both thermal and electrical needs are evaluated in literature [7, 8], to decrease the global energy demand of the building sector. In fact, buildings play a critical role in achieving sustainable development since the buildings sector contributes 36% to final energy use [9–11]. In a recent work [12], exergy and energy analysis are carried out to evaluate the behavior of a microgrid connected to residential, commercial and office buildings, supplied by RES, in different geographical locations. The aforementioned analysis aimed to investigate the benefits coming from the district heating (DH) powered by photovoltaic systems based on different climate conditions. The DH powered by renewable sources is still considered a sustainable future energy infrastructure for cities to face environmental changes and pollution [13].

Furthermore, literature shows that combined heat and power (CHP) plants offer the best performance in terms of energy generators for DH systems [14, 15], providing approximately 56% of heat. As mentioned in [16], reducing CO_2 emissions and reducing primary energy needs are also compelling features of CHP plants. In several countries, CHP plants are applied to DH systems, pointing out their potential and weaknesses. Among the CHP energy sources, the amount of waste disposed on landfills can be used, applying the concept of Waste-to-Energy (WtE) [17]. As a result, waste and renewable heat potentials have been individuated [18] and then investigated in different countries [19]. In a scientific study [20], combustion of waste for baseload and biomass for peak demand is analysed, while in other research [21] the use of WtE technologies alongside energy storage can support higher installation costs. Another work [22] is focused on enhancing the advantages of WtE and district cooling (DC) integration in existing Gas-CHP based district heating system. There is a 33% higher potential for energy-from-waste during summer as opposed to winter, which is counter to heating demand and in line with the annual distribution of energy needs for cooling. As many works underlined [22, 23], the residential sector was explored especially for the heating needs, since its final electricity use is lower compared to public buildings ones.

In line with this, the current Italian districts still need many efforts to recover, in terms of energy and sustainable ways, the existing residential buildings [24, 25].

In this context, the aim of this study is to model an energy system that guarantees the energy independence of an Italian district of Rome, both for the electric and thermal load, covering the energy needs of citizens and taking advantage of the waste through a waste-to-energy plant. Moreover, the surplus of energy obtained is used to produce and store the hydrogen, a clean energy vector. Consequently, a group of public vehicles for the waste collection could use the hydrogen stored, increasing the share of green mobility in the district. Although the presented work is a case study, the model of the energy systems developed could be replicable in a different context.

2 Materials and Methods

2.1 Case Study

Garbatella is one of the most attractive urban zones in Rome. Located in the south part of the city, at 3 km from the historic center, Garbatella is probably the last example of ancient popular district, inhabited by old families and not yet invaded by hotels and trendy clubs, unlike other historic areas (Trastevere, Testaccio, Monti etc.) that becoming very touristic and nightlife spots. From the demographic point of view Garbatella counts 43 thousand of inhabitants, with a population density of 14,300 each kilometre, more than seven times the average of Rome.

Regarding the energy field, Garbatella is almost unprovided of distributed renewable energy production. The total of the electric load of the district is covered by the national grid, wherein each kWh has a 19.4% of renewable source. With an annual demand of 130 GWh, the CO_2 emission amounted to 33.3 kTon, while the average PUN, indicator of the electric bill cost, has been in 2019 more than 50 €/MWh [25].

The thermal needs as well as domestic hot water is provided almost entirely by natural gas boilers, generally on condominium-scale. Evaluating the district thermal load of 200 GWh/year, the CO_2 produced by the boilers amounts of 47.3 kTon, with an average cost of 20 €/MWh [26]. The dependence of the district to supply its thermal and electric load from the fossil fuel, in particular the natural gas (NG), determines not only the emission of greenhouse gases but also an unpredictable fluctuation of the costs, being the NG flow affected by geopolitical events (especially in the current times).

2.2 Plant Description

The energy plant consists in three different components, that will be analysed first separately and then as a unique system (Fig. 1).

- Photovoltaic panels (PV) distributed plant. To increase the renewable amount of energy consumed in the district and guarantee the energy independence, the roofs of both public and private buildings will host solar photovoltaic panels, obviously where there isn't any architectural constraint.
- Waste-to-Energy plant (WtE) plant and District Heating (DH). The PV system alone is not able to supply the entire electric load of Garbatella; moreover, to reduce the dependence from fossil fuel and the bill costs, the project foresees the implementation of a district heating network, needing a central plant to produce the required thermal power. A relevant solution is entrusting the thermal production and part of the electric one to a Waste to Power plant. Garbatella waste production is not even close enough to supply the plant of the size needed, so will be required a part of the waste collected in the rest of Rome.
- Hydrogen Production and Storage (H2). Using the renewable energy in excess, hydrogen could be produced, through an electrolyze process, to fuel sustainable vehicles.

2.3 Case Study Energy Requirements

Firstly, it was necessary to calculate the energy requirements, both thermal and electrical ones, of the entire district. Following are briefly reported the method and the results obtained. The evaluation of electrical loads started from data furnished by Terna [27], that reported the actual power demand, metered every 15 min for the entire year (2019).

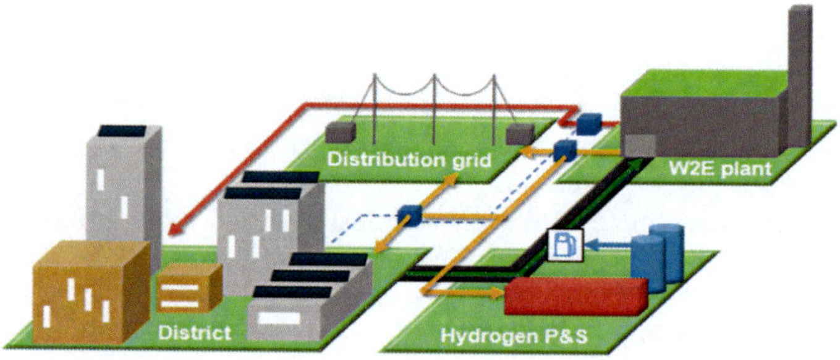

Fig. 1. Qualitative scheme of the energy plant proposed.

Using then the IEA statistics [28], the residential and service electric load is attested to 52.36% of the total (the rest is consumed by the industry). This value was therefore scaled on the district population, obtaining for the electrical needs a peak power of 22.94 MW and an annual energy of 122.46 GWh. Moving to thermal loads, their evaluation was developed based on the Italian energy use for space heating [x] and the number of district's inhabitants. Domestic hot water load was also considered. The final thermal need is about 195.202 MWh. Regarding the operating schedules, three periods can be identifying. The first one, named "high demand", in the winter months, and a second one named "medium demand" (half of the high one), in the weeks before December the 21st and after March the 20th. The load is constant during the day; indeed, even if the domestic heating plants can be switched-on only 12 h every day and there are fluctuations caused by the users' daily activities, the WtE plant will supply only the baseload of the thermal demand, while the peaks will continue to be covered by the condominium heater. The third and last period, when is not request energy for space heating, the thermal load is constituted only by the energy for domestic hot water.

3 Results

3.1 PV Plant Results

The panel selected, crystalline silicon with high performances, has a surface of 1.86 m^2 and a peak power of 400 W. At the latitudes of Rome, the maximum yield from the panel is characterized by an angle of 30° and an orientation of 0° South; the surface occupied by one panel is 3.56 m^2. The calculation for the irradiation is a complex step. The goal of the design is to obtain a curve with the power produced by the PV plant every 15 min in line with the electric load. Starting from the daily maximum irradiation along the year, recovered from a climate metering station near the district, it is possible to model the values through the time of sunrise and sunset. The annual productivity resulting amounts to 1.584 kWh/m^2. Through an approximated calculation, the total surface of the district is equal to 3.05 km^2; according to statistic data, the 30% of the total area is occupied by buildings, fixing the half of those buildings under architectural constraints, the available surface is about 0.46 km^2. Then, evaluating the real surface a quarter of the one calculated (the roofs are not entirely available), the final available surface is 114.375

m^2. Based on the available area of roofs calculated, the production of the PV plant is 18.41 GWh and its trend is showed in Fig. 2.

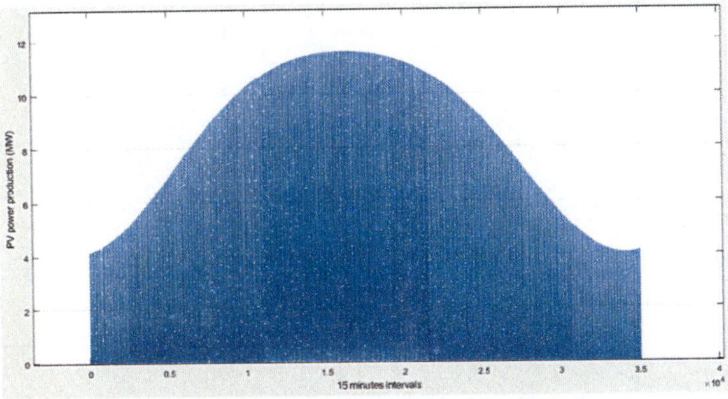

Fig. 2. PV power production obtained.

3.2 WtE Plant Results

After the PV plant installation, the district electric load is varied. The maximum power demand decreases about the 10% and more than the 15% of the annual energy load is covered by the photovoltaic production.

It is well-known that the thermal load will be more energy-intensive compared to the electric one, therefore the design process of WtE starts imposing the nominal electric power of the plant equal to the maximum power of the electric load, about 20.8 MW. WtE plant is a conventional steam plant, that uses as fuel the waste properly treated to produce a steam flow at determinate temperature and pressure. The cogeneration configuration guarantees a relevant efficiency, providing both thermal and electrical energy to the final users. In this case a part of the steam flow is drawn off at a fixed stage in the turbine and used in a heat exchanger to deliver the thermal energy to the users. The operating variables are evaluated starting from existing similar plants, and all the thermodynamics calculation are also modelled in Simulink. Summarizing, the thermal power required in the steam generator, considering a combustion efficiency of 0.9, is 84.05 MW and the electric efficiency of the plant at the nominal power is attested to 0.247. The next step is to evaluate the fuel flow needed to generate the thermal power in the steam generator. The municipal unrecyclable waste is composed by different elements, usually of four types: biological fraction, plastic, paper and metals and inert materials. It is well known that the lower heating value (LHV) of waste increases when the fraction of paper and plastic increases. Common values of LHV are around 10–12 MJ/kg, with an ash percentage of 35% and a humidity of 30% (evaluated on the dry weight) [29]. To increase the LHV and reduce both humidity content and ash production, the waste is treated through a series of thermo-physical process. The final product, called CSS (solid secondary fuel), has a higher LHV and good environmental properties. To set an accurate analysis, it was used a typical composition of a CSS, and through the oxidation energies of the different elements (carbon, hydrogen and sulphury), it was therefore possible to assess the LHV. Based on the LHV of the fuel obtained, the CSS flow rate (M_{CSS}) needed to generate the necessary thermal power in the steam generator was evaluated. Finally, assuming an annual operation of 7800 h, the annual request of municipal waste (M_{waste}) of the plant was obtained (Table 1).

Table 1. Annual request of municipal waste and fuel properties.

Fuel flow characteristics		
LHV	20.26	MJ/kg
M_{CSS}	14.93	ton/h
M_{waste}	232.440	ton/y

Table 2 summarizes the main properties of the WtE according to the three periods of the year (high demand, medium demand, and lower demand), as pointed out in Sect. 2.3. During the period of maximum thermal load, to cover the demand of the district, a considerable flow of steam will be extracted from the turbine, which will reduce the production of electricity, as well as the electrical efficiency of the plant. On the other hand, by reducing the thermal load in the autumn months, and even more markedly in the summer months, the amount of steam used in electricity production increases, as does the efficiency, until it reaches its maximum value in the summer months. Therefore, during the summer months there is the peak of electricity demand, which can be satisfied by the combined production of the PV plant and the WtE.

Table 2. Characteristics of the WtE according to the three periods of the year.

Period	P_{el} (MW)	η_{el} (%)	η_{th} (%)
High demand	14.21	16.9	66.54
Medium demand	17.52	20.8	33.27
Lower demand	20.57	24.48	2.5

To complete the technical analysis of the WtE plant, it is also reporting the final data about its production. The final annual electric energy produced is about 136.8 GWh, with an average global efficiency of 46.55%. In addition, the CO_2 emission every year will be 157.446 ton/y, wherein the emission factor evaluated in the fuel analysis is 1.349 tonCO_2 each ton of CSS. Based on IRENA calculations, the CO_2 emission obtained must decreased of 50%, therefore 78.723 ton/y is the final value.

3.3 H2 Plant Results

Considering the production of electricity from the WtE plant, the trend of the district electric load is changed (Fig. 3).

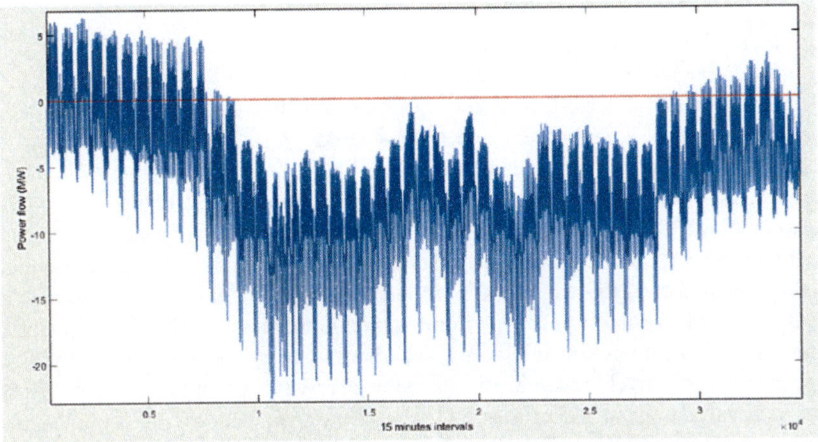

Fig. 3. Trend of the district electric load considering the presence of PV system.

Due to maintain a high level of production to guarantee the thermal load during the winter, the power flow is for the 89.1% of the year negative, so the electric power produced in the district is higher than the load requests. The excess of energy is therefore employed to feed a public vehicle park and also to sell the remain amount of energy. The first step is the sizing and the manage process of the hydrogen production. The H2 structure is constituted by three main elements: electrolyser, electrochemical machine that supplied by electrical energy produced H2 through an electrolyte process. Among the storage, two types are involved, or high-pressure cylinders (around 200 bar), which need a compressor and several security devices, or using metal hydrides, a new technology that permits to store high amount of hydrogen at atmospheric pressure. The main limit to the plant sizing is the imposed limit of the storage capacity in the filling stations, fixed by the regulatory body equal to 6000 Nm^3, that considering a density of 0.089 kg/Nm^3, correspond to 550 kg of hydrogen. Knowing this value and the power flow, through a simple MatLab script, it's possible to evaluate the maximum energy storable in a year. Moving to the results, the Electrolyser Power is 47 kWe, the energy stored is 368 MWhe/y and the storage capacity is 491 kg.

3.4 Economic Analysis

The environmental benefits that derive from the establishment of an energy community are supported by the economic ones, given by the reduction of bill costs. Considering the district as an isolated electric system, the Levelized Cost of Electricity (LCOE) of the energy system is calculated. Based on statistical data [30], the total investment cost amounts to about 245 million euros, from which the revenues of selling the super plus of electricity and using incentives provide a considerable cut for the investment [31]. The total cost for the electrical and thermal MWh consumed by the user will thus be € 25.7 and € 20.1 respectively, obtaining considerable savings for the users compared with the current costs (above € 125/MWh [25]).

4 Discussions and Conclusion

This research aims to investigate an energy system able to create an energy community, based on the use of RES [32, 33] and DH networks. Waste-to-Energy (WtE) process is chosen to supply the energy demands of the Italian district, knowing its impact on energy savings and environmental sustainability [34]. Moreover, a photovoltaic system is also employed. The production of extra electricity is used for producing hydrogen and the remaining ones for selling to the national grid. Moreover, this work highlights the positive impact of this kind of the proposed energy system on both economic and social aspects. In addition to the environmental and financial implications, social and cultural tasks play a relevant role on the decarbonization goals. Today more than ever, it is necessary to actively involve citizens in the energy transition process, which arises from political actions and decisions, but it also has an impact on the life of citizens. The energy community, therefore, lays its foundations on the inclusion of the citizens, enhancing the energy transformation process of the city.

Acknowledgement. This work was carried out within the research projects number AR12117A8A48EA00 and AR12117A8A8E6664 promoted by Sapienza University of Rome and DIAEE (Department of Astronautical Electrical and Energy Engineering).

References

1. United Nations: General Assembly of 6 December 1988, A/RES/43/53. https://www.un.org/documents/ga/res/43/a43r053.htm. A history of international climate change policy, overview
2. United Nations: The Sustainable Development Goals Report 2020. United Nations Publications, New York, NY, USA (2020)
3. BP: BP Statistical Review of World Energy 2021. https://www.bp.com/content/dam/bp/business-sites/en/global/corporate/pdfs/energy-economics/statistical-review/bp-stats-review-2021-full-report.pdf (2021). Consulted on 18 Oct 2021
4. UNFCC: The Paris Agreement. https://unfccc.int/process-and-meetings/the-paris-agreement/the-paris-agreement (2015)
5. Nardecchia, F., Groppi, D., Astiaso Garcia, D., Bise-gna, F., de Santoli, L.: A new concept for a mini ducted wind turbine system. Renew. Energy **175**, 610–624 (2021)
6. Nardecchia, F., Groppi, D., Lilliu, I., Astiaso Garcia, D., De Santoli, L.: Increasing energy production of a ducted wind turbine system. Wind Eng. **44**(6), 560–576 (2020)
7. Pompei, L., Mattoni, B., Bisegna, F., Blaso, L., Fumagalli, S.: Evaluation of the energy consumption of an educational building, based on the UNI EN 15193-1:2017, varying different lighting control systems. In: 2020 IEEE International Conference on Environment and Electrical Engineering and 2020 IEEE Industrial and Commercial Power Systems Europe, EEEIC / I and CPS Europe 2020, p. 9160588 (2020)

8. Mattoni, B., Mangione, A., Pompei, L., Spinelli, F., Zinzi, M., et al.: An alternative method for the assessment of the typical lighting energy numeric indicator for different outdoor illuminance conditions. In: Building Simulation Conference Proceedings, vol. 2, pp. 1224–1230 (2019)
9. International Energy Agency, United Nations-Environment Programme: 2019 Global Status Report for Buildings and Construction: Towards a Zero-Emission, Efficient and Resilient Buildings and Construction Sector (2019)
10. Cumo, F., Nardecchia, F., Agostinelli, S., Rosa, F.: Transforming a historic public office building in the Centre of Rome into nZEB: limits and potentials. Energies **15**(3), 697 (2022)
11. Pompei, L., Nardecchia, F., Mattoni, B., Bisegna, F., Mangione, A.: Comparison between two energy dynamic tools: the impact of two different calculation procedures on the achievement of nZEBs requirements. In: Building Simulation Conference Proceedings, vol. 6, pp. 4259–4266 (2019)
12. Pompei, L., Nardecchia, F., Mattoni, B., Gugliermetti, L., Bisegna, F.: Combining the exergy and energy analysis for the assessment of district heating powered by renewable sources. In: 2019 IEEE International Conference on Environment and Electrical Engineering and 2019 IEEE Industrial and Commercial Power Systems Europe, p. 8783426 (2019)
13. Zhang, L., Li, Y., Zhang, H., Xu, X., Yang, Z., Xu, W.: A review of the potential of district heating system in Northern China. Appl. Therm. Eng. **188**, 116605 (2021)
14. Puschnigg, S., Jauschnik, G., Moser, S., Volkova, A., Linhart, M.: A review of low-temperature sub-networks in existing district heating networks: examples, conditions, replicability. Energy Rep. **6–9** (2021)
15. Werner, S.: International review of district heating and cooling. Energy **137**, 617–631 (2017). https://doi.org/10.1016/j.energy.2017.04.045
16. Ravina, M., Panepinto, D., Zanetti, M.C., Genon, G.: Environmental analysis of a potential district heating network powered by a large-scale cogeneration plant. Environ. Sci. Pollut. Res. **24**(15), 13424–13436 (2017). https://doi.org/10.1007/s11356-017-8863-2
17. Matak, N., Tomić, T., Schneider, D.R., Krajačić, G.: Integration of WtE and district cooling in existing Gas-CHP based district heating system—Central European city perspective. Smart Energy **4**, 100043 (2021). https://doi.org/10.1016/j.segy.2021.100043
18. Persson, U., Möller, B., Werner, S.: Heat Roadmap Europe: identifying strategic heat synergy regions. Energy Policy **74**, 663–681 (2014)
19. Dénarié, A., Fattori, F., Spirito, G., Macchi, S., Cirillo, V.F., Motta, M., et al.: Assessment of waste and renewable heat recovery in DH through GIS mapping: the national potential in Italy. Smart Energy **1** (2021)
20. Benedek, J., Sebestyén, T.-T., Bartók, B.: Evaluation of renewable energy sources in peripheral areas and renewable energy-based rural development. Renew. Sustain. Energy Rev. **90**, 516–535 (2018)
21. Rentizelas, A.A., Tolis, A.I., Tatsiopoulos, I.P.: Combined Municipal Solid Waste and biomass system optimization for district energy applications. Waste Manag. **34**, 36–48 (2014)
22. Islam, S., Ponnambalam, S.G., Lam, H.L.: Energy management strategy for industries integrating small scale waste-to-energy and energy storage system under variable electricity pricing. J. Clean. Prod. **127**, 352–362 (2015)
23. Pompei, L., Blaso, L., Fumagalli, S., Bisegna, F.: The impact of key parameters on the energy requirements for artificial lighting in Italian buildings based on standard EN 15193-1:2017. Energy Build. **263**, 112025 (2022)
24. Mangialavori, M., Pompei, L., Nardecchia, F., Tronchin, L., Schibuola, L., et al.: Towards the definition of a sustainable Smart Model for the suburbs redevelopment. In: 2020 IEEE International Conference on Environment and Electrical Engineering and 2020 IEEE Industrial and Commercial Power Systems Europe, p. 9160674 (2020)

25. Pastore, L.M., Sforzini, M., Lo Basso, G., de Santoli, L.: H2NG environmental-energy-economic effects in hybrid energy systems for building refurbishment in future National Power to Gas scenarios. Int. J. Hydrog. Energy **47**, 11289–11301 (2022)
26. GME (Gestore dei servizi Energetici): https://www.mercatoelettrico.org/it/Statistiche/ME/DatiSintesi.aspx
27. Terna: https://www.terna.it/it/sistema-elettrico/transparency-report/total-load
28. IEA (International Energy Agency): https://www.iea.org/
29. Sever Akdag, A., Atımtay, A., Sanin, F.D.: Comparison of fuel value and combustion characteristics of two different RDF samples. Waste Manag. (2016)
30. European Commission Hydrogen Generation in Europe, overview of costs and key benefits (2020)
31. Procedura per la gestione degli incentivi ex certificati verdi. GSE (2017)
32. Pastore, L.M., Sforzini, M., Lo Basso, G., de Santoli: Heading towards 100% of renewable energy sources fraction: a critical overview on smart energy systems planning and flexibility measures. E3S Web Conf. **197**, 01003 (2020)
33. Mauri, L., Vallati, A., Ocłoń, P.: Low impact energy saving strategies for individual heating systems in a modern residential building: a case study in Rome. J. Clean. Prod. **214**, 791–802 (2019)
34. Vollaro, A.D.L., Galli, G., Vallati, A., Romagnoli, R.: Analysis of thermal field within an urban canyon with variable thermophysical characteristics of the building's walls. J. Phys.: Conf. Ser. **655**(1), 012056 (2015)

Renewable Energy System Applied to Social Housing Building in Mediterranean Climate

Andrea Vallati[1], Stefano Grignaffini[1], Costanza Vittoria Fiorini[1], Simona Mannucci[2], and Miriam Di Matteo[1](✉)

[1] DIAEE Department of Astronautical, Electrical and Energy Engineering, "Sapienza" University of Rome, Via Eudossiana 18, 00184 Rome, Italy
{andrea.vallati,miriam.dimatteo}@uniroma1.it

[2] DICEA Department of Civil, Constructional and Environmental Engineering, "Sapienza" University of Rome, Via Eudossiana 18, 00184 Rome, Italy

Abstract. The article analyses a hybrid heat pump system integrated with renewable energy sources both to produce heat and electricity to optimize energy consumption in public residential buildings. The proposed case study is a medium-small sized (13 apartments) building in reinforced concrete built in 1980, site in the province of Rome owned by ATER. This type of building is extremely popular on the outskirts of the city. In the system analysed, the existing boiler and the newly installed air-water-electric heat pump, and hybrid photovoltaic panels are used in a combined way. Using this approach can reduce energy consumption in buildings with low envelope performance, especially where, for different reasons, it is difficult to intervene. Furthermore, the interventions are punctual and require almost no work inside the apartments. The results show that the production of heat from PV-T panels achieves 79% of the thermal energy demand, and 75% reduction in electricity consumption followed. Additionally, electrical energy consumption has a renewable coverage of 100% annually. Finally, primary energy savings can reach 85%.

Keywords: Renewable energy · Hybrid heat pump system · Energy saving

1 Introduction

The building sector accounts significantly for primary energy demand and contributes to about 40% of energy-related carbon dioxide (CO_2) emissions [1]. Existing residential buildings are largely in charge of these issues [2, 3], and due to their widespread in the Italian stock, applying to them defossilization and adoption of renewable based technologies represents a primary concern for meeting European 2030 and 2050 Green Deal targets. In such cases, primary energy consumption can be reduced thanks to HVAC retrofit, wherever difficult to intervene on the building envelope. In residential buildings where a single device provides both space heating, cooling, and hot water production the air-to-water heat pump (HP) has several benefits, including low investment costs and easy installation. During the cold season, the COP of this device decrease when the

J. Littlewood and R. J. Howlett (Eds.): SEB 2022, SIST 336, pp. 407–417, 2023.
https://doi.org/10.1007/978-981-19-8769-4_38

building's maximum energy demand occurs, moreover it is generally designed to meet the maximum load while operating mainly with partial load, oversizing which results in a significant reduction in its seasonal performance [4]. To ease these drawbacks, hybrid systems coupling air-source heat pumps with back-up heaters connected in parallel or in series are demonstrated to be promising. The replacement in buildings by a HHP system would decrease emissions from the direct use of fossil fuel, but due to its electricity-driven technology cause a shift in emissions to the electricity sector [5]. This will lead to higher emissions, as long as the energy demand of buildings is not met by the production of renewable energy [6]. However, the RES unitability and dependency on the climate environment make it worthwhile thermal and electrical energy storage. Heat pumps, powered by photovoltaic (PV) systems [7] are the main technological pillars of the renewable electrification pathway to decarbonize buildings. PV assisted heat pumps results more profitable than solar thermal ones, due to the flexible nature of electricity as an energy carrier as opposed to heat. Furthermore, only 10–20% of solar energy can be converted directly into electrical power, with the remainder being dissipated as thermal energy, resulting in power loss and depreciation [8]. This means also high operating temperature, hesitating in cell structural damage and lifetime reduction, therefore, a cooling method for these units is vital [9]. In photovoltaic/thermal (PV/T) systems, which combine solar thermal collectors with PV panels to produce both thermal energy and electrical power [10], the fluid running in PV/T system collects excess heat and reduces the temperature of the PV panels, and once heated is employed for domestic hot water or space heating [11]. PV/T assisted direct-expansion heat pump (PV/T-DEHP) involves PVT panels to directly heat the heat pump working fluid; a cooled PV/T panel based on this system can have greater electrical efficiency than an uncooled PV panel and a comparatively high COP [12]. To solve the drawbacks of the low output temperature of PV/T and the lack of electrical production from solar thermal collectors the combination of a PV/T system with solar thermal collectors coupled in series with a heat pump is suggested in [13]. Public social building stock represents a challenging reality, quite obsolete in terms of energy and with poorly insulated envelopes, due to different economic and technical factors. Very often it is impossible to act on the building envelope and the distribution and emission system of thermal energy for the distrust of tenants [14, 15].

The present study focuses on an existing building of this residential typology, site in Palombara Sabina (Italy). Based on a detailed survey of the construction and of both energy and domestic hot water systems based on natural gas boiler (NGB) source for heat, a model of the current configuration of the building was created in TerMus and validated thanks to the data gathered during on-site measurements. Thermal loads and energy demand are then assessed. Starting from these results, two improvement scenarios were developed. First of all, the modification of the monovalent heat generation system with a hybrid solution that provides for the addition of a heat pump to the existing boiler gas was assessed, thus allowing the introduction of summer cooling. Secondly, the optimization of this configuration was proposed, through the integration of a PV-T system, which converts solar primary energy into heat and electricity. The aim is therefore the use of the ATER apartment house as a training case for the improvement of the HVAC centralized systems of the existing buildings, with the intent of suggesting a common methodology for energy retrofitting [16] applicable to all social housings built around 1980.

2 Methodology

2.1 Case Study

This study presents a residential complex owned by ATER, built in 1980–85 and located in Palombara Sabina (DD 2012), an urban center 30 km northeast of Rome.

It is a reinforced concrete three-storey building divided into thirteen flats connect by stairs, A and B, and a basement floor with technical rooms and cellars. The apartments can be grouped into four typologies, depending on their area and room number: (A) about 60 m^2; (B) about 70–80 m^2; (C) about 100 m^2; (D) about 100–120 m^2. Nevertheless, it was impossible to identify a standard configuration from the architectural and thermal point of view, as all thirteen houses differ from each other. Therefore, every apartment was characterized and studied in detail in the analysis phase. As for the building envelope, it is under the construction techniques of the historical period. The external walls consist of two layers of hollow bricks separated by air and a thin layer of insulating material ($U = 0.80$ W/m^2·K). The surfaces of the walls are composed of plaster. For what concerns transparent elements features, these are made by single glazed windows with metal frame without a thermal break ($U = 0.85$ W/m^2·K, g-value = 0.8). While the horizontal structure is composed of reinforced concrete and brick without thermal insulation ($U =$ 1.20 W/m^2·K), except for the floor adjacent to the attic ($U = 0.65$ W/m^2·K). Furthermore, it has a central heating radiator system served by a multi-stage air-blown methane boiler (NGB). The maximum heating power input is 69 kW, while the maximum and minimum heating capacity at the output are 65 kW and 51.8 kW, respectively, and the efficiency at the boiler is 0.94. The power is modulated according to the boiler outlet temperature of 80 °C. The heating system works from November 1st to April 15th. Different dimensions characterize steel radiators; due to this, for the modelling, the corresponding powers were calculated using UNI 10200, Annex D. The summer air conditioning system is absent, while to produce DHW each apartment is equipped with an independent electric boiler.

2.2 Building Validation

A preliminary survey of both architectural and technical aspects of the building and its HVAC systems was performed. A monitoring campaign from 19 to February 23rd 2021 inside an apartment with three dispersant surfaces was carried out: two bordering the external environment to the north-west and south-east and one facing the common distributive to acquire the heat flux (north-west exterior wall) and indoor and outdoor data. For this purpose, the TESTO 435 heat flow meter was used installed according to the good measurement practices suggested by the manufacturer. Heat fluxes are acquired by employing a heat-flow plate. Moreover, the internal and external air temperatures were measured using two thermal probes placed in correspondence with the heat-flow plate, one for each side of the wall. The device automatically applies the Heat-Flow Meter (HFM) method, following the ISO 9869-1 standard, recording thermal transmittance (U,

W/m^2·K) derived from heat fluxes and indoor/outdoor air temperature measurements. The measuring device saves thermal transmittance values for each data acquisition step (equal to 10 min). Then, the progressive average method was applied to determine the stationary U. In Fig. 1, the transmittance (U) measured for each step (green line) and the average U (red line) are reported. It can be observed that the measured U is equal to 0.804 W/m^2·K, which is 3% higher than the value calculated applying ISO 6946 to the stratigraphy of the wall, equal to 0.78 W/m^2·K. The percentage difference between HFM and the theoretical values is less than 20%, and then the ISO 9869-1 criterion is satisfied. The envelope modelled on Termus provide transmittance values compliant with the measured values with a delta of 0.02 W/m^2·K.

The model was validated by the comparison of heating energy needs between the Termus model and the previously validated TRNSYS (Delta 0.16).

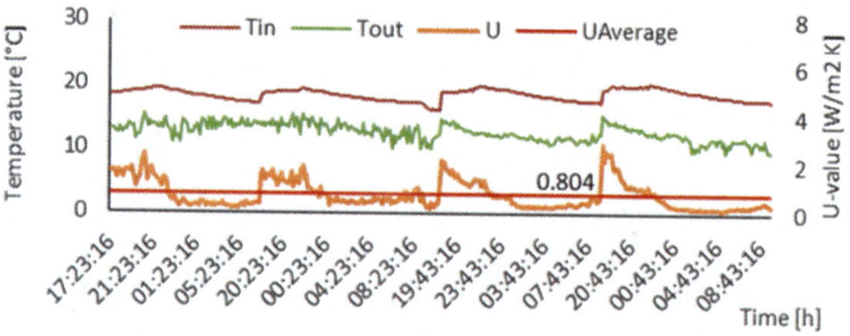

Fig. 1. Processing of the data collected by the TESTO 435 multifunction station.

2.3 Proposal for a Hybrid Heat Pump System with RES Integration

The current building, Ante Operam (AO), was analysed. After that, the installation of a hybrid heat pump (HHP) and its integration with photovoltaic panels (RED) were studied. The hybrid system consists of an inverter air-to-water heat pump rated at 51 kW and the existing natural gas boiler. Heat pumps and gas boilers are connected in series and can operate independently. Compared with the current system, the HHP heating, and cooling plans were changed as follows: the heating system is served by a bivalent system (HHP), leaving the current distribution and supply system unchanged. The sanitary water system became centralized, replacing the home-installed electrical boiler with the hybrid system above mentioned. Finally, a cooling system was placed, and the heat pump (cooling capacity 75 kW) is connected with a newly installed fain-coil. For the new DHW centralized system was considered a daily hot water demand equal to 50 l/person (Uni TS 9182, appendix E), with 50 residents. The building is oriented at 25° north, with an azimuth of 25°. The panels are placed on the double pitch roof of the building, which has a north-west/south-east direction. The panels are placed on the southeast pitch following the 30° inclination of the roof. There are 54 PV-T panels installed, each measuring 2.19 m^2 and 420 Wp, with a total area of 113.7 m^2 and a peak power of 21.84 kWp. Figure 2 shows the final system layout of the final system with the hybrid heat pump and the cooling photovoltaic panels.

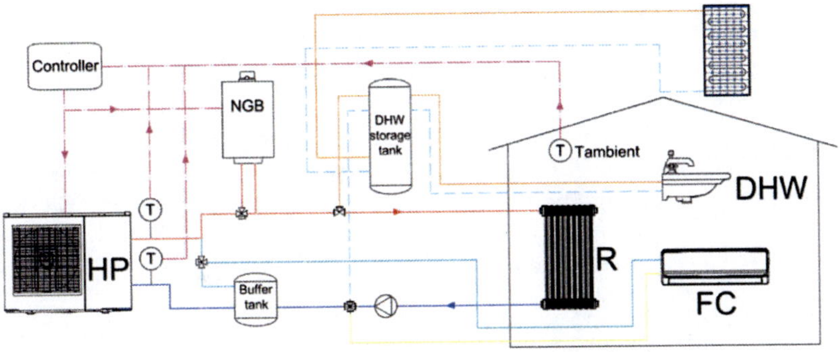

Fig. 2. Layout of the hybrid heat pump and the cooling photovoltaic panels.

2.4 Energy Background

The consumptions of the main energy carriers of the three scenarios, methane and electricity, and the corresponding primary energy requirements were then analyzed. The analyses were carried out with a breakdown according to final consumption (Heating, Cooling and Domestic hot water) and the energy vector used (Electrical energy or natural gas). The primary energy requirement per generator ($E_{p,HP}$, $E_{p,NGB}$) and total ($E_{p,tot}$), are defined by Eqs. 2–4 as the product between the input energy to the generator for the period considered and the primary energy conversion factor (*PEF* i) (UNI TS 11300-2 and DM 26/06/2015).

$$E_{P,NGB} = LHV\,Nm^3\,PEF_{\mathrm{gas}} = E_{in,HP}PEF_{gas} \tag{1}$$

$$E_{P,HP} = E_{in,HP}PEF_{ee} \tag{2}$$

$$E_{P,tot1} = E_{p,NGB} + E_{p,HP} \tag{3}$$

where PEF_{gas} = Natural gas = 1.05; PEF_{ee} = electricity = 2.42.

$E_{in,NGB}\,edE_{in,HP}$ are the energy request by the gas boiler (methane) and the heat pump (electricity) respectively, LHV is the lower heating value of methane.

The integration in the system of renewable energy sources, part of the electric energy supply requested by HP will derive from on-site production. The portion of electricity supplied by photovoltaic panels and not directly used is fed into the grid

$$E_{ee,grid,del} = E_{ee,PV,tot} - E_{ee,PV} \tag{4}$$

$$E_{ee,grid,net} = E_{ee,grid} - E_{ee,grid,del} \tag{5}$$

$E_{ee,grid,net}$ percentage of net electricity absorbed by the grid; $E_{ee,grid}$ percentage of total electricity absorbed by the grid; $E_{ee,grid,del}$ percentage of electricity fed into the grid.

The energy entering the heat pump will be distributed as follows

$$E_{in,HP,RES} = E_{in,HP}(\%_{ee,grid} + \%_{ee,PV}) \tag{6}$$

the primary energy expressed by Eq. 2 becomes:

$$E_{P,HP,RES} = E_{ee,grid,net} PEF_{ee} + E_{ee,PV} PEF_{RES} \tag{7}$$

where: $PEF_{RES} = 1$ electricity by renewable energy source.

By UNI TS 11300-4 and CTI 9: 2012, the final primary energy requirement is given by the difference between primary energy and primary energy fed into the grid (Eq. 6).

$$E_{P,\mathrm{tot},2} = E_{p,NGB} + E_{P,HP,RES} \tag{8}$$

3 Results and Discussion

3.1 Current Situation: Thermal Behaviour and Energy Consumption

The model therefore allowed the assessment of thermal loads and energy demand of the building. The thermal energy demand for a typical meteorological year (TMY). The monthly trend is shown in Fig. 3. The maximum thermal power required during the winter is equal to 67.8 kW, while the cooling power is 83.7 kW and per the DHW is 12.2 kW. When all the heat loads are summed up, the building has a heating, cooling and DHW energy demand per year are respectively: 100 MWh, 8 MWh and 15.1 MWh. Currently, the annual consumption of methane by heating system is 17,303 Nm^3. While the electricity requirement for domestic hot water is equal to 23.3 MWh. Consequently, the primary energy demand is 127 MWh for the heating system and 56 MWh to produce DHW.

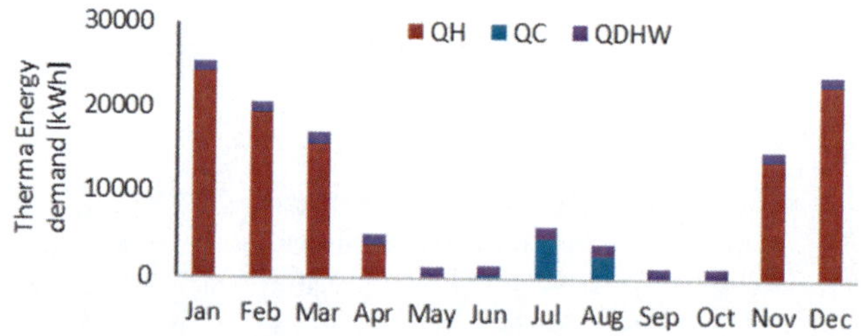

Fig. 3. Monthly heating (Q_H), cooling (Q_C), and DHW (Q_{DHW}) energy demand.

3.2 Energy Saving Potentials of Bivalent and PV-Assisted Hybrid Systems

When the current monovalent heating generation system is converted to a bivalent system (HHP), the consumption of the energy vectors reduces and, consequently, the primary energy requirement decreases.

The yearly methane consumption from AO to HHP decreases by 98%, from 17,303 to 408 Nm^3, useful to satisfy the heating demand when the capacity of the heat pump proves insufficient. However, the electrical consumption increases. The annual electrical

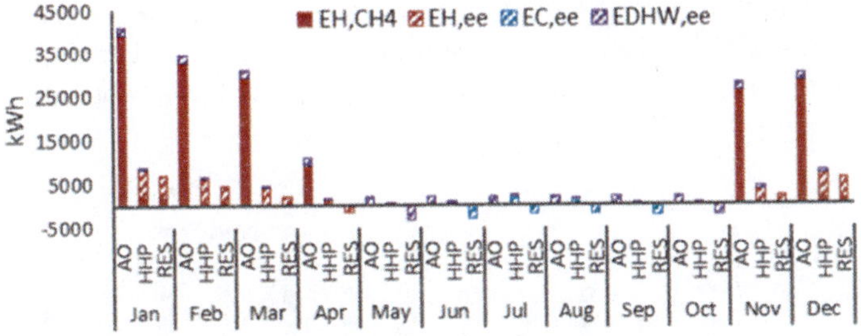

Fig. 4. Monthly electrical and methane consumption for each scenario

consumption to produce DHW decreased by 82%, going from 23.3 MWh by domestic electric boiler, to 4.2 MWh by the centralized heat pump. To these is added the electricity consumption for summer cooling (Fig. 4).

Primary energy consumption by the heating system drops to 61% and by the hot water system drops to 82%. To them was added 8.6 MWh by the cooling system. Overall, there was a reduction of 62% in the yearly primary energy consumption from the initial 228 to 85.8 MWh (Fig. 5).

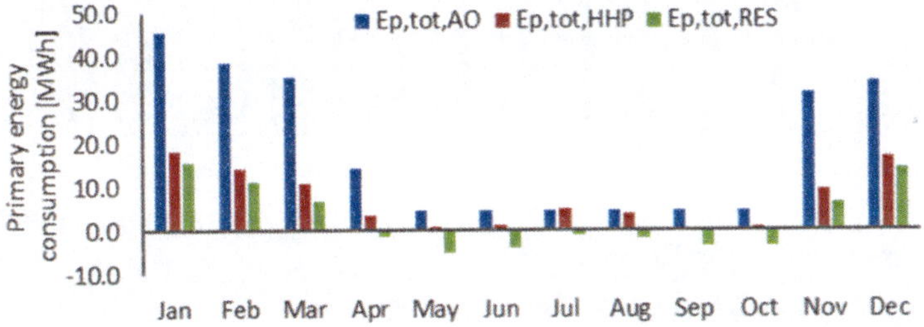

Fig. 5. Primary energy demand ante opera ($E_{p,tot,AO}$); hybrid system, HHP ($E_{p,tot,HHP}$), PV-T integration ($E_{p,tot,RES}$).

3.3 Benefits of RES Integration

The electrification of the energy demand allows the integration with a renewable energy source, in particular with the installation of cooling photovoltaic panels (PV-T) for the combined heating (at the service of DHW), and power (at the service of HP) production. DHW has an annual thermal energy demand of 25.35 MWh, of these the solar energy contributes is 20.02 MWh, with a solar fraction of 79% (Fig. 6). Due to the panels' thermal integration leads to a reduction of (DHW) electric demand by the HP equal to 75%. Annual consumption decreased from 4.18 to 1.05 MWh.

The annual electrical production by PV-T panels, net of auxiliary consumption, is 30.9 MWh, 86% of the electrical consumption of the heating, cooling, and hot water production. As there is no electrical storage, the energy produced is either consumed directly or fed into the grid. Almost 72% of the electricity production is injected into the grid, while only 28% is used for self-consumption. In particular, in the months where neither heating nor cooling is required, the energy fed into the network also represents 97% of production (Fig. 7).

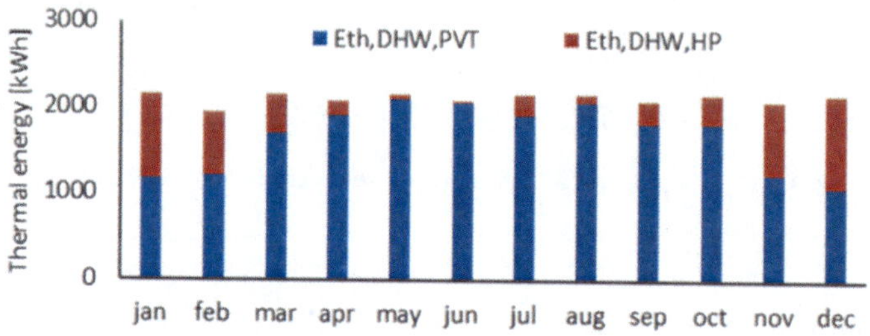

Fig. 6. Thermal energy production by PV-T ($E_{th,DHW,PVT}$) panels and by HP ($E_{th,DHW,HP}$)

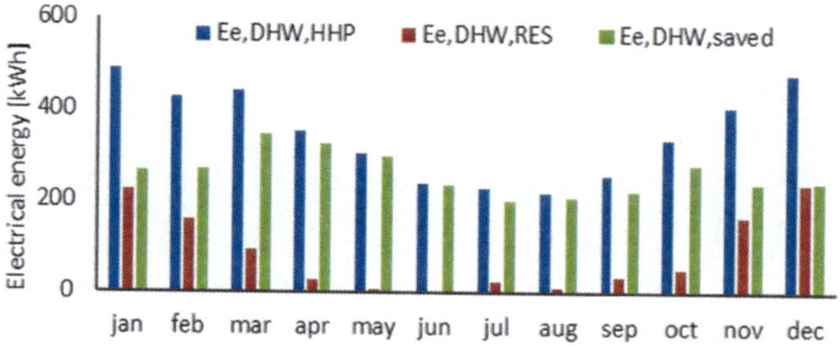

Fig. 7. DHW electrical consumption by HHP scenario ($E_{e,DHW,HHP}$) and RES scenario ($E_{e,DHW,RES}$), energy saved ($E_{e,DHW,saved}$).

In the heating period there is an increase in electrical demand with a total of 27.5 MWh and a PV-T production decrease with a total of 11.5 MWh, consequently, electrical absorption from the grid is 16.0 MWh. For the remainder of the year, electricity production exceeds consumption for domestic hot water and cooling, and 15.0 MWh are delivered to the grid. The results of the balance between energy delivered "from" (+) and "to" (−) the grid is 0 MWh. A photovoltaic generator supplies 100% of the electricity needed (Fig. 8).

The primary energy demand was calculated by Eq. 8. Witch considers the thermal and electrical contribution of the on-site renewable energy source and the electricity delivered to the grid. This aspect provides a double benefit on the primary energy saving. Firstly, due to the use of a solar thermal energy source, there is an avoided consumption of electrical energy. Therefore, there is an avoided consumption of primary energy, which is 7.58 MWh. Second, because the annual balance between electricity consumption from the grid and to the grid is zero, the energy withdrawn from the grid (PEF = 2.42) is not included in the primary energy account, only the 30.9 MWh of primary energy related to renewable sources (PEF = 1) is considered. In addition, 4.0 MWh (no-renewable) must be added the methane consumption during the heating period. The annual total

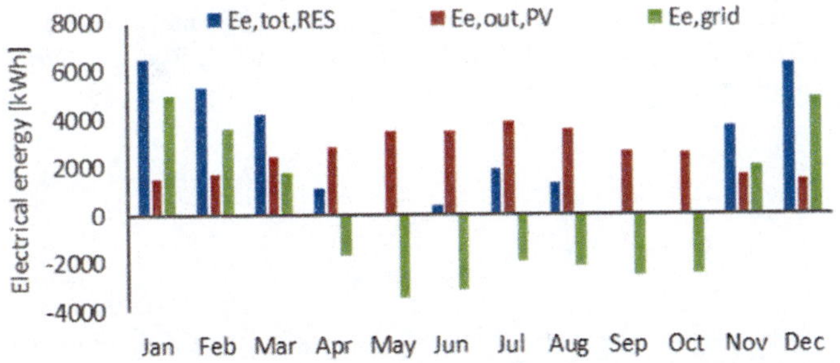

Fig. 8. Electrical energy demand by RES system (Ee,tot,RES); electrical energy by PV panels (Ee,out,PV); electrical energy delivered from the grid (positive) and to the grid (negative).

primary energy demand is 34.3 MWh. The saving compared to the previous scenario is 85% (AO) and 60% (HHP).

4 Conclusion

The explained work deals with a hybrid heat pump system integrated with cooled photovoltaic panels both to combine heating, cooling and power production to optimize energy consumption in public residential buildings. Three different scenarios were analysed each as a further improvement of the previous one. Starting to analyse the current system (AO). It was compared with a hybrid system (HHP) and successively with an HHP integrated with PV-T panels (RES).

In the HHP system part of the thermal energy demand move from the NGB to the HP, with a consequently decreased of methane consumption equal to 98%. A 62% decrease in primary energy follows.

Through the electrification of the heating generation, it is possible satisfy the HP electrical demand by an on-site power generator, specifically insert a hybrid photovoltaic panel (PV-T). The PV-T panels integration carries out several additional energy benefits. By the heating production the thermal energy demand for DHW to the heating generator, achieve the 79% of solar fraction. A 75% reduction of electricity consumption followed. Furthermore, electrical energy production and not directly used and delivered in the grid is equal to the electricity delivered by the grid in a TMY, therefore there is an annual renewable coverage equal to 100%. Finally comparing the primary energy demand of the RES scenario with AO and HHP, the reduction is of, respectively: 85% (AO-RES), 60% (HHP-RES).

Specifically, the article explores how an improved hybrid system with high-efficiency solar integration can be applied to a public residential building for the combined generation of heating, cooling, and electrical energy. With this approach, it is possible to reduce energy consumption in buildings with low envelope performance, especially whenever it is difficult to intervene for several reasons.

Acknowledgments. The research was supported by European Commission and is a part of the HORIZON 2020 project RESHeat. This project received funding from the European Union's Horizon 2020 program in the field of research and innovation on the basis of grant agreement No. 956255.

References

1. Energy statistics: an overview. Eurostat Statistics Explained. European Commission. https://ec.europa.eu/eurostat/statistics-explained/index.php?title=Energy_statistics_-_an_overview#Final_energy_consumption. Accessed 7 July 2022
2. Calama-González, C.M., Symonds, P., León-Rodríguez, A.L., Suárez, R.: Optimal retrofit solutions considering thermal comfort and intervention costs for the Mediterranean social housing stock. Energy Build. **259**, 111915 (2022)
3. Vakalis, D., Diaz Lozano Patino, E., Opher, T., Touchie, M.F., Burrows, K., MacLean, H.L., Siegel, J.A.: Quantifying thermal comfort and carbon savings from energy-retrofits in social housing. Energy Build. **241**, 110950 (2021)
4. Treichel, C., Cruickshank, C.A.: Greenhouse gas emissions analysis of heat pump water heaters coupled with air-based solar thermal collectors in Canada and the United States. Energy Build. **231**, 110594 (2021)
5. Bagarella, G., Lazzarin, R., Lamanna, B.: Cycling losses in refrigeration equipment: an experimental evaluation. Int. J. Refrig. **36**, 2111–2118 (2013)
6. Dongellini, M., Naldi, C., Mori, L.: Influence of sizing and control rules on the energy saving potential of heat pump hybrid system in a residential building. Energy Convers. Manag. (2021)
7. Schreurs, T., Madani, H., Zottl, A., Sommerfeldt, N., Zucker, G.: Techno-economic analysis of combined heat pump and solar PV system for multi-family houses: an Austrian case study. Energy Strat. Rev. **36**, 100666 (2021)
8. Li, Z., Ma, T., Zhao, J., Song, A., Cheng, Y.: Experimental study and performance analysis on solar photovoltaic panel integrated with phase change material. Energy **178**, 471–486 (2019)
9. Kalateh, M.R., Kianifar, A., Sardarabadi, M.: Energy, exergy, and entropy generation analyses of a water-based photovoltaic thermal system, equipped with clockwise counter-clockwise twisted tapes: an indoor experimental study. Appl. Therm. Eng. **215**, 118906 (2022)
10. Chow, T.T.: A review on photovoltaic/thermal hybrid solar technology. Renew. Energy **87**(2), 365–379 (2020)
11. Zhou, J.Z., Zhao, X.D., Yuan, Y.P., Li, J., Yu, M., Fan, Y.: Operational performance of a novel heat pump coupled with mini-channel PV/T and thermal panel in low solar radiation. Energy Built Environ. **1**, 50–59 (2020)
12. Dai, N., Xu, X., Li, S., Zhang, Z.: Simulation of hybrid photovoltaic solar assisted loop heat pipe/heat pump system. Appl. Sci. **7**(2), 197 (2017)
13. Sajid, A., Yanping, Y., Ataza, H., Jinzhi, Z., Chao, Z., Min, Y., Bisengimana, E.: Experimental and numerical investigation on a solar direct-expansion heat pump system employing PV/T & solar thermal collector as evaporator. Energy **254-B**, 124312 (2022)
14. Mauri, L., Vallati, A., Ocłoń, P.: Low impact energy saving strategies for individual heating systems in a modern residential building: a case study in Rome. J. Clean. Prod. **214**, 791–802 (2019)

15. Vollaro, A.D.L., Galli, G., Vallati, A., Romagnoli, R.: Analysis of thermal field within an urban canyon with variable thermophysical characteristics of the building's walls. J. Phys.: Conf. Ser. **655**(1), 012056 (2015)
16. Vallati, A., Grignaffini, S., Romagna, M., Mauri, L.: Effects of different building automation systems on the energy consumption for three thermal insulation values of the building envelope. In: EEEIC 2016 - International Conference on Environment and Electrical Engineering, p. 7555731 (2016)

Cyclic Lateral Load Test of a Wall with Timber Frame Structure and Lightearth Envelope

G. Becerra[1(✉)], S. Onnis[1], G. Meli[1], M. Wieser[1], and J. Vargas-Neumann[2]

[1] Architecture Department, Pontificia Universidad Católica del Perú PUCP, Lima, Perú
german.becerra@pucp.pe

[2] Civil Engineering Department, Pontificia Universidad Católica del Perú PUCP, Lima, Perú

Abstract. To address the housing and equipment deficit in Peru, it is necessary to have more and better construction system alternatives that prioritize not only cost and safety aspects, but also those of comfort and sustainability. The aim of this study is to identify the mechanical properties of a system made of a timber framed structure and a prefabricated panels of lightened earth enclosure, as a preliminary step to a subsequent dynamic test. A cyclic lateral load test is described, whose procedure consists of the anchoring and fixing of the wall in a fixed base, the definition of phases controlled by a displacement, the instrumentation of the wall to control the drifts and, finally, the application of a lateral load on the superior beam of the wall. The aim is to determine the mechanical properties of the wall, such as the lateral shear resistance and the lateral elastic stiffness, as well as to develop the hysteresis diagram of the proposed system. In addition to identifying the required mechanical properties, the wall proved to be quite rigid, with a high elasticity in the initial stage, and an inelastic stage of large deformation until failure. Even though the wall behaved quite auspiciously in relation to current construction standards, some weaknesses were identified that gave rise to recommendations, such as reducing the density of the cladding or reducing the spacing between the diagonal wooden laths.

Keywords: Structural analysis · Prefabricated system · Natural materials

1 Introduction

This article exposes and analyzes a structural test carried out on a wooden framed wall with an envelope made of prefabricated panels of lightened earth with natural fibers. The specific proposal of the panel has been the result of a broader research that aims to technically validate an accessible, earthquake-resistant, bioclimatic, and environmentally friendly construction system. Additionally, for the design of the proposal, other no less important conditions were considered, such as prefabrication and dry construction, modular design, and constructive flexibility [1].

The importance of researching constructive alternatives in Peru arises from the recognition of the great quantitative and qualitative deficit of housing and existing equipment

J. Littlewood and R. J. Howlett (Eds.): SEB 2022, SIST 336, pp. 418–430, 2023.
https://doi.org/10.1007/978-981-19-8769-4_39

in the country. In the specific case of housing, various authors agree on the seriousness of the situation due to the scale and complexity of the problem [2, 3]. In this sense, it is necessary to recognize that the solution necessarily involves expanding the available alternatives in the choice of construction systems that go beyond the conventional ones. Even if it is usual to prioritize low costs and earthquake resistance, aspects as sustainability and comfort provided to users should not be neglected.

The present structural test is considered preliminary, since its objective is to approach the mechanical properties of the proposed mixed system, in addition to see its weaknesses to take it to a next phase of dynamic structural testing. This structural validation, together with the verification of the good thermal performance already carried out [4] will finally allow the construction of a pilot building (25 m^2) that demonstrates its viability.

2 Background

Traditional architecture reflects, above all, a capacity to adapt to the natural environment and to the cultural conditioning factors of the place and its inhabitants. Conceived in pre-industrial times, the difficulty in extracting and managing resources determined ways of building in which efficiency, relevance and originality were essential. Safety, utility, comfort, economy, and a sense of beauty itself have been the determining factors that have shaped this apparently simple architecture, but extremely valuable and sophisticated.

In almost all regions of the world, even in those where earthquakes are recurrent, natural materials such as wood, stone or raw earth have been considered since ancient times for the construction of buildings. In the case of the Peruvian coast, evidence of mixed systems such as *quincha*, which use frameworks or fabrics with wood or cane as a structure and raw earth as a covering, can even be found in the pre-Hispanic constructions of Caral, which are almost 5,000 years old [5]. The *quincha* continued to be used and evolved during the colony and the republic until the middle of the 20th century. A particular interest in this construction technique reappeared with the study and dissemination of prefabricated *quincha* in the 1970s because of the Ancash earthquake in 1970 and Moyobamba earthquakes in 1968 and 1971 [6, 7]; the interest has continued to the present day, although often limited to the academic and experimental field [8].

The most important advantages of using a mixture of raw earth and natural fibers as a construction material are its low cost, its good thermal performance, and its low environmental impact. On the other hand, the limited structural capacity of the lightened earth requires the use of reinforcements such as wood, whose costs are much higher, so they are limited precisely to a structural function. The proposed wall system, whose structural test is presented in this article, is based on this principle. It is important to mention the growing interest that exists in this type of structural systems in the academic and commercial fields due to the advantages already exposed, which is reflected in the publications of various authors [9–11].

Based on a study of the Peruvian context, the state of the art of mixed techniques in seismic zones, and some previous successful experiences [12], the team designed a prefabricated proposal that can be assembled with dry elements (see Figs. 1 and 2). The constructive proposal that was tested is composed of a light wooden framework structure, made up of composite columns formed by two or four 2 × 2 × 230 cm pieces of wood and arranged every 80 cm, which are braced by a 2 × 8 cm horizontal crossbar each 45 cm. The wall enclosure is made up of 45 × 45 × 4 cm prefabricated lightened earth panels, with a density between 600 and 800 kg/m^3, screwed to the wooden structure. As an auxiliary structure, the wall has been stiffened with diagonal wood laths at 45° (4 × 1 cm each 25 cm) on both sides of the wall and in opposite directions, which end up covered with a layer of thick earth render mixed with sand and 3 cm long fibers (Fig. 1). The wood used was the "Tornillo" species (structural classification under Peruvian standards: C, HC = 9%) and the connections were made with 2″ and 3″ screws.

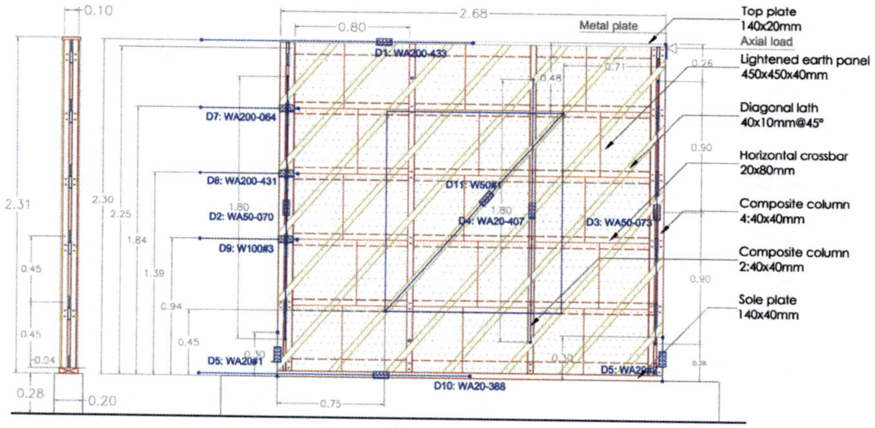

Fig. 1. Construction detail of the proposed wall.

Two images of the construction system are shown below in which the prefabricated composite columns, the horizontal crossbars, and the diagonal laths of sawn wood, as well as the prefabricated panels of lightened earth can be identified (Fig. 2). The joints between columns and plates (sole and top) were made with steel screws, between the crossbars and columns with steel nails and between the earthen panels and the wooden crossbars with screws and 35 mm diameter steel sheets.

Fig. 2. Assembly and reinforcement of the tested wall.

3 Methodology

The cyclic lateral load test is performed to quantify the properties and shear strengths of vertical elements, such as walls, against cyclic lateral loads, such as those caused by earthquakes. The test procedure consists of anchoring and fixing of the wall to a base, restraining the specimen to prevent displacements out of the plane of the wall, defining phases controlled by a seismic demand—either force or displacement, instrumenting the wall to control displacements, applying a vertical load simulating the weight of a roof and finally applying a lateral load on the top of the wall. In the analysis of results, mechanical properties such as pure shear strength (Vpeak) and elastic lateral stiffness (Ke) must be determined. The specimen tested was a wall with the construction characteristics proposed and described in the previous chapter, 2.68 m long, 2.31 m high and 0.14 m thick.

The conditions of ASTM 2126-11 [13], as well as the recommendations of the Federal Emergency Management Agency (FEMA) [14], were followed to perform the test. The test consisted of the slow and incremental application of a pattern of demands to the specimen, which can be a load or displacement, to assess the progression of damage. The FEMA procedure recommends defining damage limit states associated with some parameter that can be quantified. The procedure chosen to evaluate the damage limit states was Method B: ISO 16670 protocol [14]. The load test was carried out at the Laboratory of Antiseismic Structures at the PUCP. Table 1 shows the equipment used in the cyclic lateral load test. The procedure consisted in the application of a horizontal force by a hydraulic jack (in cycles of loading and unloading) that pushes and pulls the top beam of the wall with controlled displacements.

Table 1. Equipment of the Laboratory of Antiseismic Structures—LEDI.

Equipment	Features and functions
Reaction frame	Metallic frame in charge of containing the dynamic actuator and the wall, allowing only the displacements in the coplanar direction of the wall
200 kN hydraulic jack	Hydraulic manual equipment that applies a lateral load with controlled displacements. Model MTS 204.81 with a double-acting hydraulic actuator, 200 kN capacity and a displacement amplitude of ±100 mm
LVDT displacement sensors	These are instruments that electronically measure the relative displacement between 2 points. The LVST range goes from: ±10, ±20, ±50, ±100 mm
Load cell	Equipment that electronically records the load applied by the actuator. The cell is connected to a charge amplifier and a computer translates the measured voltage to mm

The test was conducted in four stages. The first consisted of the preparation and assembly of the reaction frame, the installation of the dynamic actuator and the subsequent lifting, moving, and positioning of the wall at the exact point where the load was applied. Additionally, a 50 kN load cell was placed during the transfer process to obtain the actual weight of the wall (see Table 2), so that it can be compared with the estimated theoretical weight.

Table 2. Specimen real and theoretical weight.

Real weight	Theoretical weight	Difference
W_R	W_T	$(W_T - W_R)/W_R$
857 kg	893 kg	4%

When the wall was placed in the reaction frame, the foundation beam was fixed to the ground using four hydraulic jacks, which applied a vertical force of 0.5 and a horizontal force of 1.3 ton at both ends of the beam. Subsequently, two rails were placed on each face of the wall to prevent out-of-plane displacement of the specimen. In a second stage, the phases of the test were defined, based on the limit states of damage of the structure, which translates into certain levels of displacement or load that induce the degradation of the structure's resistance. Experience has shown that the level of damage that a structure suffers after a seismic movement is related to the maximum displacement that the structure reaches [15], so it was decided to control the phases with a displacement time history.

The displacement time history is made up of different incremental phases. Each phase consists of 1 cycle (pushing and pulling) with an amplitude equal to a percentage of the reference parameter (Δm). The table of damage levels and structural behavior for a timber wall structure published by HAZUS and elaborated by FEMA was taken as reference [16]. Table 3 details a relationship between drift and the state of damage of a building, which is divided into three levels: immediate occupancy, life safety and collapse prevention. In addition, the maximum service displacement established by the Peruvian Technical Standards E-010 [17] was considered. The ultimate demand displacement was chosen as the maximum life-safety, (from FEMA 356, $\Delta m = 72$ mm), related to the temporary or permanent collapse prevention displacement of a structural system composed of timber walls. Table 4 describes the ten control phases for the test, each corresponding to an amplitude related to a percentage of the predefined ultimate displacement. The maximum and service displacements suggested by the Peruvian Standards E-010 [17] and E-030 [18] were included in the test, so that the behavior of the system and the level of damage in those phases could also be identified.

The application of the load in each phase comprises (1) the first load cycle, where the wall is pushed to the predetermined displacement, (2) the pulling back to its original position, (3) the pulling to the same displacement in the opposite direction and (4) the pushing back to its original position (Fig. 3). The development of each phase consists of 2 cycles of movement, or until the hysteretic loops of a force-displacement diagram are stabilized.

Table 3. Damage levels for a timber wall structure.

Structural performance levels	Drift	Displacement (h = 2.40 m)
Immediate occupation	1% temporary 0.25% permanent	24 mm temporary 6 mm permanent
Life safety	2% temporary 1% permanent	48 mm temporary 24 mm permanent
Collapse prevention	3% temporary or permanent	72 mm temporary or permanent
Service displacement (NTP.010)	1/1200	2 mm

Table 4. Phases of the incremental cyclic test.

Fase	# Cycles	Cycles amplitude (%Δm)	Δu = 72 mm	Drift	Speed (mm/min)
1	2	1.8%	1.30	0.54	1.30
2	2	2.8%	2.00	0.83	2.00
3	2	5.0%	3.60	1.50	3.60
4	2	9.5%	6.84	2.85	6.84
5	2	18.6%	13.39	5.58	10.00
6	2	26.0%	18.72	7.80	10.00
7	2	33.3%	24.00	10.00	15.00
8	2	50.0%	36.00	15.00	20.00
9	2	66.7%	48.00	20.00	25.00
10	2	100.0%	72.00	30.00	30.00

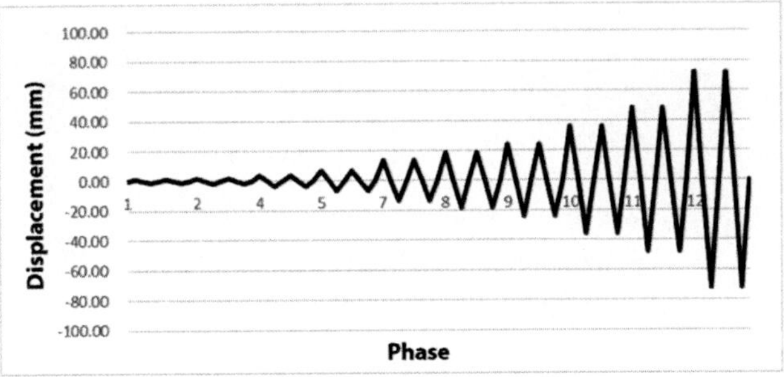

Fig. 3. Cyclic displacement schedule.

The third stage comprised the instrumentation process of the wall, which included fixing the anchor points for the motion sensors, the placement of the LVDTs and the final calibration of the instruments and equipment. The final location of the LVDT's (see Fig. 1) was defined with the criteria of determining the results of the lateral behavior, anchorage fixation and to control that the test is carried out in the correct way.

The fourth and final stage consist of the execution of the test itself, with the application of the lateral cyclic load. In the first stage of the test, a static vertical load was applied to the wall, by the installation of metal beams arranged to simulate the weight of the roof. The weight of the roof was estimated by considering the short columns the short columns needed to support an additional lightweight roof, the joists, the wooden cladding and rafters, the panels of lightened earth protected by a layer of earth covering, and a light, ventilated, sloped roof with solar panels; the total load resulted in approximately 190 kg/m^2 (Table 5). For the case of a minimum equipment module of approximately 30 m^2, it is possible to calculate the distributed load that the wall will take in real conditions to its surroundings. Assuming a tributary area of 3.80 m^2, a distributed load of approximately 280 kg/m was calculated, which is equivalent to about 720 kg for a 2.70 m long wall.

Table 5. Estimated vertical roof weight.

Roof	Unit weight	Unit
Columns	2.00	kg/m^2
Joists 2″ × 8″	11.50	kg/m^2
Wooden cladding (e = 3/4″ 1.5 cm	13.50	kg/m^2
Layer of mud lining over the insulation panels (e = 2 cm)	24.00	kg/m^2
Insulation panels of lightearth, e = 10 cm (800 kg/m^3)	80.00	kg/m^2
Joists 2″ × 6″ (roof)	9.00	kg/m^2
Rafters 2″ × 2″ @80 cm	2.00	kg/m^2
Calamine roof	13.00	kg/m^2
Solar panels type VISION 60P	5.00	kg/m^2
Overload of use (sloped roof)	30.00	kg/m^2
Weight	190.00	kg/m^2

The application of the lateral load was carried out with the constant pumping of the hydraulic jack at a constant speed, which is controlled by the 10 phases of displacement on the axis of the upper beam of the wall. The hydraulic jack applied the load in cycles of loading (pushing) and unloading (pulling) during the 10 phases, until the wall resistance was reduced by a 20%.

4 Results

The hysteresis diagram obtained during the test (relationship between force and displacement in each phase) was systematized for the results analysis. When the vertical load was applied, no representative results were obtained for the compression of the wall. This is because the applied stresses were much lower than those admissible by the wood. As proof of this, the sensors intended to measure the vertical deformation of the columns registered very small deformations, outside the measuring range of the instruments.

4.1 Hysteretic Behavior

During the test, three states of behavior were identified, which are detailed in Table 6: (i) an initial elastic stage up to phase 3; (ii) a second stage up to phase 6, where the first 45° diagonal cracks began to appear (Fig. 4); (iii) finally, a last stage up to phase 10, where the wall started to fail and to overturn as a rigid solid.

Table 6. Mechanisms of failure and damage in the wall during the test.

Elastic stage F1–F3	This stage described the elastic behavior of the wall, since there were no cracks showing evidence of damage. It was not until phase 2 that the first crack occurred at the joint between the specimen and the base of the plinth, due to bending traction
Maximum strength F4–F6	This stage described the overstrength behavior, until phase 6, when the wall showed failure by overturning, in the negative direction of load application (V = −1371.24 kg; D = −18.80 mm) and a reduction of stiffness close to 70%, where the wall anchorage at the plinth began to detach and the wall began to rise at its ends. Cracks started to occur in the plaster, at 45°, in the same direction as the diagonal wooden laths. This is because the stress began to bend the diagonals
Ultimate failure stage F7–F10	This stage described the inelastic behavior of the wall, until its failure in phase 10. The failure occurred when the wall began to overturn from one side to another, evidencing the failure of the anchorage of the column to the bottom sill, even before the failure of the anchorage of the bottom sill to the foundation

As for the hysteretic behavior, the hysteresis curves presented similar results in the elastic stage (up to phase 3) and, subsequently, they presented different results in each direction, registering much lower resistance in the pulling direction (−V). The hysteresis diagram of the wall in the 10 displacement phases is presented below (Fig. 5), which describes loops that are not very thin but oriented towards the origin, indicating a flexible structural system, with degrading lateral stiffness and high energy dissipation. According to ASTM standards [13], it is not uncommon for different results to occur in each direction, because the damage generated in the first excursion (push), can weaken the response in the next excursion (pull).

Fig. 4. View of the cracks coinciding with the diagonal laths.

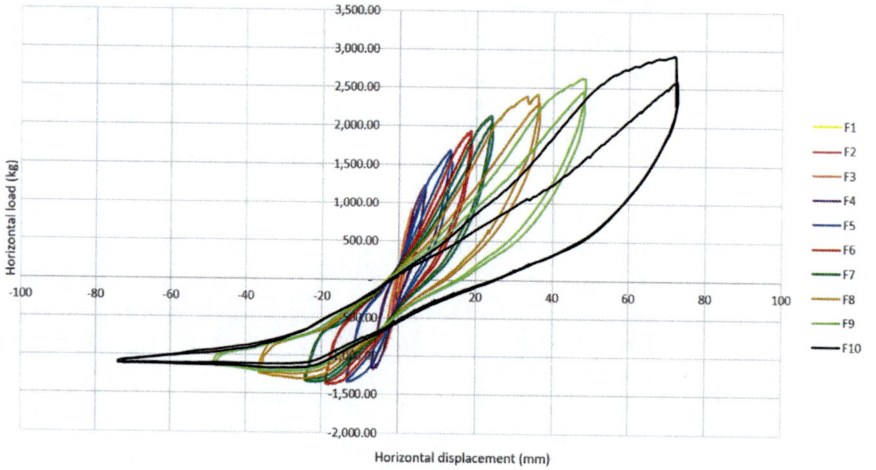

Fig. 5. Hysteresis diagram of sensor D1 (force - lateral displacement).

4.2 Envelope Curve

In relation to the envelope curve of the hysteretic behavior, which represents the maximum load and displacement values of the stable cycles of each phase, it can be observed that higher resistance values were reached in the positive direction of the load application rather than in the negative direction (Fig. 6). The envelope curve is a tool that helps to determine a relationship between the maximum allowable drifts and the progression of damage in the system. It can be observed that the positive and negative curves diverge from phase 3 onwards, where a rather short and pronounced elastic range is identified, demonstrating a high initial stiffness of the wall. Subsequently, a drop in resistance is observed for large displacements, demonstrating a great ductility and energy dissipation of the wall. The last state of ductility is due to the plinth uplift and failure of the wall anchorage.

Since the positive and negative curves differ significantly, it was decided to use the negative load values (pull) to define the values of the damage states of the wall. The behavior of the wall can be summarized in three damage states (Table 7). The first is the yield limit state, where the wall remains in its elastic range until the first significant crack in the First Major Event (FME), which occurred in phase 3, and defines the yield load and displacement (Vy – Δy). The second is the limit state of resistance, which occurred in phase 6, and defines the maximum load that causes the failure of the wall (Pm). Finally, the third is the limit failure state, which corresponds to the last point of the envelope curve, which occurred in phase 10, where the ultimate load (Pu) is approximately 80% of the limit load (Pm).

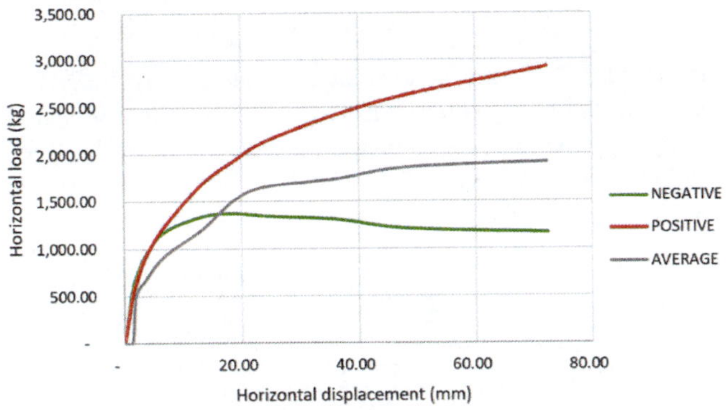

Fig. 6. Envelope curve. Positive, negative, and average capacity (force - lateral displacement).

Table 7. Maximum load and displacement values of the negative hysteresis curve

	Displacement	P peak	V peak
Yield limit state	Δy = 3.6 mm	Py = 928.2 kg-f	Vy = 346.6 kg/m
Resistance limit state	Δm = 18.7 mm	Pm = 1371.2 kg-f	Vm = 511.7 kg/m
Failure state limit	Δu = 72 mm	Pu = 1162.5 kg-f	Vu = 433.8 kg/m

4.3 System Allowable Capacities

The allowable capacities of the system were calculated to systematize a seismic-resistant design of the proposal. The allowable shear capacity and the allowable lateral stiffness are two properties that provide information on the maximum lateral resistance and the allowable displacement that the wall can have under different lateral demands. Standard safety factors were applied, such as those proposed by JUNAC tests [19], which determine the lateral allowable load as the maximum unit load (Vpeak) reduced by a safety factor of 1.5 and multiplied by a reduction factor of 0.7, which considers the reduced number of tests; see Eq. (1). It was determined that the allowable shear load of the system is equivalent to 73% of the actual weight of the wall (Table 8).

$$V_{PEAK} = \frac{V_{max-aver}}{1.5} * 0.7 = \frac{1371.2}{1.5} * 0.7 = 238.8\frac{kg}{m} \quad (1)$$

Table 8. Allowable wall shear load.

Allowable load (kg/m)	Allowable lateral resistance L = 2.68 m (kg)	Wall actual weight (kg)	% of weight (%)
235	629	857	73

For the calculation of the system's elastic lateral modulus of stiffness (Ke), see Eq. (2), it was used the phase at a deformation of h/1200, equivalent to a service displacement of 2 mm, for a wall approximately 2.40 m high. For this purpose, the second hysteresis cycle of the second phase was used to obtain the slope of the force-displacement line in the elastic range of the wall, when significant cracks were not yet present.

$$Ke = \frac{\Delta V}{\Delta D} = 346830.4\frac{kg}{m}$$

$$Keu = \frac{K}{L} = \frac{346830.4}{2.68} = 1294.1\frac{kg}{cm}/m \quad (2)$$

5 Conclusions

The test performed achieved the objective of approaching the mechanical properties of the component and allowed us to identify the weaknesses of the mixed system as a preliminary step to a subsequent dynamic test. The results showed a sufficiently stiff wall, with damage in the elastic range allowed by the service displacements established by the Peruvian seismic-resistant and timber regulations.

The hysteretic behavior of the wall show three stages of damage: (i) an elastic stage up to phase 3, where the push and pull curves presented similar results and the loops were thin and oriented towards the origin, indicating an elastic and flexible structural system; (ii) a second stage of over-resistance up to phase 6, where diagonal cracks were produced due to the buckling of the laths and the lifting of the plinths due to the failure of the anchorage, in this stage the positive loops (push) were greater than the negative ones (pull) due to the damage that was generated in the first excursion; and finally, (iii) a limit failure stage up to phase 10, where the wall began to fail and overturn as a rigid solid, and the hysteresis loops exhibited high degrading stiffness and energy dissipation.

With the results obtained, the allowable capacities of the system were calculated from the negative hysteresis curves, as they were the most conservative. It was calculated a maximum shear strength per unit of the system of 235 kg/m, which represents the maximum static lateral resistance of the proposal, considering a safety factor of 1.5 and a reduction coefficient of 0.7. The lateral elastic stiffness calculated for service displacement was 1294.1 kg/cm/m. The system demonstrated an elastic behavior for a service displacement of 2 mm, as indicated by NTP-010, and did not reach its ultimate strength for a displacement of 24 mm, as stablished by NTP-030. Even though the results were auspicious, it is important to consider some recommendations for the next stage of dynamic testing: namely, reduce the density of the earth plaster layer, reduce the spacing of the diagonal wooden laths, improve the anchorage of the wall in the plinth and use nominal measurements on the timber sections.

Acknowledgments. The authors would like to thank the National Training Service for the Construction Industry (SENCICO), the National Council for Science, Technology and Technological Innovation (CONCYTEC) and the Pontifical Catholic University of Peru (PUCP), for the funding that made possible the development of this research.

References

1. Meli, G., Onnis, S., Wieser, M.: Introducción en el contexto peruano de un nuevo sistema constructivo con madera y tierra alivianada. In: Actas del 19th SIACOT, pp. 605–613. Oaxaca (2019)
2. Blanco, A., et al.: Un espacio para el desarrollo: Los mercados de vivienda en América Latina y el Caribe. Inter American Development Bank, New York (2012)
3. Romero, J.Q., Ávila, T.A., Makedonski, P.M.: El problema de la vivienda en el Perú, retos y perspectivas. Rev. INVI **20**(53) (2005)
4. Wieser, M., Onnis, S., Meli, G.: Desempeño térmico de cerramientos de tierra alivianada. Posibilidades de aplicación en el territorio peruano. Rev. Arquit. (Bogotá) **22**(1), 164–174 (2020). https://doi.org/10.14718/RevArq.2020.2633
5. Blondet, M., Vargas, J., Tarque, N., Iwaki, C.: Construcción sismorresistente en tierra: la gran experiencia contemporánea de la Pontificia Universidad Católica del Perú. Inf. Constr. **63**(523), 41–50 (2011). https://doi.org/10.3989/ic.10.017
6. Kuroiwa, J.: Quincha modular prefabricada. Universidad Nacional de Ingeniería, Lima (1991)
7. Kuroiwa, J.: Reducción de desastres: Viviendo en armonía con la naturaleza. OPS (2002)
8. Tejada, U.: Buena tierra: apuntes para el diseño y construcción con quincha. Consideraciones sismorresistentes. CIDAP, Lima (2001)
9. Marcom, A.: Construire en terre-paille. Terre Vivante (2011)
10. Volhard, F.: Light Earth Building: A Handbook for Building with Wood and Earth. Birkhäuser, Basel (2016)
11. Placitelli, C.: Autoconstrucción ecológica con B.T.A.: (Bloque de tierra alivianada). Editorial Académica Española, España (2016)
12. Wieser, M., Onnis, S., Meli, G.: Conductividad Térmica de la tierra alivianada con fibras naturales en paneles de quincha. In: SIACOT Proceedings, pp. 199–208. La Antigua Guatemala (2018)

13. ASTM: Standard test methods for cyclic (reversed) load test for shear resistance of vertical elements of the lateral force resisting systems for buildings. ASTM E2126 (2018)
14. Federal Emergency Management Agency: Interim testing protocols for determining the seismic performance characteristics of structural and nonstructural components. FEMA Report 461, Washington (2007)
15. Pari Quispe, S.E., Manchego Meza, J.A.: Análisis experimental de muros de albañilería confinada en viviendas de baja altura en Lima, Perú (2017)
16. Ordoñez García, P.K., Lugo Chávez, Y.K.: Estructuras de madera aplicadas al sector de la construcción en el Perú (2016)
17. MVCS, Ministry of Housing, Construction and Sanitation: Norma E.010. Madera. Reglamento Nacional de Edificaciones, Perú (2014)
18. MVCS, Ministry of Housing, Construction and Sanitation: Norma E.030. Diseño sismorresistente. Reglamento Nacional de Edificaciones, Perú (2016)
19. JUNAC: Manual de diseño para maderas del grupo Andino. Junta del Acuerdo de Cartagena, Lima (1984)

Takagi-Sugeno Fuzzy Control of an Interleaved DC-DC Boost Converter

M. Nachidi[1(✉)], K. Khennoune[1], I. Ouachani[2], and A. Rabhi[3]

[1] Ecole d'Ingénieurs, Icam, Site de Grand Paris Sud, 34 rue points de vue, 77127 Lieusaint, France
meriem.labourel@icam.fr

[2] Polymont Engineering, 15 rue de la gare, 78640 Villiers-Saint-Frederic, France
rd@polymont.fr

[3] Modeling, Information and Systems Laboratory, University of Picardie Jules Verne, Amiens, France
abdelhamid.rabhi@u-picardie.fr

Abstract. This paper presents a Takagi-Sugeno (TS) fuzzy model-based controller for an interleaved DC-DC boost converter. Firstly, a TS model is developed using the bilinear model of the interleaved boost converter. Then, a TS fuzzy controller is designed to maintain the converter output voltage at the desired value with disturbance rejection. More precisely, the fuzzy controller gains are provided by solving a set of linear matrix inequalities (LMIs) based on the Lyapunov stability theory. These gains are used in the feedback control to determine the correct switch duty ratio to cause the converter output to reach the control objective even in the presence of load variations. Simulation results are given to show the applicability of the proposed control system for the interleaved DC-DC boost converter.

Keywords: Boost interleaved boost converter · Takagi-Sugeno fuzzy control · Lyapunov stability analysis

1 Introduction

Renewable and sustainable energy sources such as photovoltaic modules require a power conditioning system. Indeed, the voltages produced by photovoltaic cells are relatively low and need to be boosted to a high voltage level for a number of applications. Typically, a DC-DC boost converter is used for this purpose. However, this architecture has many drawbacks for these applications, especially in terms of reliability in case of power switch faults. Based on a 2011 survey [1] to determine industrial requirements and expectations for reliability in DC-DC converters, power switches are ranked as the most sensitive components. Indeed, more than 30% of reported failures in these converters are due to power switches. Moreover, this survey highlighted the main causes of power switch failure that are environment, transients, and electrical constraints. The reliability of a boost converter can be improved by using dynamic redundancy [2],

J. Littlewood and R. J. Howlett (Eds.): SEB 2022, SIST 336, pp. 431–440, 2023.
https://doi.org/10.1007/978-981-19-8769-4_40

which consists of adding a redundant arm. However, the use of dynamic redundancy increases the price and complexity of the converter. To avoid the use of dynamic redundancy, interleaved boost converter (IBC) topologies are presented as efficient solutions. Many research articles have emphasized the advantages of these topologies compared to other topologies developed for photovoltaic applications. Indeed, IBCs present many advantages in terms of compactness, input current ripple reduction, energy efficiency and predisposition to degraded mode operation in case of power switch faults [3,4]. It is worthy to point out that these kind of performance is conditioned by the quality of the transistor duty cycle control system. Although classical control techniques based on proportional-integral(PI) algorithm could satisfy the stability around an equilibrium [5], it is necessary to design a controller so that the stability can be meet for a wide range of system operating points. In general, nonlinear controllers are often used to solve similar problems for power control systems. [6] proposed a control system for four phases floating IBC that consists of an external voltage loop and an inner current loop to achieve good performance while rejecting disturbances. A robust dual loop control strategy has been proposed in [7] for output series IBC. The design is based on flatness theory and extended state observer for output voltage regulation and inductor current tracking of the IBC. Backstepping control law has been designed for a four-phase IBC in [8] by using a particle swarm optimization algorithm for controller parameter selection. A table comparing different control schemes for IBC across factors such as stability, controller parameter derivation has been provided by [8].

In the literature, the nonlinear sector approach is one of the most used approaches for the transition from a nonlinear control model to a Takagi-Sugeno model [9]. This method is intended to represent complex nonlinear systems on a large range of operation. It is based on a convex polymorphic transformation of the nonlinear terms of the dynamical system and guarantees that the obtained TS model exactly represents the nonlinear system in a compact space of the system state variables. In other words, the TS model is a collection of linear models blended through the nonlinear functions called normalized membership functions. The closed-loop system stability is ensured in the Lyapunov sense by solving a set of Linear Matrix Inequalities (LMIs) [10]. This methodology has been widely proven successful in many works, especially those dealing with power converters [11,12].

This study focuses on an interleaved boost DC-DC converter (Fig. 1). In order to guarantee a high level of reliability of this converter, a TS fuzzy-model-based H_∞ control design is proposed. The control objective is to regulate the output voltage of an IBC while ensuring zero steady state error, low overshoot, fast settling time and rejection of external disturbances.

First, a TS fuzzy model of the studied IBC is given using the topology of Fig. 1. Then, a smooth nonlinear state-feedback controllers, using the so-called parallel distributed compensation (PDC), is designed including the integrating action [9,13]. The gains of the controller are parametrized in terms of an LMI problem, which can be solved using the Matlab toolbox Yalmip. The proposed controller guarantees the system stability and good dynamic behavior.

This paper is organized as follows. Section 2 gives the averaged model of the IBC, from which the corresponding TS fuzzy model is derived. Section 3 presents the design of an integral action fuzzy controller. Section 4 presents simulations results to demonstrate the merits of the proposed fuzzy control approach. Finally, some conclusions are given in Sect. 5.

2 T-S Fuzzy Model of an Interleaved Boost Converter

Figure 1 shows the studied IBC topology. It consists of two conventional boost converters connected in parallel and sharing the same filter capacitor c at the output. V_{in} is the input source voltage, R is the load resistance and V_o is the output load voltage. Each branch of the IBC comprises an inductor L_i, an equivalent series resistances r_{Li}, a power switch M_i and a diode D_i ($i = 1, 2$). The controllable input of the IBC is the duty ratio d_i of the active switch. Since the converter operates 180^{deg} out of phase, the total input current ripple of the interleaved converter is reduced compared with a single converter. It is assumed that the parameters of the two converters are identical.

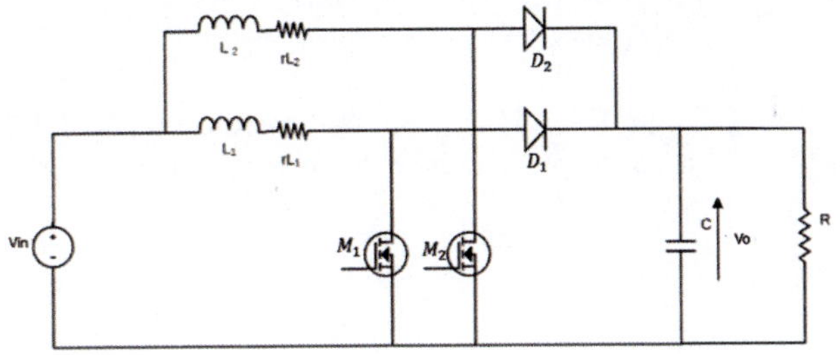

Fig. 1. Circuit diagram of the investigated IBC

According to the Kirchhoff voltage/current law, and the switching modes of the IBC, the following state-space model describes the four operating modes of Fig. 1:

$$\dot{x}(t) = \begin{cases} A_1 x(t) + E_1 V_{in}, (M_1 = 1, M_2 = 1), \\ A_2 x(t) + E_2 V_{in}, (M_1 = 1, M_2 = 0), \\ A_3 x(t) + E_3 V_{in}, (M_1 = 1, M_2 = 1), \\ A_4 x(t) + E_4 V_{in}, (M_1 = 0, M_2 = 1), \end{cases} \tag{1}$$

where $x(t) = \left[iL_1(t)\ iL_2(t)\ V_o(t) \right]^T$ and V_{in} is the input voltage and

$$A_1 = A_3 = \begin{bmatrix} -\frac{r_{L1}}{L_1} & 0 & 0 \\ 0 & -\frac{r_{L2}}{L_2} & 0 \\ 0 & 0 & -\frac{1}{Rc} \end{bmatrix},$$

$$A_2 = \begin{bmatrix} -\frac{r_{L1}}{L_1} & 0 & 0 \\ 0 & -\frac{r_{L2}}{L_2} & -\frac{1}{L_2} \\ 0 & \frac{1}{c} & -\frac{1}{Rc} \end{bmatrix}, A_4 = \begin{bmatrix} -\frac{r_{L1}}{L_1} & 0 & -\frac{1}{L_1} \\ 0 & -\frac{r_{L2}}{L_2} & 0 \\ \frac{1}{c} & 0 & -\frac{1}{Rc} \end{bmatrix}, \tag{2}$$

$$E = \begin{bmatrix} \frac{1}{L_1} \\ \frac{1}{L_2} \\ 0 \end{bmatrix}.$$

The averaged model can be derived from the four states over whole switching period as follows:

$$\begin{cases} \dot{x}(t) = Ax(t) + B(x(t))u(t) + EV_{in}, \\ y(t) = Cx(t), \end{cases} \tag{3}$$

where $x(t) = \left[iL_1(t)\ iL_2(t)\ V_o(t) \right]^T$, is the state vector, $u(t) = \left[d_1(t)\ d_2(t) \right]^T$ is the control vector, $y(t)$ is the output vector i.e. the output voltage.

$$A = \begin{bmatrix} -\frac{r_{L1}}{L_1} & 0 & -\frac{1}{L_1} \\ 0 & -\frac{r_{L2}}{L_2} & -\frac{1}{L_1} \\ \frac{1}{c} & \frac{1}{c} & -\frac{1}{Rc} \end{bmatrix},$$

$$B(x(t)) = \begin{bmatrix} \frac{V_o(t)}{L_1} & 0 \\ 0 & \frac{V_o(t)}{L_2} \\ \frac{-i_{L1}(t)}{c} & \frac{-i_{L2}(t)}{c} \end{bmatrix}, \tag{4}$$

$$C = \left[0\,0\,1 \right], E = \begin{bmatrix} \frac{1}{L_1} \\ \frac{1}{L_2} \\ 0 \end{bmatrix}.$$

The system is a nonlinear system, since the matrix $B(t)$ contains the product of state variables and input variables (control signals). To deal with this problem, the fuzzy modeling approach is applied to the model of IBC given in (3). The T-S fuzzy model is considered as an accurate representation of the nonlinear systems. It is supposed to obtain global dynamics of the system to guarantee the overall control performance.

In this paper, the fuzzy models of the investigated IBC is obtained considering $z(t) = [i_{L1}(t), i_{L2}(t), V_o(t)]$ the vector of fuzzy variables and assigning an interval $\left[\underline{z}_j(t),\ \overline{z}_j(t) \right]$ for each fuzzy variable $z_j(t)$. This leads to the following 8 fuzzy rules describing the fuzzy model of the IBC:

Rule i: If $z_1(t)$ *is* M_1^p *and* $z_2(t)$ *is* M_2^p *and* $z_3(t)$ *is* M_3^p,

$$\textit{Then} \begin{cases} \dot{x}(t) = Ax(t) + B_i u(t) + EV_{in} \\ y(t) = Cx(t), \quad i = 1, 2, ..., 8, \end{cases}$$

where $M_j^p, j = 1, 2, 3, p = 1, 2$, are the corresponding fuzzy sets, A, C, and E are system matrices given in (4) and

$$B_1 = \begin{bmatrix} \frac{\underline{V}_o}{L_1} & 0 \\ 0 & \frac{\underline{V}_o}{L_2} \\ \frac{-\underline{i}_{L1}}{c} & \frac{-\underline{i}_{L2}}{c} \end{bmatrix}, \quad B_2 = \begin{bmatrix} \frac{\underline{V}_o}{L_1} & 0 \\ 0 & \frac{\underline{V}_o}{L_2} \\ \frac{-\underline{i}_{L1}}{c} & \frac{-\bar{i}_{L2}}{c} \end{bmatrix},$$
$$B_3 = \begin{bmatrix} \frac{\underline{V}_o}{L_1} & 0 \\ 0 & \frac{\underline{V}_o}{L_2} \\ \frac{-\bar{i}_{L1}}{c} & \frac{-\underline{i}_{L2}}{c} \end{bmatrix}, \quad B_4 = \begin{bmatrix} \frac{\underline{V}_o}{L_1} & 0 \\ 0 & \frac{\underline{V}_o}{L_2} \\ \frac{-\bar{i}_{L1}}{c} & \frac{-\bar{i}_{L2}}{c} \end{bmatrix},$$
$$B_5 = \begin{bmatrix} \frac{\overline{V}_o}{L_1} & 0 \\ 0 & \frac{\overline{V}_o}{L_2} \\ \frac{-\underline{i}_{L1}}{c} & \frac{-\underline{i}_{L2}}{c} \end{bmatrix}, \quad B_6 = \begin{bmatrix} \frac{\overline{V}_o}{L_1} & 0 \\ 0 & \frac{\overline{V}_o}{L_2} \\ \frac{-\underline{i}_{L1}}{c} & \frac{-\bar{i}_{L2}}{c} \end{bmatrix}, \tag{5}$$
$$B_7 = \begin{bmatrix} \frac{\overline{V}_o}{L_1} & 0 \\ 0 & \frac{\overline{V}_o}{L_2} \\ \frac{-\bar{i}_{L1}}{c} & \frac{-\underline{i}_{L2}}{c} \end{bmatrix}, \quad B_8 = \begin{bmatrix} \frac{\overline{V}_o}{L_1} & 0 \\ 0 & \frac{\overline{V}_o}{L_2} \\ \frac{-\bar{i}_{L1}}{c} & \frac{-\bar{i}_{L2}}{c} \end{bmatrix},$$

Thus, the global T-S model is expressed as an interpolation of all subsystems of (5) via the nonlinear normalized membership functions as follows [9, 12, 13]:

$$\begin{cases} \dot{x}(t) = \sum_{i=1}^{8} h_i(z(t)(Ax(t) + B_i u(t) + EV_{in}), \\ y(t) = Cx(t), \end{cases} \tag{6}$$

where $h_i(z(t))$ is defined by:

$$h_i(z(t)) = \frac{\prod_{j=1}^{l} \mu_j^p(z_j(t))}{\sum_{i=1}^{8} \prod_{j=1}^{l} \mu_j^p(z_j(t))}, \tag{7}$$

and satisfy the convex property: $\sum_{i=1}^{8} h_i(z(t)) = 1$ with $0 < h_i(z(t)) \leq 1$. $\mu_j^p(z_j(t))$ is the membership function that corresponds to the fuzzy set $M_j^p(z_j(t))$, and are given by:

$$\mu_j^1(z_j(t)) = \frac{z_j(t) - \underline{z}_j(t)}{\overline{z}_j(t) - \underline{z}_j(t)}, \mu_j^2(z_j(t)) = \frac{\overline{z}_j(t) - z_j(t)}{\overline{z}_j(t) - \underline{z}_j(t)}, \tag{8}$$

3 T-S Fuzzy Controller Synthesis

In this section, we study the control problem for the IBC represented by the T-S fuzzy system (6) to regulate the output voltage error to zero. The main technique to achieve this purpose is to add the observable variable $e(t) = \int (V_{ref} - V_o(t))dt$ to the measured state variables, iL_1, iL_2 and V_o in order to form a new state vector.

The augmented T-S Fuzzy model is expressed as follows:

$$\begin{cases} \dot{\bar{x}}(t) = \sum_{i=1}^{8} h_i(z(t))(\bar{A}\bar{x}(t) + \bar{B}_i\bar{u}(t) + \bar{E}\bar{w}(t)), \\ y(t) = \bar{C}\bar{x}(t), \quad i = 1, 2, ..., 8, \end{cases} \tag{9}$$

where

$$\bar{A} = \begin{bmatrix} A & 0 \\ -C & 0 \end{bmatrix}, \quad \bar{B}_i = \begin{bmatrix} B_i \\ 0 \end{bmatrix},$$

$$\bar{E} = \begin{bmatrix} \frac{1}{L1} & 0 \\ \frac{1}{L2} & 0 \\ 0 & 0 \\ 0 & 1 \end{bmatrix} \overline{w}(t) = \begin{bmatrix} V_{in} \\ V_{ref} \end{bmatrix}$$

$$\overline{x}(t) = \begin{bmatrix} x(t) \\ e(t) \end{bmatrix} \text{ and } \bar{C} = [0\,0\,1\,0]$$

The fuzzy controllers is performed through the parallel distributed compensation [9,12,13]:
Rule i
$If\ i_{L_1}(t)$ is M_{1i}, $i_{L_2}(t)$ is M_{2i} and $V_c(t)$ is M_{3i},
then

$$u(t) = K_i\bar{x}(t), i = 1, \ldots, 8, \tag{10}$$

where K_i are linear control gains to be designed. The overall fuzzy controller is represented as

$$u(t) = \sum_{i=1}^{8} h_i(z(t))K_i\bar{x}(t) \tag{11}$$

By replacing the control law (11) in the fuzzy model (6), the closed-loop system is given as follows:

$$\dot{\bar{x}}(t) = \sum_{i=1}^{8}\sum_{j=1}^{8} h_i(z(t))h_j(z(t))(\bar{A} + \overline{B}_iK_j)\bar{x}(t) + \bar{E}\bar{w}(t). \tag{12}$$

The purpose is to design the controller (11) such that the equilibrium solution $V_o(t) = V_{ref}$ of the closed-loop fuzzy system (12) with $\bar{w}(t) = 0$ is asymptotically stable and the following H_∞ is satisfied:

$$\int_0^{t_f} \bar{x}^T(t)\bar{x}(t)dt \leq \rho^2 \int_0^{t_f} \bar{w}^T(t)\bar{w}(t)dt \tag{13}$$

for all $t_f > 0$ and $\bar{w}(t) \in L_2(0, \infty)$ under zero initial conditions.

In order to solve the H_∞ fuzzy control problem of system (12), the next lemma provides a sufficient condition in the form of LMIs. The detailed proof can be easily established using standard Lyapunov technique [9] and it is omitted here.

Lemma 1. *The feedback gains K_i that stabilize the boost converter represented by the closed-loop fuzzy model (12) and minimize ρ in (13) can be obtained by solving the following minimization problem based on LMIs:*

$$\underset{X,Y_1,\ldots,Y_8}{minimize}\ \rho^2$$

$$\begin{bmatrix} \begin{pmatrix} \bar{A}X + X\bar{A}^T + \frac{1}{2}(\bar{B}_iY_j + \\ Y_j{}^T\bar{B}_i{}^T + \bar{B}_jY_i + Y_i{}^T\bar{B}_j{}^T) \end{pmatrix} & \bar{E} & -X \\ \bar{E}^T & -\rho^2 I & 0 \\ -X & 0 & -I \end{bmatrix} \leq 0, \tag{14}$$

$$i \leq j, i = 1, \ldots, 8,$$

$$X > 0,$$

where

$$Y_i = K_iX, \quad i = 1, \ldots, 8,$$

4 Simulation and Experimental Results

In order to validate the effectiveness of the proposed control strategy, simulation studies were carried out on Matlab/Simulink environment. Specifications of the investigated IBC are shown in Table 1. First, a feasible solution of the optimization problem of Lemma 1 is obtained using Yalmip toolbox, the controller gains are given as follows:

Table 1. Parameters of the studied IBC.

Parameter	Symbol	Value	Unit
Inductor	L_1, L_2	0.001	H
Inductor resistor	r_{L1}, r_{L2}	0.2	Ω
Capacitor	c	$2200 * 10^{-6}$	F
Load resistor	R	24	Ω

$$K_1 = \begin{bmatrix} -0.6693 & -0.6312 & -0.5017 & 16.8099 \\ -0.5760 & -0.9854 & -0.5958 & 19.9548 \end{bmatrix}, K_2 = \begin{bmatrix} -0.9892 & -0.7969 & -0.6917 & 22.9412 \\ -0.6557 & -1.6555 & -0.9032 & 29.5935 \end{bmatrix},$$

$$K_3 = \begin{bmatrix} -2.1665 & -1.9120 & -1.6639 & 52.9592 \\ 0.6751 & 0.4235 & 0.4728 & -14.4234 \end{bmatrix}, K_4 = \begin{bmatrix} -0.5160 & -0.2981 & -0.2853 & 10.0588 \\ -0.1227 & -0.4477 & -0.2717 & 7.5250 \end{bmatrix},$$

$$K_5 = \begin{bmatrix} -0.7917 & -0.6431 & -0.5082 & 18.1187 \\ -0.4848 & -0.7668 & -0.4641 & 15.7256 \end{bmatrix}, K_6 = \begin{bmatrix} -1.6847 & -1.5848 & -1.3142 & 42.3520 \\ 0.3953 & -0.0591 & 0.1276 & -4.2421 \end{bmatrix},$$

$$K_7 = \begin{bmatrix} -1.6830 & -1.5324 & -1.2520 & 41.3414 \\ -0.0828 & -0.6836 & -0.2513 & 9.4149 \end{bmatrix}, K_8 = \begin{bmatrix} -0.9263 & -0.7511 & -0.6611 & 21.4159 \\ -0.7149 & -1.2572 & -0.8221 & 25.7760 \end{bmatrix}.$$

Then, the following control law is applied in the simulation model.

$$u(t) = \sum_{i=1}^{8} h_i(z(t)K_i\bar{x}(t) \tag{15}$$

Various scenarios were implemented to evaluate the efficiency and dynamic performance of the proposed control approach. In the first scenario, a constant reference voltage was considered. Figure 2 shows the IBC response using the conventional PI controller and the proposed TS controller. In this case, even the PID if it is well calibrated can be sufficient.

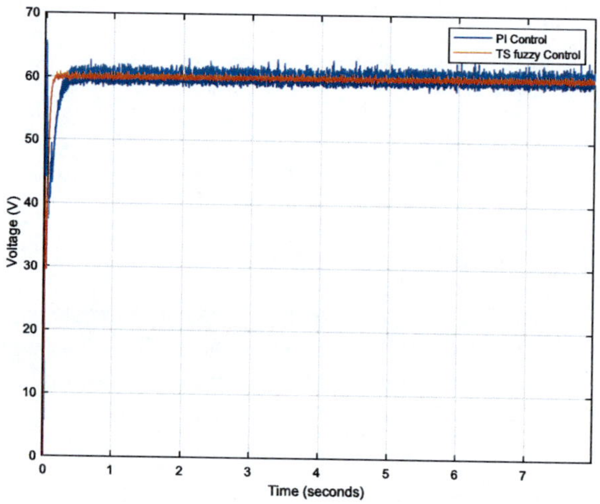

Fig. 2. Simulation results of the output voltage with the circuit parameters of Table 1.

In the second scenario, and as shown in Fig. 3, at $t = 2s$ the reference voltage changes from $60V$ to $30V$ and at $t = 5s$ the reference voltage changes from $40V$ to $30V$. We can clearly see that the proposed approach maintains the output voltage at the desired value, providing a strong tracking performance unlike the PI controller. At start-up, the converter operates at a the desired voltage $60V$. At 2s, the output voltage changes from $60V$ to $30V$. As shown in Fig. 3, with the proposed TS controller, the desired output voltage is obtained with a short response time and without any overshoot. However, for the PI controller, the output voltage is restored to the desired values ($40V$, $30V$) with a large overshoot and large response time.

The last scenario consists in testing the robustness of the proposed control strategy under the load variations. Figure 4 shows the obtained results. It can be seen that the output voltage track its reference value under the load variations with good tracking performances.

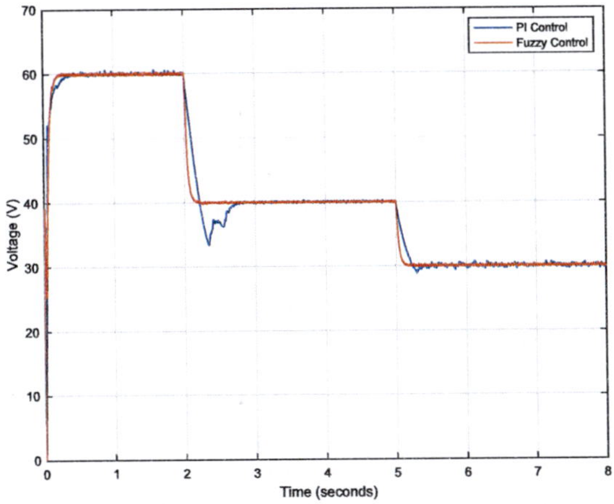

Fig. 3. Simulation results of the output voltage under reference voltage variations.

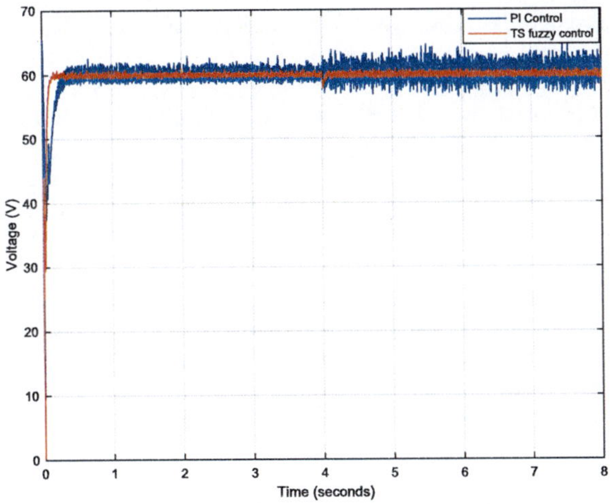

Fig. 4. Simulation results of the output voltage under the load variation: from 24Ω to 13.33Ω at 4 s.

5 Conclusions

This paper focused on T-S fuzzy control of two-arm interleaved boost converters. The control gains are obtained by solving a set of LMIs. All suggested scenarios confirm that the proposed fuzzy controller is efficient and can achieve better control performance defined by good transient response and less sensitivity to the effect of reference variations compared to the PI controller. As future work, the proposed control strategy will be implemented and tested in real time on test bed.

References

1. Yang, S., Bryant, A., Mawby, P., Xiang, D., Ran, L., Tavner, P.: An industry-based survey of reliability in power electronic converters. IEEE Trans. Ind. Appl. **47**(3), 1441–1451 (2011)
2. Isermann, R.: Fault-diagnosis applications. In: Actuators, Drives, Machinery, Plants, Sensors, and Fault-tolerant Systems, Model-Based Condition Monitoring: Actuators, Drives, Machinery, Plants, Sensors, and Fault-tolerant Systems. Springer, Berlin, Heidelberg (2011)
3. Zhang, S., Yu, X.: Design considerations of the interleaved boost converter in photovoltaic/fuel cell power conditioning system. Intelec 1–6 (2012)
4. Magne, P., Liu, P., Bilgin, B., Emadi, A.: Investigation of Impact of Number of Phases in Interleaved DC-DC Boost Converter, pp. 1–6 (2015)
5. Guruswamy, K., Divya, M.: Design, modelling and implementation of interleaved boost DC-DC converter. Int. J. Innov. Sci. Res. Technol. 2456–2165
6. Saadi, R., Hammoudi, M., Kraa, O., Ayad, M., Bahri, M.: A robust control of a 4-leg floating interleaved boost converter for fuel cell electric vehicle application. In: Mathematics and Computers in Simulation, vol. 167, pp. 32–47 (2020); International Conference on Emerging and Renewable Energy: Generation and Automation, held in Belfort, France on 4–6 July, 2017
7. Huangfu, Y., Li, Q., Xu, L., Ma, R., Gao, F.: Extended state observer based flatness control for fuel cell output series interleaved boost converter. IEEE Trans. Ind. Appl. **55**(6), 6427–6437 (2019)
8. Hao, X., Salhi, I., Laghrouche, S., Ait-Amirat, Y., Djerdir, A.: Backstepping supertwisting control of four-phase interleaved boost converter for PEM fuel cell. IEEE Trans. Power Electron. **37**(7), 7858–7870 (2022)
9. Wang, H., Li, J., Niemann, D., Tanaka, K.: T-S fuzzy model with linear rule consequence and PDC controller: a universal framework for nonlinear control systems. In: Ninth IEEE International Conference on Fuzzy Systems. FUZZ- IEEE 2000 (Cat. No.00CH37063), vol. 2, pp. 549–554 (2000)
10. Boyd, S., Ghaoui, L.E., Feron, E., Balakrishnan, V.: Linear Matrix Inequalities in System and Control Theory. SIAM, Philadelphia, PA (1994)
11. Seo, S.-W., Choi, H.H., Kim, Y.: Takagi-sugeno fuzzy model-based approach to robust control of boost DC-DC converters. J. Electr. Eng. Technol. **10**, 925–934 (2015)
12. Nachidi, M., El Hajjaji, A., Bosche, J.: An enhanced control approach for DC-DC converters. Int. J. Electr. Power Energy Syst. **45**(1), 404–412 (2013)
13. Wang, H., Tanaka, K., Griffin, M.: Parallel distributed compensation of nonlinear systems by Takagi-sugeno fuzzy model. In: Proceedings of 1995 IEEE International Conference on Fuzzy Systems, vol. 2, pp. 531–538 (1995)

Author Index

A
Ahmadian, Ehsan, 1
Albatici, Rossano, 249
Alonso-Montolio, Carlos, 261
Amara, Mohamed, 140
Angelosanti, M., 347
Arenas, Rolando Biere, 228
Arnone, D., 12
Artino, A., 12
Aversa, Patrizia, 218
Azrague, Kamal, 97

B
Barbolini, Fausto, 324
Barelli, Linda, 130
Bavarsad, Fatemeh Salehipour, 281
Becerra, G., 418
Beckers, Benoit, 175
Bernabei, Letizia, 197
Bernardini, Gabriele, 109, 197
Bernardo, Enrico, 47
Biere-Arenas, Rolando, 185

C
Cabillo, Isabel Crespo, 261
Cadena, Juan Diego Blanco, 197
Calderoni, Daniele, 47
Callegaro, Nicola, 249
Caponetto, Rosa, 164
Capozzoli, Alfonso, 313
Cascone, Stefano, 66
Cecere, Carlo, 301
Čekon, Miroslav, 291
Chatzigeorgiou, Nikolas G., 86
Chicco, Gianfranco, 371
Contrafatto, Loredana, 47
Costa Costa, César, 228
Costanzo, V., 12
Costanzo, Vincenzo, 37, 164
Cuomo, Massimo, 164
Čurpek, Jakub, 291
Currà, E., 347
Currà, Edoardo, 197

D
D'Amico, A., 347
D'Amico, Alessandro, 197
D'Orazio, Marco, 109, 120, 271, 281
Daniotti, Bruno, 218
de Bort, Inès, 175
de Santoli, L., 397
Derradji, Lotfi, 140
Detommaso, Maurizio, 37
Di Giuseppe, Elisa, 120, 271, 281
Di Loreto, Samantha, 120
Di Matteo, Miriam, 407
Di Perna, Costanzo, 120
Distefano, D. L., 12
Dotelli, Giovanni, 218
Duarte, Carlos Marmolejo, 228

E
El Youssef, Mohammad, 359
Elnolaky, Amira, 1
Endrizzi, Luca, 249
Enescu, Diana, 152, 371
Evans, N. I., 386
Evola, G., 12

J. Littlewood and R. J. Howlett (Eds.): SEB 2022, SIST 336, pp. 441–443, 2023.
https://doi.org/10.1007/978-981-19-8769-4

F
Fenili, Lorenzo, 313
Fiorini, Costanza Vittoria, 407
Fjellheim, Kristin, 97

G
Gagliano, Antonio, 66
Gallo, Antonio, 313
Garcia-Nevado, Elena, 261
Gazzo, Salvatore, 47
Georghiou, George E., 86
Giuffrida, Giada, 164
Grant, P., 208
Grignaffini, Stefano, 407
Guardigli, Luca, 324
Guarino, Francesco, 359

H
Haddad, Mohammed Amin Nassim, 140
Hale, C., 386
Hawkins, R. J. M., 386
Hong, Jin, 56

I
Imessad, Khaled, 140

J
Junaid, Muhammad Faisal, 291

K
Karani, G., 337
Khennoune, K., 431

L
Lanza, V., 397
Latini, Arianna, 120
Lazzaro, M., 12
Lekhal, Mohammed Cherif, 140
Lin, Alin, 76
Littlewood, J. R., 208, 240, 337, 386
Liu, Jing, 22
Liu, Yujing, 56
Lo Faro, Alessandro, 37
Lombardo, Grazia, 37
Longo, Michela, 130
Lori, Valter, 120
Lou, Jiankun, 76
Lu, Ming, 22
Luprano, Vincenza A. M., 218

M
Mannucci, Simona, 407
Maracchini, Gianluca, 271, 281
Margani, G., 12
Marmolejo-Duarte, Carlos, 185
Marzo, Anna, 218
Mazza, Andrea, 371
Meli, G., 418
Miño, Jairo Acuña Paz y, 175
Mochi, Giovanni, 197
Moletti, Chiara, 218
Moschella, Angela, 37

N
Nachidi, M., 431
Nardecchia, F., 397
Nocera, Francesco, 37, 164

O
Onnis, S., 418
Ouachani, I., 431

P
Paolini, C., 347
Pastore, L. M., 397
Peduzzi, Arianna, 301
Pelosi, Dario, 130
Pepperell, R., 208
Piscitelli, Marco Savino, 313
Pompei, L., 397
Pugnaletto, M., 347

Q
Quagliarini, E., 347
Quagliarini, Enrico, 109, 197

R
Rabhi, A., 431
Roca-Musach, Marc, 261
Romano, Guido, 109
Rosato, Antonio, 359
Rosso, Federica, 197
Rouag-Saffidine, D., 140
Roura, Helena Coch, 261
Russo, M., 347
Russo, Martina, 197

S
Sabbadini, Sergio, 218
Salemi, Angelo, 37
Salvalai, Graziano, 197
Sanna, F., 208, 240
Sciuto, Gaetano, 37
Semmari, Hamza, 140
Semprini, G., 12
Serpilli, Fabio, 120
Sibilio, Sergio, 359
Šikula, Ondřej, 291
Sodagar, Behzad, 1
Song, Di, 22

Stevens-Wood, K., 240
Suul, Jon Are, 97

T
Tripepi, Concetta, 218

V
Vallati, Andrea, 407
Vargas-Neumann, J., 418
Verhaert, Ivan, 1

W
Wang, Lu, 22
Wieser, M., 418
Wiik, Marianne Kjendseth, 97

X
Xing, Jun, 22

Y
Yue, Ran, 76

Z
Zaniboni, Luca, 249
Zaninelli, Dario, 130
Zhang, X., 337